Global Climate Change

Global Climate Change

Edited by

Suruchi Singh

Pardeep Singh

S. Rangabhashiyam

K.K. Srivastava

ELSEVIER

Elsevier
Radarweg 29, PO Box 211, 1000 AE Amsterdam, Netherlands
The Boulevard, Langford Lane, Kidlington, Oxford OX5 1GB, United Kingdom
50 Hampshire Street, 5th Floor, Cambridge, MA 02139, United States

Notices
Knowledge and best practice in this field are constantly changing. As new research and experience broaden our understanding, changes in research methods, professional practices, or medical treatment may become necessary.

Practitioners and researchers must always rely on their own experience and knowledge in evaluating and using any information, methods, compounds, or experiments described herein. In using such information or methods they should be mindful of their own safety and the safety of others, including parties for whom they have a professional responsibility.

To the fullest extent of the law, neither the Publisher nor the authors, contributors, or editors, assume any liability for any injury and/or damage to persons or property as a matter of products liability, negligence or otherwise, or from any use or operation of any methods, products, instructions, or ideas contained in the material herein.

Library of Congress Cataloging-in-Publication Data
A catalog record for this book is available from the Library of Congress

British Library Cataloguing-in-Publication Data
A catalogue record for this book is available from the British Library

ISBN: 978-0-12-822928-6

For information on all Elsevier publications visit our website at https://www.elsevier.com/books-and-journals

Publisher: Candice Janco
Acquisitions Editor: Marisa LaFleur
Editorial Project Manager: Sara Valentino
Production Project Manager: Sruthi Satheesh
Cover Designer: Victoria Pearson-Esser

Typeset by TNQ Technologies

Contents

Contributors

Ganesh Kumar Agrawal Research Laboratory for Biotechnology and Biochemistry (RLABB), Kathmandu, Nepal; Global Research Arch for Developing Education (GRADE) Academy Pvt. Ltd., Birgunj, Parsa, Nepal

Madhoolika Agrawal Laboratory of Air Pollution and Global Climate Change, Department of Botany, Institute of Science, Banaras Hindu University, Varanasi, Uttar Pradesh, India

Shashi Bhushan Agrawal Laboratory of Air Pollution and Global Climate Change, Department of Botany, Institute of Science, Banaras Hindu University, Varanasi, Uttar Pradesh, India

K.N. Amruthesh Department of Studies in Botany, University of Mysore, Mysuru, Karnataka, India

Shalini Anand Center for Fire, Explosive and Environment Safety, DRDO, New Delhi, India

Iqbal Ansari Academy of Scientific and Innovative Research, CSIR-Central Institute of Mining and Fuel Research, Dhanbad, Jharkhand, India

Charu Arora Department of Chemistry, Guru Ghasidas Vishwavidyalaya, Bilaspur, Chhattisgarh, India

Dhruv Arora Department of Food Technology, Gautam Budha University, Greater Noida, Uttar Pradesh, India

Dipti Bharti Department of Applied Sciences, Greater Noida Institute of Technology, Greater Noida, Uttar Pradesh, India

Shankar Bhujbal Nuclear Agriculture and Biotechnology Division, Bhabha Atomic Research Centre, Mumbai, Maharashtra, India

S. Chandrashekar Department of Studies in Biotechnology, Shivagangotri, Davangere University, Davangere, Karnataka, India

Moharana Choudhury Voice of Environment (VoE), Guwahati, Assam, India

Maha M. El-Kady Self-Pollinated Vegetable Crops, Department of Horticulture Research Institute, Agriculture Research Centre, Giza, Giza Governorate, Egypt

Annesha Ghosh Laboratory of Air Pollution and Global Climate Change, Department of Botany, Institute of Science, Banaras Hindu University, Varanasi, Uttar Pradesh, India

H.V. Girish Department of Studies in Botany, University of Mysore, Mysuru, Karnataka, India

N.K. Hemanth Kumar Department of Studies in Botany, University of Mysore, Mysuru, Karnataka, India

Shobha Jagannath Department of Studies in Botany, University of Mysore, Mysuru, Karnataka, India

Durgesh Kumar Jaiswal Institute of Environment and Sustainable Development, Banaras Hindu University, Varanasi, Uttar Pradesh, India

Bhavna Jaiswal Laboratory of Air Pollution and Global Climate Change, Department of Botany, Institute of Science, Banaras Hindu University, Varanasi, Uttar Pradesh, India

Sanjay J. Jambhulkar Nuclear Agriculture and Biotechnology Division, Bhabha Atomic Research Centre, Mumbai, Maharashtra, India

Aarif Khan Department of Environmental Science, Central University of Rajasthan, Ajmer, Rajasthan, India

Abhishek Kumar Department of Civil & Environmental Engineering, Birla Institute of Technology, Mesra, Ranchi, Jharkhand, India

Vijay Kumar Department of Applied Sciences and Humanities, Faculty of Engineering and Technology, Rama University, Kanpur, Uttar Pradesh, India

Deepak Kumar United Nation Developmental Program (UNDP), New Delhi, India

Raviteja Machanuru Centre for Environment Science and Climate Resilient Agriculture (CESCRA), Nuclear Research Laboratory (NRL), ICAR-Indian Agricultural Research Institute, New Delhi, India

Deblina Maiti Academy of Scientific and Innovative Research, CSIR-Central Institute of Mining and Fuel Research, Dhanbad, Jharkhand, India

Srishti Mishra Laboratory of Air Pollution and Global Climate Change, Department of Botany, Institute of Science, Banaras Hindu University, Varanasi, Uttar Pradesh, India

M. Murali Department of Studies in Botany, University of Mysore, Mysuru, Karnataka, India

Shilpi Nagar Department of Environmental Studies, University of Delhi, New Delhi, India; Center for Fire, Explosive and Environment Safety, DRDO, New Delhi, India

Sunayana Nath Department of Environmental Studies, Institute of Home Economics, University of Delhi, New Delhi, India

Yash Nath Department of Geology and Geophysics, Indian Institute of Technology, Kharagpur, West Bengal, India

Divya Pandey Leibniz Centre for Agricultural Landscape Research (ZALF), Müncheberg, Germany

S. Pandian E Department of Environmental Science and Engineering, IIT(ISM), Dhanbad, Jharkhand, India

Archana Rai Nuclear Agriculture and Biotechnology Division, Bhabha Atomic Research Centre, Mumbai, Maharashtra, India

Kshama Rai Laboratory of Air Pollution and Global Climate Change, Department of Botany, Institute of Science, Banaras Hindu University, Varanasi, Uttar Pradesh, India

Sanchayita Rajkhowa Department of Chemistry, Jorhat Institute of Science & Technology (JIST), Jorhat, Assam, India

Randeep Rakwal Faculty of Health and Sport Sciences, University of Tsukuba, Tsukuba, Ibaraki, Japan; Research Laboratory for Biotechnology and Biochemistry (RLABB), Kathmandu, Nepal; Global Research Arch for Developing Education (GRADE) Academy Pvt. Ltd., Birgunj, Parsa, Nepal

Deeksha Ranjan Department of Applied Sciences & Humanities, Faculty of Engineering & Technology, Rama University, Kanpur, Uttar Pradesh, India

Manisha Sarkar Laboratory of Applied Stress Biology, Department of Botany, University of Gour Banga, Malda, West Bengal, India

Abhijit Sarkar Laboratory of Applied Stress Biology, Department of Botany, University of Gour Banga, Malda, West Bengal, India

Jyotirmoy Sarma Department of Chemistry, Kaziranga University, Jorhat, Assam, India

Gereraj Sen Gupta Department of Botany, Institute of Science, Banaras Hindu University, Varanasi, Uttar Pradesh, India

Anu Sharma Department of Environmental Science, Government Degree College Bhaderwah, University of Jammu, Jammu, India

Manoj Shrivastava Centre for Environment Science and Climate Resilient Agriculture (CESCRA), Nuclear Research Laboratory (NRL), ICAR-Indian Agricultural Research Institute, New Delhi, India

Ritis Kumar Shyanti School of Life Sciences, Jawaharlal Nehru University, New Delhi, India

Pratichi Singh Department of Environmental Science and Engineering, IIT(ISM), Dhanbad, Jharkhand, India

Renu Singh Centre for Environment Science and Climate Resilient Agriculture (CESCRA), Nuclear Research Laboratory (NRL), ICAR-Indian Agricultural Research Institute, New Delhi, India

Rahul Singh G.L. Bajaj Institute of Technology and Management, Greater Noida, Uttar Pradesh, India

Bhupinder Singh Centre for Environment Science and Climate Resilient Agriculture (CESCRA), Nuclear Research Laboratory (NRL), ICAR-Indian Agricultural Research Institute, New Delhi, India

Pardeep Singh Department of Environmental Studies, PGDAV College University of Delhi, New Delhi, India

Rakesh Kumar Sinha Department of Civil Engineering, Indian Institute of Technology-Bombay, Mumbai, Maharashtra, India

M.Y. Sreenivasa Department of Studies in Microbiology, University of Mysore, Mysuru, Karnataka, India

Muniyan Sundararajan Academy of Scientific and Innovative Research, CSIR-Central Institute of Mining and Fuel Research, Dhanbad, Jharkhand, India

Supriya Tiwari Department of Botany, Institute of Science, Banaras Hindu University, Varanasi, Uttar Pradesh, India

Khushaboo Verma Faculty of Education, Banaras Hindu University, Varanasi, Uttar Pradesh, India

Poonam Yadav Institute of Environment and Sustainable Development, Banaras Hindu University, Varanasi, Uttar Pradesh, India

Poonam Yadav Centre for Environment Science and Climate Resilient Agriculture (CESCRA), Nuclear Research Laboratory (NRL), ICAR-Indian Agricultural Research Institute, New Delhi, India

Deepak Yadav Department of Chemical Engineering, HBTU, Kanpur, Uttar Pradesh, India

Durgesh Singh Yadav Laboratory of Air Pollution and Global Climate Change, Department of Botany, Institute of Science, Banaras Hindu University, Varanasi, Uttar Pradesh, India

Biographies

Dr Suruchi Singh

Dr Suruchi Singh is presently working as Faculty in the Department of Botany, Sunbeam College for Women, India. She obtained her PhD from Banaras Hindu University, India, in Botany with Environmental Science as major field. Dr Singh received her postdoctoral experiences in the Department of Botany, Banaras Hindu University under the Council of Scientific and Industrial Research and Department of Science and Technology. She has published articles in various international journals and book series (>40 peer-reviewed articles). In her research, she has identified processes and activities where environmental efficiencies of selected crop plants must be increased against UV-B and tropospheric ozone.

Dr Pardeep Singh

Dr Pardeep Singh is presently working as an assistant professor in the Department of Environmental Science, PGDAV College University of Delhi, New Delhi, India. He obtained his master's degree from the Department of Environmental Science, Banaras Hindu University, Varanasi, India. He obtained his doctorate degree from the Indian Institute of Technology (Banaras Hindu University), Varanasi. He has also published more than 65 articles in national and international journals and edited 20 books in the field of environmental sciences.

Dr Rangabhashiyam S

Dr Rangabhashiyam is currently working as Faculty in the School of Chemical and Biotechnology, SASTRA Deemed University, India. He received his Doctor of Philosophy degree from the National Institute of Technology Calicut, India, in 2015. His major research interests are bioremediation and wastewater treatment. As first author/corresponding author, he has published more than 27 research and review articles. Altogether, he has published 48 articles in peer-reviewed national and international journals.

K.K. Srivastava

K.K. Srivastava is an associate professor, PGDAV College, University of Delhi, New Delhi, India. KKS started his professional life with the communication agency Niche Communications, where his responsibilities included providing strategic inputs for marketing communication of brands. Later, he worked with his colleague and partner to provide backgrounders on Indian economy, polity, and society to media groups such as the New York Times, German TV, BBC, and so on. KKS has more than 60 articles and research papers to his credit on diverse topics in economics, marketing, advertising, human resource management, and environmental studies.

CH_4, and N_2O, has been the forerunner cause of global warming (Kweku et al., 2018). Each gas has a different global warming potential. If CO_2 is considered to have a potential of 1.0, the global warming potentials of CH_4, N_2O, HCFC-22, and CFC-12 are 10, 180, 410, and 3700, respectively (Lashof and Ahuja, 1990). The US Environmental Protection Agency had reported 6677 million metric tons of CO_2 equivalent greenhouse gases emitted into the environment in 2018. Carbon dioxide contributed to around 81% of the total emissions, while methane, nitrous oxide, and fluorinated gases (hydrofluorocarbons and sulfur hexafluoride) were ranked second (10%), third (7%), and fourth (2%), respectively, on the emission list. The various sources and their contributions to release of these gases are represented in Table 1.1 (Ward, 2015).

Table 1.1 Various sources and their contribution in production of greenhouse gases.

Greenhouse gases	Source	Contribution (%)
Carbon dioxide	Electricity	37
	Transportation	31
	Industry	15
	Residential and commercial areas	10
	Non−fossil fuel combustion	6
Methane	Natural gas and petroleum	29
	Enteric fermentation	26
	Landfills	18
	Coal mining	10
	Manure management	10
	Others	8
Nitrous oxides	Agricultural soil management	74
	Stationary combustion	6
	Industry or chemical production	5
	Transportation	5
	Manure management	5
	Others	4
Fluorinated gases	Substitution of ozone-depleting substances	90
	Electrical transmission and distribution	3
	Production and processing of aluminum and magnesium	3
	HCFC-22 production	2
	Semiconductor manufacture	2

The data have been sourced from the National Environmental Education Foundation.

Unsurprisingly, global warming was left unchecked and gave rise to problems such as increment in sea level, acidification of oceans, and boosted frequency of disasters (Kweku et al., 2018). A study has reported that the frequency of hot and sweltering days would double in the period from 2013 to 60, when compared with 1961−90, if we consider the rise in temperature to be 2°C. Interestingly, as could be expected, the cold days would decrease by 50% for the same period (Giannakopoulos et al., 2005). Additionally, the 2°C rise would alter the rainfall pattern, and the number of dry days would increase. They also reported that the precipitation would be more intense, whenever it decides to show up (Giannakopoulos et al., 2005). Such major changes would have a significant impact on several factors in consideration. The energy demand would rise because of the need to heat (or cool) workplaces, transportation, and houses (Rodríguez, 2005). The increase in hotter and drier summer would result in greater wildfire occurrences, which might be difficult to control (Moriondo et al., 2006). There would be a rise in water shortage, which might affect plants, animals, and humans. Drought periods would increase manifold (Giannakopoulos et al., 2005).

Correspondingly, the increase in temperature caused because of global warming would result in a drastic reduction in human productivity. The creation of heat stress would reduce the cognitive ability and working hours (i.e., working potential), by imposing biological constraints on working intensity (Carleton and Anne, 2018). There would be an accompanied reduction in labor productivity, affecting the economy holistically (Day et al., 2019). Additionally, the increase in temperature would increase the sea level and trigger biodiversity loss, which has been discussed in later sections.

Surprisingly, a global rise in temperature can affect agriculture in contrasting ways (Melillo et al., 2014). Depending upon the optimal growth condition for various crops, temperature rise might affect it in varying ways. Some areas may get benefitted from warmer temperatures and vice versa. An increase in CO_2 levels may be conducive to some plants, for better growth. However, alterations in the weather pattern and increase in extreme events such as drought or flood and pest attack would harm the crops (Gornall et al., 2010). Crop yields are bound to decrease, intensifying the persistent problem of food security (Backlund et al., 2008). Unsurprisingly, the crop yield response was negative toward CO_2 rise and temperature increase (Giannakopoulos et al., 2005). Additionally, the warmer temperature would have an adverse effect on livestock and fisheries (Melillo et al., 2014).

Several cities have initiated innovative approaches to tackle rising temperatures. Paris Agreement was signed under the aegis of the UN Framework Convention on Climate Change. It was determined to limit the temperature rise to 1.5°C above preindustrial levels to reduce the risks and damages associated with climate change. A climate leadership group named C40 cities are leading in handling urban heat via the cool city network (Cippec et al., 2019). Several countries have pledged to fund to decelerate the temperature increase apart from implementing climate-friendly and climate-smart technologies such as cool roofing and cool neighborhood systems. South Africa commenced a detailed study of the causes and effects of urban heat. France, Italy, and Spain have mapped heat prevalent in their cities and the vulnerable populations, and they are working on providing cool places to their citizens. Japan has launched the construction of cool pavements, which reflects sun rays. Many countries have built cool roofs to decrease infrastructural warming (Li et al., 2017; Shickman, 2017).

4. Melting of glaciers and polar icecaps

Snow, a low-temperature product, would vanish with an enhancement of temperature. As a result, glaciers and permafrost in the mountains and the polar ice caps become very sensitive to atmospheric warming. Additionally, mass wasting is sensitive to slope steepness at a higher altitude, aggravating glacial loss (Haeberli and Beniston, 1998). The later part of the 20th century and 21st century has seen significant changes in the cryosphere world. Striking differences have been seen in similar decadal pictures depicting glacial changes (Haeberli and Beniston, 1998; Haeberli, 2010).

It is essential to give special emphasis to the protection of glaciers and polar ice caps. They are involved in the water cycle, and alterations would affect the process as a whole. The mountains are involved in orographic rainfall, a very critical process in the formation of clouds and associated precipitation. This holistic process has an indirect role in the maintenance of heat balance and moisture transfer. The mountains are involved in large-scale airflow patterns, and perturbations to such cycles would be devastating (Haeberli et al., 1993; Fitzharris et al., 1996). With a reduction in the cryosphere coverage, albedo (i.e., surface reflectivity of sunlight) gets diminished, initiating a positive feedback to melting and a negative feedback to thawing (Beniston et al., 1997).

Climate change is modifying the permafrost distribution patterns, and the natural adjustment of glacial processes with time is getting impossible (Haeberli and Beniston, 1998). The Alps have seen nearly 40% loss of surface area and 50% loss of volume of the glacial belts, when compared with the preindustrial timeline, referred to as "Little Ice Age" in the report. The process of glacial loss boomed in the latter part of the 20th century (Haeberli and Beniston, 1998).

Glaciers are a source of freshwater in the form of streams and rivers, for the communities through which they flow. Such freshwater is utilized for domestic use, irrigation of farmlands, and hydropower generation (Eriksson et al., 2006). Although melting of these glaciers would boost water availability for a short period, detrimental effects are destined in the long run. The ecosystem would get uprooted. Additionally, there would be an increase in sea level, which is a disturbance to the equilibrium, and ill-effects would be bound to follow (Mei'e et al., 1985).

It was reported that 30%, 50%, and 50% of water resources are derived from ice melting in the eastern, central, and western Himalayas, respectively, and the ratio goes as high as 80% for the Karakoram. The presence of such large mountain ranges is, therefore, critical for sustenance of livelihoods of a humongous human population and the complete ecosystem. The mountains play a critical role in the weather pattern for the areas surrounding it (Beniston et al., 1997; Xu et al., 2009). It has been reported that the process of deglaciation would result in changes in hydrological characteristics of the streams and rivers, increment in the size and volume of glacial lakes, and a rise in risk associated with floods. Furthermore, agriculture, industries, animal and plant health (including human health), and biodiversity would get adversely affected (Shrestha et al., 2000; Eriksson et al., 2006).

It was proposed to impinge chemicals into the atmosphere to block incoming solar radiation. Some also proposed to pump seawater on polar ice caps, with the idea of accelerated freezing (Cox, 2018). Nevertheless, these ideas were rejected based on the risks and humongous costs involved. Lately, it has been proposed to build underwater walls beneath glaciers using robots and pumping chilled water beneath ice via tunnels to thicken the layers and prevent their ocean fall (Cox, 2018).

5. Rise in sea level, sea shape, and sea composition

The rise in sea level caused by cryosphere melting has worldwide, long-lasting, and irreversible effects. It was reported that, by 2018, global mean sea level had increased by 21–24 and 8.1 cm compared with 1880 and 1993, respectively (Lindsey, 2019). It was also stated that the rise in sea level was three times quicker during 2005–15 compared with 1993–2014 (Yi et al., 2017).

The sea level rise is changing the coastlines across the globe and would result in flooding of coastal areas and a number of small islands. The existence of the Maldives, a small archipelago located to the southwest of the Indian subcontinent, is hugely threatened by the rise in sea level and is predicted to be entirely submerged by 2100 (Gagain, 2012). Apart from the Maldives, a number of islands would get submerged, such as Tuvalu, Nauru, and the Marshall Islands located in the Pacific Ocean (Allen, 2004; Tim Lister, 2012; Letman, 2018). Fiji, Sri Lanka, Kiribati, and Mauritius are also reported to be at extreme threat (Leatherman and Beller-Simms, 1997).

The rise in sea level would result in saltwater intrusion of surface waters, erosion of coasts, and destruction of arable land near the coasts. Furthermore, there would be accompanied loss of wetlands and an increase in the water table (Noone, 2013). Damage to arable land and freshwater supply would affect agriculture, and the households living in the vicinity of the coasts would be devastated (Gornitz, 1991).The frequency of hazards, such as storm surges, would increase due to sea level rise (Allison et al., 2015). The increase in storm surges could also be an outcome of the increased occurrences of cyclones, hurricanes, and typhoons (Kossin et al., 2017). Climate change–induced increment of sea surface temperatures is the possible cause of intensification of cyclone-related storms (Gillett et al., 2008). The size of storms would worsen the damage caused because of storm surges. However, it is interesting to note that some regions may be at a greater risk depending on the storm intensity, the number of people involved, and the built-up infrastructure under consideration (Rahmstorf, 2007, 2017). There would be episodic flooding of the coasts, which may penetrate inland to a great extent (Gornitz, 1991). Additionally, the infrastructure built across the coastal areas, such as roads, houses, bridges, power plants, and others, would be damaged because of the rise (Lindsey, 2019). Correspondingly, there would be high costs involved in the construction of infrastructure to prevent damage to the existing infrastructure.

Unsurprisingly, the marine biodiversity would be negatively affected and result insignificant biodiversity loss (Barnes and Kaiser, 2009). There would be alterations in the habitat of the species thriving in the coastal regions to such an extent that these habitats may get completely destroyed. Species of fauna belonging to arthropods, mollusks, chordates, and so on and floral species such as mangroves, grasses, and so on are bound to get affected by a rise in sea level. The habitat loss for such a large number of species would result in drastic alterations to the existing biodiversity (Barnes and Kaiser, 2009; Cheung et al., 2009).

Interestingly, the addition of surplus freshwater, via the melting of glaciers and polar ice caps, would cause significant changes in the composition of salty seawater. There would be changes in density, conductivity, pH, salt concentration, and freezing point of water, which would have a significant effect on a global scale. The thermohaline circulation would see significant alterations due to the changes in the density of water (EIS, 2020). Thermohaline circulation, also called ocean conveyor belt, plays a vital role in maintaining a heat balance across the planet. Alterations in the circulation would affect the equilibrium and may initiate central heating and polar cooling (Marotzke, 2000; Berwyn, 2018). Weather patterns may change worldwide due to changes in the ocean conveyor belt. This would inevitably strengthen the problems associated with climate change.

Oceans have been described as great reservoirs of carbon dioxide and have acted as a natural control mechanism for the greenhouse effect. With the increase in global temperatures, warming of oceans would occur. The heating up of water bodies would cause a release of the stored carbon dioxide (Tsai and Yau, 1990). This would reinforce the greenhouse effect, creating a positive feedback loop. However, it was reported that an increase in carbon dioxide concentrations might boost ocean acidification. There would be a production of carbonic acid and depletion of carbonate ions. Such depletion would prevent many marine organisms from building skeletons and shells. Coral bleaching, i.e., removal of algae from the coral polyps, would occur. Coral reefs are said to be the rainforests of the oceans. As a result of coral bleaching, marine biodiversity loss would occur on a very large scale (Hoegh-Guldberg et al., 2007; Guinotte and Fabry, 2008). Sadly, at deeper levels, i.e., between a depth of 250 and 450 m, the carbon dioxide levels are increasing twice in comparison with the surface waters (Glick, 2020), and this would be devastating for the animals and plants thriving at greater depths in the oceans.

Various countries are trying to cope with rising sea levels. Coastal cities in the United States are building seawalls and drainage systems, while people living in the island nations are migrating to higher locations. The city of Boston has set up an elliptical tower to withstand strong winds and storms. A flood wall has been built in Manhattan. Miami Beach is increasing the height of its roads by 2 feet to prevent flooding. Many cities are building pumps to drain out stormwater. Florida has upgraded sewage systems to prevent stormwater seeping into prevailing drainage systems. However, these are options to cope with a rise in sea level and do not help prevent it (Amadeo, 2020; Sea Level Rise organisation, 2020). Many countries are trying to decrease the warming of the atmosphere and hydrosphere. They are reducing greenhouse gas emissions and trying to capture the existing greenhouse gases persistent in the atmosphere. These methods include carbon sequestration via protecting the forests, wetlands, and peatlands from destruction, and carbon capture and storage technologies (Amadeo, 2020). Natural infrastructure is being upgraded in several coastal cities by upgrading existing beaches, mangroves, coral reefs, and salt marshes (Sea Level Rise organisation, 2020).

6. Hazards of climate change

The number and frequency of hazards and extreme weather events such as heat waves, forest fires, droughts, floods, cyclones, and tornadoes caused by climate change are expected to rise shortly.

7. Forest fires

The previous decade saw incidences of humongous wildfires. This could be exemplified by the fact that wildfires burned more than 5.5 million acres of forests in 2016, in the United States alone. This destruction was greater than 10 million acres in 2015 (Daley, 2017). Forest fires of Australia have become everyday news (Beer and Williams, 1995; Lucas, 2005). The year 2019 saw one of the most heart-breaking forest fires of Amazon rainforests.

It has been reported that, among several other reasons, increased warmer and drier periods caused by climate change have resulted in forest fires. They are further boosted by the drought-like situation and a longer fire engulfing season (Wehner et al., 2017). It has been projected that an increase of 1°C in temperature would pump an enhancement of 600% of burned forests in the western part of the United States (Vose et al., 2012). The increase in risks related to fire and an extended fire season, in the southeast United States, would ignite 30% more than fires ignited by lightning, by 2060 (Vose et al., 2018). The involvement of human beings reportedly causes 84% of wildfires. Once the fire is put on, the warmer temperature and drier conditions help spread the fire making it extremely tough to put off (Daley, 2017).

Fire is significantly affected by four factors—fuel, ignition, weather, and anthropogenic activity (Johnson, 1992; Swetnam, 1993). The weather pattern is dynamic because of Earth's rotation, solar input, and atmospheric variations. However, the increase in anthropogenic GHGs in the atmosphere has been intensifying the alterations in the weather pattern and boosting fire activities (Thornes, 2002; IPCC, 2007).

It has been reported that the spatial extent of extreme fire conditions would increase extraordinarily, even in countries lying in the temperate zones, such as Russia and Canada (Stocks et al., 1998). They also stated that forest fires would start earlier every year, intensifying the damage caused. Furthermore, it may expand in its extent in the months of June and July (Stocks et al., 1998). The increase in the burned area may increase by 74%–118%, by the end of the 21st century, in Canada alone (Flannigan et al., 2006). It was suggested that fires caused by lightning would rise by 44%, and there would be an enhancement of burned area by 80% by the end of this century in the United States alone (Price and Rind, 1994). Additionally, they stated that lightning-induced ignition would add to atmospheric warming.

Fire intensity is measured in kW/m and is calculated by the amount of energy released at the fire front per unit length of the line of fire (Flannigan et al., 2006). Climate change is likely to affect the fire intensity by decreasing moisture content in trees (fuel to fire), increasing wind speed (a boost to fire), and increasing fuel load (Flannigan et al., 2005). Determination of fire intensity becomes vital in the scenario of enhanced efforts involved in putting out fires and to make the fire extinguishing crew safe from dangers of exceptionally intense fire. Additionally, fire intensity could also help in the estimation of vegetation mortality in case of occurrence of forest fire (Hely et al., 2003). A tremendous enhancement in fire intensity was reported in preliminary work in some fuel types (Johnston, 2001).

Forest fires are likely to affect biodiversity by altering species habitat, enhancing the problem of air pollution, loss of significant forest resource, damaged ecosystem, loss of infrastructure found in the vicinity to forests, and displacement of communities located in the forests. Correspondingly, it becomes essential to protect the ecosystem and people living near the fires to minimize the damage. A lot of funding may go into putting out the fires, neglecting the low success to failure ratio. For example, Canada alone spends more than 500 million dollars to suppress these fires. In the scenario of an increase in forest fires, the costs would rise unprecedentedly (Flannigan and Van Wagner, 1991; Flannigan et al., 2005).

Considering the immensity of damage that forest fires could cause, Integrated Forest Protection Scheme was implemented in various states of India. It included constructing fire lines and watchtowers, provisioning of fire-resistant clothes and wireless sets, and training the various stakeholders involved along with the firefighters (NIDM, 2012). European Forest Fire Information System has been used to facilitate early warning in the event of a forest fire. European Forest Fire Tactical Reserve aids in coordinating the response to the forest fire relief. European Civil Protection Mechanism involves firefighting via ground and aerial routes (Wilkinson and Zimmermann, 2014).

8. Heat waves

The frequency and intensity of hot days are increasing on a yearly basis. In the previous decade, record hot days were seen twice in comparison with record cold days across the United States, the occurrence of which was even in the 1950s. The heat waves have been increasing since the 1960s (Abatzoglou and Barbero, 2014). Heat waves have become a common occurrence, and we cannot deny the role of climate change (Diffenbaugh and Scherer, 2013; Vose et al., 2017). It has been reported that the number of days with temperatures above 100°F would double, and the number of days above 105°F would triple in the days to come (Dahl et al., 2019). The heat waves have become super long-lasting since the previous decade (Habeeb et al., 2015). There could be a 4–20 times increase in extreme heat conditions by the end of the 21st century (Dahl et al., 2019). Extreme heat could also be affected by other factors. For example, Dust Bowls are considered to intensify daytime warming throughout the central part of the United States (Abatzoglou and Barbero, 2014). Cropland intensification has contributed to a cooling effect in the daytime temperatures (Mueller et al., 2017).

Extreme heat poses several threats for the animals and plants thriving in the different ecosystems (Sherwood and Huber, 2010). There could be an outbreak of numerous disasters. Drought and wildfires may follow in the situation of excess heat. Heat island effect could be a possible outcome in different cities and towns. The infrastructure built across the human habitations would add on to the problem of heat, pushing it toward extremities. Interestingly, humid conditions aggravate heat stress (Raymond et al., 2017). In the situation of skin humidity crossing 90%, it would be difficult for humans to cool down by releasing sweat (Sherwood and Huber, 2010). This would pose a threat to the people living in such conditions and the associated ecosystem (Oudin Åström et al., 2011). Human health would be at a great risk. Diseases and deaths may follow (Anderson and Bell, 2011; Xu et al., 2012; Schmeltz et al., 2015). Additionally, it would decrease the productivity of human beings, which would affect the economy adversely. Apart from health, there would be extreme pressure on the energy sector to mitigate excess heat. Energy demand would increase, enhancing pressure on fossil fuels, which would consequently strengthen global warming. Technological advancement and economic development have somehow handled the problem of excess heat, but it is difficult for the poor to handle excess heat (Davis et al., 2003). The role of government becomes critical to increase timely communication of extreme heat-related events and provide mitigation techniques to those unable to afford costlier technologies. Reduction in GHGs, afforestation, and development of energy-efficient technology are the needs of the hour. France brought about a national action plan for tackling heat waves. The plans are inclusive of heat forecasting and warning system, health levels monitoring, infrastructural changes, and special care of the senior citizens. Lately, many European countries have implemented plans to tackle issues related to heat stress (Michelozzi, 2004; Michelon et al., 2005).

9. Drought

A drought is a natural event involving an extended period of a water shortage caused because of the absence of adequate precipitation for long (Wilhite and Glantz, 1985; Mishra and Singh, 2010). Droughts could be characterized by their duration, frequency of occurrence, and intensity (Mishra and Singh, 2010). Droughts have been classified into meteorological, hydrological, or agricultural droughts, depending upon their impact on the hydrological cycle (Sheffield and Wood, 2008; Duan and Mei, 2014; Leng et al., 2015).

Lack of rainfall for an extended period (weeks to years) is called meteorological drought. Reduction in groundwater and a decrease in streamflow causes hydrological drought and could be a consequence of meteorological drought. Loss of soil moisture over a period of time, caused because of the absence of precipitation, is known as agricultural drought. All kinds of droughts could be deleterious for nature, including humans. This could be exemplified by the increase in water scarcity in the scenario of below-average water availability in water bodies. It is accompanied by an increase in water demand. The need for irrigation and other household activities propels humans to extract more water from deeper levels, damaging the ecosystem (Wisser et al., 2010; Leng et al., 2015). Droughts have been given the topmost position among the natural hazards taking into consideration the extensive coverage, the large population affected, and the monetary losses involved (Wilhite and Glantz, 1985; Wilhite and Buchanan-Smith, 2005).

Climate change could prove to be a great contributor to increased occurrences of drought (Dai, 2013; Duan and Mei, 2014). An increase in global temperatures would enhance the evaporation of water molecules. This, combined with a scarcity of water, makes the problem severe and threatening (Sheffield and Wood, 2008). There could be the creation of a vicious circle where drought dries up soil and reduces flora cover, and is succeeded by a decrease in precipitation, further resulting in drought. Droughts are, therefore, bound to increase in frequency and intensity, with the occurrence of climate change (Leng et al., 2015). Determination of the impact of climate change on different regions becomes vital to identify the type of drought that may occur soon (IPCC, 2007; Leng et al., 2015).

It has been estimated that droughts may increase in the frequency of occurrence and affect the people for extended periods. An increase of temperature by 3°C may strengthen the intensity, duration, and frequency of agricultural droughts. An increase in 1°C in the river basins may result in strengthening of intensity and duration of hydrological and meteorological drought (Leng et al., 2015).

The droughts of 2008–09 in China affected more than 10 million people. They struggled with a shortage of water, and monetary loss was calculated to be greater than $2 billion (Yang et al., 2012; Zhang et al., 2012). There were far-reaching consequences in terms of ecosystem damage, agricultural loss, and disruption of societies, apart from economic disaster (Zhang et al., 2012). 81% of the United States had to face drought-like situations in 2012. Droughts are likely to affect agriculture and enhance food security problems. This would result in famine, forced migration, and social unrest. Droughts may give a boost to the occurrence of forest and bush fires across the globe. Energy demand would increase during droughts, putting pressure on the already struggling energy sector. Droughts might decrease forest productivity when trees fail to face high stress, and their resilience levels diminish. Additionally, forests facing drought stress might be damaged by insects and diseases, the life cycle of which incorporates warm and dry conditions (Volney and Hirsch, 2005).

India implemented the "drought-prone area program" to tackle the persistence of drought in a number of districts across the country. The various measures included watershed management program, soil moisture conservation, afforestation, water harvesting, restructuring cropping patterns, diversifying agricultural system, implementing minor irrigation, and constructing canals (Mwinjaka et al., 2010).

10. Floods

Accumulation of water on lands, which are generally dry, is called as flood. It is normally caused by river water overflow, tidal water overflow or excess of precipitation (Smith and Ward, 1998; Brázdil et al., 2006; Ranger, 2011). Floods could be classified into river floods, coastal floods, and flash floods. River floods are an outcome of water flowing above its natural bank, causing inundation of nearby inhabited lands. Coastal flood is an outcome of a coastal storm or storm surge. Flash floods are caused by an excess of rain in a very short span of time. Additionally, urban floods could arise by any of the reasons causing floods, and poor drainage aggravates the situation. Recently, climate change has been a significant contributor to flood occurrences (Rodríguez, 2005; Seneviratne et al., 2012; IPCC, 2014a).

Climate change has affected water-related factors such as precipitation and snow melting in glaciers, causing floods (Haeberli, 1983; IPCC, 2007; Kay et al., 2009). The link between climate change and floods can be confirmed by an increase in the frequency and intensity of floods over the previous decade (Milly et al., 2002; Hirabayashi et al., 2008). A number of factors could be involved in climate change, causing floods (Christensen and C, 2003; Lehner et al., 2006). The increase in frequency, intensity, and duration of precipitation enhanced flood occurrences (Dankers and Feyen, 2008; Raff et al., 2009). Higher temperatures heat the air, and warmer air has been reported to hold a more considerable amount of moisture. Consequently, the air has been reported to hold 4%–27% more of moisture in the United States compared with the beginning of the 20th century (Douglas et al., 2000; Easterling et al., 2017). A study by the National Oceanic and Atmospheric Administration (NOAA) has determined that rains have been 10%–40% more intense as a result of climate change (NOAA, 2016). Substantial precipitation occurrences might triple by the end of the century. It would increase the potential for flood events. A rise in temperature would boost the melting of snow from glaciers. Additionally, such snow melting would be boosted by frequent rains coming from warmer atmospheric conditions. Rains accompanied by melting snow would cause floods.

Climate change is enhancing the frequency of storms, hurricanes, typhoons, and cyclones across the world because of the ability of warmer parcel of air to hold more moisture (Seneviratne et al., 2012; McClearn, 2018). Sadly, the frequency of these destructive hurricanes and cyclones would rise by 80% by the end of the century. Such storms would bring more substantial precipitation initiating catastrophic floods (USGS, 2019). Additionally, more energetic storms may aggravate storm surges and high tides, which help them, penetrate farther inland. As already stated, climate change would bring about a rise in sea level, enhancing storm surges and high tide floods (NOAA, 2018). These floods have doubled over the past 30 years in the United States alone. These are predicted to increase by more than 10 times in the next few decades (Carter et al., 2018).

Multiple countries have implemented efficient strategies to tackle the problem of floods. An Integrated Model for Flood Management was implemented in Nigeria with the help of which the authorities have facilitated information related to flood warning and flood forecasting to the public. They have also incorporated safety and rescue initiatives. Additionally, they brought about a three-step strategy of retaining, storing, and draining rain and floodwater. The precautionary principle was taken into account when considering flood prevention. Dams, levees, embankments, and flood diversion systems were incorporated to decrease the incidences of flooding. Flood plain zonation and regulation was performed to conserve the natural resources in flood plains. A compensation system for the flood victims was kept in mind while framing the prevention strategy (Joshua Oyebode, 2018). Interestingly, reusing the flood water to tackle water scarcity could be taken into consideration during the formulation of an integrated approach for floodwater management.

11. Cyclones, hurricanes, and typhoons

Although scientists are inconclusive about the relation between climate change and incidences of cyclone/hurricane, there has been growing acceptance that climate change leads to an increase in the occurrence of cyclones. In the present climatic condition, more than 90% of damage is caused by 10% of the total storms, and the remaining 90% causes less than 10% damage (Mendelsohn et al., 2012). The excessive release of GHGs has caused an increment of sea surface temperatures resulting in greater hurricane frequency and hurricane intensity by increasing the wind speeds of the tropical storms (Gillett et al., 2008). The damage is enhanced by a higher population near the coasts and infrastructure established in the vicinity of coasts (Pielke, 2008). The larger the population and the higher the infrastructure to uproot, colossal is the damage caused by the hurricane. An increase in damage has been reported in the recent past (Watson, 2001; Pielke, 2008).Climate change has induced intensification of cyclones in the tropics in the past four decades (Webster et al., 2005; Emanuel et al., 2008). It has been estimated that global damage would increase $53 billion every year. Tropical nations have been stated to be at higher risk (Mendelsohn et al., 2012).

The frequency of tropical cyclones is likely to increase from 6% to 34% on a global scale and might go beyond 50% by the end of the 21st century. The intensity of wind speeds in the cyclones would increase by 2%−11% by the late 21st century, a direct result of global warming. Rainfall, as already stated, would strengthen and increase in frequency (Bender et al., 2010; Easterling et al., 2017; Knutson et al., 2019). Interestingly, the areas facing hurricane threat by tropical cyclones may shift away from the equatorial region, moving toward the poles (Kossin et al., 2017). This could be an outcome of expansion in a tropical area with a rise in global temperatures. NOAA suggests that an increase in 2°C in global temperature would strengthen wind speed of a tropical cyclone by nearly 10% and rainfall increment by 15% (Knutson et al., 2010, 2019).

A rise in sea level would increase the submergence of coasts and smaller islands (Mimura, 2007; Nicholls and Cazenave, 2010). The combination of sea level rise and an increase in cyclone occurrences would have devastating effects. There would be an increment in storm surges resulting from such a combination (Walsh et al., 2012; Knutson et al., 2019). Interestingly, the frequency and intensity of cyclones would increase, but their effects could be different in different areas. Some regions may face more damage, and some may face less (Bender et al., 2010; Lavender and Walsh, 2011).

Several initiatives were implemented in the state of Odisha in India to tackle cyclones. It included the creation of multipurpose shelters, cyclone response forces, expert integration, capacity building, technology incorporation, community involvement, nongovernmental organization participation, and mock drills. Additionally, the Indian meteorological department provides early accurate cyclone warnings. The Government of India facilitates robust communication among the various stakeholders, timely evacuation of people and livestock, and provisioning of essential services and coordinates resilience in the event of a cyclone. Furthermore, the National Disaster Response Force is brought into action to enhance the response to the cyclone. Media participates actively to act as an appropriate channel between the public, government, and nongovernment agencies (Mommen et al., 2014; Jha et al., 2016).

12. Loss of biodiversity and impact on flora and fauna

As already described in previous sections, climate change has produced devastating outcomes, inclusive of extremes of temperatures, glacier melting, polar ice disappearing, enhanced climatic variations, increased sea level, and boosted oceanic acidification (Parmesan et al., 2000; Pereira et al., 2010; Bellard et al., 2012; Seneviratne et al., 2012). The animals have to migrate and find suitable places to live and survive (Leatherman and Beller-Simms, 1997; Downing, 2002; Malcolm et al., 2006). Animals incapable of migration would have to adapt to the newly created environmental conditions such as warmer oceans. Additionally, fragmentation of habitat and its size reduction has come into the picture because of agriculture diversification and deforestation (Kappelle et al., 1999). Changes in the distribution of plant species, on a global level, are predicted with change in climatic conditions (Maley, 1996; Thuiller et al., 2008). Melting of ice has given rise to the growth of grass and moss in the Antarctica peninsula, which is unprecedented and could be threatening indications of global warming (Agrawal, 2011). Furthermore, the crustacean population has been decreasing over the years, with a significant reduction in their fecundity. This has affected the species distribution and food chains involving these crustaceans, thereby affecting the mammals on a large-scale thriving in these cold environments (Agrawal, 2011). Sea ice has a critical role in the survival of multiple fauna species since they provide breeding, nursing, and resting spaces. The tundra vegetation is reported to shift northward, thereby affecting the reindeer and caribou population (Arctic Climate Assessment, 2004).

Threats to biodiversity, leading to its losses, could be assigned to a number of factors. The rise in temperature is one such vital factor (Pacifici et al., 2017). It would affect the physiology of animals, thereby altering their homeostasis. Prolonged heat waves and forest fires would kill a number of animals and cause massive destruction of plants and trees (Bush, 1997; Maclean and Wilson, 2011; Pacifici et al., 2015). An increase in carbon dioxide levels would affect the C/N ratio in plants, which could significantly alter the insect population and the food chains associated with these insect populations. Some plants may consume more of the plants, thereby changing the flora distribution. Rising temperatures could affect the frequency and length of diapause stages in insects (Hickling et al., 2006; Agrawal, 2011). Climate change could modify the pest population and their range could expand, threatening our food security (Estay et al., 2009). Furthermore, climate change could affect the insect population across the globe, modifying the avian population and their distribution, which could be deleterious for nature (Agrawal, 2011). Warmer oceanic temperatures could alter the pisces, amphibian, and reptile population and their distribution. Interestingly, there could be alterations in the gender distribution, i.e., some species may experience higher male populations, and others may observe higher female numbers (Agrawal, 2011).

The process of ocean acidification would affect marine life (Fabry et al., 2008; Guinotte and Fabry, 2008). Planktons living in these oceans would get affected by a rise in acid levels in oceans (Joint et al., 2010). Dependence of marine food chains on planktons would cause large-scale damage to the marine ecosystem. An increase in carbon dioxide levels would reduce the oxygen levels in oceans creating a number of dead zones, i.e., oxygen-less zones, which cannot support life systems (Diaz and Rosenberg, 2008). Furthermore, the transitional zone of mangroves may get affected because of ocean acidification (Hansen et al., 2010; Beaumont et al., 2011). Interestingly, the stratification of oceans would be enhanced by ocean warming (Palacios, 2004). The upper layers would be warmed more than the tolerance of planktons. The inability of planktons to feed, in the absence of nutrients, would lead to a decline in their growth and abundance. The consequence of such a decline would be deleterious for marine food chains and the ecosystem as a whole (Dobson et al., 2006; Doney et al., 2009).

Another outcome of higher temperatures and ocean acidification would be coral bleaching (Hoegh-Guldberg, 1999; Hughes et al., 2003). Corals would expel the inhabiting algae, i.e., zooxanthellae, because of the decrease in carbonate content in oceans. The existence of a symbiotic relationship makes algae necessary for the survival of corals. The resulting "coral bleaching" would be devastating for the marine ecosystem, because of the dependence of a huge marine life on the corals (Guinotte and Fabry, 2008; Hofmann et al., 2010). These corals are rightly called "rainforests of the oceans."

The increase in extreme events and disasters, such as fire, cyclone, drought or flood, caused because of human-induced climate change, would result in biodiversity loss (Hogg and Bernier, 2005; Keith et al., 2008; Kuiper et al., 2014; Wehner et al., 2017). The vegetation cover would get damaged, limiting its ability to support life system. Invasive or alien species might increase in number, adding to the stress on the existing life (Walther et al., 2009). Storm surges arising out of the rise in sea levels would be devastating for the coastal ecosystem, uprooting them exceptionally.

The rainfall pattern is bound to change in the scenario of climate change (Vittoz et al., 2013). Plants and animals, adapted to live in a particular weather pattern, might be affected (Bellard et al., 2012; El-Keblawy, 2014). An excess or lack of rainfall would bring about alterations in the life cycle of a number of plants and animals; an inability to adapt may result in their death (Maggini et al., 2011). Unsurprisingly, acidic rains would lead to additional deleterious effects (Schindler, 1988; Likens et al., 1996).

In situ and ex situ conservation methods have been incorporated in a number of countries across the planet. In situ methods include natural parks, wildlife sanctuaries, and biosphere reserves creation to conserve flora and fauna in their natural habitats. Ex situ methods include setting up of gene banks, seed banks, zoos, botanical gardens, and captive breeding of fauna (Al-Farabi, 2013). Several laws have been formulated by the government of India to reverse the biodiversity losses. These laws include Wildlife Protection Act, 1972, Environment Protection Act, 1986, Biological Diversity Act, 2002, and Forest Conservation Act, 1980 (Jayakumar, 1987; Mohanraj and Veenakumari, 1996; Sajwan and Kala, 2007).

13. Health effects

Climate change would be associated with adverse irreversible impact on the health of animals, including humans, and plants, and there could be a massive loss of life (Houghton et al., 1995). The increase in temperature causes excess heat, which would be unbearable for organisms, killing them on multiple occasions (Kalkstein, 1995). Dependence on cooling technology, for relief, could be counterproductive in the longer run. Additionally, such technologies produce gases polluting the atmosphere, affecting other parts of the ecosystem and causing deleterious effects to each individual (Burnett et al., 1994, 1997; Morris et al., 1995; Schwartz and Morris, 1995).

The increase in extreme, unpredictable weather events, such as floods, cyclones, droughts, and forest fires, would damage to life and property massively. Additionally, they bring about innumerous diseases in plants and animals (Patz et al., 1996; Epstein, 2001, 2002). Furthermore, there would be damage to ecosystem services, especially the loss of medicinal plants. Extreme weather events would damage the productivity of soil, resulting in inhibition of plant growth (Last et al., 1998). An increase in plant diseases would be associated with climate change. Pathogens affecting the plants could spread in their geographical extent with a warming of the climate and increase in humidity with greater precipitation (Crawford et al., 1991; Luo et al., 1995, 2003; Brasier, 1996; Carter et al., 1996).

Storm surges and inland intrusion of saline seawater would make the agricultural soil infertile, consequentially damaging the plants (Gornitz, 1991; Rahmstorf, 2017). It could be associated with contamination of groundwater, further aggravating the damage on the people living near to the coasts (Beckman, 1997). Changes in rainfall pattern, increasing the humidity levels, are also associated with an increase in disease outbreak by increasing the number and persistence of vectors and pathogens (Jofre et al., 2009). Changes in wind patterns would also affect the rollout of diseases (Chen et al., 2010).

Warmer temperatures would expand the habitat of several species of vectors, such as mosquitoes, causing a number of diseases such as malaria, dengue, and encephalitis (Epstein, 1997, 2002). Furthermore, a rise in temperature would cause an expansion in zones of disease transmission (Watson, 1998). An increase in El-Nino events has been linked to causing tropical diseases such as cholera, and climate change would result in further outbreaks of such diseases (Bouma and Dye, 1997). Climate change would increase the incidences of infectious diseases by affecting the pathogen, vector, host, or their living environment. Warming of climate has been reported to expand the range of such infectious disease (Patz et al., 1996; Epstein, 1999, 2001; Altizer et al., 2013; Bouzid et al., 2014). Diseases such as Hantavirus, Salmonellosis, and Giardiasis have started to come into the picture because of warmer climate and extreme weather events (Chretien et al., 2015).

14. Food security

The population explosion has resulted in an increase in food demand to very high levels, eventually creating food security issues. With persistent climate change, there would be an enhancement of the problem of food shortage (Hoogenboom, 2000; Lal, 2005; Howden et al., 2007). Floods and droughts would damage a lot of arable lands apart from the destruction of standing crops. Famine could arise from such developments (Challinor and Wheeler, 2008; Challinor et al., 2009; Gornall et al., 2010). Changes in the weather pattern have affected the production of crops throughout the globe. Additionally, pests may develop, which could destroy crops hugely, aggravating the problem of food scarcity (Aggarwal et al., 2006; Ziska et al., 2011). Irrigation would be affected due to alterations in rainfall patterns or differential melting of glaciers, ultimately affecting food security (Chang et al., 2002; Wilby, 2005; Wilby et al., 2006). Additionally, warmer temperatures might increase or decrease the crop production depending upon the optimal growth requirements for each crop. It has been reported that yield of rice increases, and maize production decreases at higher temperatures (Reddy and Pachepsky, 2000; Luo et al., 2003; Popova and Kercheva, 2005; Krishnan et al., 2007; Soler et al., 2007). Furthermore, climate change might affect soil properties such as soil water balance, creating food shortages (Fischer et al., 2007). Interestingly, the growth period of crops might decrease with changes in climatic conditions (Khan et al., 2009).

15. Climate refugees

Climate change would result in an unprecedented mass migration of people (Black, 1998; Hartmann, 2010). This could be an outcome of a rise in sea level, extreme hazard events such as flood or drought, disease outbreak, or global warming (Biermann and Boas, 2010; Farbotko and Lazrus, 2012). Islands and coastal areas getting submerged underwater or the emergence of uninhabitable places due to global warming would lead to humongous human displacement (Farbotko, 2010). This would create the problem of environmental refugees (Myers, 2002), leading to the development of newer problems of unhygienic human settlement issues such as disease outbreak. Political problems could increase (Hartmann, 2010). Violent conflicts might arise because of environmental stress (Homer-Dixon and Percival, 1996; Percival and Homer-Dixon, 1996).

16. Conclusion

It has been well established that climate change has worldwide repercussions. Climate change has led to a rise in global temperature; melting of glaciers and polar ice caps; sea level increment and changes in sea composition; weather pattern alterations; increased incidences of hazards and extreme events; disease outbreaks and emergence of new diseases; and biodiversity loss. For the planet to thrive, it becomes extremely necessary to look for adaptation and mitigation methods and technologies (IPCC, 2014b). Reduction in emission of GHGs, increased use of renewable energy, afforestation, adoption of sustainable practices, enhancement of carbon sequestration, water management, practicing 3R's (reduce, reuse and recycle), and awareness generation and international cooperation could be few of the ways to prevent climate change and minimize the resulting damages. On an individual level, one needs to plant trees and stop cutting them down, decrease food wastage, consume less of meat and give preference to plant-based diet, reduce fossil fuel consumption and incorporate renewable energy sources, prefer walking and cycling to riding an automobile, and save water and electricity on an hourly basis(Raj and Singh, 2013; Amadeo, 2020). To survive, we must learn to adapt and adopt sustainable practices. If we do not, we are bound to perish.

References

Abatzoglou, J.T., Barbero, R., 2014. Observed and projected changes in absolute temperature records across the contiguous United States. Geophys. Res. Lett. 41 (18), 6501–6508. https://doi.org/10.1002/2014GL061441.

Aggarwal, P.K., et al., 2006. InfoCrop: a dynamic simulation model for the assessment of crop yields, losses due to pests, and environmental impact of agro-ecosystems in tropical environments. I. Model description. Agric. Syst. 89 (1), 47–67. https://doi.org/10.1016/j.agsy.2005.08.001. Elsevier BV.

Agrawal, A., 2011. Effect of global warming on climate change, flora and fauna. J. Ecophysiol. Occup. Health 11 (3–4), 161–174.

Al-Farabi, J.W., 2013. Biodiversity conservation needs and method to conserve the biological diversity. J. Biodiver. Endanger. Species 01 (03). https://doi.org/10.4172/2332-2543.1000113.

Allen, L., 2004. Will Tuvalu Disappear beneath the Sea? Smithsonian magazine.

Allison, I., et al., 2015. Chapter 20 - ice sheets, glaciers, and sea level. In: Shroder, J. F., Haeberli, W., Whiteman Risks and Disasters, C. B. T.-S. and I.-R. H. (Eds.), Snow and Ice-Related Hazards, Risks and Disasters. Academic Press, Boston, pp. 713–747. https://doi.org/10.1016/B978-0-12-394849-6.00020-2.

Altizer, S., et al., 2013. Climate change and infectious diseases: from evidence to a predictive framework. Science 341 (6145), 514–519. https://doi.org/10.1126/science.1239401. American Association for the Advancement of Science (AAAS).

Amadeo, K., 2020. Rising Sea Level Effects, Projections, and Solutions. Available at: https://www.thebalance.com/sea-level-rise-and-climate-change-4158037.

Anderson, B.G., Bell, M.L., 2011. Heat waves in the United States: mortality risk during heat waves and effect modification by heat wave characteristics in 43 U.S. communities. Environ. Health Perspect. 119 (2), 210–218. https://doi.org/10.1289/ehp.1002313.

Arctic Climate Assessment, 2004. Impacts of Warming Arctic: Arctic Climate Impact Assessment. Cambridge University Press, Cambridge United Kingdom and New York, NY, USS.

Backlund, P., et al., 2008. The Effects of Climate Change on Agriculture, Land Resources, Water Resources, and Biodiversity in the United States (June). Available at: http://www.epa.gov/climatechange/impacts-adaptation/water.html.

Barnes, D.K.A., Kaiser, S., 2009. Melting of polar ice caps: impact on biodiversity. Fisher. Aquacul. 5 (345). Available at: https://pdfs.semanticscholar.org/df77/179ae88a52f3b434a53689e575fa4a76f32c.pdf.

Beaumont, L.J., et al., 2011. Impacts of climate change on the world's most exceptional ecoregions. Proc. Natl. Acad. Sci. U.S.A. 108 (6), 2306–2311. https://doi.org/10.1073/pnas.1007217108. Proceedings of the National Academy of Sciences.

Beckman, L., 1997. Responding to Global Climate Change, Effects of Climate Change on Coastal Systems in British Columbia and Yukon, pp. 8–18. Environment Canada.

Beer, T., Williams, A., 1995. Estimating Australian forest fire danger under conditions of doubled carbon dioxide concentrations. Climat. Change 29, 169–188.

Bellard, C., et al., 2012. Impacts of climate change on the future of biodiversity. Ecol. Lett. 15 (4), 365–377. https://doi.org/10.1111/j.1461-0248.2011.01736.x.

Bender, M.A., et al., 2010. Modeled impact of anthropogenic warming on the frequency of intense Atlantic hurricanes. Science 327 (5964), 454–458. https://doi.org/10.1126/science.1180568.

Beniston, M., Diaz, H.F., Bradley, R.S., 1997. Climatic change at high elevation sites: an overview. Climat. Change 36 (3–4), 233–251. https://doi.org/10.1007/978-94-015-8905-5_1.

Berwyn, B., 2018. Scientists Say Ocean Circulation Is Slowing. Here's Why You Should Care. Inside Climate News. Available at: https://insideclimatenews.org/news/07052018/atlantic-ocean-circulation-slowing-climate-change-heat-temperature-rainfall-fish-why-you-should-care.

Biermann, F., Boas, I., 2010. Preparing for a warmer world: towards a global governance system to protect climate refugees. Global Environ. Polit. 10 (1), 60–88. https://doi.org/10.1162/glep.2010.10.1.60.

Black, R., 1998. Refugees, Environment and Development. Addison Wesley Longman, Harlow.

Bouma, M.J., Dye, C., 1997. Cycles of malaria associated with El nino in Venezuela. J. Am. Med. Assoc. 278 (21), 1772–1774. https://doi.org/10.1001/jama.278.21.1772. American Medical Association (AMA).

Bouzid, M., et al., 2014. Climate change and the emergence of vector-borne diseases in Europe: case study of dengue fever. BMC Publ. Health 14 (1). https://doi.org/10.1186/1471-2458-14-781. Springer Science and Business Media LLC.

Brasier, C.M., 1996. Phytophthora cinnamomi and oak decline in southern Europe. Environmental constraints including climate change. Ann. Des. Sci. For. 53 (2–3), 347–358. https://doi.org/10.1051/forest:19960217. Springer Science and Business Media LLC.

Brázdil, R., Kundzewicz, Z.W., Benito, G., 2006. Historical hydrology for studying flood risk in Europe. Hydrol. Sci. J. 51 (5), 739–764. https://doi.org/10.1623/hysj.51.5.739.

Burnett, R.T., et al., 1994. Effects of low ambient levels of ozone and sulphates on the frequency of respiratory admissions to ontario hospitals. Environ. Res. 65, 172–194.

Burnett, R.T., et al., 1997. Association between ambient carbon monoxide levels and hospitalizations for congestive heart failure in the elderly in 10 Canadian cities. Epidemiology 8 (2), 162−167. https://doi.org/10.1097/00001648-199703000-00007. Ovid Technologies (Wolters Kluwer Health).

Bush, M.B., 1997. Ecology of a Changing Planet. Prentice Hall, Upper Saddle River, NJ.

Carleton, Anne, T., 2018. Human Productivity in a Warmer World: The Impact of Climate Change on the Global Workforce.

Carter, T.R., Saarikko, R.A., Niemi, K.J., 1996. Assessing the risks and uncertainties of regional crop potential under a changing climate in Finland. Agric. Food Sci. 5 (3), 329−350. https://doi.org/10.23986/afsci.72750. Agricultural and Food Science.

Carter, L., et al., 2018. Southeast. In: Impacts, Risks, and Adaptation in the United States: Fourth National Climate Assessment. Volume II[Reidmiller, D.R., C.W. Avery.

Challinor, A.J., Wheeler, T.R., 2008. Crop yield reduction in the tropics under climate change: processes and uncertainties. Agric. For. Meteorol. 148 (3), 343−356. https://doi.org/10.1016/j.agrformet.2007.09.015. Elsevier BV.

Challinor, A.J., et al., 2009. Ensemble yield simulations: crop and climate uncertainties, sensitivity to temperature and genotypic adaptation to climate change. Clim. Res. 38, 117−127. https://doi.org/10.3354/cr00779. Inter-Research Science Center.

Chang, H., et al., 2002. Water resource impacts of climate change in southwestern Bulgaria. Geojournal 57 (3), 159−168. https://doi.org/10.1023/b:gejo.0000003611.11187.5c. Springer Science and Business Media LLC.

Chen, P.S., et al., 2010. Ambient influenza and avian influenza virus during dust storm days and background days. Environ. Health Perspect. 118 (9), 1211−1216. https://doi.org/10.1289/ehp.0901782.

Cheung, W.W.L., et al., 2009. Projecting global marine biodiversity impacts under climate change scenarios. Fish Fish. 10 (3), 235−251. https://doi.org/10.1111/j.1467-2979.2008.00315.x. Wiley.

Chretien, J.P., et al., 2015. Global climate anomalies and potential infectious disease risks: 2014-2015. PLoS Current. 7 https://doi.org/10.1371/currents.outbreaks.95fbc4a8fb4695e049baabfc2fc8289f. OUTBREAKS.

Christensen, J.H., C., O.B., 2003. Climate modeling: severe summertime flooding in Europe. Nature 421, 805−806.

Cippec, G.L., et al., 2019. Delivering National Climate Action through Decarbonized Cities.

Cox, D., 2018. Two Audacious Plans for Saving the World's Ice Sheets. NBC News. Available at: https://www.nbcnews.com/mach/science/can-these-bold-plans-keep-world-s-ice-sheets-melting-ncna877616.

Crawford, J.W., Neilson, R., Boag, B., 1991. The effect of potential climatic changes on the geographical distribution of the plant-parasitic nematodes Xiphinema and longidor us in europe. Nematologica 37 (1−4), 312−323. https://doi.org/10.1163/187529291x00312. Brill.

Crowley, T.J., 2000. Causes of climate change over the past 1000 years. Science 289 (5477), 270−277. https://doi.org/10.1126/science.289.5477.270.

Dahl, K., et al., 2019. Increased frequency of and population exposure to extreme heat index days in the United States during the 21st century. Environ. Res. Communicat. 1 (7), 075002. https://doi.org/10.1088/2515-7620/ab27cf.

Dai, A., 2013. Increasing drought under global warming in observations and models. Nat. Climat. Change 3 (1), 52−58. https://doi.org/10.1038/nclimate1633. Springer Science and Business Media LLC.

Daley, J., 2017. Study Shows 84% of Wildfires Caused by Humans. Smithsonian Magazine.

Dankers, R., Feyen, L., 2008. Climate change impact on flood hazard in Europe: an assessment based on high-resolution climate simulations. J. Geophys. Res. Atmos. 113 (19). https://doi.org/10.1029/2007JD009719.

Davis, R.E., et al., 2003. Changing heat-related mortality in the United States. Environ. Health Perspect. (111), 1712–1718.

Day, E., et al., 2019. Upholding labour productivity under climate change: an assessment of adaptation options. Clim. Pol. 19 (3), 367–385. https://doi.org/10.1080/14693062.2018.1517640. Taylor & Francis.

Diaz, R.J., Rosenberg, R., 2008. Spreading dead zones and consequences for marine ecosystems. Science 321 (5891), 926–929. https://doi.org/10.1126/science.1156401. American Association for the Advancement of Science (AAAS).

Diffenbaugh, N.S., Scherer, M., 2013. Likelihood of July 2012 U.S. Temperatures in preindustrial and current forcing regimes. Bull. Am. Meteorol. Soc. 94 (9), S6–S9.

Dignon, J., Hameed, S., 1989. Global emissions of nitrogen and sulfur oxides from 1860 to 1980. J. Air Waste Manag. Assoc. 39 (2), 180–186. https://doi.org/10.1080/08940630.1989.10466519.

Dobson, A., et al., 2006. Habitat loss, trophic collapse, and the decline of ecosystem services. Wiley Ecology 87 (8), 1915–1924. https://doi.org/10.1890/0012-9658(2006)87[1915:HLTCAT]2.0.CO;2.

Doney, S.C., et al., 2009. ocean acidification: the other CO2Problem. Ann. Rev. Marine Sci. 1 (1), 169–192. https://doi.org/10.1146/annurev.marine.010908.163834. Annual Reviews.

Douglas, E.M., Vogel, R.M., Kroll, C.N., 2000. Trends in floods and low flows in the United States: impact of spatial correlation. J. Hydrol. 240 (1–2), 90–105. https://doi.org/10.1016/S0022-1694(00)00336-X.

Downing, T., 2002. Climate, Change and Risk, Climate, Change and Risk. https://doi.org/10.4324/9780203026175.

Duan, K., Mei, Y., 2014. Comparison of meteorological, hydrological and agricultural drought responses to climate change and uncertainty assessment. Water Resour. Manag. 28 (14), 5039–5054. https://doi.org/10.1007/s11269-014-0789-6. Springer Science and Business Media LLC.

Easterling, D.R., et al., 2017. Precipitation change in the United States. In: climate science special report: fourth national climate assessment. In: Fourth National Climate Assessment, vol. I, pp. 207–230. https://doi.org/10.7930/J0H993CC.U.S. I.

EIS, 2020. Why Do Glaciers Matter?, Extreme Ice Survey. Available at: http://extremeicesurvey.org/why-do-glaciers-matter/.

El-Keblawy, A., 2014. Impact of climate change on biodiversity loss and extinction of endemic plants of arid land mountains. J.Biodiver. Endanger. Specie. 02 (01). https://doi.org/10.4172/2332-2543.1000120.

Emanuel, K., Sundararajan, R., Williams, J., 2008. Hurricanes and global warming: results from downscaling IPCC AR4 simulations. Bull. Am. Meteorol. Soc. 89. https://doi.org/10.1175/BAMS-89-3-347.

Epstein, P.R., 1997. Climate, ecology, and human health. Consequence. – The Nat. Implicati. Environ. Change 3 (2), 3–19.

Epstein, P.R., 1999. Climate and health. Science 285 (5426), 347–348. https://doi.org/10.1126/science.285.5426.347. American Association for the Advancement of Science (AAAS).

Epstein, P.R., 2001. Climate change and emerging infectious diseases. Microb. Infect. 3 (9), 747–754. https://doi.org/10.1016/s1286-4579(01)01429-0. Elsevier BV.

Epstein, P.R., 2002. Detecting the infectious disease consequences of climate change and extreme weather events. Environ. Change, Clim. Health 172–196. https://doi.org/10.1017/cbo9780511535987.008. Cambridge University Press.

Eriksson, M., Shrestha, A.B., Jianchu, X., 2006. Climate change and its impacts on glaciers and water resource management in the Himalayan region. Assess. Snow, Glacier Water Res. Asia 27 (4), 258–265.

Estay, S.A., Lima, M., Labra, F.A., 2009. Predicting insect pest status under climate change scenarios: combining experimental data and population dynamics modelling. J. Appl. Entomol. 133 (7), 491–499. https://doi.org/10.1111/j.1439-0418.2008.01380.x.

Fabry, V.J., et al., 2008. ocean acidification's effects on marine ecosystems and biogeochemistry: ocean carbon and biogeochemistry scoping workshop on ocean acidification research; La jolla, California, 9–11 october 2007. Eos, Transact. Am. Geophys. Union 89 (15), 143. https://doi.org/10.1029/2008eo150004. American Geophysical Union (AGU).

Farbotko, C., 2010. Wishful sinking: disappearing islands, climate refugees and cosmopolitan experimentation. Asia Pac. Viewp. 51 (1), 47–60. https://doi.org/10.1111/j.1467-8373.2010.001413.x.

Farbotko, C., Lazrus, H., 2012. The first climate refugees? Contesting global narratives of climate change in Tuvalu. Global Environ. Change 22 (2), 382–390. https://doi.org/10.1016/j.gloenvcha.2011.11.014.

Fischer, G., et al., 2007. Climate change impacts on irrigation water requirements: effects of mitigation, 1990–2080. Technol. Forecast. Soc. Change 74 (7), 1083–1107. https://doi.org/10.1016/j.techfore.2006.05.021. Elsevier BV.

Fitzharris, B.B., et al., 1996. The Cryosphere: Changes and Their Impacts, Climate Change 1995, Impacts, Adaptations and Mitigation of Climate Change: Scientific-Technical Analyses.

Flannigan, M.D., Van Wagner, C.E., 1991. Climate change and wildfire in Canada. Can. J. For. Res. 21 https://doi.org/10.1139/x91-010.

Flannigan, M.D., et al., 2005. Future area burned in Canada. Clim. Change 1–16. https://doi.org/10.1007/s10584-005-5935-y.

Flannigan, M., et al., 2006. Forest fires and climate change in the 21ST century. Mitig. Adapt. Strategies Glob. Change 11, 847–859. https://doi.org/10.1007/s11027-005-9020-7.

Gagain, M., 2012. Climate change, sea level rise, and artificial islands: saving the Maldives' statehood and maritime claims through the constitution of the oceans. Colorado J. Int. Environ. Law Pol. 23, 1–44. Available at: http://heinonline.org.ludwig.lub.lu.se/HOL/Page?handle=hein.journals/colenvlp23&id=79&div=&collection=journals%5Cnhttp://heinonline.org.ludwig.lub.lu.se/HOL/Page?handle=hein.journals/colenvlp23&collection=journals&page=77.

Giannakopoulos, C., et al., 2005. Climate Change Impacts in the Mediterranean Resulting from a 2oC Global Temperature Rise.

Gillett, N.P., Stott, P.A., Santer, B.D., 2008. Attribution of cyclogenesis region sea surface temperature change to anthropogenic influence. Geophys. Res. Lett. 35 (9). https://doi.org/10.1029/2008GL033670.

Gladwin, et al., 1982. A global view of CFC sources and policies to reduce emissions. Econom. Manag. Chlorofluorocarbon.: Stratos. Ozone Clim. Issue. 64–113.

Glick, D., 2020. The Big Thaw: As the Climate Warms, How Much, and How Quickly, Will Earth's Glaciers Melt?, National Geographic. Available at: https://www.nationalgeographic.com/environment/global-warming/big-thaw/.

Gornall, J., et al., 2010. Implications of climate change for agricultural productivity in the early twenty-first century. Phil. Trans. Biol. Sci. 365 (1554), 2973–2989. https://doi.org/10.1098/rstb.2010.0158.

Gornitz, V., 1991. Global coastal hazards from future sea level rise. Palaeogeogr. Palaeoclimatol. Palaeoecol. 89 (4), 379–398. https://doi.org/10.1016/0031-0182(91)90173-O.

Guinotte, J.M., Fabry, V.J., 2008. Ocean acidification and its potential effects on marine ecosystems. Ann. N. Y. Acad. Sci. 1134, 320–342. https://doi.org/10.1196/annals.1439.013.

Habeeb, D., Vargo, J., Stone, B., 2015. Rising heat wave trends in large US cities. Nat. Hazards 76 (3), 1651–1665. https://doi.org/10.1007/s11069-014-1563-z.

Haeberli, W., 1983. Frequency and characteristics of glacier floods in the Swiss Alps. Ann. Glaciol. 4, 85–90. https://doi.org/10.3189/s0260305500005280.

Haeberli, W., 2010. Accelerated glacier and permafrost changes in the alps. In: Mountain Environments in Changing Climates, p. 91. https://doi.org/10.4324/9780203424957_chapter_5.

Haeberli, W., Beniston, M., 1998. Climate change and its impacts on glaciers and permafrost in the Alps. Ambio 27 (4), 258–265. https://doi.org/10.2307/4314732.

Haeberli, W., et al., 1993. Monitoring the long-term evolution of mountain permafrost in the Swiss Alps. In: Proceedings of the 6th International Conference on Permafrost, pp. 214–219. Beijing, China.

Hansen, L., et al., 2010. Designing climate-smart conservation: guidance and case studies: special section. Conserv. Biol. 24 (1), 63–69. https://doi.org/10.1111/j.1523-1739.2009.01404.x. Wiley.

Hartmann, B., 2010. Rethinking climate refugees and climate conflict: rhetoric, reality and the politics of policy discourse. J. Int. Dev. 22 (2), 233–246. https://doi.org/10.1002/jid.1676. John Wiley & Sons, Ltd.

Hely, C., Flannigan, M.D., Bergeron, Y., 2003. Modeling tree mortality following wildfire in the southeastern Canadian mixed-wood boreal forest. For. Sci. 49.

Hickling, R., et al., 2006. The distributions of a wide range of taxonomic groups are expanding polewards. Global Change Biol. 12 (3), 450–455. https://doi.org/10.1111/j.1365-2486.2006.01116.x.

Hirabayashi, Y., et al., 2008. Global projections of changing risks of floods and droughts in a changing climate. Hydrol. Sci. J. 53 (4), 754–772. https://doi.org/10.1623/hysj.53.4.754.

Hoegh-Guldberg, O., 1999. Climate change, coral bleaching and the future of the world's coral reefs. Mar. Freshw. Res. https://doi.org/10.1071/mf99078. CSIRO Publishing.

Hoegh-Guldberg, O., et al., 2007. Coral reefs under rapid climate change and ocean acidification. Science 318 (5857), 1737–1742. https://doi.org/10.1126/science.1152509. American Association for the Advancement of Science (AAAS).

Hofmann, G.E., et al., 2010. The effect of ocean acidification on calcifying organisms in marine ecosystems: an organism-to-ecosystem perspective. Annu. Rev. Ecol. Evol. Syst. 41 (1), 127–147. https://doi.org/10.1146/annurev.ecolsys.110308.120227. Annual Reviews.

Hogg, E.H., Bernier, P.Y., 2005. Climate change impacts on drought-prone forests in western Canada. For. Chron. 81 (5), 675–682. https://doi.org/10.5558/tfc81675-5. Canadian Institute of Forestry.

Holden, E., 2020. How the Oil Industry Has Spent Billions to Control the Climate Change Conversation. Washington. Available at: https://www.theguardian.com/business/2020/jan/08/oil-companies-climate-crisis-pr-spending.

Homer-Dixon, T., Percival, V., 1996. Environmental Scarcity and Violent Conflict: Briefing Book.

Hoogenboom, G., 2000. Contribution of agrometeorology to the simulation of crop production and its applications. Agric. For. Meteorol. 103 (1–2), 137–157. https://doi.org/10.1016/s0168-1923(00)00108-8. Elsevier BV.

Houghton, J.T., Meiro Filho, L.G., Callander, B.A., Harris, N., Kattenberg, A., Maskell, K., 1995. The science of climate change. Clim. Change 584.

Howden, S.M., et al., 2007. Adapting agriculture to climate change. Proc. Natl. Acad. Sci. U.S.A. 104 (50), 19691–19696. https://doi.org/10.1073/pnas.0701890104. Proceedings of the National Academy of Sciences.

Hughes, T.P., et al., 2003. Climate change, human impacts, and the resilience of coral reefs. Science 301 (5635), 929–933. https://doi.org/10.1126/science.1085046. American Association for the Advancement of Science (AAAS).

IPCC, 2007. Intergovernmental panel on climate change. In: Fourth Assessment Report. Climate Change 2007: Synthesis Report. Summary for Policymakers, pp. 99–121, 47(1).

IPCC, 2014. Climate Change: Impact Adaptation and Vulnerability.

IPCC, 2014. Mitigation of climate change. In: Contribution of Working Group III to the Fifth Assessment Report of the Intergovernmental Panel on Climate Change, p. 1454.

Jayakumar, K., 1987. Environment Protection Act: A Critical Overview. Cochin University Law Review, pp. 33–39, 11.

Jha, A., Basu, R., Basu, A., 2016. Studying policy changes in disaster management in India: a tale of two cyclones. Disaster Med. Public Health Prep. 10 (1), 42–46. https://doi.org/10.1017/dmp.2015.116.

Jofre, J., Blanch, A.R., Lucena, F., 2009. Water-Borne Infectious Disease Outbreaks Associated with Water Scarcity and Rainfall Events, pp. 147–159. https://doi.org/10.1007/698_2009_22.

Johnson, E.A., 1992. Fire and Vegetation Dynamics: Studies from the North American Boreal Forest. Cambridge University Press, Cambridge. https://doi.org/10.1017/CBO9780511623516.

Johnston, M., 2001. Sensitivity of Boreal Forest Landscapes to Climate Change. SRC. Publication No. 11341-6E01. Prepared for the Government of Canada's Climate Change Action Fund. Saskatchewan Research Council, Saskatoon, SK.

Joint, I., Doney, S.C., Karl, D.M., 2010. Will ocean acidification affect marine microbes? ISME J. 5 (1), 1–7. https://doi.org/10.1038/ismej.2010.79. Springer Science and Business Media LLC.

Joshua Oyebode, O., 2018. Strategy for tackling flood and water scarcity for adequate water supply and public health in Nigeria. Euro. J. Adv. Eng. Technol. 5 (8), 656–664.

Kalkstein, L.S., 1995. Lessons from a very hot summer. Lancet 346, 857–859.

Kappelle, M., Van Vuuren, M.M.I., Baas, P., 1999. Effects of climate change on biodiversity: a review and identification of key research issues. Biodivers. Conserv. 8 (10), 1383–1397. https://doi.org/10.1023/A:1008934324223.

Kay, A.L., et al., 2009. Comparison of uncertainty sources for climate change impacts: flood frequency in England. Clim. Change 92 (1–2), 41–63. https://doi.org/10.1007/s10584-008-9471-4.

Keith, D.A., et al., 2008. Predicting extinction risks under climate change: coupling stochastic population models with dynamic bioclimatic habitat models. Biol. Lett. 4 (5), 560–563. https://doi.org/10.1098/rsbl.2008.0049. The Royal Society.

Khan, S., Hanjra, M.A., Mu, J., 2009. Water management and crop production for food security in China: a review. Agric. Water Manag. 96 (3), 349–360. https://doi.org/10.1016/j.agwat.2008.09.022. Elsevier BV.

Kirschke, S., et al., 2013. Three decades of global methane sources and sinks. Nat. Geosci. 6 (10), 813–823.

Knutson, T.R., et al., 2010. Tropical cyclones and climate change. Nat. Geosci. 3 (3), 157–163. https://doi.org/10.1038/ngeo779.

Knutson, T., et al., 2019. Tropical cyclones and climate change assessment. Bull. Am. Meteorol. Soc. 100 (10), 1987–2007. https://doi.org/10.1175/BAMS-D-18-0189.1.

Kossin, J.P., et al., 2017. Extreme storms. In: Wuebbles, D.J., Fahey, D.W., Hibbard, K.A., Dokken, D.J., B.C. S (Eds.), Climate Science Special Report: Fourth National Climate Assessment, Volume I, pp. 257–276.

Krishnan, P., et al., 2007. Impact of elevated CO2 and temperature on rice yield and methods of adaptation as evaluated by crop simulation studies. Agric. Ecosyst. Environ. 122 (2), 233–242. https://doi.org/10.1016/j.agee.2007.01.019. Elsevier BV.

Kuiper, J.J., et al., 2014. The impact of river regulation on the biodiversity intactness of floodplain wetlands. Wetl. Ecol. Manag. 22 (6), 647–658. https://doi.org/10.1007/s11273-014-9360-8.

Kweku, D.W., et al., 2018. Greenhouse effect: greenhouse gases and their impact on global warming. J. Sci. Res. Rep. 17, 1–9. https://doi.org/10.9734/JSRR/2017/39630.

Lal, R., 2005. Climate change, soil carbon dynamics, and global food security. In: Climate Change and Global Food Security. CRC Press, pp. 113–143. https://doi.org/10.1201/9781420028614.ch5.

Lashof, D.A., Ahuja, D.R., 1990. Relative contributions of greenhouse gas emissions to global warming. Nature 344 (6266), 529–531. https://doi.org/10.1038/344529a0.

Last, J., Trouton, K., Pengelly, D., 1998. Taking Our Breath Away: Effects, the Health Change, Climate.

Lavender, S.L., Walsh, K.J.E., 2011. Dynamically downscaled simulations of Australian region tropical cyclones in current and future climates. Geophys. Res. Lett. 38 (10) https://doi.org/10.1029/2011gl047499 n/a-n/a.

Leatherman, S.P., Beller-Simms, N., 1997. Sea level rise and small island states: an overview. Coastal Education & Research Foundation, Inc. J. Coast Res. 1–16. Available at: http://www.jstor.org/stable/25736084.

Lehner, B., et al., 2006. Estimating the impact of global change on flood and drought risks in Europe: a continental, integrated analysis. Climatic Change 75 (3), 273–299. https://doi.org/10.1007/s10584-006-6338-4.

Leng, G., Tang, Q., Rayburg, S., 2015. Climate change impacts on meteorological, agricultural and hydrological droughts in China. Global Planet. Change 126, 23–34. https://doi.org/10.1016/j.gloplacha.2015.01.003. Elsevier BV.

Letman, J., 2018. Rising Seas Give Island Nation a Stark Choice: Relocate or Elevate, Environment in National Geographic.

Li, Z., et al., 2017. Towards low carbon based economic development: shanghai as a C40 city. Sci. Total Environ. 576, 538–548. https://doi.org/10.1016/j.scitotenv.2016.10.034.

Likens, G.E., Driscoll, C.T., Buso, D.C., 1996. Long-term effects of acid rain: response and recovery of a forest ecosystem. Science 272 (5259), 244–246. https://doi.org/10.1126/science.272.5259.244.

Lindsey, R., 2019. Climate change: global sea level. Clim. Watch. https://www.climate.gov/news-features/understanding-climate/climate-change-global-sea-level. (Accessed 12 November 2020).

Lucas, C., 2005. Fire climates of Australia: past, present and future. In: 6th Symposium on Fire and Forest Meteorology. Canmore, Alberta, Canada.

Luo, Y., et al., 1995. Simulation studies on risk analysis of rice leaf blast epidemics associated with global climate change in several asian countries. J. Biogeo. JSTOR 22 (4/5), 673. https://doi.org/10.2307/2845969.

Luo, Q., et al., 2003. Quantitative and visual assessments of climate change impacts on South Australian wheat production. Agric. Syst. 77 (3), 173–186. https://doi.org/10.1016/s0308-521x(02)00109-9. Elsevier BV.

Maclean, I.M.D., Wilson, R.J., 2011. Recent ecological responses to climate change support predictions of high extinction risk. Proc. Natl. Acad. Sci. U.S.A. 108 (30), 12337–12342. https://doi.org/10.1073/pnas.1017352108. Proceedings of the National Academy of Sciences.

Maggini, R., et al., 2011. Are Swiss birds tracking climate change?. Detecting elevational shifts using response curve shapes. Ecol. Model. 222 (1), 21–32. https://doi.org/10.1016/j.ecolmodel.2010.09.010.

Malcolm, J.R., et al., 2006. Global warming and extinctions of endemic species from biodiversity hotspots. Conserv. Biol. 20 (2), 538–548. https://doi.org/10.1111/j.1523-1739.2006.00364.x. Wiley.

Maley, J., 1996. The African rain forest - main characteristics of changes in vegetation and climate from the Upper Cretaceous to the Quaternary. Proc. Roy. Soc. Edinb. B Biol. Sci. 104, 31–73. https://doi.org/10.1017/S0269727000006114.

Marotzke, J., 2000. Abrupt climate change and thermohaline circulation: mechanisms and predictability. Proc. Natl. Acad. Sci. U.S.A. 97 (4), 1347–1350. https://doi.org/10.1073/pnas.97.4.1347. The National Academy of Sciences.

McClearn, M., 2018. Rising Seas and Climate Change: Everything You Need to Know. The Globe and Mail. Available at: https://www.nrdc.org/stories/hurricanes-and-climate-change-everything-you-need-know.

Mei'e, R., Renzhang, Y., H. B, 1985. An Outline of China's Physical Geography. Foreign Language Press, *Beijing*.

Melillo, J.M., et al., 2014. Climate Change Impacts in the United States: The Third National Climate Assessment, US Global Change Research Program. https://doi.org/10.7930/j0z31WJ2.

Mendelsohn, R., et al., 2012. The impact of climate change on global tropical cyclone damage. Nat. Clim. Change 2 (3), 205–209. https://doi.org/10.1038/nclimate1357.

Michelon, T., Magne, P., Simon-Delavelle, F., 2005. Lessons of the 2003 Heat-Wave in France and action taken to limit the effects of future heat-waves. In: Extreme Weather Events and Public Health Responses, pp. 131–140. https://doi.org/10.1007/3-540-28862-7_13.

Michelozzi, P., 2004. Impact of heatwaves on mortality—rome, Italy, june-august 2003. J. Am. Med. Assoc. 291 https://doi.org/10.1001/jama.291.21.2537.

Milly, P.C.D., et al., 2002. Increasing risk of great floods in a changing climate. Nature 415 (6871), 514–517. https://doi.org/10.1038/415514a.

Mimura, M.L., 2007. Small islands. In: Parry, M.L., et al. (Eds.), Climate Change 2007: Impacts, Adaptation and Vulnerability. Contribution of Working Group II to the Fourth Assessment Report of the Intergovernmental Panel on Climate Change. Cambridge Univ. Press, Cambridge.

Mishra, A.K., Singh, V.P., 2010. A review of drought concepts. J. Hydrol. 391 (1–2), 202–216. https://doi.org/10.1016/j.jhydrol.2010.07.012. Elsevier BV.

Mohanraj, P., Veenakumari, K., 1996. Nomenclature, classification and the basis of the schedules in the Indian wildlife (protection) act, 1972. Curr. Sci. 70 (6), 428–432.

Mommen, B., Roy, D., Sethi, J., 2014. Resilient against the elements: the preparedness and response to cyclone "Phailin" by the state of Odisha, India. In: Sustainable Water and Sanitation Services for all in a Fast Changing World in 37th WEDC International Conference, pp. 1−6 (December 1999).

Moriondo, M., et al., 2006. Potential impact of climate change on fire risk in the Mediterranean area. Clim. Res. 31 (1), 85−95. https://doi.org/10.3354/cr031085.

Morris, R.D., Naumova, E.N., Munasinghe, R.L., 1995. Ambient air pollution and hospitalization for congestive heart failure among elderly people in seven large US cities. Am. J. Publ. Health 85 (10), 1361−1365. https://doi.org/10.2105/AJPH.85.10.1361. American Public Health Association.

Mueller, N.D., et al., 2017. Global relationships between cropland intensification and summer temperature extremes over the last 50 years. J. Clim. 30 (18), 7505−7528. https://doi.org/10.1175/JCLI-D-17-0096.1.

Mwinjaka, O., Gupta, J., Bresser, T., 2010. Adaptation strategies of the poorest farmers in drought-prone Gujarat. Clim. Dev. 2 (4), 346−363. https://doi.org/10.3763/cdev.2010.0058.

Myers, N., 2002. Environmental refugees: a growing phenomenon of the 21st century. Phil. Trans. Biol. Sci. 357 (1420), 609−613. https://doi.org/10.1098/rstb.2001.0953.

Nicholls, R.J., Cazenave, A., 2010. Sea-level rise and its impact on coastal zones. Science 328 (5985), 1517−1520. https://doi.org/10.1126/science.1185782.

NIDM, 2012. Forest Fire Disaster Management, p. 232.

NOAA, 2016. Climate Change Increased Chances of Record Rains in Louisiana by at Least 40 Percent | National Oceanic and Atmospheric Administration. Available at: http://www.noaa.gov/media-release/climate-change-increased-chances-of-record-rains-in-louisiana-by-at-least-40-percent?dom=prime&src=syn.

NOAA, 2018. What Is High Tide Flooding? NOAA.

Noone, K.J., 2013. Chapter 5 - sea-Level rise. In: Noone, K.J., Sumaila, U.R., Diaz, R.J.B.T.-M.O.E., a C. C. (Eds.), Managing Ocean Environments in a Changing Climate. Elsevier, Boston, pp. 97−126. https://doi.org/10.1016/B978-0-12-407668-6.00005-7.

Odjugo, P.A.O., 2010. General overview of climate change impacts in Nigeria. J. Hum. Ecol. 29 (1), 47−55. https://doi.org/10.1080/09709274.2010.11906248.

Oppenheimer, M., Anttila-Hughes, J.K., 2016. The Science of Climate Change, Future of Children. Cambridge University Press, Cambridge, UK. https://doi.org/10.1353/foc.2016.0001.

Oudin Åström, D., Bertil, F., Joacim, R., 2011. Heat wave impact on morbidity and mortality in the elderly population: a review of recent studies. Maturitas 69 (2), 99−105. https://doi.org/10.1016/j.maturitas.2011.03.008.

Pacifici, M., et al., 2015. Assessing species vulnerability to climate change. Nat. Clim. Change 5 (3), 215−225. https://doi.org/10.1038/nclimate2448.

Pacifici, M., et al., 2017. Species' traits influenced their response to recent climate change. Nat. Clim. Change 7 (3), 205−208. https://doi.org/10.1038/nclimate3223.

Palacios, D.M., 2004. Long-term and seasonal trends in stratification in the California Current, 1950−1993. J. Geophys. Res. 109 (C10) https://doi.org/10.1029/2004jc002380. American Geophysical Union (AGU).

Parmesan, C., Root, T.L., Willig, M.R., 2000. Impacts of extreme weather and climate on terrestrial biota*. Bull. Am. Meteorol. Soc. 81 (3), 443−450. https://doi.org/10.1175/1520-0477(2000)081<0443:ioewac>2.3.co;2. American Meteorological Society.

Patz, J.A., et al., 1996. Global climate change and emerging infectious diseases. J. Am. Med. Assoc. 275. https://doi.org/10.1001/jama.1996.03530270057032.

Percival, V., Homer-Dixon, T., 1996. Environmental scarcity and violent conflict: the case of Rwanda. J. Environ. Dev. 5 (3), 270–291. https://doi.org/10.1177/107049659600500302. SAGE Publications.

Pereira, H.M., et al., 2010. Scenarios for global biodiversity in the 21st century. Science 330 (6010), 1496–1501. https://doi.org/10.1126/science.1196624. American Association for the Advancement of Science (AAAS).

Pielke, R.A., 2008. Normalized hurricane damages in the United States: 1900–2005. Nat. Hazards Rev. 9. https://doi.org/10.1061/(ASCE)1527-6988(2008)9:1(29).

Polyakov, I.V., et al., 2010. Arctic ocean warming contributes to reduced polar ice cap. J. Phys. Oceanogr. 40 (12), 2743–2756. https://doi.org/10.1175/2010jpo4339.1. American Meteorological Society.

Popova, Z., Kercheva, M., 2005. CERES model application for increasing preparedness to climate variability in agricultural planning—risk analyses. Phys. Chem. Earth, Parts A/B/C 30 (1–3), 117–124. https://doi.org/10.1016/j.pce.2004.08.025. Elsevier BV.

Price, C., Rind, D., 1994. The impact of a 2 × CO2 Climate on lightning-caused fires. J. Clim. 7, 2.0.CO;2.

Raff, D.A., Pruitt, T., Brekke, L.D., 2009. A framework for assessing flood frequency based on climate projection information. Hydrol. Earth Syst. Sci. 13 (11), 2119–2136. https://doi.org/10.5194/hess-13-2119-2009.

Rahmstorf, S., 2007. Sea-level rise a semi-empirical approach to projecting future. Science 315. https://doi.org/10.1126/science.1135456.

Rahmstorf, S., 2017. Rising hazard of storm-surge flooding. Proc. Natl. Acad. Sci. Unit. States Am. 114 (45). https://doi.org/10.1073/pnas.1715895114, 11806 LP – 11808.

Raj, B., Singh, O., 2013. A study about realities of climate change: glacier Melting and growing crises. Clim. Change - Real., Impact. Over Ice Cap, Sea Level Risk. https://doi.org/10.5772/54968.

Ranger, N., et al., 2011. An assessment of the potential impact of climate change on flood risk in Mumbai. Clim. Change 104, 139–167.

Raymond, C., Singh, D., Horton, R.M., 2017. Spatiotemporal patterns and synoptics of extreme wet-bulb temperature in the contiguous United States. J. Geophys. Res.: Atmos. 122 (24) https://doi.org/10.1002/2017JD027140, 13,108-13,124.

Reddy, V.R., Pachepsky, Y.A., 2000. Predicting crop yields under climate change conditions from monthly GCM weather projections. Environ. Model. Software 15 (1), 79–86. https://doi.org/10.1016/s1364-8152(99)00011-0. Elsevier BV.

Rodríguez, J.M.M., 2005. A Preliminary General Assessment of the Impacts in Spain Due to the Effects of Climate Change. Ministerio de Medio Ambiente and Universidad de Castilla la Mancha.

Sajwan, B.S., Kala, C.P., 2007. Conservation of medicinal plants: conventional and contemporary strategies, regulations and executions. Indian For. 133 (4), 484–495.

Santer, B.D., et al., 1996. A search for human influences on the thermal structure of the atmosphere. Nature 382 (6586), 39–46. https://doi.org/10.1038/382039a0.

Satterthwaite, D., 2009. The implications of population growth and urbanization for climate change. Environ. Urbanizat. 21 (2), 545–567. https://doi.org/10.1177/0956247809344361.

Schindler, D.W., 1988. Effects of acid rain on freshwater ecosystems. Science 239 (4836), 149–157. https://doi.org/10.1126/science.239.4836.149.

Schmeltz, M.T., et al., 2015. Identifying individual risk factors and documenting the pattern of heat-related illness through analyses of hospitalization and patterns of household cooling. PloS One 10 (3). https://doi.org/10.1371/journal.pone.0118958.

Schwartz, J., Morris, R., 1995. Air pollution and hospital admissions for cardiovascular disease in detroit, Michigan. Am. J. Epidemiol. 142 (1), 23–35. https://doi.org/10.1093/oxfordjournals.aje.a117541. Oxford University Press (OUP).

Sea Level Rise organisation, 2020. Solutions Can Protect Coastal Communities, Solutions Can Protect Coastal Communities. Available at: https://sealevelrise.org/solutions/.

Seneviratne, S.I., et al., 2012. Changes in climate extremes and their impacts on the natural physical environment: an overview of the IPCC SREX report. EGU Gene. Assem. 2012 14, 12566. https://doi.org/10.1017/CBO9781139177245.006.

Sheffield, J., Wood, E.F., 2008. Projected changes in drought occurrence under future global warming from multi-model, multi-scenario, IPCC AR4 simulations. Clim. Dynam. 31 (1), 79–105. https://doi.org/10.1007/s00382-007-0340-z. Springer Science and Business Media LLC.

Sherwood, S.C., Huber, M., 2010. An adaptability limit to climate change due to heat stress. Proc. Natl. Acad. Sci. U.S.A. 107 (21), 9552–9555. https://doi.org/10.1073/pnas.0913352107.

Shickman, K., 2017. A Hot Topic: Cities Tackle Rising Temperatures. Available at: https://www.c40.org/blog_posts/cool-cities-august.

Shrestha, A.B., et al., 2000. Precipitation fluctuations in the Nepal Himalaya and its vicinity and relationship with some large scale climatological parameters. Wiley Int. J. Climatol. 20 (3), 317–327. https://doi.org/10.1002/(sici)1097-0088(20000315)20:3317::aid-joc4763.0.co;2-g.

Smith, K., Ward, R., 1998. Floods: Physical Processes and Human Impacts. Book, p. 394. Available at: http://eu.wiley.com/WileyCDA/WileyTitle/productCd-0471952486.html.

Soler, C.M.T., Sentelhas, P.C., Hoogenboom, G., 2007. Application of the CSM-CERES-Maize model for planting date evaluation and yield forecasting for maize grown off-season in a subtropical environment. Eur. J. Agron. 27 (2–4), 165–177. https://doi.org/10.1016/j.eja.2007.03.002. Elsevier BV.

Solomon, S., et al., 2007. Climate Change 2007: The Physical Science Basis: Contribution of Working Group I to the Fourth Assessment Report of the Intergovernmental Panel on Climate Change. Cambridge Univ. Press, Cambridge.

Stocks, B.J., et al., 1998. Climate change and forest fire potential in Russian and Canadian boreal forests. Clim. Change 38. https://doi.org/10.1023/A:1005306001055.

Stott, P.A., Tett, S.F.B., Jones, G.S., Allen, M.R., Ingram, W.J., Mitchell, J.F.B., 1999. Attribution of twentieth century temperature change to natural and anthropogenic causes. Clim. Dynam. 17 (1), 1–21.

Swetnam, T.W., 1993. Fire history and climate change in giant sequoia groves. Science 262. https://doi.org/10.1126/science.262.5135.885.

Tett, S.F.B., et al., 1999. Causes of twentieth-century temperature change near the Earth's surface. Nature 399 (6736), 569–572. https://doi.org/10.1038/21164.

Thornes, J.E., 2002. IPCC, 2001: climate change 2001: impacts, adaptation and vulnerability, contribution of working group II to the third assessment report of the intergovernmental panel on climate change. edited by J. J. McCarthy, O. F. Canziani, N. A. Leary, D. J. Dokken a Int. J. Climatol. 22 (10), 1285–1286. https://doi.org/10.1002/joc.775. Wiley.

Thuiller, W., et al., 2008. Predicting global change impacts on plant species' distributions: future challenges. Perspect. Plant Ecol. Evol. Systemat. 9 (3–4), 137–152. https://doi.org/10.1016/j.ppees.2007.09.004. Elsevier BV.

Tim Lister, 2012. Rising Sea Level Puts Island Nations like Nauru at Risk. CNN.

Tsai, F., Yau, J., 1990. Solubility of Carbon Dioxide in Water. Available at: https://www.metlink.org/wp-content/uploads/2013/10/solubility-of-CO2.pdf.

USGS, 2019. USGS: Florence Set at Least 28 Flood Records in Carolinas. USGS. Available at: https://www.usgs.gov/news/usgs-florence-set-least-28-flood-records-carolinas.

Vittoz, P., et al., 2013. Climate change impacts on biodiversity in Switzerland: a review. J. Nat. Conserv. 21, 154–162. https://doi.org/10.1016/j.jnc.2012.12.002.

Volney, W.J.A., Hirsch, K.G., 2005. Disturbing forest disturbances. For. Chron. 81 (5), 662–668. https://doi.org/10.5558/tfc81662-5. Canadian Institute of Forestry.

Vose, J.M., Peterson, D.L., Patel-Weynand, T., 2012. Effects of climatic variability and change on forest ecosystems: a comprehensive science synthesis for the US forest sector. Forest Serv. Gener. Tech. Rep. 1–265. PNW-GTR-87.

Vose, R.S., et al., 2017. Temperature changes in the United States. In: Climate Science Special Report: Fourth National Climate Assessment, pp. 185–206. https://doi.org/10.7930/J0N29V45.U.S.

Vose, J.M., et al., 2018. Forests. In: Reidmiller, D.R., Avery, C.W., Easterling, D.R., Kunkel, K.E., Lewis, K.L.M., Maycock, T.K., Stewart, B.C. (Eds.), Impacts, Risks, and Adaptation in the United States: Fourth National Climate Assessment, Volume II. U.S. Global Change Research Program, Wash, pp. 232–267.

Walsh, K.J.E., McInnes, K.L., McBride, J.L., 2012. Climate change impacts on tropical cyclones and extreme sea levels in the South Pacific - a regional assessment. Glob. Planet. Change 80–81, 149–164. https://doi.org/10.1016/j.gloplacha.2011.10.006.

Walther, G.R., et al., 2009. Alien species in a warmer world: risks and opportunities. Trends Ecol. Evol. 24 (12), 686–693. https://doi.org/10.1016/j.tree.2009.06.008. Elsevier BV.

Ward, M., 2015. The principal greenhouse gases and their sources. Nat. Environ. Educat. Foundat. Available at: https://www.neefusa.org/weather-and-climate/climate-change/principal-greenhouse-gases-and-their-sources.

Watson, R.T., 1998. The Regional Impacts of Climate Change. IPCC.

Watson, R.T., 2001. IPCC - Climate Change 2001: Synthesis Report. Cambridge Univ. Press. Available at: http://www.grida.no/publications/267.

Watson, R.T., et al., 1992. Greenhouse gases: sources and sinks. Clim. change 92, 25–46.

Webster, P.J., et al., 2005. Atmospheric science: changes in tropical cyclone number, duration, and intensity in a warming environment. Science 309 (5742), 1844–1846. https://doi.org/10.1126/science.1116448.

Wehner, M.F., et al., 2017. Droughts, floods, and wildfires. In: Wuebbles, D.J., Fahey, D.W., Hibbard, K.A., Dokken, D.J., Stewart, B.C., Maycock, T.K. (Eds.), Climate Science Special Report: Fourth National Climate Assessment, Volume I.

Wilby, R.L., 2005. Uncertainty in water resource model parameters used for climate change impact assessment. Hydrol. Process. 19 (16), 3201–3219. https://doi.org/10.1002/hyp.5819. Wiley.

Wilby, R.L., et al., 2006. Integrated modelling of climate change impacts on water resources and quality in a lowland catchment: river Kennet, UK. J. Hydrol. 330 (1–2), 204–220. https://doi.org/10.1016/j.jhydrol.2006.04.033. Elsevier BV.

Wilhite, D.A., Buchanan-Smith, M., 2005. Drought as hazard: understanding the natural and social context. Drought Water Crise.: Sci., Technol. Manage. Issue. 3–29. https://doi.org/10.1201/b22009-2. CRC Press.

Wilhite, D.A., Glantz, M.H., 1985. Understanding: the drought phenomenon: the role of definitions. Water Int. 10 (3), 111–120. https://doi.org/10.1080/02508068508686328. Informa UK Limited.

Wilkinson, C., Zimmermann, S., 2014. Evaluation of Civil Protection Mechanism Case Study Report - Forest Fires in Europe.

Wisser, D., et al., 2010. Reconstructing 20th century global hydrography: a contribution to the global terrestrial network- hydrology (GTN-H). Hydrol. Earth Syst. Sci. 14 (1), 1–24. https://doi.org/10.5194/hess-14-1-2010. Copernicus GmbH.

Xu, J., et al., 2009. The melting Himalayas: cascading effects of climate change on water, biodiversity, and livelihoods. Conserv. Biol. 23 (3), 520–530. https://doi.org/10.1111/j.1523-1739.2009.01237.x. Wiley.

Xu, Z., et al., 2012. Impact of ambient temperature on children's health: a systematic review. Environ. Res. 117, 120–131. https://doi.org/10.1016/j.envres.2012.07.002.

Yang, J., et al., 2012. Extreme drought event of 2009/2010 over southwestern China. Meteorol. Atmos. Phys. 115 (3–4), 173–184. https://doi.org/10.1007/s00703-011-0172-6. Springer Science and Business Media LLC.

Yi, S., Heki, K., Qian, A., 2017. Acceleration in the global mean sea level rise: 2005–2015. John Wiley & Sons, Ltd Geophys. Res. Lett. 44 (23). https://doi.org/10.1002/2017GL076129, 11,905-911,913.

Zhang, L., et al., 2012. The 2010 spring drought reduced primary productivity in southwestern China. Environ. Res. Lett. 7 (4), 45706. https://doi.org/10.1088/1748-9326/7/4/045706. IOP Publishing.

Ziska, L.H., et al., 2011. Invasive species and climate change: an agronomic perspective. Clim. Change 105 (1), 13–42. https://doi.org/10.1007/s10584-010-9879-5. Springer Science and Business Media LLC.

Impact of climate change on biodiversity and shift in major biomes

2

N.K. Hemanth Kumar[1], *M. Murali*[1], *H.V. Girish*[1], *S. Chandrashekar*[3], *K.N. Amruthesh*[1], *M.Y. Sreenivasa*[2], *Shobha Jagannath*[1]
[1]Department of Studies in Botany, University of Mysore, Mysuru, Karnataka, India; [2]Department of Studies in Microbiology, University of Mysore, Mysuru, Karnataka, India; [3]Department of Studies in Biotechnology, Shivagangotri, Davangere University, Davangere, Karnataka, India

1. Introduction

Climate change signifies one of the vital research challenges presently faced by plant biologists, agronomists, and conservation biologists. With global greenhouse gas emissions continuing to increase for the predictable future, the impact of elevated atmospheric carbon dioxide and associated shifts in temperature and rainfall are all expected to impact plant physiology, distribution, and interactions with other biotic and abiotic factors (Settele et al., 2014). Biodiversity and biodiversity-based ecosystem services are intrinsically dependent on the climate. During the 20th century, climate change has posed major threats to biodiversity, and impacts are expected to increase as climate change continues and perhaps even accelerates. However, a certain change in climatic conditions is well suited with the natural ecosystem structure and its function. The climate sometimes shifts rapidly and is detrimental to the variety of life (Heller and Zavaleta, 2009). Biodiversity provides several direct benefits to humans; around 40% of the world's economy is derived from biological resources. Maintaining biodiversity gives greater food security and provides a foundation for new pharmaceuticals and other medical advances. Ironically, maintaining biodiversity levels and functioning ecosystems is critical to ameliorating climate change; yet, climate change is expected to cause severe disruptions to Earth's ecological systems, resulting in an overall loss of biodiversity and a reduction in the goods and services provided to humans. Extinction rates in the future are exceedingly difficult to predict. However, with immediate and decisive action to mitigate climate change, losses of biodiversity can be minimized (Reed, 2012).

Climate is one of the key components of the environment that directly influences the structure and function of the ecosystems. Changing climatic factors might directly or indirectly affect the ecosystems in different ways. For example, increasing the universal temperature may force species to shift or migrate to other geographical areas or higher altitudes or latitudes depending upon their suitability, where the temperature

Global Climate Change. https://doi.org/10.1016/B978-0-12-822928-6.00007-1

was found to be more suitable to their survival (Araujo et al., 2013; Bellard et al., 2012). Likewise, as sea level increases, marine water invasion into a freshwater ecosystem may compel some important species to change place and thus take off predators or prey that are significant in the accessible food chain. From the past century, the average universal temperature has risen by 0.7°C, and it was estimated to continue rising in the future also (Bates et al., 2014; Bellard et al., 2012). The Intergovernmental Panel on Climate Change (IPCC) predicted that the average universal temperature is projected to increase by 1.1−6.4°C by the end of the 21st century (IPCC, 2012). The link between global climatic change and biodiversity has long been well established. Even though, during the history of the Earth, the climatic conditions have always been changed with types of ecosystems and species immigration and emigration, quick climate change may affect different ecosystems and the species capability to acclimatize to new climatic conditions; as a result, the biodiversity loss increases.

Plant and animal species may be unable to acclimatize to the hastily changing environmental conditions, and this is one of the most acute conditions; thus, many plant and animal species will be threatened by extinction (Bates et al., 2014). The change in the climatic condition does not only honestly affect the different ecosystem and species diversity, but it is also associated with some of the other anthropogenic stressors. Even though some stressors cause cumulative effects when acting unaccompanied, but sometimes their growing influence may lead to amazing ecological changes (Settele et al., 2014).

Due to anthropogenic activities, biodiversity is found to be under growing threats (Johnson et al., 2017). Climate change, directly and indirectly, affects the global biodiversity value because the climate factors are one of the chief components to determine the phytogeographical distribution of species (Pearson and Dawson, 2003). Therefore, in ecological areas where the environmental condition is no longer more suitable, may shift their geographical ranges to their new suitable range (Bellard et al., 2012). All these unenthusiastic impacts deteriorate the challenge of supervising and preserve biodiversity (CBD, 2019).

2. Effects of climate change on biodiversity

The global biodiversity is being gradually more susceptible through a range of human-made activities (Barlow et al., 2018). The Convention of Biological Diversity (CBD) defines "biodiversity loss" as "the long-term reduction in apparatus of biodiversity and their capability to provide goods and services." The fossil records proved that it took on approximately 1 million years to an entity vertebrate species, which became extinct (Lawton and May, 1995). Therefore, change in the climatic condition might be responsible for the change in biodiversity, which might have led to the shift in the form of organisms from one environment to another (Whiteside and Ward, 2011). The reasons behind the present-day biodiversity are due to the change in the habitat structure, climate, overexploitation, introduction of the invasive alien species, and extinction of the organisms, which are collectively considered as five evil for loss of biodiversity

(Guo et al., 2017). As mentioned in the report IPCC (2013), the global warming and rainfall are estimated to augment, the varying climate is suspected to be one of the most horrible reasons for loss of biodiversity in the next 50−100 years, and supplementary embellished the effects of previous intimidation on biodiversity loss. Nevertheless, only some isolated studies have reported honestly correlated climate change−induced biodiversity loss (Botkin et al., 2007; Callaghan et al., 2004).

The various components that are accountable for climate change in different levels of biodiversity are depicted in Fig. 2.1. They chiefly concern a range of strengths and forms of strength, which are expressed at diverse levels and have may affect the individuals in population, species, and ecosystems levels. At the basic levels of biodiversity, change in the climatic condition is maybe responsible for a decrease in genetic diversity of the populations due to different types of selection and voyage, which might, in turn, affect the ecosystem performance and adaptability (Botkin et al., 2007). Conversely, most studies are concentrated on impacts at elevated organizational levels, and climate change has explored for a small number of species (CBD, 2019).

Furthermore, the diverse belongings on populations are expected to change the "web of interactions" at the level of community. In essence, the reaction of some species to change in climatic conditions may comprise a circuitous bang on the species which are depending on them (Koh et al., 2004; Walther, 2010). In addition to this, in several species, primary effect of climate change might be through possessions on the same time with species basic requirements such as food and shelter. The change in the climatic condition has led to phonological shifts in Angiospermic plants and anemophily pollinators and wrong identification between plants and pollinators, which leads to the loss of both the plant and the associated pollinator (Kiers et al., 2010).

3. Changes in the recurring life cycle events

The climate of our Earth is at a disturbing stage; constantly increasing sea levels considerably affects the plant's growth and development. Climate change is a very diversified phenomenon faced by all the living entities. It comprises many disciplines beyond science such as society, economics, politics, and mainly moral and ethical issues; climatic fluctuations are considered as a global problem, but arise and experience from small scale (https://climate.nasa.gov/solutions/adaptation-mitigation).

For many plant and animal species, the environmental or climatic condition, where they live out part of life, significantly influences important stages of their life cycle, viz., blooming, movement, and imitation. As winters are shorter and milder in some of the places, the period of these programs has been altered in few parts of the country. For example, in the United States, the former springs have led to previous nesting of 28 traveling bird species on the East Coast (Settele et al., 2014).

Due to sessile nature, plants cannot be able to move from one climatic condition to another like all of us. Therefore, plants must adapt themselves to other alternative mechanisms. Because of changes in physiology more specifically, we can say that

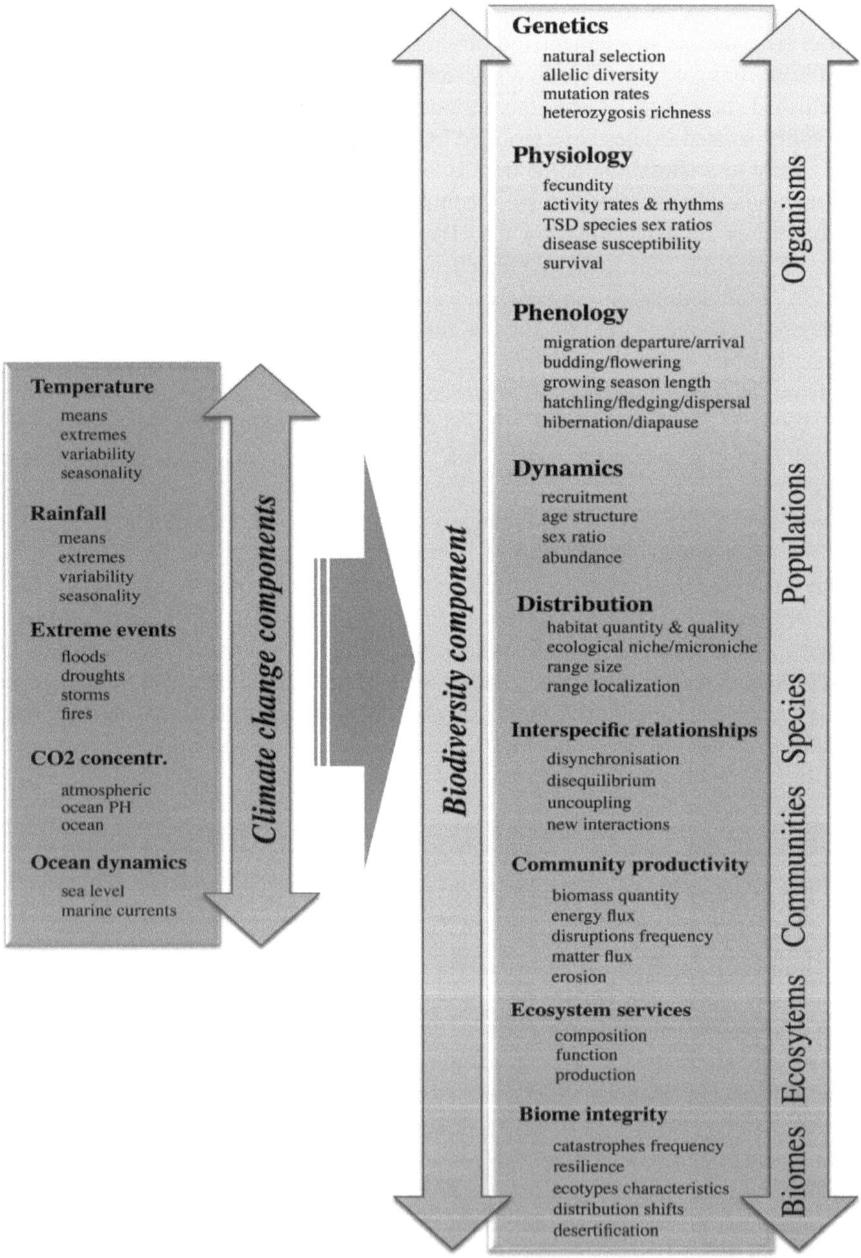

Figure 2.1 Summary of some of the predicted aspects of climate change and some examples of their likely effects on different levels of biodiversity.
Source: Bellard, C., Bertelsmeier, C., Leadley, P., Thuiller, W., Courchamp, F. 2012. Impacts of climate change on the future of biodiversity. Ecol. Lett. 15, 365–377.

variation in secondary metabolites production is considered to be one of the plant's defense mechanisms toward unfavorable conditions (Ober, 2005; Pichersky and Gang, 2000). Secondary metabolites are the compounds, which are not required for plant growth and development, but these compounds such as alkaloids, terpenes, and cyanogenic glycosides together boost the plant immune system (Hartmann, 2007).

Change in the climatic condition may directly or indirectly affect the quality of natural products and its influence on the medicinal value of plants (Gore, 2006). However, it was reported that such alterations might be positive or negative. Generally, secondary metabolite production is increased in stressed conditions; nevertheless, the production of secondary metabolites is inspired by many biotic and abiotic factors such as competition between the plants, light, soil, and humidity (Bates et al., 2014). Change in climatic conditions implicated in a significant impact on life cycles and distribution patterns of plants; therefore, many plants become endemic to geographic regions (Das et al., 2016).

As per the recent study published in the journal *Nature Climate Change*, change in the climatic conditions could lead to the globally widespread loss of plants that arose around the world. This study collected data of 50,000 familiar plant as well as animal species. Furthermore, they concluded that more than half of the plants would be affected until 2080 due to the constant rise in greenhouse gas emissions (Whiteside and Ward, 2011).

4. Climatic factors

Plants are dependent on some climatic factors such as temperature, rainfall, light, carbon dioxide (CO_2), and moisture for their growth and development. The amount of these climatic factors varies in different geographical conditions. Plant growth and its development are dependent upon the temperature; each species has its optimum temperature range to survive and flourish the environment (Hatfield and Prueger, 2015) (Table 2.1).

5. Biological responses and ecosystem health

The species in the environment were affected by the climatic conditions in different ways, counting range shifts, change in relative abundance inside species, and usage of microhabitat. The geographical distribution of different species mainly depends upon its environmental conditions. As the climatic conditions will change, the species must either adapt themselves to tolerate the change in the environment for a new existing condition. The ongoing species may increase the capacity to adjust to live in new locations or decrease the capability to persevere where they are presently located (Bates et al., 2014).

Climate is one of the key factors determining the geographic distribution of plant species that can result in the shift in species ranges. Studies have shown that, under

Table 2.1 List of major factors affecting climatic conditions.

Sl. No.	Climatic factors	Description
1	Altitude	At high altitude, climatic conditions become colder.
2	Wind patterns	There are different wind patterns in northern and southern Hemispheres although wind pattern is associated with seasonal variations.
3	Topography	The topography of the area can strongly affect the climate. Mountains ranges are known as natural barriers of air movement.
4	Geography	Geography of a zone, i.e., distant from mountains and substantial areas of water, plays an important role in determination of climate.
5	Surface of the Earth	The quantity of sunlight absorbed or reflected through surface determines the amount of atmospheric heating occurred.
6	Climate change over time	Cold and warm conditions are main variants on Earth; sometimes changes in these conditions are noticeably short, but sometimes it may take several years.

Source: https://www.climateandweather.net/global-warming/factors-that-influenceclimate.html.

upcoming climate circumstances, the proper areas for many species would change, decreasing or increasing in size or shifting in latitude and elevation. There is an interest that these changes will boost the extinction risk and affect the interaction among the biotic community. Further additional evidence proved that changes in climate are affecting both plant phenology and insect phenology, in the latter case postponing flowering periods and causing a discrepancy with the presence of key migratory pollinators. The loss of a single pollinator can interrupt an entire pollination process network by affecting patterns of specialization and floral fidelity of other pollinators. Therefore, shifts in temperature and precipitation systems under current climate change circumstances might place plant–pollinator mutualisms under a complex mix of spatial and temporal threats. With respect to the loss of pollinators, there are increased calls for the insertion of the risk factors of climate change into vulnerability evaluations of extinction risk across multiple taxa and in specific climate change assessments into the IUCN Red List process (Gómez-Ruiz and Lacher, 2019). The distributional responses of some species delay after the climate change. Such pause can happen from a variety of reasons, which may be together with species-specific physiological, ecological, behavioral, and evolutionary responses. Absence of sufficient habitat connectivity and permission to microhabitats may be linked to microclimates that are predictable to be grave in escalating experience to macroclimatic warming and extreme thermal events; thus, shifts of some species were delayed (Scheffers et al., 2014). Moreover, range shifts are frequently diverse across different geographic gradients, when the edaphic factors other than heat force species rearrangement. For example, rainfall changes can cause declined rise shifts as climate warm (Lenoir and Svenning, 2014). Even though species may be adapted to altering climates, either through morphologic plasticity or by the natural selection, all the species have their limits to their competence for adapting themselves for changing environments, and these limits are improbable to augment for species previously experiencing hot temperatures remarkably close to their tolerance limits (Araujo et al., 2013).

6. Buffer and porch effects

Ecosystems can itself act as a natural buffer from diverse actions such as wildfires, drought, and flooding. The change in the climatic condition might be responsible for the change in the ecosystems' capability to frustration the impacts of severe conditions. Therefore, that may be a boost in susceptibility to damage to the species. For example, coral reefs and barriers of islands, which, in turn, protect the coastal ecosystems from the storm; the wetland ecosystems help in the absorption of floodwaters; and recurring wildfires help in the clearing of excess forest debris, which is accumulated during the winter season (Groffman et al., 2014). When the coral reefs turn into stressed due to increased temperature in the ocean, they drive out the microorganisms that survive inside their tissues and are beneficial to their health system. This mechanism of spill-out of microorganism due to high temperature is known as coral bleaching. As the ocean temperatures increase, the acidity of the ocean will be increased (Leong et al., 2014).

7. Range shifts

Distributions of species are presently changing in response to globalization and climate change (Chen et al. 2011). For instance, global warming is shifting some of the species distribution limits toward the poles. Therefore, present range shifts may lead to shifts in biotic interactions because interacting organisms may not be able to move all at the same speed. Range expanding species are thus likely to encounter different abiotic as well as biotic environments in their new range. These novel selection pressures during range expansion could lead to the fast adaptive evolution of the species in the new range (Buckley and Bridle 2014).

As the environmental temperatures increase, the niche of many organisms will be shifted from one place to another (Fig. 2.2). The species that live in the North Pole might be moving toward higher elevations. In present years, in both terrestrial and marine environments, both animals and plants have shifted to a higher altitude at an average rate of 36 feet per decade. Whereas a range spreading out of some of the species migration is less comfortable for suitable habitat, improved competition, and range of reduction, some of the species will not migrate at all, since they have previously reached the top of the mountain suitable for their habitat. These climatic factors

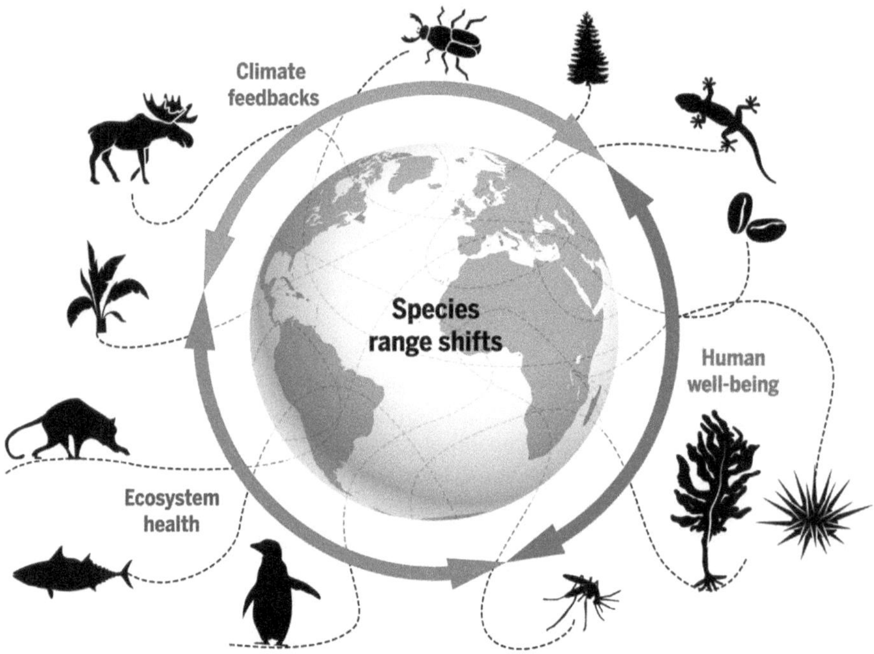

Figure 2.2 As the climate changes, human welfare, ecosystem function, and climate itself are ever more affected by the geographically shifting of life.
Source: Pecl, G.T., Araujo, M.B., Bell, J.D., Blanchard, J., Bonebrake, T.C., Chen, I.C., Clark, T.D., Colwell, R.K., Danielsen, F., Evengård, B. and Falconi, L., 2017. Biodiversity redistribution under climate change: impacts on ecosystems and human well-being. Science 355 (6332).

might be responsible for extinctions of some organisms in particular geographical areas. Therefore, the ranges of vegetative biomes are predictable to alter across 5%–20% of the Earth in the United States by 2100 (Groffman et al., 2014).

For example, in case of boreal forests invading of tundra ecosystem, there might be decreased habitat for the numerous distinctive species that mainly depend on the tundra ecosystem, such as arctic foxes, caribou, and snowy owls. Other practical changes in the United States include a range shift in the climatic temperature where the broadleaf/conifer type of forest ecosystem boundary may be shift in the shrubland/conifer forest border in New Mexico and an uphill increase shift of the temperate-mixed forest frontier in Southern California (Callaghan et al., 2004).

As rivers and temperate water fishes are increasing, their habitat areas are earlier inhabited by cold-water species. As the temperature of the water rises, the water gets slightly warm; as a result, cold-water fishes, as well as many valued trout and salmon species, are lost their habitat, with an estimation of about 47% habitat loss by 2080 (Groffman et al., 2014). In certain areas in the western United States, there may be a sufferer of western trout species that may go beyond more than 60%, whereas in other regions, fatalities of bull trout may attain about 90%. The range shifts mainly

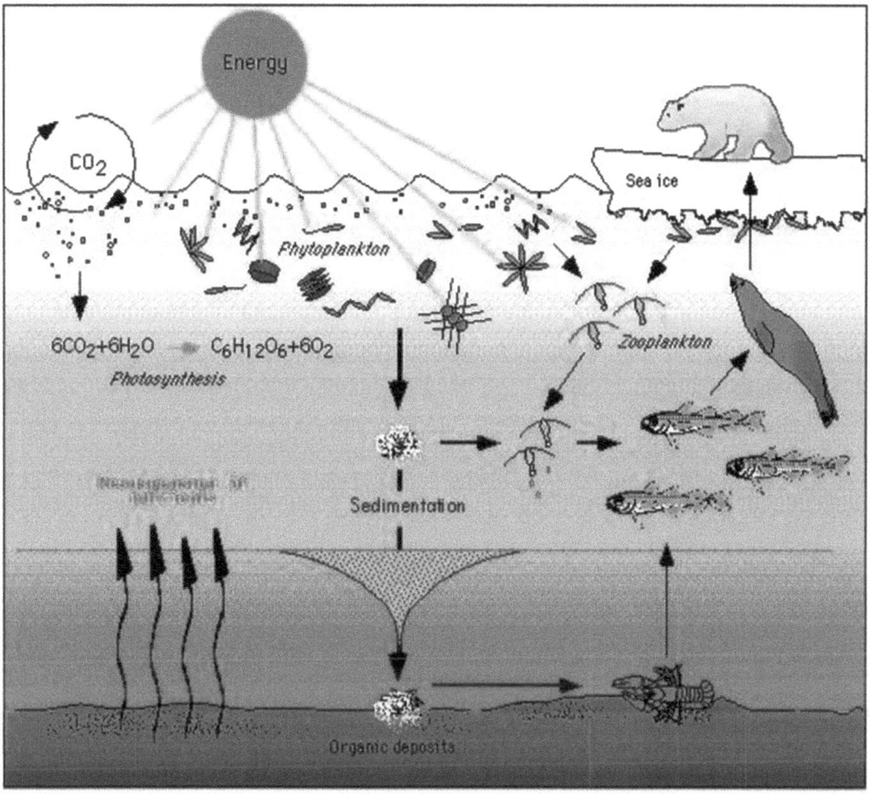

Figure 2.3 The Arctic food web is complex. The loss of sea ice can ultimately affect the entire food web, from algae and plankton to fish to mammals.
NOAA (2011).

responsible for the disturbance of the present situation of the ecosystem can edge the opportunities for fishing and hunting (Karl et al., 2009). The loss of sea ice might eventually be disturbing the complete food web, from algae and plankton to fish to mammals (Fig. 2.3).

8. Extinction risks

The change in the climatic condition, along with the pollution and habitat destruction, is one of the significant problems that can throw species into extinction. The IPCC predicted that 20%–30% of the animal and plant species studied so far with respect to climate change studies are at the line of risk of extinction if the habitat temperature reaches the maximum levels. The universal rates of species extinctions are possible to move toward the upper limit of practical natural rates of destruction in the fossil record (Settele et al., 2014). For example, some of the species exist only in the specific climatic condition; they might be at risk of considerable losses; it consists of animals that are modified to mountain climatic conditions. One of them is the pika, animals that are reliant on sea ice habitats, such as ringed seals and polar bears and salmon in the Pacific Northwest (Groffman et al., 2014).

9. Summary and conclusion

Climate is one of the chief components of the ecosystem; it determines the organism's growth and species distributions; therefore, a changing climatic condition has the possibility to change the composition of plant and soil communities and the interactions between them. Climate is one of the integral factors of the environment, which is mainly responsible for the distribution and range of habitat of all living organisms. Even though the geological range limits the species distribution and is forceful and change over time, climate change is urging a global rearrangement of life on the Earth. Some of the terrestrial species are also moving toward the cooler region, thereby moving toward cold water. Since diverse species will respond at diverse rates and to changeable degrees, key connections or communications among species are often distracted, and new interactions may be developed. These results in the development of novel biotic communities and rapid changes in functioning of ecosystem with persistent and from time to time surprising consequences that broadcast through and affect both human and biological communities.

References

Araujo, M.B., Ferri-Yanez, F., Bozinovic, F., Marquet, P.A., Valladares, F., Chown, S.L., 2013. Heat freezes niche evolution. Ecol. Lett. 16 (9), 1206–1219.

Barlow, J., França, F., Gardner, T., Hicks, C., Lennox, G.D., Berenguer, E., Castello, L., et al., 2018. The future of hyperdiverse tropical ecosystems. Nature 559, 517–526. https://doi.org/10.1038/s41586-018-0301-1.

Bates, A.E., Pecl, G.T., Frusher, S., Hobday, A.J., Wernberg, T., Smale, D.A., Holbrook, N.J., 2014. Defining and observing stages of climate-mediated range shifts in marine systems. Global Environ. Change 26, 27–38.

Bellard, C., Bertelsmeier, C., Leadley, P., Thuiller, W., Courchamp, F., 2012. Impacts of climate change on the future of biodiversity. Ecol. Lett. 15, 365–377.

Botkin, D.B., Saxe, H., Araujo, M.B., Betts, R., Bradshaw, R.H.W., Cedhagen, T., Chesson, P., Dawson, T.P., Etterson, J.R., Faith, D.P., Ferrier, S., Guisan, A., Hansen, A.S., Hilbert, D.W., Loehle, C., Margules, C., New, M., Sobel, M.J., Stockwell, D.R.B., 2007. Forecasting the effects of global warming on biodiversity. Bioscience 57, 227–236.

Buckley, J., Bridle, J.R., 2014. Loss of adaptive variation during evolutionnary responses to climate change. Ecol. Lett. 17 (10), 1316–1325.

Callaghan, T.V., Bjorn, L.O., Chernov, Y., Chapin, T., Christensen, T.R., Huntley, B., Matveyeva, N., 2004. Biodiversity, distributions and adaptations of Arctic species in the context of environmental change. AMBIO A J. Hum. Environ. 33 (7), 404–417.

CBD, 2019. Biodiversity and the 2030 Agenda for Sustainable Development. Technical note. Secretariat of the convention on biological diversity, Montréal.

Chen, I.C., Hill, J.K., Ohlemuller, R., Roy, D.B., Thomas, C.D., 2011. Rapid range shifts of species associated with high levels of climate warming. Science 333, 1024–1026.

Das, M., Jain, V., Malhotra, S.K., 2016. Impact of climate change on medicinal and aromatic plants. Indian J. Agric. Sci. 86, 1375–1382.

Gómez-Ruiz, E.P., Lacher Jr., T.E., 2019. Climate change, range shifts, and the disruption of a pollinator-plant complex. Sci. Rep. 9 (1), 1–10.

Gore, A., 2006. An Inconvenient Truth. Rodale, New York.

Groffman, P.M., Kareiva, P., Carter, S., Grimm, N.B., Lawler, J., Mack, M., Matzek, V., Tallis, H., 2014. Ch. 8: ecosystems, biodiversity, and ecosystem services. In: Melillo, J.M., Terese, T.C., Richmond, Yohe, G.W. (Eds.), Climate Change Impacts in the United States: The Third National Climate Assessment. U.S. Global Change Research Program, pp. 200–201.

Guo, D., Desmet, P.G., Powrie, L.W., 2017. Impact of the future changing climate on the southern Africa biomes, and the importance of geology. J. Geosci. Environ. Protect. 5, 1–9.

Hartmann, T., 2007. From waste products to ecochemicals: fifty years research of plant secondary metabolism. Phytochemistry 68, 2831–2846.

Hatfield, J.L., Prueger, J.H., 2015. Temperature extremes: effect on plant growth and development. Weather Clim. Extremes 10, 4–10.

Heller, N.E., Zavaleta, E.S., 2009. Biodiversity management in the face of climate change: a review of 22 years of recommendations. Biol. Conserv. 142, 14–32.

IPCC, 2013. Stocker, T.F., Qin, D., Plattner, G.-K., Tignor, M., Allen, S.K., Boschung, J., Nauels, A., Xia, Y., Bex, V., Midgley, P.M. (Eds.), Climate Change 2013: The physical science basis. Working Group I Contribution to the Fifth Assessment Report of the Intergovernmental Panel on Climate Change. Cambridge University Press, Cambridge, United Kingdom and New York, NY, USA, p. 1535.

IPCC (Intergovernmental Panel on Climate Change), 2012. Managing the Risks of Extreme Events and Disasters to Advance Climate Change Adaptation. In: A Special Report of Working Groups I and II of the Intergovernmental Panel on Climate Change. Cambridge University Press, Cambridge, UK.

Johnson, C.N., Balmford, A., Brook, B.W., Buettel, J.C., Galetti, M., Guangchun, L., Wilmshurst, J.M., 2017. Biodiversity losses and conservation responses in the Anthropocene. Science 356, 270–275.

Karl, T.R., Melillo, J.M., Peterson, T.C. (Eds.), 2009. Global Climate Change Impacts in the United States. "Climate Change Impacts by Sectors: Ecosystems." United States Global Change Research Program. Cambridge University Press, New York, NY, USA.

Kiers, E.T., Palmer, T.M., Ives, A.R., Bruno, J.F., Bronstein, J.L., 2010. Mutualisms in a changing world: an evolutionary perspective. Ecol. Lett. 13, 1459–1474.

Koh, L.P., Dunn, R.R., Sodhi, N.S., Colwell, R.K., Proctor, H.C., Smith, V.S., 2004. Species coextinctions and the biodiversity crisis. Science 305, 1632–1634.

Lawton, J.H., May, R.M., 1995. Extinction Rates. Oxford University Press, Oxford.

Lenoir, J., Svenning, J.C., 2014. Climate-related range shifts – a global multidimensional synthesis and new research directions. Ecography 38, 15–28.

Leong, J.-A., Marra, J.J., Finucane, M.L., Giambelluca, T., Merrifield, M., Miller, S.E., Polovina, J., Shea, E., Burkett, M., Campbell, J., Lefale, P., Lipschultz, F., Loope, L., Spooner, D., Wang, B., 2014. Ch. 23: Hawai'i and U.S. Affiliated Pacific islands. In: Melillo, J.M., Terese, T.C., Richmond, Yohe, G.W. (Eds.), Climate Change Impacts in the United States: The Third National Climate Assessment. U.S. Global Change Research Program, pp. 537–556.

Ober, D., 2005. Seeing double: gene duplication and diversification in plant secondary metabolism. Trends Plant Sci. 10, 444–449.

Pearson, R.G., Dawson, T.P., 2003. Predicting the impacts of climate change on the distribution of species: are bioclimate envelope models useful? Global Ecol. Biogeogr. 12, 361–371.

Pecl, G.T., Araujo, M.B., Bell, J.D., Blanchard, J., Bonebrake, T.C., Chen, I.C., Clark, T.D., Colwell, R.K., Danielsen, F., Evengård, B., Falconi, L., 2017. Biodiversity redistribution under climate change: impacts on ecosystems and human well-being. Science 355 (6332).

Pichersky, E., Gang, D., 2000. Genetics and biochemistry of secondary metabolites in plants: an evolutionary perspective. Trends Plant Sci. 5, 439–445.

Reed, D.H., 2012. Impact of climate change on biodiversity. In: Chen, W.Y., Seiner, J., Suzuki, T., Lackner, M. (Eds.), Handbook of Climate Change Mitigation. Springer, New York, NY.

Scheffers, B.R., Evans, T.A., Williams, S.E., Edwards, D.P., 2014. Microhabitats in the tropics buffer temperature in a globally coherent manner. Biol. Lett. 10, 20140819.

Settele, J., Scholes, R., Betts, R., Bunn, S., Leadley, P., Nepstad, D., Overpeck, J.T., Taboada, M.A., 2014. Terrestrial and Inland water systems. In: Field, C.B., Barros, V.R., Dokken, D.J., Mach, K.J., Mastrandrea, M.D., Bilir, T.E., Chatterjee, M., Ebi, K.L., Estrada, Y.O., Genova, R.C., Girma, B., Kissel, E.S., Levy, A.N., MacCracken, S., Mastrandrea, P.R., White, L.L. (Eds.), Climate Change 2014: Impacts, Adaptation and Vulnerability. Part A: Global and Sectoral Aspects. Contribution of Working Group II to the Fourth Assessment Report of the Intergovernmental Panel on Climate Change. Cambridge University Press, Cambridge, United Kingdom and New York, NY, USA.

Walther, G.R., 2010. Community and ecosystem responses to recent climate change. Phil. Trans. Biol. Sci. 2010 (365), 2019–2024.

Whiteside, J.H., Ward, P.D., 2011. Ammonoid diversity and disparity track episodes of chaotic carbon cycling during the early mesozoic. Geology 39, 99–102.

Further reading

https://www.climateandweather.net/global-warming/factors-that-influenceclimate.html.

https://climate.nasa.gov/solutions/adaptation-mitigation.

NRC, 2008. Ecological Impacts of Climate Change. National Research Council The National Academy Press, Washington, DC, USA.

Sintayehu, D.W., 2018. Impact of climate change on biodiversity and associated key ecosystem services in Africa: a systematic review. Ecosys. Health Sustain. 4 (9), 225–239.

Climate-resilient agriculture: enhance resilience toward climate change

3

Renu Singh, Raviteja Machanuru, Bhupinder Singh, Manoj Shrivastava
Centre for Environment Science and Climate Resilient Agriculture (CESCRA), Nuclear Research Laboratory (NRL), ICAR-Indian Agricultural Research Institute, New Delhi, India

1. Introduction

Climate change is becoming a big challenge to the productivity and profitability of agriculture. The minute change in the climatic pattern will be leading to a big change in climatic events such as change in wind and rainfall patterns, increase in global temperature, melting of polar ice caps, and so on. The frequency of the floods has now changed from every 100 years to every 5–25 years. According to *Intergovernmental Panel on Climate Change* (*IPCC*), climate change may be defined as "A change in the state of the climate that can be identified (*e.g.* by applying statistical tests) by changes in the mean and/or the variability of its properties and that persists for an extended time, may be decades or longer." Climate change can be because of natural internal or external forces or because of continual anthropogenic changes in the land use or composition of the atmosphere.

1.1 Major causes for climate change

Climate change is not new nowadays; it is happening from hundreds of millions of years. This climate change has triggered by natural as well as anthropogenic activities (Fig. 3.1).

1.1.1 The natural factors causing climate change

(a) Changes in the earth's orbit: The distance between the sun and the earth: when the earth is nearer to the sun, the climate will be warmer, and when it is farther, it is cooler.
(b) Change in the point of reference of the earth's axis of rotation: The axis of rotation has been changed from 22.1 to 24.5 degrees over about 41,000 years and back again. As the angle increases, the winter becomes much cooler, and the summer will be much warmer.
(c) Volcanic eruptions: Volcanic gases from the active volcanoes release large amount of carbon dioxide (CO_2), sulfur dioxide (SO_2), water vapor, dust, and ash particles into the atmosphere. The sulfur dioxide and dust particles cause cooling effect. But carbon dioxide causes warming effect.
(d) Change in the direction of ocean currents: Shift in the direction of ocean currents leads to warmer or cooler effect.

Global Climate Change. https://doi.org/10.1016/B978-0-12-822928-6.00016-2

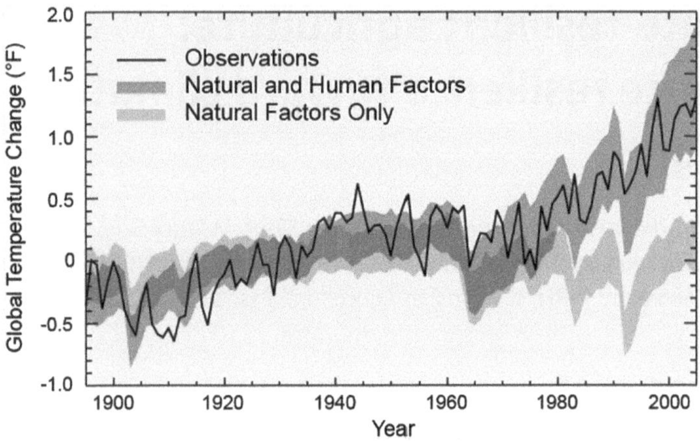

Figure 3.1 Natural and human influences on climate change.
Source: https://www.agronomy4me.org/climate-change-can-we-blame-nature.

1.1.2 Anthropogenic activities

(a) Industrialization: From the 19th century, the use of fossil fuels to run industries has contributed to global warming by release of CO_2. Burning of fossil fuels contributes to 49% of greenhouse gases (GHGs).

(b) Loss of vegetation: For the purpose of agriculture and urbanization, lots of trees were cut, thereby inhibiting natural carbon sequestration.

(c) Agriculture sector: There is increase in use of fertilizers especially nitrogenous fertilizers such as urea, from which nitrogen is lost through volatilization. Only 30% of applied urea is utilized by the plants, and rest is lost through volatilization and leaching. Methane released from livestock and paddy fields is the potential source of global warming. An estimated 23% of total anthropogenic GHG emissions (2007−16) derive from agriculture, forestry, and other land use (IPCC, 2019)

(d) As a result of microbial action in the soil, there will be release of nitrous oxide (N_2O), which is having global warming potential of 310.

1.2 Pros of climate change

In northern latitudes, because of climate change, there will be less severe winters, more precipitation in some dry areas and less in some wet places, and increased food production and forest growth. In C-3 and C-4 plants, the photosynthesis rate will increase with increase in temperature and concentration of carbon dioxide to a certain level, so there will be increase in food production in certain crop species. For example, coconut yields are projected to increase by 30% in Kerala (Rama Rao et al., 2013).

1.3 Cons of climate change

In contrast to pros, due to climate change, tropical areas experience excessive heat, lack of water for irrigation, change in rainfall pattern, decrease in food production, and so on.

(a) Impact on agriculture: There will be shift in agriculture lands, decreased crop yields, depletion of irrigation sources, and newly emerging pest and diseases.
(b) Impact on water resources: There will be melting of polar ice caps, glaciers, increased frequency of floods and droughts, and decrease in water quality due to pollution.
(c) Impact on weather extremes: There will be increased number of deaths due to heat waves and cold waves and increase in floods due to prolonged and heavy rainfall.
(d) Impact on biodiversity: There will be a decline in the species diversity due to loss of their habitat, fragmentation, and overexploitation.
(e) Impact on human health: There will be increase in mosquito-borne disease such as malaria and dengue.

2. Climate-resilient agriculture

Climate-resilient agriculture can be defined as "agriculture that reduces poverty and hunger in the face of climate change, improving the resources it depends on for future generations" (Christian Aid, Time for Climate Justice, 2015).

There have been significant negative effects found with medium-term (2010–2039) climate change, such as yield reduction by 4.5%–9%, because of variation in temperature, rainfall pattern, increase in pest incidence, and so on. Since agriculture contributes 15% of India's GDP, a 4.5%–9.0% of negative impacts on production result to the cost of climate change approximately 1.5% of GDP every year.

Hazards related to climate change have a big impact on the human life, especially the poor. Heavy rainfall or no rain, increase in temperatures, abrupt hailstorms, and increased frequency of droughts, storms, and floods are expected to rise further. The impacts are severe in developing countries having food and water scarcity along with livelihoods insecurity. Survival becomes very tough that can be addressed by promoting climate-resilient agriculture in such areas affected by climate change severity to ensure food security and income for rural communities. Communities or systems, which are prone to hazards, should be able to foresee the risks so that they can respond when disaster strikes and adapts to changing risks and situations.

3. Major programs for climate-resilient agriculture

All the major programs related to climate change in agriculture include few key drivers as shown in Fig. 3.2.

3.1 NICRA: National Innovations on Climate Resilient Agriculture

National Innovations on Climate Resilient Agriculture (NICRA) was launched during February 2011 by Indian Council of Agricultural Research (ICAR) with the funding from Ministry of Agriculture, Government of India. It is a unique project that brings all the sectors of agriculture on one platform to meet the resilience and ICAR—Central

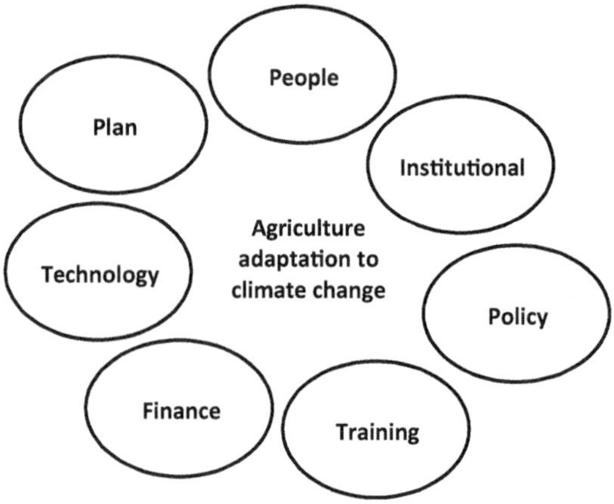

Figure 3.2 Key divers of climate change programs for agriculture.

Research Institute for Dryland Agriculture, Hyderabad (CRIDA), is the nodal institute. The megaproject has three major objectives of strategic research, technology demonstrations, and capacity building. Assessment of the impact of climate change simultaneous with formulation of adaptive strategies is the prime approach under strategic research across all sectors of agriculture, dairying, and fisheries.

3.2 *Objectives of National Innovations on Climate Resilient Agriculture*

A. To enhance the resilience of Indian agriculture covering crops, livestock and fisheries to climatic variability and climate change through development and application of improved production and risk management technologies.
B. To demonstrate site specific technology packages on farmers' fields for adapting to current climate risks.
C. To enhance the capacity of scientists and other stakeholders in climate-resilient agricultural research and its application.

Significant achievements of this project include extensive field phenotyping of germplasm of target crops (rice, wheat, pigeon pea, and tomato) to multiple abiotic stress and preparation of first ever vulnerability atlas of India at district level for 572 rural districts. Best infrastructure was established across the country at different institutes to facilitate the climate change research. Evolving climate-resilient agricultural technologies that would increase farm production and productivity vis-à-vis continuous management of natural and human-made resources constitute an integral part of sustaining agriculture in the era of climate change. The four modules of NICRA—natural resource management, improving soil health, crop production, and livestock are aimed at making the farmers' self-reliant.

3.3 Village Climate Risk Management Committee

It is a village-level committee that represents all the farmers including landless farmers and women formed under the guidance of Gram Sabha to decide on the strategies regarding interventions, farmer's participation, and harmony with ongoing Government schemes pertaining to the climate change adaptation. This project was implemented in one of the hundred selected villages, which is more prone to drought, floods, cyclones, and so on.

4. Smart practices and technologies for climate-resilient agriculture

Farmers need to wisely adapt to the changing climate to sustain crop yields and farm income. Enhancing resilience of agriculture to climate risk is of supreme importance for protecting livelihoods of small and marginal farmers. Traditionally, technology transfer in agriculture has aimed at enhancing farm productivity. However, in the context of climate change and variability, farmers need to adapt quickly to enhance their resilience to increasing threats of climatic variability such as droughts, floods, and other extreme climatic events. There have been different practices and technologies identified toward the climate-resilient agriculture. Some of resilient practices and technologies are presented in Table 3.1.

4.1 Conservation of natural resources

In situ moisture conservation techniques such as mulching, practicing conservational agriculture, rain water harvesting and using it for supplementary irrigation, artificial ground water recharge, and so on help in solving water scarcity problems that arise due to climate change. Enhancing the available water through the surface and subsurface storage structures is one of the means of creating resilience at the local level for agricultural systems. This is especially important in low to medium rainfall zones, as rainfall variability and occurrence of intense rainy events are considered to be high. A multipronged approach was adopted, which includes the creation/renovation of community tanks, check dams, individual farm ponds, and desiltation of conveyance systems for safe disposal of excess rain into community tanks, thereby leading to in situ conservation of rainfall helps in soil and land management practices.

4.2 Conservation of soil moisture by mulching

Mulching helps in the conservation of moisture in soil, which is otherwise exposed to dry winds and temperature that causes hard-baking and drying of exposed soil. Mulching provides a cooler and lesser fluctuated temperature regime to the soil in comparison

Table 3.1 Smart technologies for climate-resilient agriculture.

Technology	Technology adaptation/mitigation potential
1. Water-smart **I.** Rainwater harvesting **II.** Drip irrigation **III.** Laser land leveling **IV.** Furrow irrigated bed planting **V.** Drainage management **VI.** Cover crops method	**i.** Interventions that improve water use efficiency **ii.** Collection of rainwater not allowing to run off and use for agricultural in rain-fed/dry areas and other purposes on-site. **iii.** Application of water directly to the root zone of crops and minimize water loss **iv.** Leveling the field ensures uniform distribution of water in the field and reduces water loss (also improves nutrient use efficiency) **v.** This method offers more effective control over irrigation and drainage as well as rainwater management during the monsoon (also improves nutrient use efficiency) **vi.** Removal of excess water (flood) through water control structure **vii.** Reduces evaporation loss of soil water (also adds nutrients into the soil)
2. Energy-smart **I.** Zero tillage/minimum tillage	**i.** Interventions that improve energy use efficiency **ii.** Reduces amount of energy use in land preparation **iii.** In the long run, it also improves water infiltration and organic matter retention into the soil
3. Nutrient-smart **I.** Green manuring **II.** Leaf color chart **III.** Intercropping with legumes	**i.** Optimum supply of soil nutrients over time and space matching to the requirements of crops with right product, rate, time, and place **ii.** Site-specific integrated nutrient management **iii.** Cultivation of legumes in a cropping system. This practice improves nitrogen supply and soil quality **iv.** Quantify the required amount of nitrogen use based on greenness of crops. Mostly used for split dose application in rice but also applicable for maize and wheat crops to detect nitrogen deficiency **v.** Cultivation of legumes with other main crops in alternate rows or mixed. This practice improves nitrogen supply and soil quality.
4. Carbon-smart agro forestry (AF) **I.** Concentrate feeding for livestock **II.** Fodder management **III.** Integrated pest management	**i.** Interventions that reduce greenhouse gas emissions **ii.** Promote carbon sequestration including sustainable land use management **iii.** Reduces nutrient losses and livestock requires low amount of feed **iv.** Promote carbon sequestration including sustainable land use management **v.** Reduces use of agrochemicals

Table 3.1 Smart technologies for climate-resilient agriculture.—cont'd

Technology	Technology adaptation/mitigation potential
5. Weather-smart **I.** Climate-smart housing for livestock **II.** Weather based crop Agro-advisory **III.** Crop insurance	**i.** Interventions that provide services related to income security and weather advisories to farmers **ii.** Protection of livestock from extreme climatic events (e.g., heat/cold stresses) **iii.** Climate information-based value-added agroadvisories to the farmers **iv.** Crop-specific insurance to compensate income loss due to vagaries of weather
6. Knowledge-smart **I.** Contingent crop planning **II.** Improved crop varieties **III.** Seed and fodder banks	**i.** Use of combination of science and local knowledge **ii.** Climatic risk management plan to cope with major weather-related contingencies **iii.** Like drought, flood, heat/cold stresses during the crop season **iv.** Crop varieties that are tolerant to drought, flood, and heat/cold stresses **v.** Conservation of seeds of crops and fodders to manage climatic risks

with the nonmulched soil. It results in even plant growth at optimum temperature with fewer evaporation conditions. Under this condition, plant roots get favorable environment at the soil surface with appropriate nutrient levels and air content for plant growth.

Mulches also break the force of irrigation water and rain by preventing soil compaction, erosion, and crusting. Mulched soils soak up water at a faster rate. It prevents the splashing of mud and plant disease causing organisms on plants and flowers during overhead irrigation or rain. The mulch covering prevents light that decreases the germination of weed seeds, which in turn poses less competition for available moisture and nutrients to the crop. In fact, application of mulch for weed control reduces the dependency on herbicides.

4.2.1 Benefit of mulching

i. Mulch provides the soil cover and protects the soil from disintegration by rainwater intensity, which in turn prevents the soil loss by the erosion.
ii. Mulch protects sunlight that minimizes soil evaporation by retaining soil moisture.
iii. Organic mulch decomposes and releases nutrients for plant uptake.
iv. Weed seeds remain dormant under mulch.

4.3 Conservation agriculture for sustainable land use

In response to concerns about food security, farm profitability, and land degradation in agriculture around the world, a specific combination of measures has been identified, which is known as "conservation agriculture" (CA). CA involves the three key components, including zero tillage (or at least minimum soil disturbance), retention of crop residues for soil cover (mulching), and rotation (or sometimes intercropping) of cereals with legumes, or sometimes with other crops (Pannell et al., 2014). CA can be defined as a sustainable agriculture production system comprising of custom-made farming practices according to the crops and local conditions. CA supports farming and soil management techniques that help in improving soil quality and biodiversity, protect the soil from erosion and degradation, and preserve natural resources. CA practices adopted in many parts of the world are focused on ecological principles making the land use more sustainable. Conservation agriculture essentially relies on three concepts that are connected together and need to be considered for effective processes of design, planning, and implementation as shown in framework in Fig. 3.3 (Bhan et al., 2014).

4.4 Artificial recharging for enhancement of groundwater

Diversion of overflowing water into the defunct bore well during the peak monsoon recharges the bore wells and increases the water table. The recharged bore wells can be used for providing irrigation during critical stages of crop growth, and significant area can be brought under irrigation during the *Rabi* season. Farmers can grow

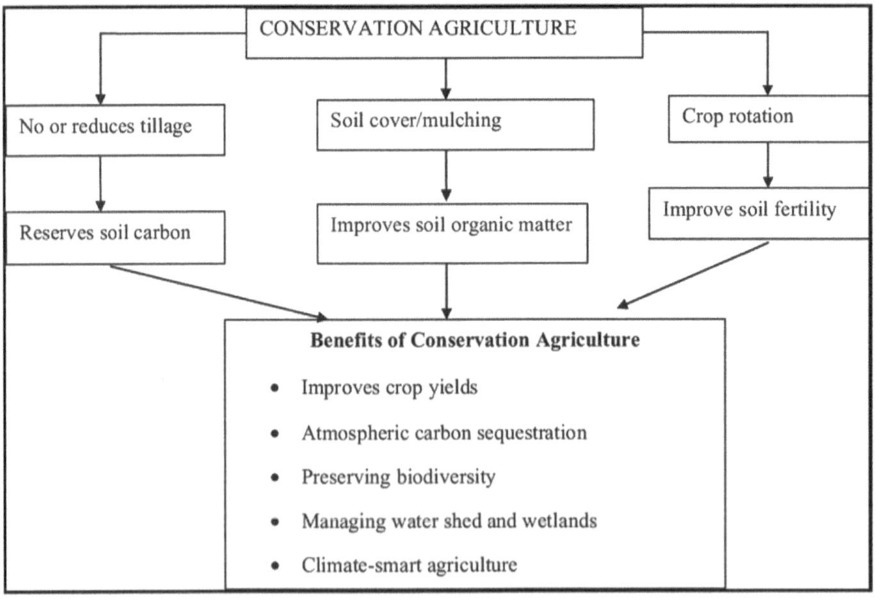

Figure 3.3 Concept of conservation agriculture.

vegetable and flower crops during *Rabi* and earn higher income who otherwise grow single crop each year. Renovation/desilting of existing rainwater harvesting structures and creation of new ponds/Jalkunds improve the facility for storage of rainwater so that it can be used in both *Kharif* and *Rabi* crops to enhance resilience and increase cropping intensity. Use of appropriate farm implements helps in the adoption of ridge furrow, broadbed furrow, and furrow-irrigated raised-bed planting methods. Crop residues can be used for mulching in high-value crops such as winter vegetables in rice fallows. Zero till sowing in several rabi crops helps to attain advancement in sowing date to avoid high-temperature stress at flowering and maturity stages, conserve soil moisture, and reduce the number of irrigations (Srinivasa Rao et al., 2015). For delayed monsoon and deficit rainfall conditions during *Kharif*, rainwater harvesting and its efficient use enable farmers to provide lifesaving and critical irrigations to rain-fed crops and use the stored water for increasing the cropping intensity in *Rabi*. Renovation of cement plugs and farm ponds of different water storage capacity helps in recharging the surrounding open wells. Desilting of these structures increases percolation of impounded water, leading to increase in water table.

4.5 Community approach for soil and water conservation

Due to recharge of aquifer and water level in open wells, farmers can irrigate their crops in both *Kharif* and *Rabi* seasons. Increase in number of wells helps farmers to adopt microirrigation systems for efficient use of irrigation water.

4.6 Sustainable crop production under climate change scenario

Crop production under climate change scenario can be increased by introducing drought/temperature-resistant varieties, advancing the planting dates of *Rabi* crops in areas having terminal heat stress. Methods for water-saving paddy cultivation, viz., System of Rice Intensification (SRI), direct seeding, aerobic rice, fumigation for frost control in horticulture, delayed monsoon community nurseries, and custom recruiting centers for timely seeding are useful under adverse climatic condition.

Components of sustainable agriculture:

- **i.** Management of soil salinity and sodicity
- **ii.** Management of energy used in agriculture
- **iii.** Soil fertility management
- **iv.** Integrated nutrient management
- **v.** Integrated pest management
- **vi.** Management of soil and water erosion
- **vii.** Water and irrigation management
- **viii.** Postharvest losses management
- **ix.** Pastureland management
- **x.** Livestock management
- **xi.** Wetland management
- **xii.** Fishery and aquaculture management

4.6.1 Selection of appropriate crop varieties

Short-duration and drought-tolerant varieties are useful for rain-fed crops. The varieties include soybean (JS-95-60, JS 93-05), pigeon pea (ICPL-88,039, BDN-711, NDA-1, TJT-501, PRG-159, BRG-2 & 4, ICPL-87, Pusa Arhar 16) maize (GM-6, HQPM-1 & 7, JM 216, Birsa Makka-1), short-duration pulses (Samrat, Vishal, Azad-3, BBN-3, TGM-3, RNG-344, TARM-1), finger millet (ML-365, GPU-48, GPU28, A 404), and paddy (Naveen, Sahbhagi dhan, Basanti, Birsa Vikas Dhan, MTU-1010, NRC-7, MAS-26, Pusa Basmati 1509) (Srinivas Rao et al., 2015).

4.6.2 System of Rice Intensification

Two major challenges for agriculture are food security of exponentially growing world population; along the situations such as scarce availability of new land for agriculture and increasing scarcity of water resources for rice (*Oryza sativa* L.)-based cropping system. By the year 2025, the world's farmers would be producing about 60% additional rice to meet the food demands of growing world population by that time. It is imperative to efficiently utilize the water and land resources with increased productivity to meet future need. Over 30 years ago, SRI was developed in Madagascar with an opportunity for reducing water demand without compromising rice crop yield, being practiced in more than 50 countries. SRI comprises of several core components as per local conditions. The generic agronomic practices for growing transplanted rice, i.e., nursery raising, transplanting, irrigation, weed control, and nutrient management, all the steps are there in SRI by keeping an eye on water and resources conservation.

There are six principle components of rice intensification (SRI) (Uphoff, 2007).

- **i.** Seedlings get transplanted at a much younger age
- **ii.** Only single seedling, instead of a handful of seedlings that get planted in each hole
- **iii.** Increased use of organic fertilizer to enhance soil fertility
- **iv.** Intermittent water application to increase wet and dry soil conditions, instead of continuous flood irrigation
- **v.** Plants are spaced wider apart instead of close, and the seed rate is reduced by 80%–90%
- **vi.** Rotary weeding to control weeds and promote soil aeration

4.6.3 Aerobic rice cultivation

Aerobic rice is grown like an upland crop such as wheat, under nonflooded conditions in nonpuddled and unsaturated (aerobic) soil. With appropriate management, the system aims for yield of at least 4–6 t/ha. The usual establishment method is dry direct seeding that can be rain-fed or irrigated. Irrigation can be applied through flash-flooding, furrow irrigation (or raised beds), or sprinklers. Aerobic rice can save as much as 50% of irrigation water in comparison with lowland rice (Lal et al., 2014). Weed infestation is the most severe constraint to widespread adoption of aerobic rice. Weed pressure in dry direct-seeded aerobic rice is significantly greater than that recorded in transplanted rice. Weeds in plots with a lower seeding rate have more chances to emerge, grow, and build up a strong population and thus pose a serious crop–weed competition.

4.6.4 Intensive mixed farming system

Intensive mixed farming with multiple enterprises is advantageous because it allows farmers to diversify and spread risk, enables recycling and reduces dependence on external inputs, reduces the spread of diseases and pests, leads to high profits, guarantees work throughout the year, and enhances farmer's social status. The utilization of equipment, labor, and land is very efficient in these systems. The flow of nutrients within intensive farming system is effectively controlled and balanced. The fertility of soil is also maintained in farming systems due to recycling of residues. Mixed farming promotes water conservation and minimizes soil erosion.

4.6.5 Soil enrichment using organic manure/amendments and biofertilizers

Biofertilizers are the substances that contain living microbes and, when applied to soil either by soil application or seed priming, colonize the rhizosphere. These applied microbes interact with plants and improve growth and yield with their different growth-promoting mechanisms. Generally, the organic inputs such as well-decomposed organic manure/vermicompost and green manure are used for soil application in appropriate quantity. These organic inputs require the soil nourishment and improve soil fertility as well as soil physical condition. Crop nutrient requirement can be met only with healthy soil, which is rich in microflora and microfauna. Organic manure and biofertilizers help in improving root development and nutrient uptake, in increasing water holding capacity, and in reducing the incidence of diseases and nematodes along with possible reduction in GHG emissions and global warming potential of agriculture.

4.6.6 Livestock and fisheries

Community lands can be used for fodder production during droughts/floods, with improved fodder/feed storage method. Preventive vaccination, improved shelters for reducing heat stress in livestock, management of fish ponds/tanks during water scarcity and floods, and so on can be adopted in climatic stress conditions.

4.6.6.1 Fodder production

The number of livestock is growing rapidly, but the grazing lands are gradually diminishing due to pressure on land for agricultural and nonagricultural uses. Most of the grazing lands have either been degraded or encroached upon restricting its availability for grazing. The area under fodder cultivation is limited to about 4% of the cropping area, and it has remained static for the past four decades. Owing to the importance of food crops and other cash crops, it is not possible that the area under fodder cultivation would increase substantially. Therefore, the need of the time is to adopt the practice of land use with multiple crops in a sustainable manner. Adopting silvipastoral and hortipastoral models suitable to the area can help in substantially enhancing the availability of forage for the livestock. About 29 million ha area in the country falls under the category of open forests with less than 0.4 canopy density, which can be developed with fodder trees. This huge land resource can be utilized for growing fodder, as an understory on the partially shaded ground without affecting standing trees. Similar development is also possible in the area under horticulture orchards. While the forest department can undertake silvipastoral plantations through the Joint Forest Management Committees, the hortipastoral activities can be initiated by incentivizing the farmers who are the owners of the orchards. Emphasis should also be laid on the non-cropped areas in the agricultural land which are not cultivated, viz., bunds, pond embankments, slopes of nala course, basins of plantation and horticultural crops, hedges with fodder crops, and so on.

4.6.6.2 Aquaculture and coping measures

Farmers consider breach of pond embankment, mortality of fishes due to saline water ingression as extreme risk; escape of fish stock and diseases as high risk; entry of unwanted species, retardation of growth, and deterioration of water quality as medium risk; and damage of pond environment as low risk. To reduce the threats against saline water ingression, farmers are taking some coping measures such as increase in pond dyke height; repair and strengthening of dyke; plantation on dyke; dewatering and addition of fresh/rainwater; application of chemicals/lime/dung; addition of tree branches in pond for hide outs, and so on. Climate change events and the physical and economic impacts are perceived by the aquaculture farmers in coastal areas of India. Livestock–crop integration with aquaculture as Climate Adaptive Integrated Farming (CIAF) has been tried among the marginal fish farmers of Sundarban in participatory mode. Incorporation of salt-tolerant fruits and horticultural crops with salt-tolerant aquacultural species has resulted in better economic returns, capacity enhancement, and reduced risk. This has been achieved through land shaping, reclamation, reexcavation of ponds including three-tier step-cutting, or terracing on inward slopes of the ponds. Plantation on the top of the dyke has resulted in its strengthening. This has also given additional (70%–90% more) income to the farmers through agri–horticultural crop. The intervention showed encouraging results with increased production and profitability through recycling of waste, better utilization of space and manpower, and crop diversification. Farmers are facing difficulties in procurement of quality stock in required quantity, and the transportation of fish seed leads to poor survival and increased cost on fingerling size seed. Due to the variability in rainfall, there is uncertainty in the availability of water in tank, which results in uncertainty in the release of fish seed in time. To overcome these constraints, rearing of early-stage fish seed (spawn/fry) to fingerling stage is done in captive condition, which reduces the cost on seed.

5. Institutional interventions

Improving the existing agricultural practices by institutional interventions by either strengthening the existing ones or initiating new ones related to fodder bank, seed bank, custom hiring center, commodity groups, collective marketing, weather index–based insurance, and climate literacy through a village-level weather forecast station (Srinivas Rao et al., 2019).

5.1 Preserving the genetic diversity

A seed bank stores seeds to safeguard genetic diversity; hence, it is a type of gene bank. The seeds can be preserved to protect the genes that plant breeders need to increase yield, disease resistance, taste, nutritional quality, drought tolerance, and so on of crops. Conservation efforts such as seed banks are probably to play an important role as climate change progresses. Seed banks provide communities a supply of climate-resilient seeds to withstand altering local climates. The challenges arise from climate change; community-based seed banks can improve access to a climatically diverse selection of local crops and also enhance indigenous understandings of plant management such as seed selection, treatment for disease protection, storage, and distribution in the market (Prasad et al., 2014).

5.2 Role of fodder bank in improving the quality of fodder

Livestock play an important role in most small-scale farming systems throughout the world. They provide traction to plow fields, manure that maintains crop productivity, and nutritious food products for human consumption. In most small-scale farming systems, livestock graze in pastures or woodlands feeding on grass and other herbaceous plants. During the wet season, these lands provide adequate forage to maintain productive animals. In the dry season, however, the quantity and quality of forage greatly decreases and is generally low in nutritional value. Livestock sustained on such diets often lose weight and productivity (Venugopalan et al., 2016). Fodder banks are valuable crops that support productive farming systems. They should be managed intensively. Thorough site preparation is necessary to establish fodder banks. This can be accomplished by mechanical or chemical means; use of methods practiced locally for other high-value crops; and seed bed, in which plant growth and survival will be maximized. As with all tree planting activities, fodder bank establishment must coincide with the rainy season that assures high plant survival.

5.3 Custom hiring centers for increasing farm mechanization

Indian agriculture is undergoing a gradual shift from dependence on human power and animal power to mechanical power because of increasing cost for upkeep of animal and growing scarcity of human labor. Furthermore, use of mechanical power has a direct bearing on the productivity of crops apart from reducing the drudgery and facilitating timeliness of agricultural operations. The farm power availability for small/marginal land holdings is the lowest. As the small/marginal holdings constitute 80% of total land holdings, the potential for custom hiring centers (CHCs), which will cater to the farm machinery requirement of such a vast area, is quite huge. To bring farm machinery available within the reach of small/marginal holdings, collective ownership or CHCs needs to be promoted in a big way. This model scheme is prepared to demonstrate the banks that financing for establishment of CHCs is a financially viable unit.

CHCs for farm implement empowering farmers to tide over the shortage of labor and improve efficiency of agricultural operations. A committee of farmers' nominated by the *gramsabha* manages the custom hiring center. The rates for hiring the machines/implements are decided by the Village Climate Risk Management Committee.

5.3.1 Benefits from custom hiring centers

i. Access to small and marginal farmers to costly farm machinery
ii. Timeliness in farm operations and efficient use of inputs
iii. Adoption of climate-resilient practices and technologies by farmers because of availability of appropriate machines at reasonable hiring charges
iv. Reduces drudgery
v. Increase in cropping intensity wherever feasible
vi. Crop residue recycling and prevents burning of residues
vii. Reduction in cost of cultivation
viii. Work opportunities to skilled labor and small artisans

Trained agriculture professionals establish agri-business centers that are commercial units of agriventures. Such ventures include maintenance and custom hiring of costly farm equipment, sale of farming inputs, and other services in agriculture and allied areas, such as postharvest management and market linkages for more income and entrepreneurship development. The scheme covers full financial support for training and handholding, provision of loan to the framers, and credit-linked composite subsidy.

5.4 Weather Based Crop Insurance

Weather Based Crop Insurance is to alleviate the hardship of the insured farmers against the chances of financial loss on account of expected crop loss due to incidence of adverse conditions of weather parameters such as temperature, rainfall, frost, humidity, and so on. Weather Based Crop Insurance Scheme (WBCIS) works on the concept of "Area Approach," i.e., for the purposes of compensation; a "reference unit area (RUA)" is deemed to be a homogeneous factor of insurance. This RUA is notified in advance of the season by the state government, and all the insured cultivators of a specific insured crop in the area will be eligible to be on par in the assessment of the claims. Each RUA is linked to a reference weather station (RWS), which is the basis of current weather data and the claims. Insured payout is entitled according to the adverse weather incidences, which is subjected to the weather triggers defined in the "payout structure" and the terms and conditions of the scheme. WBCIS gives protection to the insured cultivators in the event of loss in crops yields resulting from the variable weather incidences, such as unseasonal/excess rainfall, frost, heat (temperature), relative humidity, and so on. To capture the adverse incidence of weather parameters on crop yield, these triggers are broadly fixed.

6. Village-level weather forecasting

The India Meteorological Department aims to establish a village-level weather forecasting network, covering an area of 2–3 km, to reduce crop and livestock losses due to climate change. There are various and varied sources of weather and climate-related threats in agriculture: limited water supplies, drought, desertification, soil depletion, deforestation, hail, floods, early frosts, and many more. Good weather and climate knowledge and consulting services will inform farmers, which helps in decision-making and enhances their management of associated agricultural risks. These services can help to build sustainable and economically viable farming systems, boost productivity and quality, reduce losses and risks, reduce costs, increase water capacity, labor and energy, conserve natural resources, and decrease pollution by agricultural chemicals or other agents that contribute to the degradation of the environment. Thus, the importance of the Agromet Advisory Services that have now been established at district levels in India (Rathore and Chattopadhyay, 2016) can be realized. These services meet farmers' real-time needs and contribute to weather-based crop/livestock management strategies and operations to improve crop production and food health. They can make a huge difference in agricultural production by helping farmers benefit from benevolent weather and reduce the adverse effects of malevolent weather.

To address the issues related to climate change, Dickie et al. (2014) suggested some strategies, which are well suited in different countries including India:

i. Shift consumption patterns: It suggests the majors to be taken to combat agricultural GHG emissions, which are due to the increased pressure on the production system. Rising demand of food production on the limited land resources is the root cause of unsustainable consumption of chemical fertilizers. Dietary shift and reduction of the food item wastage can mitigate 3 gT of CO_2 equivalents.

ii. Focus is more on important agricultural producers that have the potential to achieve productivity to meet demand. It includes the strategies such as providing such diet to the animals that not only help in gaining the weight quickly but also reduce the enteric fermentation. This change may help in reducing CH_4 emissions. In this line, interventions in manure storage conditions, better control on irrigation water, and more tailor-made plans on case-by-case and country basis.

iii. Pursue catalytic, cross-cutting interventions: Agribusinesses, agricultural ministries, and financial institutions need to initiate, promote, and adopt suitable integrated climate change combating strategies to achieve higher productivity from agriculture.

iv. Rational approach for agricultural carbon sequestration by supporting long-term management and preservation of soil carbon.

7. Conclusion

To attain the sustainable crop yields and farm income, farmers need to rationally adapt the changing climate. Enhanced resilience of agriculture with respect to the climatic risk is of vital importance for protecting livelihoods of small and marginal farmers. Farmers need to adapt promptly to enhance their resilience to increasing threats of climatic variability such as floods, droughts, and other extreme climatic events during climate change and variability. Efficient resource management, environmental and social safeguards, sustainable approaches, and long-term vision in agriculture are greatly required. The major concern of resilient agriculture is to increase the agricultural productivity sustainability, to reduce the GHG emissions, and to enhance the resilience to tackle the consequences of climate change. By practicing resilient agriculture, it is possible to manage the natural resources sustainably, improve the soil health, and increase the crop production and livestock. Thereby, it helps the farmer to be self-reliant and improve their socioeconomic status.

References

Bhan, S., Behera, U.K., 2014. Conservation agriculture in India—Problems, prospects and policy issues. Int. Soil Water Conserv. Res. 2 (4), 1—12.

Dickie, A., Streck, C., Roe, S., Zurek, M., Haupt, F., Dolginow, A., 2014. Strategies for Mitigating Climate Change in Agriculture: Abridged Report. Climate Focus and California Environmental Associates, Prepared with the Support of the Climate and Land Use Alliance. https://www.agronomy4me.org/climate-change-can-we-blame-nature.

IPCC, 2019. Summary for policymakers. In: Shukla, P.R., Skea, J., Calvo Buendia, E., Masson-Delmotte, V., Pörtner, H.-O., Roberts, D.C., Zhai, P., Slade, R., Connors, S., van Diemen, R., Ferrat, M., Haughey, E., Luz, S., Neogi, S., Pathak, M., Petzold, J., Portugal Pereira, J., Vyas, P., Huntley, E., Kissick, K., Belkacemi, M., Malley, J. (Eds.), Climate Change and Land: An IPCC Special Report on Climate Change, Desertification, Land Degradation, Sustainable Land Management, Food Security, and Greenhouse Gas Fluxes in Terrestrial Ecosystems, ISBN 978-92-9169-154-8.

Lal, B., Nayak, A.K., Gautam, P., Tripathi, R., Singh, T., Katara, J., 2014. Aerobic rice: a water saving approach for rice production. Popular Kheti 1, 1—4.

Pannell, D.J., Llewellyn, R.S., Corbeels, M., 2014. The farm-level economics of conservation agriculture for resource-poor farmers. Agric. Ecosyst. Environ. 187, 52—64.

Prasad, Y.G., Maheswari, M., Dixit, S., Srinivasa Rao, C., Sikka, A.K., Venkateswarlu, B., Sudhakar, N., Prabhu Kumar, S., Singh, A.K., Gogoi, A.K., Singh, A.K., Singh, Y.V., Mishra, A., 2014. Smart Practices and Technologies for Climate Resilient Agriculture. Central Research Institute for Dryland Agriculture (ICAR), Hyderabad, 76 pp.

Rama Rao, C.A., Raju, B.M.K., Subba Rao, A.V.M., Rao, K.V., Rao, V.U.M., Kausalya, R., Venkateswarlu, B., Sikka, A.K., 2013. Atlas on Vulnerability of Indian Agriculture to Climate Change. Central Research Institute for Dryland Agriculture, Hyderabad, p. 116.

Rathore, L.S., Chattopadhyay, N., 2016. Weather and climate services for farmers in India. WMO Bull. J. World Meteorol. Org. 65 (2), 40—43.

Srinivasa Rao, C., Maheswari, M., Srinivasa Rao, M., Sharma, K.L., Vanaja, M., Rao, V.U.M., Ramana, D.B.V., Rama Rao, C.A., Vijay Kumar, P., Prasad, Y.G., Prasad, J.V.N.S., Sikka, A.K., 2015. National Innovations on Climate Resilient Agriculture (NICRA), Research Highlights (2014–15). Central Research Institute for Dryland Agriculture, Hyderabad, 120pp.

Srinivasa Rao, C., Prasad, R.S., Mohapatra, T., 2019. Climate Change and Indian Agriculture: Impacts, Coping Strategies, Programmes and Policy. Technical Bulletin/Policy Document 2019. Indian Council of Agricultural Research, Ministry of Agriculture and Farmers' Welfare and Ministry of Environment, Forestry and Climate Change, Government of India, New Delhi, p. 25.

Uphoff, N., 2007. The System of Rice Intensification: using alternative cultural practices to increase rice production and profitability from existing yield potentials. Int. Rice Comm. Newsl. 140pp. No. 55, Food and Agriculture Organization, Rome.

Venugopalan, V., Mathyam, P., Gopinath, K.A., Pankaj, P., Srinivas Rao, C., 2016. Community fodder banks for addressing forage shortage in rainfed areas. Indian Farming 66, 21–25.

Influence of anthropocene climate change on biodiversity loss in different ecosystems

4

Sunayana Nath [1], *Ritis Kumar Shyanti* [2], *Yash Nath* [3]
[1]Department of Environmental Studies, Institute of Home Economics, University of Delhi, New Delhi, India; [2]School of Life Sciences, Jawaharlal Nehru University, New Delhi, India; [3]Department of Geology and Geophysics, Indian Institute of Technology, Kharagpur, West Bengal, India

1. Introduction

The most remarkable component of Earth is the presence of life and its phenomenal diversity. Energy transfers through trophic levels to maintain the nutrient cycling so that various communities coexist in conjunction. Roughly, 9 million varieties of plants, animals, protists, and fungi together with 7 billion humans coexist on the planet (Cardinale et al., 2012). Biodiversity is very intrinsic to the structure of ecosystem (Kumar and Yadav, 2009). An exceptional feature of life, biodiversity, is today continuously suffering the threat of climate change at alarming rate (Brondizio et al., 2019; O'Connor et al., 2020). Nearly, past three decades ago, convention on biological diversity was the uttermost decision of the "Earth Summit" that pursue conservation of habitats and ecosystem from extinction. In the continuation, the current year 2020 is being tagged as superyear for nature and biodiversity by the UN Environment Programme. The environmental concerns in public for climate change threats are interlinked with the biodiversity loss (Skogen et al., 2018). Altered climatic patterns can exert profound structural and functional impacts on the sensitive ecosystems and biodiversity of planet. The magnitude of global climate change in anthropocene continue unprecedentedly, affecting atmospheric surface warming, glacial retreat, fluctuating precipitation patterns, rising sea levels, and ocean acidification. Climate change has significant impact on biodiversity in terms of species interactions, primary productivity, animal migration, foraging behavior, disease emergence, and plant reproduction (Zuckerberg, et al. 2020). The major objective of this chapter is to underline the climate change factors that impacted biodiversity deleteriously and climate change–induced species responses in various terrestrial and aquatic ecosystems.

Global Climate Change. https://doi.org/10.1016/B978-0-12-822928-6.00001-0

2. Drivers of climate change

The global emissions of carbon dioxide (CO_2) and other greenhouse gases (GHGs) such as methane (CH_4), and nitrous oxide (N_2O) have been dramatically elevated since industrial era (Roche et al., 2016; Shakoor et al., 2020). Many anthropogenic activities including agriculture, livestock management, and associated land use reported to contribute 10%–14%, whereas deforestation leads to 12%–17% of total GHG emissions (Tubiello et al., 2015; Paustian et al., 2016). The elevated radiative forcing due to human-induced emissions of CO_2 together with other non-CO_2 GHGs, mentioned under Kyoto Protocol including hydrofluorocarbons (HFCs), perfluorocarbons (PFCs), and sulfur hexafluoride (SF_6) as well as Montreal gases chlorofluorocarbons (CFCs), have caused detrimental effects on climate. GHGs are expected to further intensify in future decades and will decide the changing amplitudes of global climate (Montzka et al., 2011; Wuebbles et al., 2017). Fossil fuels consumption in the energy sectors has been noted to contribute around 35% of GHG emissions (Victor et al., 2014; Iordan et al., 2018). The concentrations of CO_2, a major GHG during preindustrial era, was around 280 parts per million (ppm) that continues to cross 400+ ppm levels and projected to hit 1000 ppm by the end of current century (Fig. 4.1) (Wuebbles et al., 2017; Woodward, 2019). Similarly, CH_4, the second major contributor after CO_2, plays a significant role in GHG release. Its potential of global warming is 28–34 times higher than CO_2. However, CH_4 emissions are known from preindustrialization times; atmospheric CO_2 was produced enormously during industrialization period and recognized as a prime source of GHG. Since industrial revolution, the levels of CO_2, CH_4, and N_2O have raised by 41%, 160%, and 20% as depicted in Fig. 4.1 (IPCC, 2013).

To combat climate change, several international efforts have been initiated. Roughly two decades back in 1997, Kyoto Protocol to UN Framework Convention on Climate Change (UNFCCC) commits industrialized countries to reduce the GHG emissions at average of 5% below 1990 levels over the periods of 2008–12 (Breidenich et al., 1998; Hargrove et al., 2019; Miyamoto and Takeuchi, 2019). The

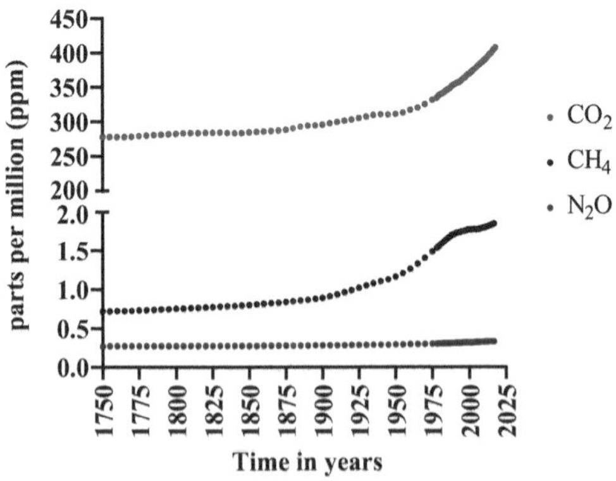

Figure 4.1 Trends in atmospheric concentration of carbon dioxide (CO_2), methane (CH_4), and nitrous oxide (N_2O) in ppm since industrial revolution.
Adapted from https://www.eea.europa.eu.

international multilateral landmark, Paris Agreement of UNFCCC targeted to keep the global temperature below 2°C with low carbon future even appear challenging to meet (Ove,et al. 2019; O'Connor et al., 2020). A continual increment of 1°C in global atmospheric temperature during 1901–2016 has been recorded and projected to strike near 5°C or more by 21st century compared with period 1850–1900 (Stocker et al., 2013; Wuebbles et al., 2017; Woodward, 2019). Quaternary global climate will exert several inches increments in sea level in next future years and threaten to rise nearly 1–4 feet by next century (Wuebbles et al., 2017). Climate drivers including temperature, precipitation, and CO_2 are the key factors that have affected surface warming, constricting ice caps, rising sea level, ocean acidification, and extreme weather patterns that are directly or indirectly responsible for biodiversity crisis at planetary scale.

3. Climate change–induced species response

Ecological biodiversity is highly sensitive to changing climate. Thus, the forecast of the species response toward climate change is highly associated with ecology, climatology, and evolutionary biology (Jackson et al., 2009). Elimination of climate specific species and biological traits is apparently coupled with stressed environmental conditions that have affected the functions and sustainability of both terrestrial and aquatic ecosystems, leading to biodiversity decline (Hooper et al., 2012; Häder and Barnes, 2019). The anthropocene climatic tolerance of the species has been evident collectively as disturbed primary production, population dynamics (Worm and Lotze, 2009), and distribution pattern (Lenoir and Svenning, 2015; Hidasi-Neto et al., 2019). The climate-driven reshuffling of species has been noticed in the temporal and spatial levels (Kappelle et al., 1999; Rosenzweig et al., 2007; Bartomeus et al., 2011). Many studies of mountain ecosystems have recorded the range shifts of taxa toward higher altitude and latitude associated with temperature and moisture availability (Crimmins et al., 2011; Ehrlén and Morris, 2015; Fei et al., 2017). Montane tree line commencing elevation gradients indicated resilience toward climate warming (Shi et al., 2020). Forest ecosystems are facing compositional variations and regional extinctions due to climatic variability (Fei et al., 2017; Kuhn and Gégout, 2019). Spatial shifts of flowering phenology along elevation gradient have been observed with elevation specific changes in temperature and precipitation (Foster et al., 2019; Rafferty, et al. 2020). Insects providing ecological services such as bees, which have a mutualistic relationship with the host plant, have been reported to emerge with temperature-associated shifts in flowering phase of host plant (Bartomeus et al., 2011). Temperature effect on physiology and behavior of ecotherms has shown changed species interactions and community dynamics (Boukal et al., 2019). Tropical birds are regarded as a reliable indicator of climate change. Birds extinction due to climate change–driven habitat loss has been reported 50% (Şekercioğlu et al., 2012). Atmospheric surface warming of 3.5°C by the year 2100 may cause extinction of 600–900 land bird species out of which 89% would be tropical birds. According to an estimation of future habitat loss, approximately 100–500 bird extinction may occur with per degree of surface warming (Şekercioğlu et al., 2012). Wetland ecosystems are facing reduced soil moisture and warmer climates affecting riparian vegetations to compromise with conifers for climate resilience (Dwire et al., 2018). Increasing heat wave events because of the increasing thermal marine conditions have damaged coral reef ecosystems to drive out algae (Hughes et al., 2017; Leggat et al., 2019). Invasive species introduction cause devastating ecological and economic impacts on native one. Invasion may have negative impact on both flora and fauna populations as a result of direct competition, habitat degradation, disease spread, or dislocation (Bergman et al., 2000; Brain et al., 2020; Gillard et al., 2020). Optimal environmental conditions, behavioral plasticity, reproductive biology, and food requirement are basic amenities for animals to achieve better reproducing functions essentially for their survival. The annual breeding season for individual species primarily lies in maintaining energy requirement, time for foraging, and production of offspring that are in sync with the environment. It is therefore necessary to cope with the variable climatic conditions and balance the risk–benefit parameters to gain maximum reproductive fitness (Reimer et al., 2019). Unfavorable condition occurring in the environment creates evolutionary pressure on plant and animal species so that they can adapt to the critical daylength changes failing, which would lead to destructive life cycle for individual species. The seasonal adaptation to changing photoperiod for the optimal reproduction was observed to be more adjustable by the insects and plants than birds and mammals (Walker et al., 2019).

4. Biodiversity loss in terrestrial environment

Around 30% of the planet's surface area that is covered under forest ecosystems comprised of 80% of the terrestrial biodiversity pool (Aerts and Honnay, 2011; Keenan et al., 2015; Morales-Hidalgo et al., 2015). Forest ecosystem provides essential ecological services including climate regulation, nutrient cycling, and wildlife habitat conservation. Quaternary climatic pressures have caused severe vulnerability to the quality of forest ecosystem (Sharma et al., 2017). Studies have suggested a kind of empirically positive correlations between biodiversity and the provisioning of services offered by forest ecosystems (Isbell et al., 2011; Hooper et al., 2012; Brockerhoff et al., 2017). Rich biodiversity contributes higher productivity and sustainability of the forest assisting resilience against extreme climate (Isbell et al., 2015; Augustynczik et al., 2020).

It is likely that warmer temperature, precipitation pattern, increased CO_2 concentration, and occurrence of extreme climatic events have exerted physiological stresses to the tree species (Cramer et al., 2004; Field et al., 2012; Sharma et al., 2017). Tree line distributions are observed to high-latitude shifts for favorable landscape expansion (Kelly and Goulden, 2008; Ehrlén and Morris, 2015; Kumar et al., 2019). Climate-driven changes in tree species composition reported to affect forest productivity (Morin et al., 2018). Tree species distribution range shifts under current climate change regime have been projected to accelerate in future (Morin et al., 2008; Anderegg et al., 2013). Flowering phenology—a governing reproductive process—serves as a key indicator of climate change. Climate is driving phenological and spatial reshuffling along local elevation gradient among flowering phenology (Rafferty et al., 2020). Study revealed that frost and snowmelt environmental sequences could have large effects on natural populations of wild flower community (Inouye, 2008). Temperate species responses to climate change are more likely to be associated with reshuffling of time while tropics alter their space or evolved (Pau et al., 2011). Extinctions of forest species or key species loss threatened of extreme events including droughts and heat wave have occurred round the globe (Allen et al., 2010; Anderegg et al., 2013).

4.1 Montane and subalpines ecosystems

Mountain landscapes harbor 22% of Earth's terrestrial surface (Romeo et al., 2015; Foggin, 2016). These rugged systems offer home and vital services such as nutrient retention, carbon storage, the provision of water, to 13% inhabitants and 26% population in immediate proximity (Foggin, 2016; Peters et al., 2019). Montane ecosystems are key centers for rich biodiversity worldwide supporting the global biodiversity hot spots (Noroozi et al., 2018; Peters et al., 2019). The mountain ranges are identified core of endemic richness (Noroozi et al., 2018). Mountain ecosystems are generally considered as sensitive indicators of perturbations and climate change (Rogora et al., 2018; Winkler et al., 2019). Alpine communities represent 4% of higher plants; above the tree line in mountains are biotic zones constituting 3% of land area (Körner, 2003; Winkler et al., 2019). Higher elevations are found to be strongly correlated to atmospheric warming compared with lower elevations (Pepin et al., 2015; Silveira et al., 2019). Alpines exhibited altered community composition, and primary production is largely determined by soil moisture under climate change (Winkler et al., 2016). The habitats of tree line species in the Qinling Mountains of North–Central China migrated upslope along the elevation gradient as a response to climate warming (Shi et al., 2020). Variability in temperature, precipitation, and snow pattern during 30 years period concluded vegetation cover shifts along elevation gradient in Southern California's Santa Rosa Mountains (Kelly and Goulden, 2008). Another long-term ecological research in tropical snow free mountains predicted altered biodiversity and ecosystem of the area (Silveira et al., 2019). High elevation herbaceous plants have been projected to suffer habitat loss under climate change stress (Engler et al., 2011; Rogora et al., 2018). Occurrence of thermophilization, a reduction in cold adapted species, and increase in warm in European mountain system suggests a progressive decline of cold mountain habitats (Gottfried et al., 2012).

4.2 Dryland ecosystems

Globally, dryland accounts approximately 41% of the Earth's terrestrial land surface and provides a net primary productivity of around 40% (Reynolds et al., 2007). Grasslands, shrublands, and savannas are the major ecosystems representing dryland (Yao et al., 2020). In view of the inadequate soil water and supplements just as the fragile vegetation spread, drylands are especially under threatened conditions to changes in climate variables (Maestre et al., 2012; Huang et al., 2020). Dryland is characterized by aridity index (AI) having value less than 0.65. AI is defined by the formula of annual precipitation to potential evapotranspiration (PET) ratio. There are major categories of dryland on the basis of AI value such as hyperarid ($AI \leq 0.05$), arid ($0.05 < AI \leq 0.2$), semiarid ($0.2 < AI \leq 0.5$), and dry subhumid ($0.5 < AI < 0.65$). Global climate change due to enhanced warming and droughts has severe impact on dryland expansion, degradations, and conversions, which lies in between the historical drylands ($AI < 0.65$) and the humid lands ($AI \geq 0.65$) (Yao et al., 2020). These changes as a result of future climatic conditions can damage dryland ecosystem structures and functions in terms of nutrient cycling, plant productivity, and microbial communities and thereby gross primary productivity (Berdugo et al., 2020; Huang et al., 2020). According to IPCC (2013) report, a predicted temperature rise to 4°C along with increased precipitation variability can cause a substantial shift in species richness and ecological resilience in dryland ecosystems (Stocker et al., 2013; Huang et al., 2016, 2020). The impact of climate change on dryland ecosystem was noticed in a large-scale conversion of vegetated grassland into bare land over 30 years of time span (Song et al., 2018). The reduced precipitation rate during growing season in Australia, introduction of invasive species in southwestern United States, and overgrazing along with elevated temperatures and reduced rainfall in Mongolian steppe are few examples of dryland conversions. Impact of uneven warming on biological ecosystem function and structure is very complicated. Asymmetric nocturnal temperatures have been reported to accelerate egg development in grasshoppers, intrude woody species in grassland-woodland ecotones, and disturb prey–predator relationship by altering plant diversity (Wu et al., 2012; Barton and Schmitz, 2018; Huang et al., 2019). According to one latest report, Berdugo et al. (2020) have described progression of AI and its resulting effects on dryland in three phases. In the very first "vegetation decline phase," the AI value is greater than 0.54 with declined vegetation. Further increase in AI value more than 0.7 is "soil disruption phase" attributing decreased soil quality, lowered organic carbon and nitrogen content, abundant fungal species, and reduced positive plant–plant interactions. Reduced soil nutrient and plant productivity trigger vegetation shifts during soil disruption phase, which leads to conversion of grasslands and savannahs to shrublands as adaptive measures to limited nutrient and sandy soils (D'Odorico et al., 2012; Phillips et al., 2019; Berdugo et al., 2020). The third and final phase at the extreme aridity condition with AI value more than 0.8 is known as "ecosystem breakdown phase." This is the drastic condition for different species inhabiting in such an ecosystem facing severe water scarcity as well as nutrient. This leads to strong inhibition of plant species richness and their productivity. This kind of environmental conditions allow replacement of vegetation with stress tolerant summer deciduous shrub species where sparse and unpredictable rain events are common (Berdugo et al., 2020).

5. Biodiversity loss in aquatic environment

5.1 Marine and coral reef ecosystems

The consequences of anthropocene climate change have significant effects on physical and chemical properties of the ocean directing the aquatic ecosystems. Oceans are consider as key absorber of the 93% heat generated during GHG emissions predicting addition of 1−10°C to the seawater temperature in the following century (IPCC, 2013; Baltar et al., 2019). Marine systems act as an active sinks for atmospheric CO_2 accredited to lengthy seawater residence time, stretched surface area along with abundance of autotrophic organism (Sabine et al., 2004; Hasler et al., 2016). Increased water temperatures, ocean acidification, water column circulation, and stratification have caused migrations, change in habitats, and altered productivity of marine life from local to global scale (Worm and Lotze, 2016). Sea level rise is apparent to accelerated in coming future by several inches (Wuebbles et al., 2017; He and Silliman, 2019). The coastal zones that represent the world's most populated areas are the most threatened landscapes to rising sea surface (Harley et al., 2006; He and Silliman, 2019). Biodiversity of marine communities has been observed responding variously toward climate change effects. These responses of marine biota include physiochemical changes, species compositional alterations, phenology variations, and range shifts toward higher latitudes or adaptation from near water to greater depths (Poloczanska et al., 2016; Bestion and Cote, 2018; He and Silliman, 2019). These climate-driven changes have been studied mostly in fish and plankton species distribution and diversity pattern. Tropical ocean experienced elevated heat stress compelling marine species to shift their biotic range toward temperate regions (Worm and Lotze, 2016). Studies have reported temperature-driven substitution of productive kelp forest habitats and coral reef ecosystems to seaweed turfs or sea urchin barrens habitats (Wernberg et al., 2016). Oceanic warming, decreased pH, lowered levels of dissolved oxygen, nutrient stress, and availability have misbalanced the phytoplankton assemblage and have favored the frequent occurrence of harmful algal blooms events (Barton et al., 2016; Bestion and Cote, 2018; Griffith and Gobler, 2020; Hennon and Dyhrman, 2020). Repeated incidence of extreme weather episodes such as storm and tropical cyclone is also known to promote harmful algal blooms (Paerl et al., 2018).

Ocean absorbs around one-third of atmospheric CO_2 released from anthropogenic sources. Absorbed CO_2 alters ocean carbonate budget and lowers water pH leading to acidification (IPCC, 2013; Baltar et al., 2019). Acidifications tend to dissolution and erosion of calcium carbonate stocks causing coral bleaching (Hoegh-Guldberg et al., 2019). The coral reef communities are diverse and productive ecosystems that are highly vulnerable to anthropocene climate change effects. Even if coral ecosystems stretched over only 0.5% of seabed, they are the habitats for approximately 1 million species including nearly 30% of the world's marine fish species (Heron et al., 2017). The loss of coral reef is threatening to biodiversity and the livelihoods of nearly 400 million fisheries and allied communities across 100 countries (Morrison et al., 2019). Almost 90% of coral reefs listed in World Heritage sites worldwide have been reported to mass bleaching due to heat waves episodes in past two decades (Heron et al., 2017). Climate-driven severe heat waves have caused loss of 50% corals population in the world's largest reef system, Great Barrier Reef, during 2016 and 2017 (Hughes et al., 2019).

5.2 Freshwater and wetland ecosystems

Freshwater environments including rivers, lakes, ponds, streams, wetlands, and reservoirs have restricted spatial extent. Despite covering less than 1% of Earth's surface and constituting 2.3% of global water, these ecosystems embrace extraordinary biodiversity of planet. Riparian habitats are one of the most biological diverse ecosystems on Earth (Albert et al., 2020). Freshwater serves habitats for 9.5% of animal species including nearly one-third of all vertebrate species (Reid et al., 2019; Tickner et al., 2020). Considering the Living Planet Index of World Wide Fund for Nature's, freshwater population decreased up to 83% from 1970 to 2014 (Reid et al., 2019). According to IUCN Red List, 27% of the freshwater species referred threatened with extinction. Among these, the species at risk are turtles (62%), followed by gastropods (47%), mammals (42%), amphibians (33%), decapod crustaceans (30%), fishes (28%), and birds (20%) (Jones, 2019; Tickner et al., 2020). Freshwater habitats are crucial to global biodiversity and providing ecosystem services to environment and society (Albert et al., 2020; Hassan et al., 2020). Freshwater ecosystems are vulnerable to consequences of climate change that may lead to undesirable and irreversible deterioration of these habitats (Hassan et al., 2020; Znachor et al., 2020). Freshwater communities are severely threatened to extinction more than terrestrial or marine species. The climate change–induced impacts on freshwater habitats could be minimized by strengthening its biota (Hassan et al., 2020).

The synergistic effects of climate change variables could have pronounced effects on ecosystem productivity. A study conducted on plankton productivity in nutrient deficient oligotrophic condition with elevated temperature and high CO_2 levels revealed the differential ecological equilibrium and species dominance in natural freshwater phytoplankton community (Cabrerizo et al., 2020). The interactive effects of warming and nutrients scarcity in freshwater bodies could lead to dysfunctional ecosystems and biodiversity (Verbeek et al., 2018).

Climate change and land use change collectively are proposed as principal drivers of stress on freshwater environments (Bayramoglu et al., 2020). These factors are likely to raise dissolved organic carbon and iron concentrations of freshwater ecosystems resulting in browning of freshwater environments. Browning is associated with severe environmental degradation by lowering drinking water quality standards and limiting light penetration in lakes leading to reduced fish production (Kritzberg et al., 2019). Measurements of freshwater biodiversity using fish-based index indicated lower biodiversity in agricultural and pasture areas compared with forests. The differences of biodiversity among these land use types will be exacerbating under climate change in future (Bayramoglu et al., 2020). Distribution and phenology of freshwater biota have been affected by impacts of climate change. A study of freshwater ecosystem in Iran showed 10 out of 15 endemic fish species will manage to gain new suitable habitats, while rest of them will lose their current dwelling range partially under climate change stress (Yousefi et al., 2020). The disruptions of freshwater ecosystems have devastating effects on ecoservices including biogeochemical cycles, global scale declines in biodiversity, and reduced ecosystem productivity (Grantham et al., 2019).

There are 2303 wetlands having significant recognition worldwide where Europe has highest in numbers while Asia holds largest sizes. Wetland ecosystems are primarily affected by four parameters, namely, pollution, changing natural systems, biological resources utilization, and activities related to agriculture and aquaculture. Among different types of wetlands, lakes and coastal areas are the most susceptible to impacts of climate change, whereas the river wetlands are more vulnerable to biodiversity loss. The wetland ecosystem has two important GHG contributors: CH_4, as a natural source (30%–40% of total emission) and CO_2 as a carbon sink. The major climatic parameters directly influencing the wetland methanogenesis process are temperature, precipitation, and soil–water content. Gaseous exchange processes of CO_2 and CH_4 with atmosphere are likely to be altered by climate change or anthropogenic factors. It is projected that global warming will tend to increase the optimum temperature for methanogenesis process above 37°C worldwide (Jerman et al., 2009; Koffi et al., 2020). Under climate change cumulative stress of rising temperature and precipitation, the highly productive wetland habitats will be endangered to wetland expansion (Zhang et al., 2017; Koffi et al., 2020). Therefore, acquiring crucial implementation via management and policy, monitoring, restoration, knowledge, and funding are very much essential for saving and expanding wetlands (Fig. 4.2).

6. Conclusions

In this chapter, we have highlighted the current state of information on the drivers of anthropocene climate change and its impact on declining global biodiversity of Earth. The anthropocene interference in climate directly or indirectly influences global biodiversity at both terrestrial and marine levels (Malhi et al., 2020). Species and their habitats in montane tree lines, alpines, and drylands have been evident to respond negatively across the planet. Similarly, the ocean biodiversities in aquatic ecosystems of marine species, coral reefs, freshwater, and wetland environments provide provisional and regulatory services that are being suppressed under quaternary climatic stress (Falkenberg et al., 2020). Terrestrial biodiversity loss in major biomes and ecosystems can be minimized by policy interventions (reducing disturbances in landscape or within forest), effective management of protected areas by IUCN, sustainable use reserves, legal stipulation to maintain forest cover on private lands, and proper agricultural intensification on deforested lands (Falkenberg et al., 2020). As far as aquatic biosystems, ocean acidification has turned into a major threat to marine ecosystems that needs to be effectively managed. Marine disturbance not only is affecting the ocean life but also possesses severe risk to human health in several ways. To combat the ocean acidification challenge, combined strategies are required to regulate the CO_2 emission, GHGs, decarbonization in energy-providing sectors, and efficient carbon sequestration techniques either naturally or engineered. Additional measures can be opted by creating marine-protected areas (MPAs), restoration programs, or ecosystem-based management, restricting the pollutants, which can develop resistance toward acidification process and related biodiversity loss (West and Salm, 2003;

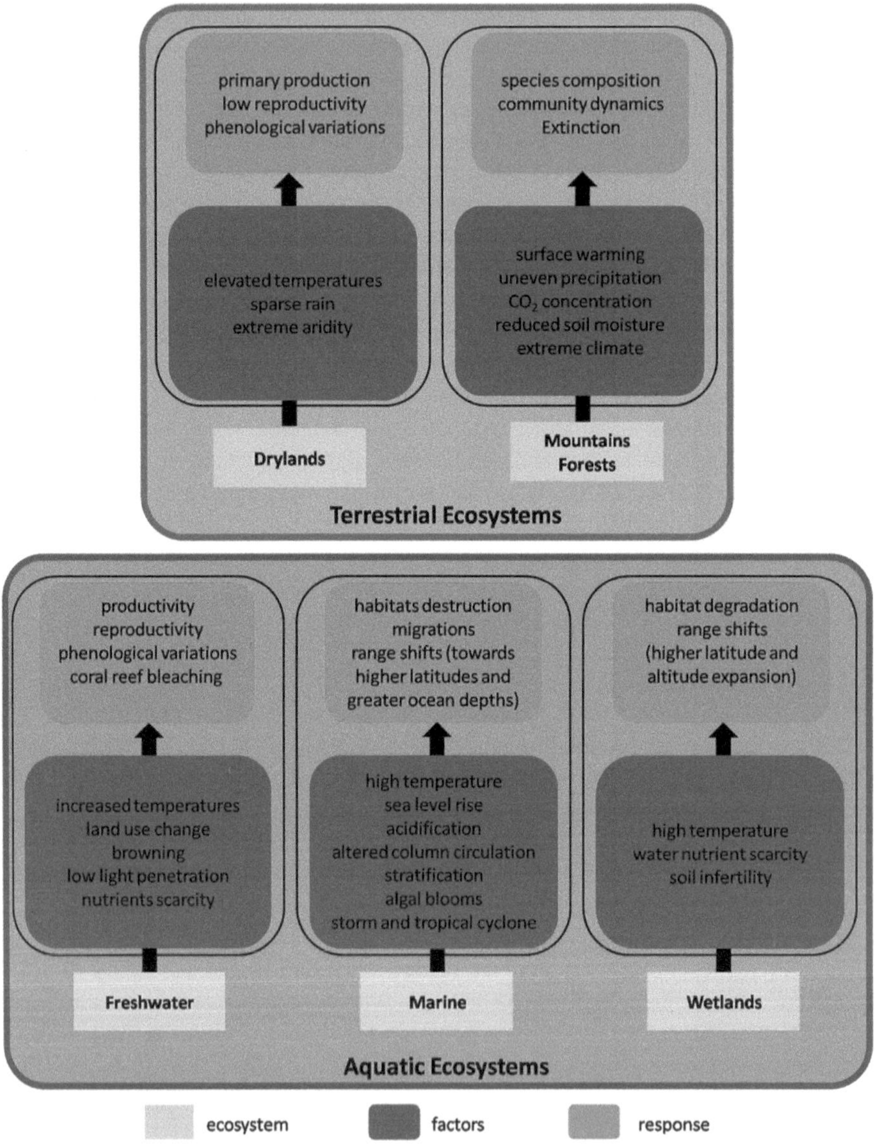

Figure 4.2 Effects and ecological response of different ecosystems under anthropocene climate change. Elevated temperature, high CO_2 concentrations, and reduced precipitation are the primary climate change drivers in both terrestrial and aquatic ecosystems. Changes in species composition, primary production, range shifts, phenology, and habitat degradations are the major common responses of species in both the ecosystems against climate change.

McLeod et al., 2009; Hendriks et al., 2014; Ussiri and Lal, 2017; Falkenberg et al., 2020). This will restore or improve the quality and quantity of oceanic biodiversity and provide benefits to human being. Predicting the future course of reaction of biodiversity to environmental change has turned out to be an operational and, to some extent, an unavoidable field of research. Though international organizations such as IPCC, IUCN, UNEP, EPA, WWF, and others are continuously involved in researching the comprehensive models of climate change, it is really not very accurate when predicted on biodiversity loss over a long time. There is a need of making committed global partners from government and societies to come under a common platform to beware of potential future dangers, giving a way to reinforce the coordinated policies and advancement of technological proactive systems to combat the challenge of quaternary climate change and biodiversity.

References

Aerts, R., Honnay, O., 2011. Forest restoration, biodiversity and ecosystem functioning. BMC Ecol. 11 (1), 29.

Albert, J.S., et al., 2020. Scientists' warning to humanity on the freshwater biodiversity crisis. Ambio 1−10.

Allen, C.D., et al., 2010. A global overview of drought and heat-induced tree mortality reveals emerging climate change risks for forests. For. Ecol. Manag. 259 (4), 660−684.

Anderegg, W.R., et al., 2013. Consequences of widespread tree mortality triggered by drought and temperature stress. Nat. Clim. Chang. 3 (1), 30−36.

Augustynczik, A.L.D., et al., 2020. Socially optimal forest management and biodiversity conservation in temperate forests under climate change. Ecol. Econ. 169, 106504.

Baltar, F., et al., 2019. Towards integrating evolution, metabolism, and climate change studies of marine ecosystems. Trends Ecol. Evol. 34 (11), 1022−1033.

Bartomeus, I., et al., 2011. Climate-associated phenological advances in bee pollinators and bee-pollinated plants. Proc. Natl. Acad. Sci. U S A 108 (51), 20645−20649.

Barton, A.D., et al., 2016. Anthropogenic climate change drives shift and shuffle in North Atlantic phytoplankton communities. Proc. Natl. Acad. Sci. U S A 113 (11), 2964−2969.

Barton, B.T., Schmitz, O.J., 2018. Opposite effects of daytime and nighttime warming on top-down control of plant diversity. Ecology 99 (1), 13−20.

Bayramoglu, B., et al., 2020. Impacts of land use and climate change on freshwater ecosystems in France. Environ. Model. Assess. 25 (2), 147−172.

Berdugo, M., et al., 2020. Global ecosystem thresholds driven by aridity. Science 367 (6479), 787−790.

Bergman, D.L., et al., 2000. The Economic Impact of Invasive Species to Wildlife Service-s'cooperators, p. 21.

Bestion, E., Cote, J., 2018. Species responses to climate change: integrating individual-based ecology into community and ecosystem studies. In: The Encyclopedia of the Anthropocene, pp. 139−147.

Boukal, D.S., et al., 2019. Species interactions under climate change: connecting kinetic effects of temperature on individuals to community dynamics. Curr. Opin. Insect Sci. 35, 88−95.

Brain, R.A., et al., 2020. Anthropogenic factors affecting wildlife species status outcomes: why the fixation on pesticides? Environ. Sci. Pollut. Res. 1−24.

Breidenich, C., et al., 1998. The Kyoto protocol to the United Nations framework convention on climate change. Am. J. Int. Law 92 (2), 315–331.

Brockerhoff, E.G., et al., 2017. Forest Biodiversity, Ecosystem Functioning and the Provision of Ecosystem Services. Springer.

Brondizio, E., et al., 2019. Global Assessment Report on Biodiversity and Ecosystem Services of the Intergovernmental Science-Policy Platform on Biodiversity and Ecosystem Services. IPBES Secretariat.

Cabrerizo, M.J., et al., 2020. Warming and CO_2 effects under oligotrophication on temperate phytoplankton communities. Water Res. 173, 115579.

Cardinale, B.J., et al., 2012. Biodiversity loss and its impact on humanity. Nature 486 (7401), 59–67.

Cramer, W., et al., 2004. Tropical forests and the global carbon cycle: impacts of atmospheric carbon dioxide, climate change and rate of deforestation. Phil. Trans. Roy. Soc. Lond. B Biol. Sci. 359 (1443), 331–343.

Crimmins, S.M., et al., 2011. Changes in climatic water balance drive downhill shifts in plant species' optimum elevations. Science 331 (6015), 324–327.

D'Odorico, P., et al., 2012. A synthetic review of feedbacks and drivers of shrub encroachment in arid grasslands, 5 (5), 520–530.

Dwire, K.A., et al., 2018. Potential effects of climate change on riparian areas, wetlands, and groundwater-dependent ecosystems in the Blue Mountains, Oregon, USA. Climate Serv. 10, 44–52.

Ehrlén, J., Morris, W.F., 2015. Predicting changes in the distribution and abundance of species under environmental change. Ecol. Lett. 18 (3), 303–314.

Engler, R., et al., 2011. 21st century climate change threatens mountain flora unequally across Europe. Glob. Chang. Biol. 17 (7), 2330–2341.

Falkenberg, L.J., et al., 2020. Ocean acidification and human health. Int. J. Environ. Res. Publ. Health 17 (12), 4563.

Fei, S., et al., 2017. Divergence of species responses to climate change. Sci. Adv. 3 (5), e1603055.

Field, C.B., et al., 2012. Managing the Risks of Extreme Events and Disasters to Advance Climate Change Adaptation: Special Report of the Intergovernmental Panel on Climate Change. Cambridge University Press.

Foggin, J., 2016. Conservation Issues: Mountain Ecosystems.

Foster, A.C., et al., 2019. Importance of tree-and species-level interactions with wildfire, climate, and soils in interior Alaska: implications for forest change under a warming climate. Ecol. Model. 409, 108765.

Gillard, M.B., et al., 2020. Seed source regions drive fitness differences in invasive macrophytes. Am. J. Bot. 107 (5), 749–760.

Gottfried, M., et al., 2012. Continent-wide response of mountain vegetation to climate change. Nat. Clim. Chang. 2 (2), 111–115.

Grantham, T.E., et al., 2019. Shifting currents: managing freshwater systems for ecological resilience in a changing climate. Water Secur. 8, 100049.

Griffith, A.W., Gobler, C.J., 2020. Harmful algal blooms: a climate change co-stressor in marine and freshwater ecosystems. Harmful Algae 91, 101590.

Häder, D.-P., Barnes, P.W., 2019. Comparing the impacts of climate change on the responses and linkages between terrestrial and aquatic ecosystems. Sci. Total Environ. 682, 239–246.

Hargrove, A., et al., 2019. Global governance for climate justice: a cross-national analysis of CO_2 emissions. Glob. Trans. 1, 190–199.

Harley, C.D., et al., 2006. The impacts of climate change in coastal marine systems. Ecol. Lett. 9 (2), 228–241.
Hasler, C.T., et al., 2016. Freshwater biota and rising pCO_2? Ecol. Lett. 19 (1), 98–108.
Hassan, B., et al., 2020. Impact of climate change on freshwater ecosystem and its sustainable management. In: Fresh Water Pollution Dynamics and Remediation. Springer, pp. 105–121.
He, Q., Silliman, B.R., 2019. Climate change, human impacts, and coastal ecosystems in the Anthropocene. Curr. Biol. 29 (19), R1021–R1035.
Hendriks, I.E., et al., 2014. Photosynthetic activity buffers ocean acidification in seagrass meadows. Biogeosciences 11 (2), 333–346.
Hennon, G.M., Dyhrman, S.T., 2020. Progress and promise of omics for predicting the impacts of climate change on harmful algal blooms. Harmful Algae 91, 101587.
Heron, S.F., et al., 2017. Impacts of Climate Change on World Heritage Coral Reefs: a First Global Scientific Assessment.
Hidasi-Neto, J., et al., 2019. Climate change will drive mammal species loss and biotic homogenization in the Cerrado Biodiversity Hotspot. Perspect. Ecol. Conserv. 17 (2), 57–63.
Hoegh-Guldberg, O., et al., 2019. People and the changing nature of coral reefs. Reg. Stud. Mar. Sci. 30, 100699.
Hooper, D.U., et al., 2012. A global synthesis reveals biodiversity loss as a major driver of ecosystem change. Nature 486 (7401), 105–108.
Huang, H., et al., 2020. CAM plant expansion favored indirectly by asymmetric climate warming and increased rainfall variability. Oecologia 1–13.
Huang, H., et al., 2019. The competitive advantage of C4 grasses over CAM plants under increased rainfall variability. Plant Soil 442 (1–2), 483–495.
Huang, J., et al., 2016. Accelerated dryland expansion under climate change. Nat. Clim. Chang. 6 (2), 166–171.
Hughes, T.P., et al., 2017. Global warming and recurrent mass bleaching of corals. Nature 543 (7645), 373–377.
Hughes, T.P., et al., 2019. Ecological memory modifies the cumulative impact of recurrent climate extremes. Nat. Clim. Chang. 9 (1), 40–43.
Inouye, D.W., 2008. Effects of climate change on phenology, frost damage, and floral abundance of montane wildflowers. Ecology 89 (2), 353–362.
Iordan, C.M., et al., 2018. Integrating impacts on climate change and biodiversity from forest harvest in Norway. Ecol. Indicat. 89, 411–421.
IPCC WGI, 2013. Climate Change 2013: The Physical Science Basis: Working Group I Contribution to the Fifth Assessment Report of the Intergovernmental Panel on Climate Change. IPCC WGI. (Accessed 14 May 2014).
Isbell, F., et al., 2011. High plant diversity is needed to maintain ecosystem services. Nature 477 (7363), 199–202.
Isbell, F., et al., 2015. Biodiversity increases the resistance of ecosystem productivity to climate extremes. Nature 526 (7574), 574–577.
Jackson, S.T., et al., 2009. Ecology and the ratchet of events: climate variability, niche dimensions, and species distributions. Proc. Natl. Acad. Sci. U S A 106 (Suppl. 2), 19685–19692.
Jerman, V., et al., 2009. Wetland restoration and methanogenesis: the activity of microbial populations and competition for substrates at different temperatures. Biogeosci. Discuss. 6 (1).
Jones, J.B., 2019. [IUCN] International Union for Conservation of Nature. The IUCN Red List of Threatened Species, version 2019-1. IUCN.

Kappelle, M., et al., 1999. Effects of climate change on biodiversity: a review and identification of key research issues. Biodivers. Conserv. 8 (10), 1383–1397.
Keenan, R.J., et al., 2015. Dynamics of global forest area: results from the FAO global forest resources assessment 2015. For. Ecol. Manag. 352, 9–20.
Kelly, A.E., Goulden, M.L., 2008. Rapid shifts in plant distribution with recent climate change. Proc. Natl. Acad. Sci. U S A 105 (33), 11823–11826.
Koffi, E.N., et al., 2020. An observation-constrained assessment of the climate sensitivity and future trajectories of wetland methane emissions. Sci. Adv. 6 (15) eaay4444.
Körner, C., 2003. Alpine Plant Life: Functional Plant Ecology of High Mountain Ecosystems; with 47 Tables. Springer Science & Business Media.
Kritzberg, E.S., et al., 2019. Browning of freshwaters: consequences to ecosystem services, underlying drivers, and potential mitigation measures. Ambio 1–16.
Kuhn, E., Gégout, J.C., 2019. Highlighting declines of cold-demanding plant species in lowlands under climate warming. Ecography 42 (1), 36–44.
Kumar, V., Yadav, S.K., 2009. "Plant-mediated synthesis of silver and gold nanoparticles and their applications. J. Chem. Technol. Biotechnol. Int. Res. Process Environ. Clean Technol. 84 (2), 151–157.
Kumar, M., et al., 2019. Assessing vulnerability of forest ecosystem in the Indian Western Himalayan region using trends of net primary productivity. Biodivers. Conserv. 28 (8–9), 2163–2182.
Leggat, W.P., et al., 2019. Rapid coral decay is associated with marine heatwave mortality events on reefs. Curr. Biol. 29 (16), 2723–2730 e2724.
Lenoir, J., Svenning, J.C., 2015. Climate-related range shifts—a global multidimensional synthesis and new research directions. Ecography 38 (1), 15–28.
Maestre, F.T., et al., 2012. It Is Getting Hotter in Here: Determining and Projecting the Impacts of Global Environmental Change on Drylands. The Royal Society.
Malhi, Y., et al., 2020. Climate Change and Ecosystems: Threats, Opportunities and Solutions. The Royal Society.
McLeod, E., et al., 2009. Designing marine protected area networks to address the impacts of climate change, 7 (7), 362–370.
Miyamoto, M., Takeuchi, K., 2019. Climate agreement and technology diffusion: impact of the Kyoto Protocol on international patent applications for renewable energy technologies. Energy Policy 129, 1331–1338.
Montzka, S.A., et al., 2011. Non-CO_2 greenhouse gases and climate change, 476 (7358), 43–50.
Morales-Hidalgo, D., et al., 2015. Status and trends in global primary forest, protected areas, and areas designated for conservation of biodiversity from the Global Forest Resources Assessment 2015. For. Ecol. Manag. 352, 68–77.
Morin, X., et al., 2018. Long-term response of forest productivity to climate change is mostly driven by change in tree species composition. Sci. Rep. 8 (1), 5627.
Morin, X., et al., 2008. Tree species range shifts at a continental scale: new predictive insights from a process-based model. J. Ecol. 96 (4), 784–794.
Morrison, T.H., et al., 2019. Save Reefs to Rescue All Ecosystems. Nature Publishing Group.
Noroozi, J., et al., 2018. Hotspots within a global biodiversity hotspot-areas of endemism are associated with high mountain ranges. Sci. Rep. 8 (1), 1–10.
O'Connor, B., et al., 2020. Monitoring global changes in biodiversity and climate essential as ecological crisis intensifies. Ecol. Inf. 55, 101033.
Ove, Hoegh-Guldberg, et al., 2019. The human imperative of stabilizing global climate change at 1.5°C. Science 365 (6459).

Paerl, H.W., et al., 2018. Mitigating the Expansion of Harmful Algal Blooms across the Freshwater-To-Marine Continuum. ACS Publications.

Pau, S., et al., 2011. Predicting phenology by integrating ecology, evolution and climate science. Glob. Chang. Biol. 17 (12), 3633–3643.

Paustian, K., et al., 2016. Climate-smart soils. Nature 532 (7597), 49–57.

Pepin, N., et al., 2015. Elevation-dependent warming in mountain regions of the world. Nat. Clim. Chang. 5 (5), 424–430.

Peters, M.K., et al., 2019. Climate–land-use interactions shape tropical mountain biodiversity and ecosystem functions. Nature 568 (7750), 88–92.

Phillips, M.L., et al., 2019. Differences in root phenology and water depletion by an invasive grass explains persistence in a Mediterranean ecosystem, 106 (9), 1210–1218.

Poloczanska, E.S., et al., 2016. Responses of marine organisms to climate change across oceans. Front. Mar. Sci. 3, 62.

Rafferty, N.E., et al., 2020. Changing climate drives divergent and nonlinear shifts in flowering phenology across elevations. Curr. Biol. 30 (3), 432–441.

Reid, A.J., et al., 2019. Emerging threats and persistent conservation challenges for freshwater biodiversity. Biol. Rev. 94 (3), 849–873.

Reimer, J.R., et al., 2019. Modeling optimal responses and fitness consequences in a changing Arctic. Glob. Chang. Biol. 25 (10), 3450–3461.

Reynolds, J.F., et al., 2007. Global desertification: building a science for dryland development. Science 316 (5826), 847–851.

Roche, L., et al., 2016. Impact of fertiliser nitrogen formulation, and N stabilisers on nitrous oxide emissions in spring barley. Agric. Ecosyst. Environ. 233, 229–237.

Rogora, M., et al., 2018. Assessment of climate change effects on mountain ecosystems through a cross-site analysis in the Alps and Apennines. Sci. Total Environ. 624, 1429–1442.

Romeo, R., et al., 2015. Mapping the Vulnerability of Mountain Peoples to Food Insecurity. FAO, Rome.

Rosenzweig, C., et al., 2007. Assessment of Observed Changes and Responses in Natural and Managed Systems.

Sabine, C.L., et al., 2004. The oceanic sink for anthropogenic CO_2. Science 305 (5682), 367–371.

Şekercioğlu, Ç.H., et al., 2012. The effects of climate change on tropical birds. Biol. Conserv. 148 (1), 1–18.

Shakoor, A., et al., 2020. Biogeochemical transformation of greenhouse gas emissions from terrestrial to atmospheric environment and potential feedback to climate forcing. Environ. Sci. Pollut. Res. Int. 1–24.

Sharma, J., et al., 2017. Vulnerability of forests in India: a national scale assessment. Environ. Manag. 60 (3), 544–553.

Shi, H., et al., 2020. Disparity in elevational shifts of upper species limits in response to recent climate warming in the Qinling Mountains, North-central China. Sci. Total Environ. 706, 135718.

Silveira, F.A., et al., 2019. Tropical mountains as natural laboratories to study global changes: a long-term ecological research project in a megadiverse biodiversity hotspot. Perspect. Plant Ecol. Evol. Systemat. 38, 64–73.

Skogen, K., et al., 2018. Concern about climate change, biodiversity loss, habitat degradation and landscape change: embedded in different packages of environmental concern? J. Nat. Conserv. 44, 12–20.

Song, X.-P., et al., 2018. Global land change from 1982 to 2016. Nature 560 (7720), 639–643.

Stocker, T.F., et al., 2013. Climate change 2013: the physical science basis. In: Contribution of Working Group I to the Fifth Assessment Report of the Intergovernmental Panel on Climate Change, p. 1535.

Tickner, D., et al., 2020. Bending the curve of global freshwater biodiversity loss: an emergency recovery plan. Bioscience 70 (4), 330–342.

Tubiello, F.N., et al., 2015. The contribution of agriculture, forestry and other land use activities to global warming, 1990–2012. Glob. Chang. Biol. 21 (7), 2655–2660.

Ussiri, D.A., Lal, R., 2017. Carbon Sequestration for Climate Change Mitigation and Adaptation. Springer.

Verbeek, L., et al., 2018. Warming and oligotrophication cause shifts in freshwater phytoplankton communities. Glob. Chang. Biol. 24 (10), 4532–4543.

Victor, D., et al., 2014. Climate change 2014: mitigation of climate change. Clim. Chang. 1454.

Walker, W.H., et al., 2019. Global climate change and invariable photoperiods: a mismatch that jeopardizes animal fitness. Ecol. Evol. 9 (17), 10044–10054.

Wernberg, T., et al., 2016. Climate-driven regime shift of a temperate marine ecosystem. Science 353 (6295), 169–172.

West, J.M., Salm, R.V.J.C.B., 2003. Resistance and resilience to coral bleaching: implications for coral reef conservation and management. Conserv. Biol. 17 (4), 956–967.

Winkler, D.E., et al., 2016. Soil moisture mediates alpine life form and community productivity responses to warming. Ecology 97 (6), 1553–1563.

Winkler, D.E., et al., 2019. Responses of alpine plant communities to climate warming. In: Ecosystem Consequences of Soil Warming. Elsevier, pp. 297–346.

Woodward, A., 2019. Climate change: disruption, risk and opportunity. Glob. Transit. 1, 44–49.

Worm, B., Lotze, H.K., 2009. Changes in marine biodiversity as an indicator of climate change. Clim. Chang. 263–279. Elsevier.

Worm, B., Lotze, H.K., 2016. Marine biodiversity and climate change. Clim Chang. 195–212.

Wu, T., et al., 2012. Specificity responses of grasshoppers in temperate grasslands to diel asymmetric warming. PLoS One 7 (7).

Wuebbles, D.J., et al., 2017. Climate Science Special Report: Fourth National Climate Assessment, Volume I.

Yao, J., et al., 2020. Accelerated dryland expansion regulates future variability in dryland gross primary production. Nat. Commun. 11 (1), 1–10.

Yousefi, M., et al., 2020. Using endemic freshwater fishes as proxies of their ecosystems to identify high priority rivers for conservation under climate change. Ecol. Indicat. 112, 106137.

Zhang, Z., et al., 2017. Emerging role of wetland methane emissions in driving 21st century climate change. Proc. Natl. Acad. Sci. U S A 114 (36), 9647–9652.

Znachor, P., et al., 2020. Changing environmental conditions underpin long-term patterns of phytoplankton in a freshwater reservoir. Sci. Total Environ. 710, 135626.

Zuckerberg, B., et al., 2020. Climate dipoles as continental drivers of plant and animal populations. Trends Ecol. Evol. 35 (5), 440–453.

Link between air pollution and global climate change

5

Pratichi Singh [1], *Deepak Yadav* [2], *S. Pandian E* [1]
[1]Department of Environmental Science and Engineering, IIT(ISM), Dhanbad, Jharkhand, India; [2]Department of Chemical Engineering, HBTU, Kanpur, Uttar Pradesh, India

1. Air pollution

Whole world is facing critical air pollution as a result of industrialization, population growth, construction for housing and infrastructure, vehicular traffic, congested roads, poorly maintained motor vehicles, less access to clean fuel and deficiency of effective pollution control programs (WHO, 2016). In India, the National Capital Territory of Delhi, the capital of India is among the sixth greatest polluted metropolitan in the world (WHO, 2016). One of our era's greatest banes is air pollution, on account of its impact on climate change and public and individual health. A report by Health Effects Institute mentioned that ~1.2 million people died in India due to air pollution in 2017 (HEI, 2019). It is important to understand their interdependence to mitigate their effect. These short-lived climate-forcing pollutants (SLCPs) include methane, black carbon, ground-level ozone, and sulfate aerosols.

Air pollution means the deleterious effects on environment due to any sources, which alters the basic composition of the atmosphere. Pollutants occur in all the phases of the materials such as solid, liquid, and gas phases (Vallero, 2007; Manisalidis et al., 2020). More than 25,000 forms of chemicals, toxins, and contaminants are present in air, which has been identified; most of them are organic in nature emitted due to combustion of fossil fuel, especially in motor vehicle engines (Lefol et al., 2013). Many air pollutants and greenhouse gases (GHGs) have common sources and diffuse globally, interact in the atmosphere, and jointly affect ecosystems. Ozone is particularly relevant for the linkages between climate change and air pollution. Increasing background ozone concentrations affect climate change because ozone is a potent GHG itself and indirectly influences the lifetime of other GHGs such as methane (Schaub and Paoletti, 2007). Conversely, climate change influences ozone concentrations through dynamic and chemical changes in the atmosphere. Climate change results from net radiative forcing (RF) caused due to heating or cooling effect of GHGs and aerosols. For example, black carbon, emitted from combustion has warming effect on the earth, while particulate sulfates contribute to the cooling of the earth's atmosphere (IPCC, 2018). The changing reflectivity of Earth's surface due to land-use change also exists. All these lead to atmosphere's changing heat structures, which directly drive the massive weather systems.

Global Climate Change. https://doi.org/10.1016/B978-0-12-822928-6.00009-5

Table 5.1 Most polluted countries worldwide (World Air Quality Report, 2019).

SN.	Countries	PM2.5 concentration (μg/m^3)
1	Bangladesh	83.3
2	Pakistan	65.8
3	Mongolia	62.0
4	Afghanistan	58.8
5	India	58.1
6	Indonesia	51.7
7	Bahrain	46.8
8	Nepal	44.5
9	Uzbekistan	41.2
10	Iraq	39.6

High in the atmosphere, the mix of warm and cool air leads to extra tropical cyclones and everyday thunderstorms. While climate change has intensified hurricanes, and increased sea levels; the circulation of these huge weather systems has been weakening. The result is cities swathed for days in pollution and whole regions more vulnerable to sudden torrential storms (De Sario et al., 2013). IPCC's Fifth Assessment Report (AR5) clearly states that the recent CO_2 emission is due to anthropogenic activity, and it is the highest in the history. The atmosphere and ocean have warmed, the amounts of snow and ice have diminished, and the sea level has risen (Climate Change, 2013; IPCC).

Poor air quality is a tenacious problem that leads to public health and environmental issues. Globally, authorities are looking for measures to address and control it. Countries emitting the most air pollution worldwide are shown in Table 5.1.

1.1 Air quality standards

Air Quality Standards (AQS) set limits of air pollutants in the atmosphere to avoid the adverse effects of these pollutants on public health. In turn, Air Quality Index (AQI) measures the difference in level of pollutants from standard level, which further helps to estimate the Global Burden of Disease due to increased pollution. Estimation of air pollutants of vehicle emissions involves laboratory tests. These laboratory tests are highly precise measurement but under controlled conditions and therefore are not the representative of the real-world conditions (Tzirakis et al., 2006). To get accurate and repeatable measurements of vehicle emissions, laboratory-based emission tests are typically commenced by experts or mechanical drivers over a consistent testing procedure and driving cycle (Council of European Union, 2001; EEA, 2005; ICCT, 2019a,b). But factors like age of vehicle, cold starts emissions, vehicle payload, type of tire and its pressure, wheel-driving systems like four wheel driving, and engine remapping are seldom tested as the cost of laboratory tests will escalate. Funding to test novel vehicle technologies are always available, but the testing of the in-service vehicles is rarely done, which are deterioration and lead to more pollution. Laboratory tests do not account for the increased load on the engine that road gradient introduces. In Indian estimation of NOx emissions from vehicles is done at monitoring stations, which is not adequate to get accurate database. Rather to fill the gap of data for vehicular emissions, India requires 1600–4000 monitors at an estimated cost of US $212–540 million. Therefore, remote sensing seems to be viable option for monitoring (ICCT, 2018).

Two kinds of air quality standards are there—primary standards are meant to set the AQS limits to protect public health, especially of "sensitive" populations such as children and elderly having respiratory illness. Secondary standards are meant to set the AQS limits to protect community welfare, such as protection against reduced visibility, damage to statues, monuments, buildings, and flora and fauna (Kelishadi et al., 2009). National Air Quality Standards are shown in Table 5.2.

Table 5.2 National ambient air quality standards (Kelishadi et al., 2009).

Pollutants	Time weighted average	Concentration in ambient air	Ecologically sensitive area (notified by central government)	Methods of measurement industrial, residential, rural, and other areas
Sulfur dioxide (SO_2), μg/m^3	Annual* 24 h**	50 80	20 80	- Improved West and Gaeke method - Ultraviolet fluorescence
Nitrogen dioxide (NO_2), μg/m^3	Annual* 24 h**	40 80	30 80	- Jacob and Hochheiser modified (NaOH– $NaAsO_2$) method - Gas phase chemiluminescence
Particulate matter (PM) (size less than 10 μm) or PM10, μg/m^3	Annual* 24 h**	60 100	40 60	- Gravimetric - TEOM - Beta attenuation
Ozone (O_3) μg/m^3	8 h* 1 h**	100 180	100 180	UV photometric - Chemiluminescence - Chemical method

Continued

Table 5.2 National ambient air quality standards (Kelishadi et al., 2009).—cont'd

Pollutants	Time weighted average	Concentration in ambient air	Ecologically sensitive area (notified by central government)	Methods of measurement industrial, residential, rural, and other areas
Lead (Pb) μg/m^3	Annual* 24 h**	0.50 1.0	0.50 1.0	AAS/ICP method after sampling on EPM 2000 or equivalent filter paper—ED-XRF using Teflon filter
Carbon monoxide (CO), mg/m^3	8 h* 1 h**	02 04	02 04	Nondispersive infrared (NDIR) spectroscopy
Ammonia (NH_3), μg/m^3 Annual*	24 h** 100	400 100	400 —	Chemiluminescence - Indophenol blue method
Benzene (C_6H_6), μg/m^3 —	Annual* 24 h**	05	05	Gas chromatography (GC)−based continuous analyzer—Adsorption and desorption followed by GC analysis
Benzo(a) pyrene (BaP) Particulate phase only, ng/m^3	Annual* 24 h**	01	01	Solvent extraction followed by HPLC/ GC analysis
Arsenic (As), ng/m^3 —	Annual* 24 h**	06	06	AAS/ICP method after sampling on EPM 2000 or equivalent filter paper
Nickel (Ni), ng/m^3	Annual* 24 h**	20	20	- AAS/ICP method after sampling on EPM 2000 or equivalent filter paper

* Annual arithmetic mean of minimum 104 measurements in a year at a particular site taken twice a week 24 hourly at uniform intervals.

**24 hourly or 8 hourly or 1 hourly monitored values, as applicable, shall be complied with 98% of the time in a year. 2% of the time, they may exceed the limits but not on two consecutive days of monitoring.

1.2 Air quality index

Air quality is a mirror image of public health and environment. Ambient air quality is synonym of air pollution index (AQI). AQI measures the deviation in quantity of pollutants in the atmosphere from standard. Human health in a particular area is determined by the level of pollution in that area, which is expressed in terms of AQI (Senthilnathan, 2007). AQI is devised by EPA (2000) and other pollution control agencies (Goyal, 2001). It is meant to keep the public well informed on local air quality on regular basis. The higher value (above 125) of an index refers to a great level of air pollution (Shukla et al., 2010). The AQI is calculated using the following formula:

$$\text{AQI} = 100 \times \frac{\text{Observed mean concentration of a pollutant}}{\text{Srandard for the respective pollutant}}$$

The Pollutant Standard Index (PSI) is used to facilitate risk assessment due to the level of pollutants. Its value is expressed between 0 and 500 (USEPA, 1996). Thom and Ott in 1974 first introduced PSI as a standard for reporting air quality (Thom and Ott, 1976). Hence, it estimates the relative contribution of each pollutant to total risk (Cairncross et al., 2007; Pryor et al., 2015). The calculation of PSI is based on the concentration of five major air pollutants including PMs, SO_2, NO_2, CO, and O_3 in the air. PSI is an indice AQI, which is obtained by averaging data collected for the past 24 h.

According to Johnson et al., "AQI is defined as a quantification of the purity of air as compared to the pure air required for sustenance of human being and other living organisms" (Johnson et al., 1997). AQI is divided into various ranges, each range having numerical value and corresponding color coding. The numbers ranges from 0 to 500, which indicates the level of health risk associated with air quality as shown in Table 5.3.

Table 5.3 Color-coded air quality index (USEPA, 2020).

AQI Value	AQI Category	AQI Color
0 - 50	Good	Green
51 - 100	Moderate	Yellow
101 - 150	Unhealthy for Sensitive Groups	Orange
151 - 200	Unhealthy	Red
201 - 300	Very Unhealthy	Purple
301 - 500	Hazardous	Maroon

2. Sources of air pollution

The sources of pollution can vary from as small as from cigarettes burning to large natural and anthropogenic phenomenon like volcanic eruptions, emission from motor engines of automobiles and industrial chimneys and furnaces (Robinson, 2005; Habre et al., 2014). Major sources are the use of fossil fuels for heating and cooling for transportation, for industry, and for energy conversion (responsible for majority of air pollution on a global scale); incineration of various forms of industrial, municipal, and private wastes; certain chemicals (e.g., perchloroethylene from dry cleaners, methylene chloride used as solvent and paint stripper), and materials (e.g., asbestos used as fireproofing materials) used in different processes and purposes also contribute to air pollution.

2.1 *Classification of major air pollution sources*

(1) Mobile sources/transportation: motor vehicle, rail, ship, aircraft.
(2) Stationary sources: utility, industrial, institutional, and commercial facilities. Examples are power plants, heating plants, paper-pulp industries, petroleum refineries, municipal waste combustors.
(3) Area sources: many individually small activities, for example, gasoline service stations, small paint shops, open burning associated with solid waste, agriculture and forest management, cooking in slum areas.
(4) Incineration/burning of waste: household and commercial waste; agricultural burning; industrial and hazardous waste incineration.
(5) Miscellaneous: resuspension from road; domestic fuel, wood burning; forest fire, volcanic eruptions, pollen grains, certain bacteria, viruses (natural); chemicals and materials used in different processes (perchloroethylene, methyl chloride).

There are several natural and anthropogenic sources of air pollution shown in Fig. 5.1.

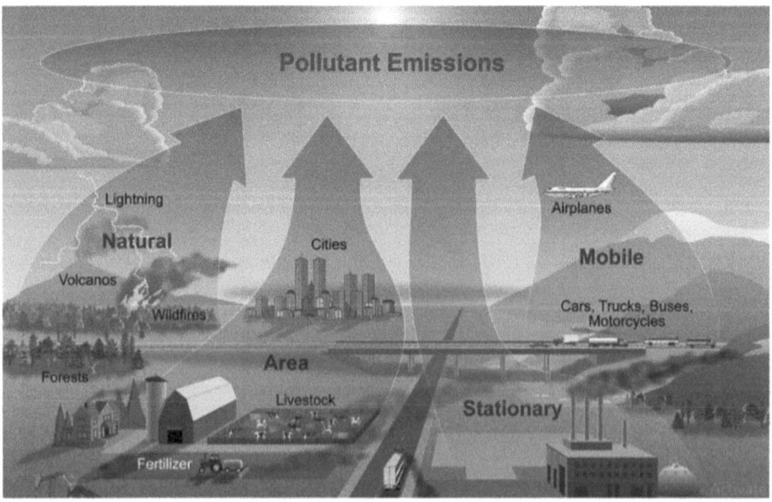

Figure 5.1 Natural and manmade air pollution sources.

Table 5.4 Major anthropogenic sources of air pollution (Popescu and Ionel, 2010).

Source	CO (%)	SO_2 (%)	HC(%)	NO - NOx (%)
Transportation	92	4	65	42
Industries	4	32	26	21
Power plants	–	48	–	32
Space heating	3	12	3	5
Combustion	1	4	6	0

The major sources of air pollution and their contribution are shown in Table 5.4.

Road transport is the largest source of GHGs within the transport sector (Sarath et al., 2003). The transport sector, therefore, affects the global climate (Weiss et al., 2000; Johansson, 2003). In turn, climate change impacts on air quality by perturbing ventilation rates (wind speed, mixing depth, convection and frontal passages in addition to precipitation, dry deposition, the production of various chemicals and the rate of loss, natural emissions and persistent natural background concentrations of emissions. According to the Kyoto Agreement, the causes of CO_2 emissions have been allocated to the transport sector (Bates et al., 2001).

2.2 *Types of pollutants*

Six major criteria air pollutants identified as causing health effects at concentrations above thresholds established at levels known to be safe. These are CO, NO_2, O_3, SO_2, particulate matter (PM), and Pb (Ott and Hunt, 1976) shown in Table 5.5.

(i) Particulate matters (PM2.5 and PM10)

PMs are respirable tiny particles composed of sulfate, nitrates, ammonia, sodium chloride, black carbon, mineral dust, and water. Particles having less than 10 microns in (PM_{10}) diameter are harmful. But the fine particles that are less than 2.5 microns ($PM_{2.5}$) are capable of penetrating into the lungs ultimately entering the bloodstream pose. This causes very high risk to human health as well as environment. Sources of PM include combustion engines (both diesel and petrol), solid-fuel (coal, lignite, heavy oil, and biomass) combustion for energy production in households and industry, as well as other industrial activities (building, mining, manufacture of cement, ceramic and bricks, and smelting) (Jeffrey et al., 1998). They may be produced naturally or as a direct or indirect result of human activities. Atmospheric aerosol is characterized by these relatively small particles. The formation and increase of atmospheric aerosol by both natural and anthropogenic sources is a major air quality concern because aerosol particles may scatter light, reducing visibility, pose an inhalation hazard to humans and animals, and affect climate on a regional and global scale.

(ii) Black carbon

Black carbon is a major component of PM2.5 and a driver of climate change, also known as a short-lived climate pollutant (SLCPs) exist for a shorter period compared with CO_2 in the atmosphere. Notwithstanding, black carbon plays the major role in promoting global warming after CO_2, even with its short atmospheric lifetime. It is responsible for reduction in agricultural productivity and increase melting of glaciers.

Table 5.5 Classification of pollutants.

Classification of pollutants	Types	Remarks
According to origin	Primary pollutants	Emitted directly into the atmosphere and are found in form in which they were emitted, for example, SOx, NOx, hydrocarbon (HC)
	Secondary pollutants	Derived from the primary pollutants by chemical or photochemical reactions in the atmosphere, for example, ozone, peroxyacetyl nitrate
According to chemical composition	Organic	HCs, aldehydes, and ketones (HCO), volatile organic compounds (VOCs), polychlorinated biphenyls (PCBs), polycyclic aromatic hydrocarbons (PAHs)
	Inorganic	NOx, SOx, CO, HCl, H_2SO_4, H_2S, NH_3
According to state of matter	Gaseous	CO, NOx, SOx (inorganic), benzene, methane (organic)
	Particulates/ aerosols	Dust, smoke, fume, fly ash (solid), mist, spray (liquid), pollen, bacteria, virus (natural)
Air quality determinants	Major indirect greenhouse gases	• Carbon monoxide (CO) • Nitrogen oxides (NOx) • VOCs, for example, ammonium
	Aerosols	• Sulfur dioxide (SO_2) • Secondary sulfate particles • Particulate matter (PM10)/nitrate • Organic carbon • Black carbon (soot)
	Main greenhouse gases	• Carbon dioxide (CO_2) • Methane (CH_4) • Nitrous oxide (N_2O) • Hydrofluorocarbons (HFCs) • Perfluorocarbons • Sulfur hexafluoride (SF_6) • Chlorofluorocarbons • Hydrochlorofluorocarbons

(iii) Tropospheric ozone

Ground-level ozone is the third largest global mean RF (An Editorial Essay, 2004), and lifetime of other GHGs like methane is influenced by it. Tropospheric ozone is formed when gaseous pollutants like NOx, CO, methane, and volatile organic compounds (VOCs) react in the atmosphere in the presence of sunlight, specifically in the ultraviolet (UV) spectrum. It is not directly emitted, it is formed due to the presence of other pollutants; therefore, and it is a secondary pollutant. CO, VOCs, methane, and NOx are not only ozone precursors but also hazardous air pollutants themselves. Tropospheric ozone in turn leads to the formation of further more deleterious photochemical smog. Major sources of CO, methane, NOx, and VOCs are releases from vehicle exhausts, industrial emissions, and chemical solvents. Sources of methane are emitted from burning of natural gas, waste and the fossil fuel, livestock, and agriculture. This poses health risks linked to breathing problems, increase in asthmatic attacks, lung dysfunction, and respiratory ailments. Apart from its adverse health impacts, ground level ozone is a short-lived climate pollutant and one of the most important GHGs.

The photolysis of ozone yields electronically excited oxygen atoms (Eq. 5.1), which are of major importance in the tropospheric chemistry (Gaffney and Marley, 2003):

$$O_3 + h\nu(290 - 336\ \text{nm}) \rightarrow O_2 + O(^1O) \tag{5.1}$$

and this formed O (^{1}D) react vigorously with water vapor (H_2O) to form hydroxyl radical (OH), and hydroxyl radical is the main component in the formation of O_3 and other secondary pollutants.

(iv) Nitrogen dioxide

Major source of nitrogen dioxide emissions is from power generation followed by industrial and traffic sources. NOx is very harmful because it is an important component of PM and ozone formation. Many authors have reported that NOx alone can have effects on respiratory conditions causing inflammation of the airways at high levels, bronchitis, and asthma (Lal and Patil, 2001). Long-term exposure can increase the risk of respiratory conditions and the response to allergens and decreases the function. NOx also contributes to the formation of fine particles (PM) and ground level ozone, both of which are associated with adverse health effects. Evidence also suggests that NO_2 may be responsible for a large disease burden, with exposure linked to premature mortality and morbidity from cardiovascular and respiratory diseases.

Nitrogen dioxide gets photolyzed to form ground-state oxygen atom with molecular oxygen (Eq. 5.2)

$$NO_2 + h\nu(290 - 423\ \text{nm}) \rightarrow NO + O(^3P) \tag{5.2}$$

Which further react in the presence of M (an energy-absorbing molecule) to form 1 molecule of O_3 in the troposphere (Eq. 5.3):

$$O(^3P) + O_2 + M \rightarrow O_3 + M \tag{5.3}$$

(v) Sulfur dioxide

Sulfur dioxide (SO_2) is primarily produced due to burning of fossil fuels such as coal and oil containing sulfur. Coal-fired thermal power plants are the largest sources of SO_2. Smelting of mineral ores that contain sulfur is also one of the major reasons behind the emission of SO_2. Other natural and anthropogenic sources of SO_2 emissions are trains, large ships, processing industry (e.g., pulp and paper and metal industry) and some diesel equipment that burns high sulfur fuel, and by volcanic eruptions. SO_2 exposure has adverse effect on respiratory tract and the functioning of the lungs and causes irritation of the eyes. Whenever emissions of SO_2 level increase in the atmosphere, it leads to increased hospital admissions and visits to emergency rooms due to in the inflammation of the respiratory tract and aggravated asthma and bronchitis cases. SO_2 has an acidifying effect on the environment. Acidification impacts both plants and animals living on land as well as in water. When there is acidification on ground, leaching of vital nutrients from the soil occurs leading to eventual decrease in forest growth. Furthermore, metals are released in the ground that can harm decomposing organisms in the ground as well as birds and mammals higher up in the food chain, including humans. Most deleterious effect is the formation of acid rain as SO_2 combines with water in the air to form sulfuric acid, a major component of acid rain.

(vi) Carbon monoxide (CO)

Carbon monoxide (CO) is a colorless and odorless toxic air pollutant produced largely from vehicle emissions. There are several natural and anthropogenic sources of CO. The three main anthropogenic sources of CO emissions are residential, industrial, and transportation (Prasad and Singh, 2012). CO intoxication is one of the most common types of poisoning in the modern world especially in the United States. According to USEPA, breathing CO at high levels can be harmful to humans by impairing oxygen transport by hemoglobin, which has health effects that include impaired reaction timing, headaches, lightheadedness, nausea, vomiting, weakness, clouding of consciousness, coma, and, at high enough concentrations and long enough exposure, death. CO is difficult to detect and can mimic other common disorders. Therefore, the true incidence of CO poisoning is unknown, and many cases probably go unrecognized. CO influences the atmospheric concentrations of both OH^- radicals and O_3 and, thus, affects the oxidizing potential of the atmosphere (Eq. 5.4).

$$CO + OH^- \rightarrow CO_2 + H^- \qquad (5.4)$$

Subsequent reactions of H^- and O_2 produce the highly reactive hydroperoxyl radical, HO_2^-(Eq. 5.5)

$$H^- + O_2 + M \rightarrow HO_2^- + M \qquad (5.5)$$

where M is an energy-absorbing molecule, e.g., nitrogen N_2 or O_2. From those to reactions we can conclude (Eq. 5.6):

$$CO + H^- + O_2 \rightarrow CO_2 + HO_2^- \qquad (5.6)$$

In the presence of photons of light energy another reaction will occur (Eq. 5.7):

$$CO + O_2 + h\nu \rightarrow CO_2 + O_3 \qquad (5.7)$$

where O_2 is ground-state atomic oxygen and hv is a photon of light energy. As a result of these reactions, CO is converted to CO_2, with one molecule of O_3 produced for each CO molecule oxidized. As such, CO oxidation in the troposphere is an important (but low-level) source of O_3. Because CO and CH_4 both compete for OH^- radical, CO indirectly affects tropospheric concentrations of CH_4 and stratospheric H_2O vapor (derived from CH_4 oxidation).

3. Greenhouse gases

Any gas that has infrared (IR) absorption bands can be a GHG. Natural occurring GHGs in the atmosphere are water vapor, CO_2, methane, nitrous oxide, and ozone.

Anthropogenically produced GHGs are sulfur hexafluoride, perfluorocarbons (PFCs), hydrofluorocarbons (HFCs), and they are ozone-depleting. They have high global warming potential (GWP) (USEPA, 2010). GHGs can be better recognized as longer-lived climate pollutant (LLCPs) in the troposphere such as chlorofluorocarbons, methane, and nitrous oxide, which are considered trace GHG species and short-lived climate pollutants (SLCPs) such as carbonaceous soots whose pronounced heating effect has been addressed in the Northern Hemisphere by global-scale modeling (Jacobson, 2002).

The GHGs listed by the IPCC (CO_2, methane, nitrous oxide, HFCs, PFCs, and sulfur hexafluoride) are discussed below. Water vapor, although the most abundant GHG, is not discussed because it is impossible to predict its concentration as its natural occurrence and fluctuations far outweigh anthropogenic influences. Ozone is an indirectly RF GHG. Other ozone-depleting substances are HFCs, PFCs, and sulfur hexafluoride (IPCC, 2010).

(i) Carbon dioxide (CO_2)

CO_2 is the most important anthropogenic GHG and accounts for more than 75% of all anthropogenic GHG emissions. It has long atmospheric lifetime (on the order of decades to centuries) and thus remains in the atmosphere for decades after its emission. Even if very efficient GHG mitigation efforts to reduce GHG concentrations are promulgated, it will take decades long time to control its concentration in the atmosphere (IPCC, 2007a). CO_2 increase in the atmosphere is mainly due to emissions from the burning of fossil fuels, gas flaring, cement production, iron and steel industry, and land use changes. Three-quarters of the current RF is likely caused by anthropogenic CO_2 emissions. Fossil fuel combustion and the global manufacture of cement are responsible for more than 75% of human-caused CO_2 emissions (IPCC, 2007b); approximately one-quarter of RF results from land-use changes (IPCC, 2007b).

Since the Industrial Revolution, the concentration of anthropogenic emissions of CO_2 has increased most notably from ~280 to 379 ppm (more than 35%) (IPCC, 2007b). IPCC estimates that the present atmospheric concentration of CO_2 has not been exceeded in the last 650,000 years and is likely to be the highest ambient concentration in the last 20 million years (IPCC, 2007c).

(ii) Methane

Methane is the second largest GHG with a GWP of 21 compared with CO_2 (IPCC, 2007b). It is the main component of natural gas. The anthropogenic activities, which contribute to the emissions of such a large quantity of methane, are cattle raising, growing rice, burning natural gas, and coal mining. Atmospheric methane has increased from a pre-industrial concentration of 715 ppb to 1775 ppb in 2005 (IPCC, 2001; NOAA, 2008).

Methane reacts with hydroxyl radical to form the methyl radical (CH_3) and subsequently the methylperoxy radical, which can react with nitric oxide to form nitrogen dioxide (Eqs. 5.8–5.10):

$$OH + CH_4 \rightarrow CH_3 + H_2O \tag{5.8}$$

$$CH_3 + O_2 \rightarrow CH_3O_2 \tag{5.9}$$

$$CH_3O_2 + NO \rightarrow CH_3O + NO_2 \tag{5.10}$$

The methoxy radical (CH_3O) then can react with molecular O_2 leading to the formation of formaldehyde and hydroperoxyl radical (Eq. 5.11):

$$CH_3O + O_2 \rightarrow HCHO + HO_2 \quad (5.11)$$

Hydroperoxyl radical react with N_2O producing nitrogen dioxide and OH radical and thus leading to the chain reactions to form nitrogen dioxide and ozone. Formaldehyde can photolyze to form another hydrogen atom and the formyl radical (HCO) (Eq. 5.12):

$$HCHO + h\nu \rightarrow HCO + H \quad (5.12)$$

This process can be followed vice-versa, by the reaction of both products with oxygen to form carbon monoxide and hydroperoxyl radical. On the other hand, formaldehyde can react with hydroxyl radical by abstraction to form the formyl radical and water (Eq. 5.13):

$$HCHO + OH \rightarrow HCO + H_2O \quad (5.13)$$

Here the consequent reaction of the formyl radical with oxygen will lead to the formation of carbon monoxide and hydroperoxyl radical.

(iii) Nitrous oxide

Nitrous oxide is a potent GHG, with a GWP of 310 (IPCC, 2007b). The major sources of nitrous oxide are agricultural processes, nylon production, fuel-fired power plants, nitric acid production, and vehicle emissions. Nitrous oxide is also used in rocket engines and racecars and as an aerosol spray propellant. Agricultural processes that result in anthropogenic emissions of nitrous oxide are fertilizer use and microbial processes in soil and water. Nitrous oxide concentrations in the atmosphere have increased from preindustrial levels of 270−319 ppb in 2005, an 18% increase (IPCC, 2007b). It also produce free radicals (Eq. 5.14)

$$HONO + h\nu(310 - 400\ nm) \rightarrow OH + NO \quad (5.14)$$

(iv) HFCs

HFCs have high GWPs and are ozone-depleting substance. They are manmade, and they do not exist naturally in ambient conditions. These chemicals are used in commercial, industrial, and consumer products (EPA, 2010). HFCs are used in automobile air conditioners and refrigerants. The most abundant HFCs, in order from the most abundant to least, are HFC-134a (35 ppT), HFC-23 (17.5 ppT), and HFC-152a (3.9 ppT).

(v) PFCs

The most abundant PFCs are tetrafluoromethane (PFC-14) and hexafluoroethane (PFC-116). These manmade chemicals are emitted largely from aluminum production and semiconductor manufacturing processes. PFCs are extremely stable compounds that are destroyed only by very high-energy UV rays, which results in the very long lifetimes of these chemicals (EPA, 2010).

(vi) Sulfur hexafluoride

Sulfur hexafluoride, another manmade chemical, is used as an electrical insulating fluid for power distribution equipment, in the magnesium industry, and in semiconductor manufacturing and also as a trace chemical for study of oceanic and atmospheric processes (EPA, 2010). In 1998, atmospheric concentrations of sulfur hexafluoride were 4.2 ppT and are steadily increasing in the atmosphere. Sulfur hexafluoride is the most powerful of all GHGs listed in IPCC studies, with a GWP of 23,900 (IPCC, 2007b).

3.1 Global warming potential

GWP is for comparison of the global warming potential of GHGs. It is defined as a measure of energy absorbed by the emissions of 1 ton of a particular GHG over a given period of time, compared with the emissions of 1 ton of CO_2. The higher the GWP is, the more given gas warms the earth atmosphere compared with CO_2 over given time period. The time period usually used for GWPs is 100 years. Table 5.6 enlists the GWPs of chemically reactive GHGs.

3.2 Residence time

The residence time and the physical and chemical properties of air pollutants, GHGs, and aerosols are important to understand the environmental risk assessment. The atmospheric residence time of a gas is equal to the total atmospheric abundance of the gas divided by its rate of removal (Seinfeld and Pandis, 2006). The atmospheric residence time of a gas is also its half-life measurement. Half-life indicates the length of time up to which a gas can persist in the atmosphere. These data are useful when removal mechanisms such as chemical transformation and deposition are considered. The main compounds of significance in the present context and their role in promoting toxic and climate change effects are shown in Table 5.7.

4. Greenhouse gases, global warming, and global climate change

Global climate change is the deviation of the climate of the Earth due to natural as well as anthropogenic reasons. It imparts many global impacts such as sea level rise, melting glaciers, and further life-threatening meteorological occurrences due to changes in climate patterns. Climate change is not only affects environmental phenomenon by causing drought, flooding, deforestation but also distresses socio-economic aspects by causing loss of shelter and annihilation of animal species and plant species resulting in drought and epidemic (Hansen et al., 2005; Anthes et al., 2006).

There are several illustrative examples from India itself, which shows visible link between air pollutants especially GHGs, global warming, and climate change (Fig. 5.2). India is one of the most vulnerable countries in the world with regard to climate change sources and impacts. At the same time, potential climate impacts in India are severe; they include sea level rise, changes in the monsoon, increased severe storms and flooding, and more drought. The incident of biomass burning during the winter season in central India led to unprecedented increase in aerosols. Thomas et al. used 15 years (2003−17) of satellite and model reanalysis data sets over India during the dry winter season (November to February). The number of hazy days showed an

Table 5.6 Chemically reactive greenhouse gases and their precursors: abundances, budgets, lifetimes, trends, and global warming potentials (GWPs) (Seinfeld and Pandis, 2006).

		Abundance ppt		**Trend**			
	Formula	**1998**	**1750**	**1990s**	**Annual emission late 90s**	**Lifetime (year)**	**100-year GWP**
Methane	CH_4 (ppb)	1745	700	7.0	600 Tg	8.4/12	23
Nitrous oxide	N_2O (ppb)	314	270	0.8	16.4 TgN	120/114	296
Perfluoromethane	CF_4	80	40	1.0	~15 Gg	>50,000	5700
Perfluoroethane	C_2F_6	3.0	0	0.08	~2 Gg	10,000	11,900
Sulfur hexafluoride	SF_6	4.2	0	0.24	~6 Gg	3200	22,200
HFC-23	CHF_3	14	0	0.55	~7 Gg	260	12,000
HFC-134a	CF_3CH_2F	7.5	0	2.0	~25 Gg	13.8	1300
HFC-152a	CH_3CHF_2	0.5	0	0.1	~4 Gg	1.40	120

Table 5.7 Residence times in the atmosphere, toxic properties, and effects on climate change of different gases and particle types (Seinfeld and Pandis, 2006).

Compound	Residence time	Toxic properties	Climate change properties
Carbon dioxide	150 years	Acidification of sea waters, affects photosynthesis	Climate gas, long residence time
Nitrous oxide	110 years	Destruction of the stratospheric ozone layer	Climate gas, long residence time
Methane	10 years	Precursor of ground-level ozone	Climate gas, intermediate residence time
Ozone	1 month	Adverse effects on health and vegetation	Climate gas, short residence time
Sulfur dioxide	1 week	Acidification, health effects	Sulfate particles suppressing global warming
Soot	1 week	Health effects	Soot and black particles increase global warming
Nitrogen oxides	1 week	Precursor of ground-level ozone, acidification, eutrophication	Nitrate particles may suppress global warming
Ammonia	<1 week	Acidification, eutrophication	Ammonia particles may suppress global warming

increasing trend of ~2.6 days per year over Central India (Thomas et al., 2019). They reported that the anomalous increment in hazy days over Central India is associated with the relatively higher increase in biomass burning over the region.

Srivastava et al. reported analysis based on air quality data for before lockdown and after lockdown (first phase of 21 days) periods of 21 days each. The results showed significant decline in the studied air pollution indices and demonstrated improvement of air quality in both the cities (Srivastava et al., 2020). The major impact was seen in the levels of PM2.5, NO_2, and CO. The levels of SO_2 showed less significant decline during the lockdown period. Although such a short duration data for air pollution cannot be validated for climate change effects, but the whole world is experiencing pleasant weather during the lockdown. Intense deforestation activities have taken place along the foothills of Himalayas and in the Assam region, and land use patterns have undergone definite changes over parts of Rajasthan and Punjab (Northwest India). Surface cooling with significant increase in rainfall has been observed in the peripheral regions of the Rajasthan desert; increased area under irrigation may be one of the main causal factors (Lal, 2001; Brenkert and Malone, 2005).

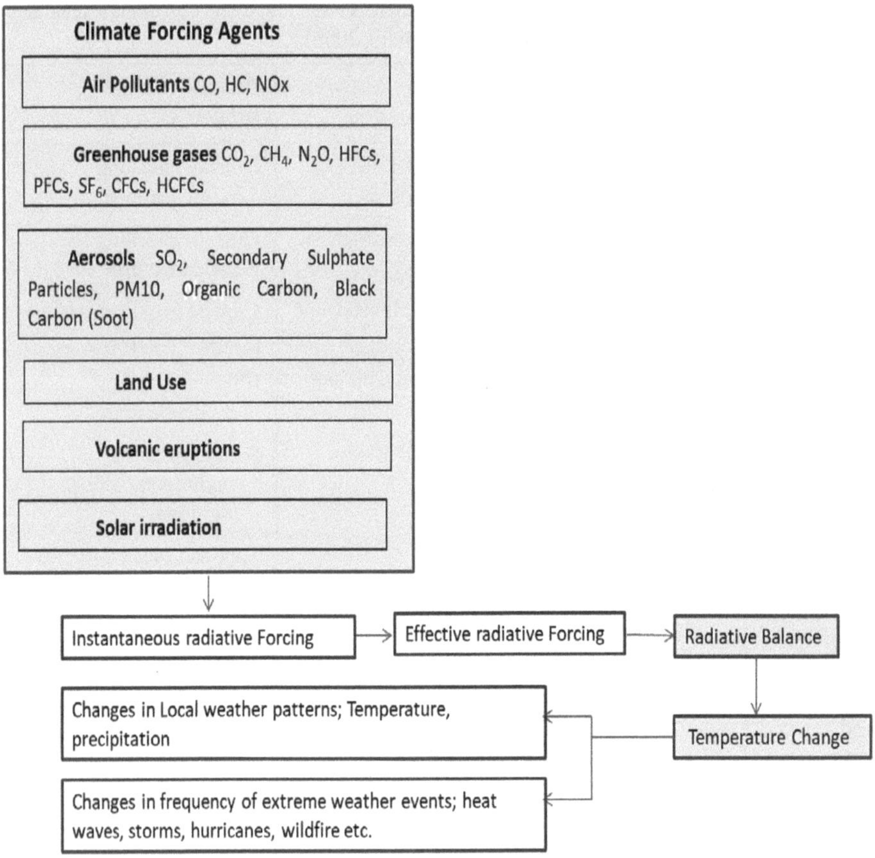

Figure 5.2 Illustration of link between greenhouse gases, global warming, and climate change.

Global warming discusses to the rise in the average global heats, melting of glaciers, and obvious rise of sea levels. GHGs' heating effect allow to absorb, discharge IR radiations wavelengths of 4.26–14.99 μm, correspondingly net rise in temperature leading to global warming (Muraza and Galadmina, 2015; Er-Rbib et al., 2012). Numerous air pollutants adds to climate change after infusion with sunlight and create secondary pollutants and global warming pollutants. Agreeing to IPCC report on the effects of 1.5°C global warming strains that declines in non-CO_2 emissions, chiefly the air pollutants CH_4 and black carbon, are crucial (IPCC, 2010). The link between global warming and climate change owing to air pollution is becoming clearer with each passing year.

4.1 Radiative forcing

RF is a degree of the alteration in energy balance as a consequence of a variation in a driving agent (e.g., aerosol, GHGs, cloud, etc.) to upset the worldwide energy balance and subsidize to climate change. It has been used as a degree to inspect the climate sympathy of an earth structure element as shown in Fig. 5.3 (IPCC, 2007a, 2018).

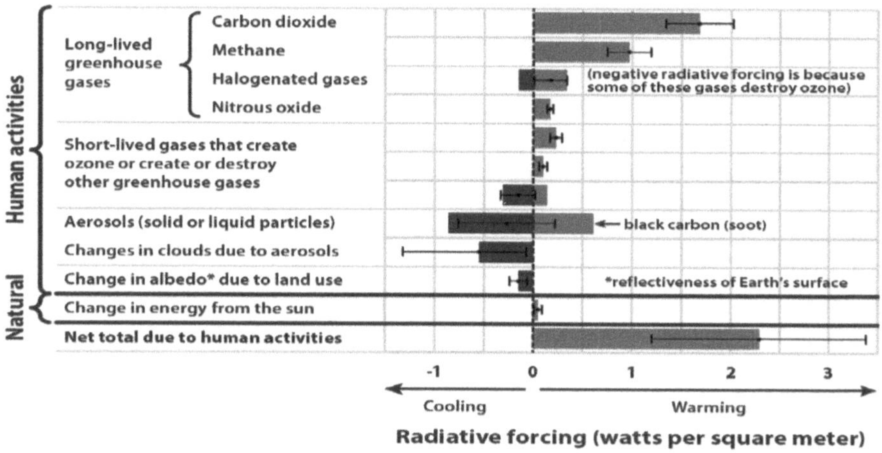

Figure 5.3 Energy balance due to natural and anthropogenic activities (IPCC, 2013).

Terrestrial surface deviations could affect lasting global weather over physical devices with deviations in evapotranspiration, roughness, and surface albedo. Among these aspects, land shallow albedo panels the supply of solar radiation at the edge of external and troposphere, which can pointedly move weather and long-term global climate.

4.2 Net radiative forcing

The RF owing to CO_2, CH_4, N_2O, and various halocarbons is due to absorption of IR radiation. These gases persist in the air long sufficient to be well mixed and have little spatial variability. Their concentrations have increased substantially since preindustrial times, and they are the utmost suppliers to total anthropogenic RF. The entire net radiative making of GHGs, aerosols, and whole anthropogenic radiative driving is shown in Fig. 5.4, the IPCC estimate of the radiative driving due to well-mixed GHGs is $+2.43\ W/m^2$ from 1750 to 1998 (present), comprising CO_2 ($1.46\ W/m^2$), CH_4 ($0.48\ W/m^2$), N_2O ($0.15\ W/m^2$), and halocarbons ($0.34\ W/m^2$) (IPCC, 2001).

From Fig. 5.4, it is obvious that although improved GHG concentrations have surely led to a positive active radiative driving of environment and later a heating result, the effect of amplified concentrations of anthropogenic aerosols is the

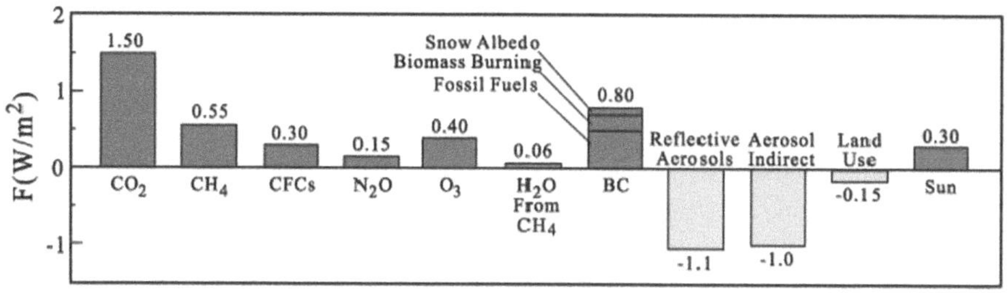

Figure 5.4 Energy balance due to greenhouse gaseous, aerosol, cloud, and surface albedo (IPCC, 2001).

prospective climate cooling. The net active RF is an optimistic net driving of the climate scheme and later a warming, but the worries in the probability density function of aerosols feeds done to source a far better ambiguity in the net RF of environment than if individual GHGs were measured (Table 5.7).

Many aerosols such as sulfate and nitrate can indirectly affect radiative balance. They are hydroscopic in nature and can form cloud condensation nuclei. Increases in size of condensation nuclei will also increase cloud formation and thus, light scattering or absorption by liquid water, another important greenhouse species (Marley et al., 1994). Therefore, a cloud can cool or heat the atmosphere depending on light scattering or absorption. Aerosols are important in the global hydrologic processes and they effect precipitation also. Aerosols transport to long-range and their interactions must be taken into account while considering radiative balance and consequently are tied to climate change.

5. Nitrogen oxide emission an indirect greenhouse gas

Nitrogen oxide (NOx) emissions lead to the formation of ozone, which is a significant GHG. In spite of its significance, slight importance was before set on authenticating methods to analyze gifts of distinct emissions to ozone and therefore toward climate change. Identification of the favorable moderation methods involves a good empathetic specific anthropogenic emissions, for example, from road transportation, shipping, and aviation. Atmospheric concentration of NOx controls nonlinear mode of atmospheric O_3 chemistry (Liu et al., 1987; Lin et al., 1988; Grooβ et al., 1998): At low NOx concentrations, ozone mitigation is imperfect by the disposal of NOx molecules. At high NOx concentrations, new complexes bound ozone creation, for example, HOx (hydrogen oxide) radicals (Lin et al., 1988). Additional ozone devastation by reactions with NOx becomes gradually significant at the premier NOx concentrations. Finally, it indicates to negative net ozone creation rates (Dahlmann et al., 2011). Ozone is a significant GHG; therefore, NOx-linked ozone changes have an effect on environment that is complex than environment effects of CO_2 increases (Grewe et al., 2012). NOx pioneer releases from road transportation maxim a big growth in the 1970s and 1980s, then later reduced owing to the wide-spread execution of catalytic converters (Uherek et al., 2010).

Anthropogenic discharges of transitory, chemically sensitive gases, for example, CO and NOx, are identified to effect environment through changing the chemistry of the universal atmospheric and thus plenty of the GHGs CH_4, CO_2, and HFCs. This learning practices the features of the normal modes of the tropospheric organic system to decay the greenhouse result NOx and CO emissions into (i) short-lived manners connecting mainly tropospheric O_3 and (ii) the prolonged approach linking a global CH_4-CO-O_3 perturbation. Merging these two segments of greenhouse perturbations—short-lived, large, regional O_3 rises and reduced, long-term, universal declines in O_3 and CH_4 we find that supreme types of anthropogenic NO discharges lead to a negative radiative driving and earth cooling (Table 5.8).

Table 5.8 Other chemically active gases directly or indirectly affecting radiative forcing (Seinfeld and Pandis, 2006).

	Formula	Abundance ppt		Trend 1990s	Annual emission late 1990s	Lifetime (year)	100-year global warming potential
		1998	1750				
Tropospheric ozone	O_3 (DU)	34	25	–	Depends on abundances of NOx, CH_4, CO, VOC, O_3, and H_2O	0.01–0.05	–
Tropospheric NOx	$NO + NO_2$	5–999	–	–	~52 TgN	<0.01–0.03	–
Carbon monoxide	CO (ppb)	80	–	6	~2800 Tg	0.08–0.25	Very sensitive
Stratospheric water	H_2O (ppm)	3–6	3–5	–	–	1–6	–

The identification of RF caused by emission of short-lived, chemically reactive gases with both short-lived and long-lived chemical modes is extremely useful. This separation of the problem can greatly reduce the computational effort for tropospheric chemical modeling, enabling an extensive choice of sensitivity studies such as the examples here. Further, it provides both insight and a clear quantitative separation of RF that is regional (short-lived modes) from that which is global (the long-lived CH_4-CO-O_3 mode). Since these two different patterns of RF may not cancel, even with opposite signs, it may become important to diagnose them separately. Anthropogenic surface emissions of NO produce a net RF, owed to deviations in O_3 and CH_4, that is, consistently negative but can vary by an order of magnitude depending on region. Combined industrial emissions of CO and NO, however, yield a positive RF. Thus choices to switch global tropospheric O_3 and therefore global warming by critical NO emissions only would yield the reverse effect once the long-term, global changes to CH_4 and O_3 are considered (Oliver et al., 2001).

6. Parameters affecting air pollution

Air movements affect the outcome of air pollutants. So several studies of air pollution should comprise a revision of the meteorology (local climate patterns). If the air is quiet and contaminants cannot diffuse, then concentration of pollutants increases. Conversely, once strong, wild winds shock, pollutants diffuse rapidly, ensuing in minor pollutant concentrations (Kumaravel et al., 2013). Meteorological data help to recognize the source of contaminants, expect air pollution events, for example, downturns and high-pollutant absorption days as well as simulate and predict air quality using computer models. Once studying air quality, it is significant to measure the following factors as they can help us understand the chemical reactions that arise in the atmosphere:

(i) Wind speed and direction

Once high-pollutant concentrations arise at an observing room, wind data accounts can govern the overall course and extent of the discharges. Identifying the foundations incomes forecasting to diminish the influences on air quality can take place. An instrument called an anemometer measures wind speed.

(ii) Temperature

Computing temperature provisions air quality calculation, modeling, and estimating activities. Temperature and solar radiation (sunlight) play an imperative part in the chemical reactions that arise in the troposphere to custom photochemical smog from supplementary pollutants. Favorable circumstances can principal to amplified concentrations of smog. The utmost collective method of calculating temperature is to use a material with a conflict that variations with temperature, for example, platinum (Pt) wire. A sensor measures this change and converts it into a temperature reading.

(iii) Humidity

Comparable solar radiation and temperature, water vapor shows an imperative part in several photochemical reactions and thermal in the atmosphere (troposphere). As water molecules are small and highly polar, they can bind strongly to many substances. If attached to particles suspended in the air, they can significantly increase the total of light dispersed by the particles. When the water molecules assign to corrosive gases, for example, SO_2 (sulfur dioxide), the gas will dissolve in water to form acid solution, which can damage health and property. Water vapor satisfied of air is reported as a percentage of the saturation vapor pressure of water at an assumed temperature. This is the relative humidity. The amount of water vapor in the troposphere is highly variable as it will be governed by physical place, how adjacent water bodies are, wind direction and air temperature. Relative humidity is generally higher during summer when temperature and rainfall are also at their highest. Humidity determination follows the absorption properties of a polymer film absorbs or drops water vapor as the relative humidity of the ambient air changes. A sensor procedures these variations and changes them into a humidity analysis.

(iv) Rainfall

Scavenging effect of rain washes PM or soot out of the atmosphere and dissolves gaseous pollutants. Eliminating particles expands reflectivity. Wherever there is regular great rain, air quality is normally enhanced. When the rain dissolves gaseous pollutants, for example SO_2, it forms acid rain causing in probable materials or vegetation damage. A common scheme to quantify rainfall is a tipping bucket rain gauge.

(v) Solar radiation

It is imperative to monitor solar radiation for use in modeling photochemical smog measures, as the sunlight intensity has a key impact on the chemical reactions rate that yields to smog. The muddiness of the sky in day time and topographical site all affect intensity of sunlight. An instrument called a pyrometer measures solar radiation from the output of a type of silicon cell sensor.

7. Coupling of air pollution with global warming process

Eruptions from volcanoes, earthquake ruptures, meteorites, and sand hurricanes terrific into the Earth crust are regular marvels that can origin air pollution and ultimate climate change. The overwhelming bulk of scientific confirmation specifies that anthropological action is the principal cause of current climate change. It is evident that the increase in carbon dioxide (CO_2) and other GHG emissions, subsequently the industrial insurgency, is the topmost source of perceived global warming. Local and year-on-year disparities are estimated within weather schemes, but the signal demonstrations heating over the past half-century that cannot be clarified by natural reasons (Schneidemesser et al., 2015).

CO_2 is currently at stages much advanced than at any period in past years, and sustained emissions are anticipated to substantial additional warming. Furthermore, the haste of warming will be quicker than through past normal climate change actions, making revision tougher. This alteration in weather is anticipated to bring deviations in provincial temperature, precipitation and to raise the regularity of heavy rainfall, heat waves, and certain other kinds of dangerous withstand measures. These will have a serious adverse effect on social comfort and the ordinary biosphere. In 1999, through 243.3 MT of carbon unconfined from the feeding and burning of fossil fuels. India is graded fifth in the domain after the Japan, the United States, Russia, and China. India's impact to world carbon releases is probable to upsurge in the pending years because of the fast stride of expansion, change from noncommercial to profitable fuels, enlarged vehicular practice, and sustained use of elder and extra incompetent fuel power-plants and coal-fired plants (Rong et al., 2010).

The heat-trapping countryside of CO_2 and other gases was verified in the mid-19th century (Hegerl et al., 1996). Their capacity to distress the transfer of IR dynamism finished the troposphere is the technical foundation for such revelation. So, it is very clear that improved heights of GHGs reason the Earth to sincere in response. Ice centers drained from Antarctica to Tropical Mountain glaciers display that the Earth's environment returns to deviations in GHG intensities. Ancient evidence (tree rings, ocean sediments) reveals that present warming is arising roughly 10 times faster than the average rate of ice-age-recovery warming (NRC, 2006). The indication for speedy climate transformation can be confirmed by the following evidences:

(i) Global temperature rise

In the late 19th century, the planet's average surface temperature has risen about 1.62°F (Fahrenheit) or 0.9°C (Celsius). An alteration obsessed chiefly by higher CO_2 and extra manmade emissions into the troposphere (NCDC). Most of the warming ascended in the earlier 35 years, counting warmest years on record since 2010. The year 2016, the warmest year on record with June among those particular (January−September) 8 months (GISS).

(ii) Glacial retreat and shrinking ice sheets

Glaciers are receding nearly all over round the world counting in the Himalayas to Alaska and Africa (National Ice Data and Snow Center). Antarctic and Greenland ice sheets have diminished in figure. NASA's Gravity Recovery and Climate Research data show during 1993−2016 period, Greenland vanished ~286 billion tons of ice/year, whereas Antarctica lost ~127 billion tons ice/year. The rate of mass loss has triplicated in Antarctica ice sheet in the past decade (NASA).

(iii) Waning Arctic Sea ice and sea level rise

Individually the degree and breadth of Arctic Sea ice has waned fast over the last few years. Global sea level increased approximately 8 inches in the past century. The degree in the previous two decades, yet, is approximately twofold that of the past century and is rushing marginally each year (Singh and Singh, 2012).

(iv) Decreased snow cover

Satellite observations reveal that the amount of spring snow cover in the Northern Hemisphere has decreased over the past five decades and that the snow is melting earlier (Cryosphere).

(v) Extreme events

Smooth, small increases in normal global heats can have vast belongings. Perhaps the main, most noticeable result in rapid melting of glaciers. The melted water drained down into oceans, instigating rise in sea altitudes and the sea water turn out to be less saline.

(vi) Warming oceans

Since 1969, the oceans have captivated much of this amplified heat, about 700 m of ocean display heating of more than 0.4°F (NOAA).

8. Design of large-scale facility for effective global warming system

Agricultural production would slow down, and consequently, food prices will rise. Intergovernmental Panel on Climate Change (IPCC) figures advice that the regularity of heat waves and exciting precipitation will increase. Climate outlays like storms and humid cyclones become strong. Storms might be sturdier, further repeated, and long-lasting. That can be monitored via storm rushes, the instant increase in sea level subsequent tornados. Tornado floods are mostly destructive to seaside regions since their erosion possessions are long-lasting. Therefore, decreasing GHG productions are the precarious pace in reducing the global warming inclination, and many governments round the globe are operational to this objective.

The main effort in terms of Kyoto Protocol has been accepted in 1997, implemented in 2005, and by the end of 2009, 187 countries signed and consented. Underneath this convention, 37 developed countries and the European Union have stanch to decrease their greenhouse emissions. Many customs can decrease GHGs on individuals, industries, and government levels. As an individual, energy efficiency can be improved in homes, institutions, and businesses. There is a regular advancement in industries the fuel proficiency of automobiles and production processes. Government agencies are promoting subsidiary schemes for the development of alternative energy (solar power and biofuels) sources.

Various scientists have dedicated their research to capture carbon dioxide and store it underground, instead of atmospheric discharge called carbon sequestration. Trees and vegetation absorb carbon dioxide for their growth. Reforesting forest protection and implanting new ones as in reforestation can support stability greenhouse fumes in the atmosphere. Alterations in unproductive farming practices can also decrease GHG emissions. For instance, farmers use huge volumes of nitrogen-based fertilizers like ammonia or ammonium nitrate (NH_4NO_3) that increase nitrogen oxide (NOx) discharges from the soil. Dropping the practice of these fertilizers would cut the sum of these gases in the troposphere. The manner farmers use animal compost can also influence on total global warming. Once compost is stocked as liquid/slurry in tanks or ponds, it discharges methane (CH_4) not in case of solid.

Decreasing GHG emissions is extremely essential. Although, the worldwide heat has now altered and will record probable remained altered continuously for upcoming years. IPCC proposes that societies discover new customs to adjust to slow global warming. Some of the recommendations for adapting are as follows:

Improved water distribution system over desalination, rainwater harvesting, management, and its reuse.

- Construction of storm flow barriers, dams, building wetlands and swamps as bumpers beside rising sea altitudes and flooding.
- Crafting global warming-health execution strategies, enhancing disaster health facilities, cultivating good health practices by disease surveillance and control.
- Diversifying tourism fascinations by good road and train infrastructure to forest sceneries.
- Amending crop positions, and diversity to increase soil fertility and cope up with warming.
- Reinforced energy infrastructure, cultivating improved energy efficacy, and decreasing dependency on single source strategy of energy.

9. Current applications and future aspects

Enriched indulgence is obligatory to monitor the growth of lengthy monitoring approaches that could best proficiently describe a future adaptive supervision method to oil sand ecological prevention of influences and monitoring (Brook et al., 2019).

Significant advances have been made in baseline evidence through air, water, and biota, as an effect, forecasting can be anticipated to be more subtle and receptive to conservational alterations linked with oil sand development.

In the future, air pollution control efforts would benefit from a more specific understanding of the source categories and chemical species that have the greatest influences on human health, including developing risk functions separately for different sources (e.g., coal combustion and traffic), allowing control actions to be better prioritized (West et al., 2016). Air pollution is rising in several developing countries, which pays to the hasty millions of global deaths each year and air quality problems. While previous policy exertions have thriven in sinking PM and trace gases in America and Europe. The adverse health effects are established at even these minor stages of air pollution, and future strategy movements will benefit from better understanding of interfaces and health effects of altered source groups. Achieving this new understanding requires air pollution scientists and engineers to work increasingly closely with health scientists. In precise, research is desired to well apprehend the physical and chemical properties of intricate air pollutant mixes and to practice new remarks delivered by dispersed innovative in-situ measurement procedures, micro monitoring systems, and satellites, to well illustrate air pollution disclosure for toxicological and epidemiological research, and to improved computation the effects of specific source sectors and mitigation strategies.

9.1 Carbon capture and storage/utilization technologies

It is essential to develop the ways to mitigate carbon emission. One of such methods is carbon capture and storage (CCS) technologies to cope with the global demand of CO_2 reduction. Numerous operative approaches of CCS, such as adsorption, membrane separation, and cryogenic separations, have been suggested to diminish the volume of released CO_2 in the troposphere, which later buried in deep underground (Cuellar-Franca and Azapagic, 2015).

For carbon capture, vastly proficient adsorbents, for example, activated carbon, zeolites, porous silica, and alkali metal oxides, are used. In the recent times, metal-organic frameworks (MOFs), N-doped carbon, porous clays, porous organic polymers (POPs), etc. are probed as flexible and rather sophisticated technique for new era of CO_2 capture. MOFs, porous and nanoporous carbon clays, and POPs are used as CO_2 adsorbents, which retain high specific surface area and facile surface chemistry for improving contact with CO_2. MOF-210 has very high surface area of 6240 m^2/g and CO_2 absorption capacity of ~71 wt% CO_2 uptake at 25°C and 50 bar, which are the peak and measured as novel and best among other solid porous materials (Sharples, 2019). Although, CCS facades a total of practical and financial obstacles that must be overwhelmed formerly it can be organized on a big scale. The key financial hindrance is the point that it is a nonprofit action that needs big investment outlay.

Additionally, newly, an associated substitute—carbon capture and utilization or CCU—has been ongoing to appeal courtesy global as it can convert surplus CO_2 discharges into valued products, for example, fuels and chemicals, whereas at the identical period funding to climate change mitigation. One of the benefits of CCU over CCS is that utilization of CO_2 is generally a lucrative action as harvests can be traded.

Successful utilization of CO_2 has the prospective to shrink CO_2 emissions by 3.7 Gt/y, which is equivalent around 10% of the annual world CO_2 emissions (Andrew, 2018). The development of new innovative processes for chemicals/products with large market demand is essential and effective way to enhance CO_2 utilization. The challenge in development of such processes is the ability to have the right balance of safety, reliability, economic feasibility, and product quality.

Much effort is currently in progress to convert CO_2 via catalytic hydrogenation processes to various value-added hydrocarbon products such as methanol or methane. The addition of H_2, produced from renewable energies, is considered a potential sustainable path toward a carbon neutral manufacturing process for hydrocarbon fuels. Thermal chemical, electrochemical, and photochemical catalysis are three interesting approaches.

Catalysts for CO_2 hydrogenation are being intensively studied in laboratories for conversion to CH_4. Heterogeneous catalysts offer several technical advantages such as low energy consumption, stability, selectivity, product separation, reprocess the catalysts, and simple reactor design. Challenges in selectivity and life still remain before commercialization can be realized.

The Sabatier reaction provides the theoretical foundation for CO_2 methanation (Eq. 5.15). This is an exothermic process with high equilibrium conversion within the temperature range from 25 to 400°C. The presence of a suitable catalyst, CO_2, methanation can be complete with over 99% CH_4 selectivity thereby avoiding product separation.

$$CO_2 + H_2 \rightarrow CH_4 + 2H_2O \quad \Delta H = -165.0 \text{ kJ/mol} \tag{5.15}$$

Other value-added products in which CO_2 can be converted.

Applications of CO_2:

- As solid CO_2 in refrigeration process
- Liquid CO_2 is needed in carbonated beverages
- Gaseous CO_2 used as a neutralizing agent and the basic raw material for Na_2CO_3, $NaHCO_3$ production
- Used in creating inert atmosphere in fire extinguishers, welding, and metal fabrication
- Supercritical CO_2 as solvent, oil recovery, and coal bed methane recovery

9.2 Concluding remarks

The choices we make now will have far reaching consequences. We need to develop mitigation and adaptation strategies to address the challenges that climate change poses. These approaches embrace emerging and organizing low carbon expertise, successful energy competence, and altering manners to permit viable growth.

The last appraisal we can mark amid air pollution and climate change is the best anticipative: together cut a collective resolution, the outline of a more justifiable energy model. Paris Agreement, 2015 will eventually aid to shrink the contaminating discharges that increase the hotness of the earth and brand the atmosphere such a contaminated atmosphere. The claim of trials from use of energy proficiency, the use of electric vehicles, less source intake, and more renewable energy.

It is very confirmatory that instantaneous variations to air pollution levels also have abrupt possessions. Quick accomplishment on dipping vastly effective, short-lived environment contaminants—CH_4, tropospheric O_3, HFCs, and black carbon—can considerably decline the risks of eliciting risky environment slanting themes, identical the permanent issue of CO_2 and CH_4 from thawing Arctic permafrost. Continued and improved networks of measurements that provide long-term data are essential to gain a more robust understanding about past and present changes in concentrations of air toxins and GHGs. Such systems comprise surface monitoring. Aircraft trials joined with study using mathematical models have verified to be mainly suitable in evolving our awareness around key physical and chemical methods in the atmosphere. Furthermore, there is a perfect need for amended emission accounts that trail varying foundations of air pollutants besides GHGs above an extensive array of sites to long periods.

Continuing investigation can offer challenges for manufacturers to select strategies that decrease GHGs consequently recover air quality and encounter health goals. From the past 3 years, important developments have been understood in satellite annotations if innovative facts on the dispersal of several gases (CH_4 and O_3) in the atmosphere. Atmospheric levels of carbon dioxide (CO_2) are rising significantly from approximately 315 ppm in 1959 to a current ~405 ppm in 2017. CO_2 is the key GHG and contributes more than 60% to global warming because of its huge emission amount. This increase in its levels has led to global warming, which is considered as the biggest environmental challenge.

Although CO_2 emission seemed to be a problem, but it can also be seen as a cheap, recyclable, and nontoxic carbon source for the formation of value-added product. Therefore, instead of discharging CO_2 into the atmosphere, the effort are put to capture CO_2 and then either store it which is known as CCS technology or after capturing, use it, called as CCU technology. CCU is favored over CCS as CO_2 conversion to fuel or other value-added products saves extra cost of transportation of captured CO_2 and storage, rather it will be a profitable affair.

References

An Editorial Essay, 2004. A good climate for clean air: linkages between climate change and air pollution. Climatic Change 66, 263–269. Water in the frequency range 3000-934 cm^{-1} Appl Optics 33, 8041.

Anthes, R.A., Corell, R.W., Holland, G., Hurrell, J.W., MacCracken, M.C., Trenberth, K., 2006. Hurricanes and global warming—potential linkages and consequences. Bull. Am. Meteorol. Soc. 87, 623–628. (Accessed 15 February 2020).

Bates, J., Brand, C., Davidson, P., Hill, N., 2001. Economic Evaluation of Sectorial Emission Reduction Objectives for Climate Change: Economic Evaluation of Emission Reductions

in the Transport Sector of the EU. AEA Technology Environment, UK. Bottom-up Analysis.
Betts, R.A., Jones, C.D., Knight, J.R., Keeling, R.F, Kennedy, J.J., Wiltshire, A.J., Aragão, L.E., 2018. A successful prediction of the record CO2 rise associated with the 2015/2016 El Nino. Philos. Trans. R Soc. Lond. B Biol. Sci. 373 (1760), 20170301.
Brenkert, A.L., Malone, E.L., 2005. Modeling vulnerability and resilience to climate change: A case study of India and Indian states. Clim. Change 72 (1–2), 57–102.
Brooks, B.E., 2019. Greening chemistry and ecotoxicology towards sustainable environmental quality. Green Chem. 21 (10), 2575–2582.
Cairncross, E.K., John, J., Zunckel, M., 2007. A novel air pollution index based on the relative risk of daily mortality associated with short-term exposure to common air pollutants. Atmos. Environ. 41, 8442–8454.
Council of European Union, 2001. EU Directive 2001/116/EC. Retrieved from: http://eur-lexeuropaeu/legal-content/en/ALL/?uri=CELEX:32001L0116.
Cryosphere. http://nsidc.org/cryosphere/sotc/snow_extent.html.
Cuéllar-Franca, R.M., Azapagic, A., 2015. Carbon capture, storage and utilisation technologies: A critical analysis and comparison of their life cycle environmental impacts. J. CO2 Util. 9, 82–102.
Dahlmann, K., Grewe, V., Ponater, M., Matthes, S., 2011. Quantifying the contributions of individual NOx sources to the trend in ozone radiative forcing. Atmos. Environ. 45 (17), 2860–2868.
De Sario, M., Katsouyanni, K., Michelozzi, P., 2013. Climate change, extreme weather events, air pollution and respiratory health in Europe. Eur. Respir. J. 42 (3), 826–843.
(EEA) European Environment Agency, Updated 2005. Emission Trends of Nitrogen Oxides NOx. https://wwweeaeuropaeu/data-and-maps/indicators/emission-trends-of-nitrogen-oxides-nox. (Accessed 15 January 2020).
Er-Rbib, H., Bouallou, C., Werkoff, F., 2012. Dry reforming of methane–review of feasibility studies. Chem. Eng. Trans. 29, 163–168. https://doi.org/10.3303/CET1229028.
Gaffney, J.S., Marley, N.A., 2003. Atmospheric chemistry and air pollution. Mini-Rev. Sci. World J. 3, 199–234. https://doi.org/10.1100/tsw.2003.18. ISSN 1537-744X.
GISS. https://www.giss.nasa.gov/research/news/20170118/.
Goyal, P., 2001. Flexibility in estimating air quality index: a case study of delhi. Glob. J. Flex. Syst. Manag. 2, 39–44.
Grewe, V., Dahlmann, K., Matthes, S., Steinbrecht, W., 2012. Attributing ozone to NOx emissions: implications for climate mitigation measures. Atmos. Environ. 59, 102–107.
Grooß, J.-U., Brühl, C., Peter, T., 1998. Impact of aircraft emissions on tropospheric and stratospheric ozone. Part I: chemistry and 2-D model results. Atmos. Environ. 32, 3152–3184.
Habre, R., Coull, B., Moshier, E., Godbold, J., Grunin, A., Nath, A., et al., 2014. Sources of indoor air pollution in New York city residences of asthmatic children. J. Expo. Sci. Environ. Epidemiol. 24, 269–278.
Hansen, J., Nazarenko, L., Ruedy, R., Sato, M., Willis, J., Del Genio, A., Koch, D., Lacis, A., Lo, K., Menon, S., Novakov, T., Perlwitz, J., Russell, G., Schmidt, G.A., Tausnev, N., 2005. Earth's energy imbalance: confirmation and implications. Science 308, 1431–1435.
Hegerl, G.C., Von Storch, H., Hasselmann, K., Santer, B.D., Cubasch, U., Jones, P.D., 1996. Detecting greenhouse-gas-induced climate change with an optimal fingerprint method. J. Clim. 9, 2281–2306.
(HEI), 2019. Health Effects Institute State of Global Air Special Report. Health Effects Institute, Boston, MA.

(ICCT) 2018 International Council on Clean Transportation, updated 2018. Real-World NOx Emissions from Remote Sensing: An Update of the TRUE Rating. https://theicctorg/blog/staff/true-rating-update-dec2018. (Accessed 19 January 2020).

International Council on Clean Transportation, Anenberg, S., Miller, J., Henze, D., Minjares, R., 2019a. A global snapshot of the air pollution-related health impacts of transportation sector emissions in 2010 and 2015. https://theicct.org/sites/default/files/publications/Global_health_impacts_transport_emissions_2010-2015_20190226.pdf.

(ICCT) 2019 International Council on Clean Transportation, Bernard, Y., German, J., Muncrief, R., updated 2019b. Worldwide Use of Remote Sensing to Measure Motor Vehicle Emissions. ICCT, Washington, DC, 2018. https://theicctorg/publications/real-world-emissions-using-remote-sensing-data.

(IPCC), 2001. Intergovernmental panel on climate change. Atmospheric chemistry and greenhouse gases. In: Climate Change 2001: Working Group I: The Scientific Basis. Available from: http://www.ipcc.ch/ipccreports/tar/wg1/pdf/TAR-04.PDF. (Accessed 4 January 2020).

IPCC, 2007a. Intergovernmental panel on climate change. Introduction. In: Climate Change 2007: Mitigation (Working Group III Fourth Assessment Report). Available from: http://www.ipcc.ch/pdf/assessment-report/ar4/wg3/ar4-wg3-chapter1.pdf. (Accessed 21 January 2020).

IPCC, 2007b. Intergovernmental panel on climate change (February). Changes in atmospheric constituents and in radiative forcing. In: Climate Change 2007: The Physical Science Basis (Working Group 1 Fourth Assessment Report). Available from: http://www.ipcc.ch/SPM2feb07.pdf. (Accessed 4 January 2020).

IPCC, 2007c. Intergovernmental panel on climate change, summary for policymakers. In: Climate Change 2007: The Physical Science Basis. Contribution of Working Group I to the Fourth Assessment Report of the Intergovernmental Panel on Climate Change. Cambridge University Press, Cambridge, United Kingdom.

(IPCC), 2010. Assessment Reports. IPCC. See: http://www.ipcc.ch/publications_and_data/publications_and_data_reports.htm#1.

(IPCC), 2018. Special Report on the impacts of global warming of 1.5 C above pre-industrial levels and related global greenhouse gas emission pathways, in the context of strengthening the global response to the threat of climate change, sustainable development, and efforts to eradicate poverty. In: Intergovernmental Panel on Climate Change.

Jacobson, M.Z., 2002. Control of fossil-fuel particulate black carbon and organic matter, possibly the most effective way of slowing global warming. J. Geophys. Res. 107 (D19), 4410.

Jeffrey, J., Ruth, P., Weiner, F., Aarne, P., Vesilind, 1998. Chapter 20 - Measurement of Air Quality Environmental Pollution and Control, fourth ed., pp. 287–295

Johansson, B., 2003. Transportation fuels - a system perspective. In: Hensher, D.A., Button, K.J. (Eds.), Handbooks in Transport 4: Handbook of Transport and the Environment. Elsevier, pp. 141–158.

Johnson, D., Ambrose, S., Bassett, T., Bowen, M., Crummey, D., Isaacson, J., et al., 1997. Meanings of environmental terms. J. Environ. Qual. 26, 581–589.

Kelishadi, Mirghaffari, N., Poursafa, P., Gidding, S.S., 2009. Lifestyle and environmental factors associated with inflammation, oxidative stress and insulin resistance in children. Atherosclerosis 203, 311–319.

Kumaravel, B., Palanivelraja, S., Chockalingam, M.P., 2013. Influence of temperature, relative humidity and seasonal variability on ambient air quality in a coastal urban area. Int. J. Atmos. Sci. 7. https://doi.org/10.1155/2013/264046.

Lal, S., Patil, R.S., 2001. Monitoring of atmospheric behaviour of NOx from vehicular traffic. Environ. Monit. Assess. 68, 37–50.
Lefol, L., Guarieiro, N., Lefol, A., Guarieiro, N., 2013. Vehicle Emissions: What Will Change with Use of Biofuel? https://doi.org/10.5772/52513.
Lin, X., Trainer, M., Liu, S.C., 1988. On the nonlinearity of the tropospheric ozone production. J. Geophys. Res. 93, 15,879–15,888.
Liu, S.C., Trainer, M., Fehsenfeld, F.C., Parrish, D.D., Williams, E.J., Fahey, D.W., Hübler, G., Murphy, P.C., 1987. Ozone production in the rural troposphere and the implications for regional and global ozone distributions. J. Geophys. Res. 92, 4191–4207.
Manisalidis, I., Stavropoulou, E., Stavropoulos, A., Bezirtzoglou, E., 2020. Environmental and health impacts of air pollution: a review. Front. Public Health 8 (14), 1–13. https://doi.org/10.3389/fpubh.2020.00014.
Marley, N.A., Gaffney, J.S., Cunningham, M.M., 1994. Lambert absorption coefficients of water in the frequency range of 3000–934 cm– 1. Applied optics 33 (34), 8041–8054.
NASA. https://www.jpl.nasa.gov/news/news.php?feature=7159.
NCDC. https://www.ncdc.noaa.gov/monitoring-references/faq/indicators.php.
NOAA National Centers for Environmental Information. Dataset. https://ds.doi.org/10.7289/V53F4MVP.
Muraza, O., Galadmina, A., 2015. A review on coke management during dry reforming of methane. Int. J. Energy Res. 39, 1196–1216.
(NOAA) National Oceanic and Atmospheric Administration, 2008. Greenhouse Gases: Frequently Asked Questions. Availabl from: http://lwf.ncdc.noaa.gov/oa/climate/gases.html. (Accessed 13 January 2020).
(NRC) 2006 National Research Council, 2006. Surface Temperature Reconstructions for the Last 2,000 Years. National Academy Press, Washington, D.C.
Oliver, W., Prather, M.J., Hajime, A., 2001. Indirect long-term global radiative cooling from emissions. Geophys. Res. Lett. 28 (9), 1719–1722.
Ott, W.R., Hunt Jr., W.F., 1976. A quantitative evaluation of the pollutant standards index. J. Air Pollut. Contr. Assoc. 26 (11), 1050–1054. https://doi.org/10.1080/00022470.1976.10470358.
Popescu, F., Ionel, I., 2010. Anthropogenic Air Pollution Sources. https://doi.org/10.5772/9751.
Prasad, R., Singh, P., 2012. A review on CO oxidation over copper chromite catalyst. Catal. Rev. Sci. Eng. 54 (2), 224–279.
Pryor, S.C., Crippa, P., Sullivan, R.C., 2015. Atmospheric chemistry. In: Reference Module in Earth Systems and Environmental Sciences. https://doi.org/10.1016/B978-0-12-409548-9.09177-6.
Robinson, D.L., 2005. Air pollution in Australia: review of costs, sources and potential solutions. Health Promot. J. Aust. 16, 213–220.
Rong, F., 2010. Understanding developing country stances on post-2012 climate change negotiations: Comparative analysis of Brazil, China, India, Mexico, and South Africa. Energy policy 38 (8), 4582–4591.
Sarath, K.G., Gregory, R.C., Giuseppe, C., 2003. The contribution of megacities to regional sulphur pollution in Asia. Atmos. Environ. 37, 11–22.
Schaub, M., Paoletti, E., 2007. Introductory Remarks to the Special Issue–XXII IUFRO World Congress, 2005 Brisbane–Air Pollution and Climate Change: A Global Overview of the Effects on Forest Vegetation.
Schneidemesser, E. von, Monks, P.S., Allan, J.D., Bruhwiler, L., Forster, P., Fowler, D., Lauer, A., 2015. Chemistry and the linkages between air quality and climate change. Chem. Rev. 115 (10), 3856–3897.

Seinfeld, J.H., Pandis, S.N., 2006. Atmospheric Chemistry and Physics: From Air Pollution to Climate Change, second ed. Wiley, New York. ISBN: 978-0-471-72018-8.
Senthilnathan, T., 2007. Analysis of concentration of air pollutants and air quality index levels in the ambient air in Chennai city. IE (I). J. Eng. 87, 3–7.
Sharples, J. 2019. LNG supply chains and the development of LNG as a shipping fuel in Northern Europe. Andrew, 2018
Shukla, V., Dalal, P., Chaudhry, D., 2010. Impact of vehicular exhaust on ambient air quality of Rohtak city, India. J. Environ. Biol. 31 (6), 929–932.
Singh, B. R., Singh, O. 2012. Study of impacts of global warming on climate change: rise in sea level and disaster frequency, Global warming—impacts and future perspective.
Srivastava, S., Kumar, A., Bauddh, K., Gautam, A.S., Kumar, S., 2020. 21-day lockdown in India dramatically reduced air pollution indices in Lucknow and New Delhi, India. Bull. Environ. Contam. Toxicol. 1.
Thom, G.C., Ott, W.R., 1976. Air Pollution Indices: A Compendium and Assessment of Indices Used in the United States and Canada, sixth ed. Ann Arbor Science, Michigan, USA.
Thomas, A., Sarangi, C., Kanawade, V.P., 2019. Recent increase in winter hazy days over central India and the Arabian sea. Sci. Rep. 9 (1), 1–10.
Tzirakis, E., Pitsas, K., Zannikos, F., Stournas, S., 2006. Vehicle emissions and driving cycles: comparison of the Athens driving cycle (ADC) with ECE-15 and European driving cycle (EDC). Global NEST J. 8 (3), 282–290.
Uherek, E., Halenka, T., Borken-Kleefeld, J., Balkanski, Y., Berntsen, T., Borrego, C., Gauss, M., Hoor, P., Juda-Rezler, K., Lelieveld, J., Melas, D., Rypdal, K., Schmid, S., 2010. Transport impacts on atmosphere and climate: land transport. Atmos. Environ. 44, 4772–4816.
U. S. Environmental Protection Agency (USEPA), 1996. Metropolitan Area Trends. US Environmental Protection Agency, Washington, DC. (Accessed 20 January 2020).
U.S. Environmental Protection Agency (USEPA), 2010. High Global Warming Potential (GWP) Gases. Available from: http://www.epa.gov/highgwp/index.html. (Accessed 2 February 2020).
(USEPA), 2020. Air Quality System (AQS) API. https://aqs.epa.gov/aqsweb/documents/data_api.html. (Accessed 10 January 2020).
Vallero, D., 2007. Fundamentals of Air Pollution, fourth ed. Academic Press, California, USA.
Weiss, M.A., Heywood, J.B., Drake, E.M., Schafer, A., AuYeung, F.F., October 2000. On the Road in 2020 - A Life Cycle Analysis of New Automobile Technologies. Energy Laboratory, Massachusetts Institute of Technology, Winter.
West, J. J., Cohen, A., Dentener, F., Brunekreef, B., Zhu, T., Armstrong, B., ... & Dockery, D. W. (2016) What we breathe impacts our health: improving understanding of the link between air pollution and health.
WHO, 2016. https://www.who.int/sustainable-development/transport/health-risks/air-pollution/en/.
World Air Quality Report Region & City PM2.5 Ranking, 2019.

Dimensions of climate change and its consequences on ecosystem functioning

6

Srishti Mishra[a], *Annesha Ghosh*[a], *Kshama Rai, Bhavna Jaiswal, Durgesh Singh Yadav, Madhoolika Agrawal, Shashi Bhushan Agrawal*
Laboratory of Air Pollution and Global Climate Change, Department of Botany, Institute of Science, Banaras Hindu University, Varanasi, Uttar Pradesh, India

1. Introduction

Generally, climate implies the overall prolonged attributes of weather at specific location, primarily determined by averages of rainfall, temperature, relative humidity, and other meteorological conditions, whereas weather reflects daily short-term deviations in natural atmospheric conditions (Climateurope, 2020). According to World Meteorological Organization (WMO), average time of 30 years has been determined for evaluation of any changes in the climatic pattern (Climateurope, 2020). However, predicting such change in climatic condition in advance requires considerable skill and involvement of several challenges (Australian Academy of Science, 2020). In recent time of Anthropocene, prolonged emissions of greenhouse gases (GHGs) enhanced global warming, which is responsible for extreme incidences of drought, heat waves, precipitation pattern, and triggering occurrences of disaster events, thereby inducing the vulnerability of whole ecological systems (NASA, 2020a). Besides, poor and developing nations have a great threat of such hazards due to their large dependency on naturally available resources and, with further variations in climate, could lead them to a devastated situation (Adedeji, 2014).

Variations in natural climate exist according to different meteorological conditions where different ecosystems function altogether as a unit. However, climate change is a catastrophic phenomenon for any ecosystem's normal functioning (Adedeji, 2014; Melillo et al., 1990). Increasing GHG concentration and temperature (Pachauri et al., 2014), frequent disaster, and extreme events displayed their rapid increase (Adedeji, 2014) along with loss of composition of species in an ecosystem, alteration of community pattern and productivity (Rogers et al., 2019; Bryndum-Buchholz et al., 2019; Grimm et al., 2013), shifting of biomes (Grimm et al., 2013; Lenoir et al., 2008), increase infestation of insects (Chattopadhyay et al., 2019; Grimm et al., 2013), change in natural cyclic process of nutrients and flow patterns of aquatic bodies, productivity, nutrient mixing pattern of oceans and stratification-related

[a] Equal contribution.

Global Climate Change. https://doi.org/10.1016/B978-0-12-822928-6.00003-4

processes (Doney et al., 2011; Melillo et al., 1990), and threatening food security (Adedeji, 2014) that are the burning issues challenging the survival of life on this planet.

Human well-being is totally dependent on the healthy ecosystem and due to human-made activities; its disruption has been common in the recent few decades. The consequences of changing climate are many in which some are rapid stratospheric O_3 depletion, desertification, deforestation, excess atmospheric nitrogen deposition, decline in freshwater resources, change in land use pattern, and biotic exchange have become conspicuous with passing time (Thuiller, 2007). Migration, shifting in structure of community, and apparent loss of sensitive organisms have now been primary challenges in natural ecosystem and also listed as major ecological consequences of climate change (Lenoir et al., 2008). Under Intergovernmental Panel on Climate Change (IPCC) Fifth Assessment Report (Pachauri et al., 2014), rate of increasing global mean temperature by 0.05–0.15°C per decade may exceed from 1.5 to 2°C by the 21st century. Furthermore, such elevation in temperature has caused a rise in sea level by 0.19 m throughout the period of 1901–2010 owing to the accelerated glaciers melting and thermal expansion of oceans. However, variations exist in between climate change at global and regional levels due to difference in source–sink, clouds patterns, and geographical locations. However, several uncertainties prevail in predicting the long-term climate change consequences, and the biggest question lies in concerning such serious undertaking is "Can these consequences be reserved if resort to preventive measures?" Although details from the past studies of climate change largely endorsed the several predictions underlying the modeling studies, imprecision in data and gaps in knowledge regarding potential factors may influence such changes, which is likely to reinforce such uncertainties (Pachauri et al., 2014; Joos and Spahni, 2008; Stott et al., 2001).

In this chapter, we have discussed about the issues of climate change and their implications on environmental and ecological functioning. We also tried to discuss the factors affecting environment and our ecosystems, which are directly or indirectly related with the survival and sustenance of life along with suggestion of some preventive measures that can be adopted to reduce the impact of global climate change in the long run.

1. Historical preview and recent updates

Earth climate has been continuously changing with times, which are evident from the past records. Cycles of interglacial and glacial periods occurred after interval of long time. In the past of Earths' climate, numerous glacial and interglacial cycles had occurred where warming and cooling of atmosphere and change in ice-cover area and sea level have been observed (Folland et al., 1990). Five main Ice age were Huronian period (2400–2100 million years ago [Mya]), Cryogenic period (850–635 Mya), Andean-Saharan period (460–430 Mya), Karoo period (360–250 Mya), and in recent time of Holocene Quaternary period started since 2.6 Mya till now (History, 2020). Little Ice Age (1500AD–1900AD) from the 16th to 19th century came after Medieval period (950–1100 AD) (Folland et al., 1990). Recent interglacial period has been started from 11,500 years to present time, maybe not given the way to next glacial period (Past Interglacial Working Group of PAGES, 2016).

An IPCC report on the "Climate Change and Land" (Shukla et al., 2019) displayed an increasing pattern for temperature, GHG emission, and land use changes. During 1850–1990, temperature ranges between 0.3 and 0.5°C but increases up to 1.6°C till 2018 (Shukla et al., 2019). GHG emissions from agriculture, forestry, and land use during 2007–16 increased by 23% and desertification that was almost absent in 1961 increased by 200% till 2016 (Shukla et al., 2019). Net CO_2 emission from anthropogenic sources in 1980 was 3.9 Gt CO_2 per year, which in 2000 rose to 5 GtCO_2 per year, and in 2016, it was 6 Gt CO_2 per year. On the other hand, N_2O emission from agriculture sector showed increment from 0.7 to 2 GtCO_2 equivalents per year during 1961–2016 (Shukla et al., 2019). Cumulative CO_2 emission has been predicted in ranges of 1000–2000 Gt CO_2 during 1980–2020 along with increment in temperature from 0.2 to 1.5°C during 1980–2017, which can be disastrous if such increase continues in future (Masson-Delmotte et al., 2018). Fifth Assessment Report (IPCC, 2014) also reported increasing trends for some of the GHG, as for CO_2 showing an increase of 100 ppm (280–380 ppm) and for CH_4 by 60 ppb and N_2O ranges in between 280 and 320 ppb during the time period 1850–2000. Fluorinated gases (F-gases) also displayed similar percentage increase (1970–2010) to that of GHGs owing to enhanced anthropogenic interventions (fossil fuel combustion, flaring, industrial processes, forestry, agriculture, and land use changes) (Pachauri et al., 2014).

2. Causes and factors of climate change

Greenhouse effect is completely a natural phenomenon. However, accelerated human interventions accelerate the emissions of GHG, resulting to an elevation of their background concentration. Volcanic activities and solar irradiance are important natural factors, whereas emissions of GHGs, changes in land use patterns, nutrient pollution, and so on are major anthropogenic factors responsible for changes in climate and represented in Fig. 6.1 (Fahey et al., 2017). Likewise, aerosols also help to alteration in radiative forcing (Earth's energy budget measurement) by either absorbing or

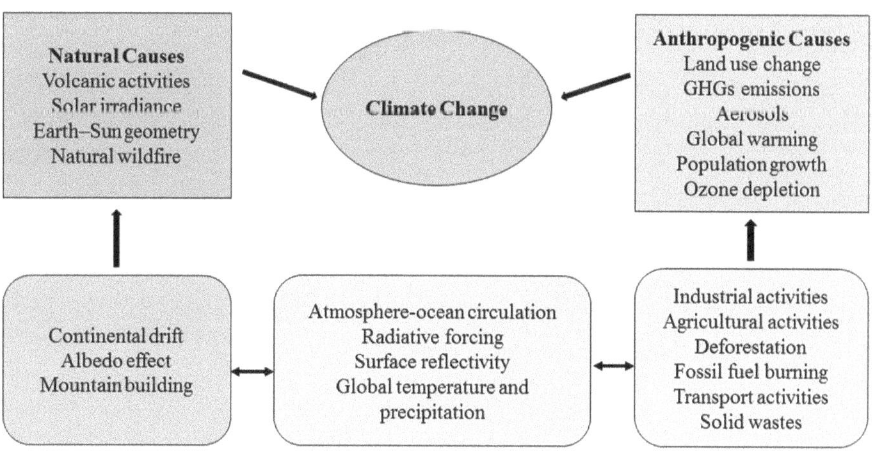

Figure 6.1 Major driving cause of climate change.

scattering of solar incoming energy, thereby resulting in either warming or cooling. However, enhancement in aerosol particle in atmosphere has aggrandized the climate change impacts. One such example is black carbon that is a black sooty particulate matter, which has emerged out to be a global environmental concern due to its implication in altering the climatic pattern by influencing the cloud formation and precipitation (NASA, 2020b).

Combustion of fossil fuel and solid wastes, deforestation, and degradation of soil are the major sources of CO_2 emission while, coal and natural gas/oil production, livestock management, land filling of solid waste produces excess CH_4. N_2O is mainly generated from industrial activities, solid/fossil fuel burning and agricultural practices (by use of nitrogenous fertilizers) on the other hand, F-gases (such as chlorofluorocarbons- CFCs) are produced from the commercial and industrial processes, domestic uses of refrigerants, domestic or automobile air conditioning, cleaning solvents which act as both ozone-depleting substances (ODSs) and GHG (US EPA, 2020; Pachauri et al., 2014; NOAA). Tropospheric O_3 is a secondary air pollutant and recognized as GHG, generally produced from photochemical reaction by oxides of nitrogen (NO_X) and volatile organic compounds (US EPA, 2018).

3. Dimensions of climate change and its consequences on ecosystem functioning

Global climate change has been observed as a dire consequence of increased anthropogenic intervention around the world. However, such conjugal relationship of climate change and anthropogenic activities has resulted in several pressing concerns in the context of ecosystem functioning (Fig. 6.2).

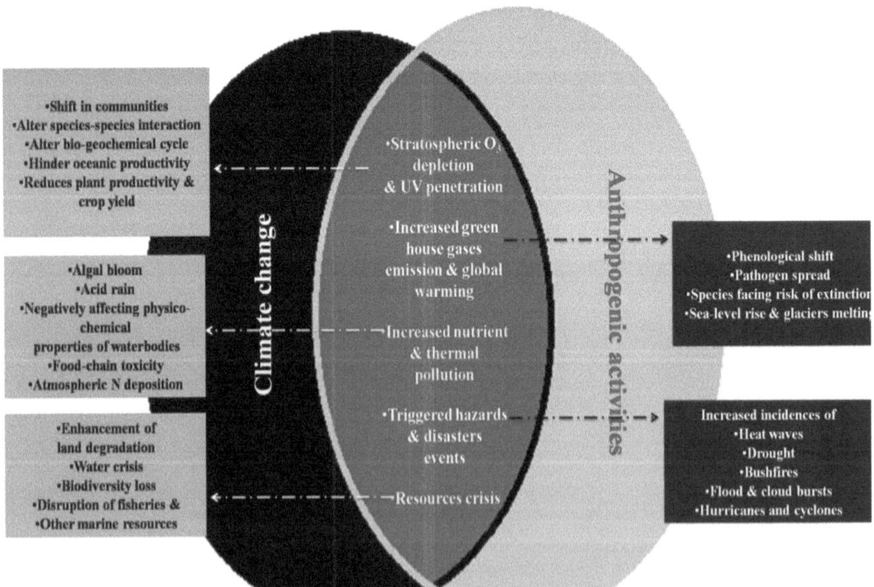

Figure 6.2 Consequences of human-induced climate change on ecosystem functioning.

2. Ozone depletion, UV-B penetration, and their impacts on different ecosystems

2.1 Stratospheric ozone depletion

Ozone (O_3) is a trioxygen compound, discovered by a German chemist C.F. Schönbein (1839) (Rubin, 2001), and the stratospheric O_3 layer was first discovered by two French physicists Fabry and Buisson (1931). However, the properties of the O_3 layer were discussed by G.M.B. Dobson, with the help of his spectrophotometer, which was known to measure the stratospheric O_3 layer thickness from the Earth. The column density of O_3 layer is represented in Dobson Unit (DU). Additionally, the average ideal O_3 layer thickness is 300 DU, which ranged between 250 and 350 DU at equatorial regions and midlatitudes. The major role of stratospheric O_3 layer is to absorb the harmful ultraviolet (UV) radiation coming to the Earth's surface and protects all the life forms. Photochemical mechanism responsible for the formation and destruction of O_3 layer was described by Chapman (1930), according to which oxygen (O_2) molecules under light (UV radiation) break into nascent O_2 atoms (O) and combine with the molecular O_2 yielding O_3. The natural process of O_3 formation and destruction occurs at high temperature, UV radiation, and O_2, so it can be said that O_3 and UV radiation are highly interconnected. In 1974, Stolarski and Cicerone (1974) observed that chlorine has capability to destroy the atomic structure of O_3, and at the same time, Molina and Rowland (1974) also described the fact that human-made chlorofluorocarbons (CFCs) have long atmospheric residential period and it could be the possible reason for the stratospheric O_3 destruction. CFCs are sparingly soluble in water, which makes them resistant to rainout and washout processes (local pollutant removal processes). The tropical regions are the primary entry point of CFCs and other gaseous pollutants into the atmosphere. From the tropical stratosphere, the gases move upward and toward the poles whence, in the presence of UV radiation, the molecular structures of CFCs are destroyed by UV-B releasing the chlorine atoms. These free chlorine atoms later attack on O_3 causing its degradation into O_2 and free chlorine.

Ozone formation:

$$O_2 + \text{UV radiation} \rightarrow 2(O)$$

$$(O) + O_2 + \text{UV radiation} \leftrightarrow O_3$$

Ozone destruction:

$$CFCl_3 + \text{electromagnetic radiation} \rightarrow CFCl_2 + Cl$$

$$Cl + O_3 \rightarrow ClO + O_2$$

$$ClO + O_3 \rightarrow Cl + 2O_2$$

Other reactions involved in ozone destruction:

$$O_3 + C_2H_4 \rightarrow CH_2O_2^{\bullet} + HCHO$$

$$CH_2O_2^{\bullet} + SO_2 \rightarrow SO_3 + HCHO$$

$$O_3 + NO \rightarrow NO_2 + O_2$$

$$NO_2 + O_3 \rightarrow NO_3 + O_2$$

Depletion of O_3 layer was first observed and documented over Antarctica by Farman et al. (1985) and was named "ozone hole." Furthermore, it was studied and observed that this massive depletion was a combined result of CFCs and other gaseous pollutants released from various homogenic activities. The extreme cold climate, long dark period, and geophysical activities (such as polar vortex, polar stratospheric clouds formations) of Antarctica boost the O_3 depletion at this region. To fill this lacuna created by O_3e depletion, more and more influx of O_3-rich air toward Antarctica takes place and efflux of O_3-depleted air toward midlatitudes leads to the low O_3-related problems in the mid- and higher latitude regions. The Brewer–Dobson circulations are involved in the rise and descend of O_3 from the tropics to mid- and higher latitudes. Soon after the discovery and alarming situation of stratospheric O_3 layer depletion, a global meeting, Montreal Protocol on limiting the emissions of ODSs, was held and signed in 1987. After the successful implementation of Montreal Protocol around the globe (Chipperfield et al., 2015), it was initially expected that O_3 hole will recover by 2050. But more recently, NASA scientists have found and predicted that stratospheric O_3 layer will return to pre-1980 levels by about 2075 (Bais et al., 2018). Recent UNEP Environmental Assessment Panel updates specified that though the O_3 levels over Antarctica have underway to recover (Bais et al., 2018; Solomon et al., 2016), the forfeiture of stratospheric O_3 layer over the Arctic region was found in the winter of 2015–16 and water vapor injection into the lower stratosphere throughout severe storms is still the reason of concern.

2.2 Impacts of ozone depletion on regional and global basis

Most of the studies advocate the climate change on the basis of the changes in the climatic events of Southern Hemisphere (SH) of the Earth, as it is the epicenter of all the stratospheric O_3-related studies. These studies are mainly manifested by the Southern Annular Mode (SAM) or Antarctic oscillations, which describes the pressure differences in between 60°S and 45°S (Abram et al., 2014; Marshall et al., 2006) and also the strength and position (latitudinal) of westerly winds (Bornman et al., 2019; Bais et al., 2018). The O_3 depletion led to increase in the pressure differences at mid- and high latitudes and narrowing the westerlies toward Antarctica (Solomon et al., 2016), which further led to changes in atmospheric and oceanic circulations, surface temperature, wind, and precipitation patterns. These changes resulted into frequent flooding and long drought, forest fires, and fluctuations in available water (Bais et al., 2019; Wu and Polvani, 2017; Bai et al., 2016; Munoz et al., 2016; Mariani and Fletcher, 2016; Holz and Veblen, 2011).

2.3 Impacts of O_3 depletion on the regional rainfall and water availability

Manatsa et al. (2016, 2013) reported that the East African variations in rainfall pattern, surface, and diurnal temperature ranges have been acknowledged and linked to the O_3 depletion induced changes in the SAM activities.

Atmospheric circulations shifting and modulations in rainfall pattern (precipitation) lead to the alterations in distribution of precipitation, some regions to receive more precipitation while others to become drier. The changes in the precipitation distribution led to disturbances in small plant diversity. For instance, the lichens are desiccation-tolerant species, but their colonizations on trees have been affected by the climatic factors.

It is well known fact that decreased precipitation induces prolong dryness to some regions (mid- to high latitude) of the Earth making them more prone to wildfire incidences, e.g., increased forest fire in the West of the Andes region (Holz et al., 2017) and the Chilean forest fire (2016–17); the drivers responsible were prolonged drought induced by SAM and El Niño-Southern Oscillation (ENSO). The areas receiving more moisture are subjected to increase agronomic and plant productivity, leading to enhancement of plant diversity (Gonzalez et al., 2014).

2.4 Impacts of O_3 depletion induced changes in surface temperature on terrestrial ecosystem

The movement of warm westerlies toward the eastern side of the Antarctic Peninsula (atmospheric circulations)–induced warming during the summer–autumn (as per 1965–2004 data) has been linked to the O_3 layer depletion and anthropogenic emissions of GHG, which was three times higher than the western side (i.e., mean annual near-surface temperature warmer by 2.7°C). The summer–autumn warming plays a crucial role in melting of ice shelves, e.g., collapse of Larsen-B Ice Shelf. However, western part of the Antarctic continent also showed surface warming (mean annual near-surface temperature warmer by 2.9°C) during the winter during the period of 1951–2004 (Van Lipziget al., 2008; Marshall, 2007). Besides the Antarctic Peninsula, other regions including Eastern Australia, Southern Africa, and South America have been under the severe influence of stratospheric O_3 layer depletion–induced surface warming during summer. The changes in biodiversity of mosses and lakes inhabiting species and dryness-induced stress in the plant communities observed in East Antarctica are some pioneer investigations that correlate the SAM, O_3 depletion, and climate-attributable changes in the terrestrial ecosystem.

3. UV radiation

It is aforementioned that the O_3 layer and UV radiation are interconnected with each other. The solar radiation is comprised of electromagnetic spectrum which ranges from cosmic waves to radio waves, and the UV spectrum is most important in case of O_3 layer context. The UV spectrum (100–400 nm) is broadly classified into three wave bands: UV-C (100–280 nm), UV-B (280–320 nm), and UV-A (320–400 nm) (Global Solar UV Index Guide, WHO, 2002). The UV-C portion of the UV spectrum is extremely harmful to the life forms because of having shorter wavelength having high energy, but, fortunately, it is fully absorbed by the stratospheric O_3 layer. While, UV-A being near to longer waves, it is not harmful, not absorbed by the O_3 layer, and reaches the Earth's surface without any diurnal, seasonal, or yearly change, UV-B (mid portion of the UV spectrum) has high energy, is harmful to all life forms, and is not fully (~95%) absorbed by the stratospheric O_3 layer. Approximately, 5% of UV-B reaches to the Earth's surface. But due to the advent of anthropogenically induced depletion of stratospheric O_3 layer, the level of UV-B penetration increased. The increased UV-B reaching to the Earth's surface affects all the life forms and processes. Plants, being immobile in nature, inevitably suffer from this potent abiotic stress; consequently, they have developed certain defense mechanisms and organs to cope up such surplus stressful conditions. Nevertheless, the response of plants toward UV-B varies with the dose of the stress perceived by the plants. Moreover, the ambient UV-B dose itself depends on several factors such as stratospheric O_3 layer thickness, solar zenith angle, time of day and year, cloud cover, altitude, latitude, prevailing atmospheric conditions of that zone, and plant characteristics (including variety, genotype, cultivar, developmental stage, etc.) (Takshak and Agrawal, 2019; Tripathi et al., 2011; Kumari et al., 2010; Jenkins, 2009; Yang et al., 2004, 2005).

3.1 Global climate change due to ozone depletion and UV-B penetration

The stratospheric O_3 depletion and UV-B penetration induce positive changes in SAM, and movements of oceanic and atmospheric circulations (poleward shift of midlatitude jet and subtropical Hadley cells) and cloud formation processes, which are further known for altering the regional weather and further are involved in inducing climatic changes too.

Past studies, during austral summer (December–February) across the midlatitudes to the tropics between 1960 and 2000, stratospheric O_3 depletion has been the major driver of alterations in the atmospheric circulations across the southern hemisphere, while GHGs play a crucial role during other seasons.

3.2 Impacts of UV-B on terrestrial ecosystem

Various researches on the past, current, and future interactions of stratospheric O_3 layer status, UV-B radiation levels, and climate changes on various ecosystems including terrestrial ecosystem are assessed from time to time. It has been found that the O_3 layer depletion and UV-B penetration induce change in the climatic conditions of the entire ecosystem (Bornman et al., 2019; Bais et al., 2018; Robinson and Erickson, 2015). These studies also described the ways, which showed evidences supporting the change in climate induced by the stratospheric O_3 layer depletion. The climatic changes along with changes in associated UV-B penetration on the Earth's surface may affect the plants' activity inducing the shifting of plant communities. These shifts might be due to the variations in the incoming UV radiations with time. For instance, the shifting of life from aquatic to terrestrial domain was also facilitated by the stratospheric O_3 layer and UV-B interactions. The encroachment of plants to other ecosystems or geographical ranges is not only the result of UV radiation alone but also a cumulative response of overall biotic and abiotic factors involved in atmospheric and meteorological changes pushing the plant communities to shift for acclimatization and adaptation. UV-B also gives hand in global warming by breaking and decomposing the dead plant materials causing the release of phytocompounds, carbon, and essential nutrients into the soil, which alters the phytoavailability of nutrients, leading to induce plant diversity change.

UV-B has potential to damage plants via damaging the genetic material, the photosynthetic apparatus, and other cellular processes. However, UV-B is also known to upregulate the plant defense responses, including genetic photorepairs processes, antioxidant activities, and UV-screening compounds. However, in the process of acclimation/adaptation to the combination of UV radiation exposure and climate change, plant modifies its morphology, anatomy, growth, development, phenology, production, yield, and geographical distribution, which hamper its role in ecosystem functioning, nutrition, and food quality. Therefore, the interactive responses of climatic factors and UV radiation are essential for assessing the consequences of UV radiation on different ecosystems.

3.3 Impacts of ozone depletion and UV radiation on ecosystem functioning

Ecosystems can be modified in several ways by the interactive effects of O_3e depletion, UV radiation, and climate change including biotic and abiotic drivers and factors of the environment too (Bornman et al., 2019; Sulzberger et al., 2019).

➢ **Species–species interactions**

O_3 depletion, UV radiation, and climatic changes have linked with the shift in the competitive balance between species (Bornman et al., 2019). In the agricultural fields, the crops and weeds species are competing with each other for available resources (Barnes et al., 2005); in the forest ecosystem and aquatic ecosystem, the main competition is for the light and food. These changes in the competitive balance are connected to differential effects of O_3 depletion–induced enhanced UV-B radiation and climate change on geographical distribution, morphology, growth and development, and interactions of species with each other. Similarly, Yang et al. (2018) reported that ambient UV-B radiation and its variable impacts on the plant height and leaf area have induced the change in species composition of plant communities in Alpines (Bais et al., 2019; Bornman et al., 2019; Williamson et al., 2019).

➢ **Herbivory and plant–pathogen interactions**

The plant response to O_3 depletion–induced enhanced UV-B radiation showed alterations in the organisms (species) at almost every trophic level from producer to consumer. UV-B radiation is known to affect reductions in herbivory. Few variations in plant chemistry elicited by UV-B radiation simulate the synthesis of some compounds important for plant communications with other organisms (Bornman et al., 2019).

➢ **Decomposition of litter**

Among the ecosystems, carbon storage is essential for the functioning and nutrient recycling, while decomposition of litter is the strong element of the environment which is fully involved in the recycling processes. However, other elements such as soil fertility and biogeochemical cycles are also involved in playing regulatory roles. Decomposition process is also driven by several factors, i.e., temperature and moisture availability, activity of decomposing microorganisms (bacteria and fungi), and plant litter (e.g., leaf chemistry, woody tissue) (Bornman et al., 2019; Ballaré and Austin 2017; Barnes et al., 2015). These alterations in the chemical composition of the plant litter help the microbes to effortlessly decompose litter; this mechanism is also controlled by the climatic factors such as cloud cover, precipitation, pollution, and canopy of the plants.

3.4 Impacts of O_3 depletion and UV-B penetration on aquatic ecosystem

Stratospheric O_3 layer depletion and associated UV radiation and climate change alter the habitat structure of most of the aquatic (marine and inland surface waters) life forms as well as the materials. The factors such as episode of irradiance, snow and ice cover, transparency of water, and the depth (to which life forms passively circulate) help in all atmosphere-driven modifications (Williamson et al., 2019).

The Aquatic organisms move to surface water of uniform temperature (commonly called as mixed layer depth) through vertical mixing (Cullen and Neale, 1994). The depths of the mixed layer in inland waters, coastal areas, and open oceans vary with seasons, regions, and characteristics of the water bodies. In polar regions, mixed layer depths are deeper during winter and shallower in coastal and tropical zones including in all other waters in the summer (Neale et al., 2003). The vertical distribution of bottom-dwelling organisms could be due to stratospheric O_3 layer depletion–induced enhanced UV radiation exposure.

There are various factors that play a significant role in controlling the penetration of UV radiation in aquatic ecosystems such as (1) incident irradiance of UV radiation which is further controlled by O_3 layer, cloud cover, and aerosols. For instance, O_3 layer depletion and climate change–driven long droughts induce forest fires, and the smoke of fire is well known to filter out the UV radiation potentially affecting the aquatic biodiversity (Bais et al., 2019; Sulzberger et al., 2019; Wilson et al., 2019). (2) Snow and ice cover are known to prevent most of the arriving UV radiation from entering the water column, but climate change–driven thinning and melting of ice sheets is a widely recognized impact on polar regions (Kirtman et al., 2013; Perin and Lean, 2004). The enhanced UV radiation exposure could be detrimental to important marine zooplanktons, which ultimately increases the emission of CO_2 by decomposing dissolved organic matter (DOM) to the atmosphere (Sulzberger et al., 2019; Cory et al., 2015) hampering the carbon cycle. (3) Transparency of surface water is regulated by the DOM, which decreases the transparency to UV radiation (Williamson et al., 2019; Monteith et al., 2007).

For any ecosystem, primary producer and primary productivity are very important. In aquatic ecosystem, phytoplankton and macroalgae are primary producers and are inevitably dependent on solar light including UV radiation, which could be detrimental to them (Gao and Häder, 2017). These primary producers are the potential sink for CO_2 (which is a GHG), and its uptakes CO_2 from the anthropogenic emissions. However, the exposure to enhanced UV radiation adversely affects the diversity of primary producers, resulting into the direct consequences for the global biogeochemical cycle of carbon and the primary productivity of the water bodies. Primary productivity is an important factor of the global carbon budget, and a significant decline in the population of the primary producers adds to the alarming issue of global warming (Williamson et al., 2019).

UV radiation along with climate change reduces the zooplankton population in various regions. Zooplankton acts as a primary consumer in trophic level, and it also plays an important role in transferring the food from primary producer (algae) to the consumers (fishes) (Häder et al., 2015). It sustainably controls the algal and fish population as well as water quality. But enhanced UV radiation exposure altered the abundance, behavior, reproduction, and mortality rate of the zooplanktons, ultimately disturbing the natural aquatic biodiversity, e.g., Red Sea displayed high sensitivity of several zooplankton species under enhanced natural UV radiation (Al-Aidaroos et al., 2015).

Various studies reported the role of the Southern Ocean (oceanic and atmospheric circulation) in regulating the climate-related activities. Williamson et al. (2019) also suggest that the Southern Ocean also plays a very important role in the global carbon budget by absorbing CO_2 from the atmosphere and sequestering it to the deep ocean. However, positive SAM reduces the atmospheric uptake of CO_2 and also reduces its sequestration (Bais et al., 2019; Bornman et al., 2019). Hence, O_3 depletion, enhanced UV radiation, and associated climate changes hinder the overall oceanic productivity, gases uptake, and aquatic biodiversity, consequently posing substantial alterations in sea birds and seals population too.

Stratospheric O_3 depletion–driven dynamics, UV radiation, and associated climate change control the exposure to UV radiation and have both positive and negative effects on underwater aquatic ecosystems. The recent reports of stratospheric O_3 layer recovery suggested that prevailing climate changes and other environmental factors will soon be the driving factors of controlling the UV radiation exposure in the aquatic ecosystem.

The mixed effects of stratospheric O_3 depletion, UV radiation, and climate change necessitate continued observance to understand and monitor the consequences, imposed on ecosystems and associated human, food, and water security.

3.5 Climate change–induced alteration in UV radiation exposure of organisms

Stratospheric O_3 layer depletion–driven climate change also employs stronger pressure on the changes in UV radiation penetration on Earth's surface (Bornman et al., 2015). However, as a result of changing cloud cover, vegetative cover, shifting in geographical ranges of the species, phenology, growth, and reproduction timing change and migration of species, climate change is exerting stronger pressure on the UV radiation received by the organisms. The natural flora and fauna ranges are encroaching and shifting to higher latitudes as a result of climate changes (Bornman et al., 2019; Pecl et al., 2017; Pachauri et al., 2014). However, their rate of migration varies from species to species, e.g., herbs and forbs generally shift more rapidly than woody species (trees and shrubs); similarly, nonnative species are also shifting with faster rates than native species. These geographical range shifts will possibly expose the organisms to combination of other stresses too. Therefore, the knowledge regarding the rate of acclimation of organisms to new environments could be beneficial in the conservation of biodiversity and ecosystem.

4. Global warming

Global warming is the first and foremost direct effect of global climate change and is the result of enhanced greenhouse effect that occurs due to uncontrolled emission of GHGs. The rate of annual increase in temperature per decade has been 0.07°C since 1880, and since 1981, it has become 0.18°C (Lindsey and Dahlman, 2020). Global warming causes a significant effect on terrestrial as well as aquatic ecosystems (Scheffers et al., 2016). Due to temperature extreme, organisms are adopting tolerance to warmer temperature. Sex ratio is also shifted due to high temperature, and metabolic costs are increased in both terrestrial and aquatic ecosystem–belonging organisms (Scheffers et al., 2016). Smaller body size is reported in many species due to warming, as large surface-to-volume ratio helps in maintaining metabolism (Sheridan and Bickford, 2011).

4.1 Impact of global warming on terrestrial ecosystems

Terrestrial ecosystems that include forests, grasslands, agricultural fields, and so on are suffered due to warming of the environment. It directly affects the productivity and biomass production either positively or negatively. Phenological changes such as budding and flowering were preponed or delayed due to climatic variations (Scheffers et al., 2016). Range expansion and range contraction have also been observed, respectively, in warm-adopted and cold-adopted species (Scheffers et al., 2016). Mendelsohn et al. (2016) reported shift in biomes due to warming. Tropical forests are shifting toward temperate forests, and temperate forests are shifting toward boreal forests.

Global warming has reported to increase the extent of biological invasion. Invasion is menace to the native communities due to predation and competition. Growth and distribution of invasive species are reported to increase in high temperature as it causes better photosynthetic assimilation, biomass accumulation, and thus high reproductive efficiency (Robakowski et al., 2018). Altitudinal and latitudinal variation plays a significant role in changing the composition and distribution of species (Wu et al., 2017; Corell, 2006; Stohlgren et al., 2005). Warm temperature increases the competitive behavior of invasive species for nutrient and space, and so the native species are bound to decline in diversity (Lu et al., 2013).

The effect of global warming on plants and their pollinator interaction is of concern and is very important from the perspective of biodiversity. Gérard et al. (2020) reported that the effect is due to spatial and temporal changes in their occurrence, morphology, and behavior (Scheffers et al., 2016).Giannini et al. (2020) reported that climate change resulted in reduction in population size of bees and their habitat in Eastern Amazon, thus affecting the productivity of crops.

Wild fires and heat waves are also the consequences of global warming (Kumar and Pooja, 2020). Australian bush fire in 2020 was the adverse effect of climate change that costs lives of thousands. Fire expansion was due to low rainfall and high temperature. Along with the destruction of habitat, lives, and homes of many, it caused increase in the concentration of particulate matters, causing health emergency to nearby populations (Yu et al., 2020). Nolan et al. (2020) reported that drought acted as a catalyst for fire as the dried Eucalyptus and other dried biomass were acting as fuel, and also the weather condition was favorable to spread the fire. Rogers et al. (2020) also considered the role of climate change in fire occurrence and change in fire regimes. Fuel availability is more during drought for forest fires (Nolan et al., 2020; Rogers et al., 2020; Yu et al., 2020).

Spread of pathogens is more common under changing climatic conditions (US EPA, 2017). Along with the geographical location, physiology and behavior of pests get affected due to global warming. Changes in distribution and vigor of pathogens and emergence of new pathogenic strains are the warnings of climate change that is directly affecting the humans and other animals and also the plant system (Scheffers et al., 2016). Extinction rate of species is increased due to adverse change in climatic variability such as ringed seals, polar bears, salmon, and so on (US EPA, 2017; Kumar and Pooja, 2020; US EPA, 2017; Corell, 2006). Adoption and survival of the organisms to warm environment is attributed to their genetic composition, inherited adoptive mechanism, and phenotypic plasticity (Ross and Adam, 2013). The organisms endemic to a particular environment are also facing risk of extinction and, being endangered such as grizzly bear, bald eagle, and moose, are endemic to boreal forests, and due to warming and shift in biome, these organisms will be threatened (Mendelsohn et al., 2016).

All these correspond to low yield and overall productivity, reduced genetic diversity, increased disease occurrence and intensity, reduced timber and other ecosystem services, poor livestock management, etc. This is greatly affecting humans and wellbeing directly or indirectly. Agriculture, forest, grass ecosystem, and so on get affected with global warming and increased concentration of GHGs. Along with these, extreme weather variables are also the adverse effects of climate change that makes the whole ecosystem to suffer with, such as heavy precipitation, drought, extreme heat, extreme cold, and so on (Kumar and Pooja, 2020).

4.2 Impact of global warming on aquatic ecosystems

Primary productivity of lakes is reported to be reduced due to high temperature (Hassan et al., 2020; Scheffers et al., 2016). Jeppesen et al. (2015) and Hassan et al. (2020) reported that global warming results in disturbance not only in nutrient cycle but also in quality of water as well as water level. The magnitude of effect depends upon the local climate variability (Jeppesen et al., 2015). Eutrophication is also a common adverse effect. Hassan et al. (2020) showed direct effect of climate change on physical and biological entities of lakes, rivers, and wetlands, including the organism's intrusion, seasonal variation, eutrophication, and so on. Mangroves are also sensitive to high temperature (Ross and Adam, 2013).

Ocean warming and ocean acidification are emerged as problems due to global warming (Chang et al., 2020; Kumar and Pooja, 2020; Nagelkerken and Connell, 2015) and affect the productivity negatively, which will in turn lead to less CO_2 absorption and thus increase in GHGs, further adding to global warming. Nagelkerken and Connell (2015) reported that ocean acidification led to shift in community structure of marine species and ecosystem processes and stability (Chang et al., 2020). Ecosystem structure appears to be influenced by both ocean warming as well as acidification (Nagelkerken and Connell, 2015). They have also reported the decline in diversity and abundance of many tropical and temperate species due to acidification of ocean. Albouy et al. (2020) also showed the risk of extinction of many marine species due to loss of habitation. Wetlands are also affected by warming and acidification of ocean (Ross and Adam, 2013). Fertilization, development of larva, and growth of microfauna such as mollusks, arthropods, and vertebrates are affected by warming and acidification (Ross and Adam, 2013). Ocean temperature plays a key role in diversity, distribution, and abundance of marine organisms. The whole process is complex and interrelated to each other; imbalance in one structure causes disturbance in entire system related to it in food web.

Increase of coral disease, dissolution of shells and corals, and higher mortality are observed as critical impacts of climate change (Scheffers et al., 2016). Spawning of fishes in both fresh and marine systems and migration of animals were affected due to climate change (Scheffers et al., 2016).

4.3 Sea level rise and glacier melting

NASA (2019) has reported 8 inches rise in sea level since the past recordings due to climate change, and it will reach to 1–4 feet by the year 2100. From the past two decades, the rate of increase in sea level is increased and is reached to 3.2 millimeters/year (Kumar and Pooja, 2020). This increase in sea level is directly associated with melting of glaciers and thermal expansion of water. This is direct effect of global warming due to climate change. Glacier covers 10% of the Earth surface and are melting with a rate that it will vanish within decade (Cauvy-Fraunié and Dangles, 2019).

State of the climate report (2018) showed that there was loss of 951 mm of water, equal to 3.1 feet, from 142 glaciers, in 2017 (Lindsey, 2020). In consequence, sea water intrusion will result in irreversible damages such as loss of fertile agricultural lands, habitation, infrastructures, tourisms, and so on. Corell (2006) has listed the impact of climate change on arctic glacier that is melting due to warming.

As reported by US EPA (2017), the whole food web gets disturbed due to climatic variation. An example cited by US EPA (2017) showed that polar bears are suffering with habitat loss not only due to ice sheets melting but also indirectly due to scarcity of food. As the duration and extent of ice are reducing, the phytoplankton population are getting affected that in turn affects the zooplanktons, which primarily feed on them. These zooplanktons are fed by arctic cod, and they, in turn, are fed by seals. Seals are the major food source for polar bears, and thus, they are deprived of food and many more organisms involved in the food web. The relationship between prey and predator is affected due to warming, thus also leading to disturbance in food web (Albouy et al., 2020).

Sea level rise causes salinization of coastal water bodies, and due to salinity level, phytoplankton population is get stressed, and thus the food web gets disturbed (Jeppesen et al., 2015; Kumar and Pooja, 2020); as phytoplanktons being the major source of primary productivity, they control the major ecosystem function and structure. Salinization affects not only the aquatic organisms but also the surrounding area, harming the ecological and socioeconomical structure. Due to warming, rate of surface evaporation will increase, and the land surface will be left with salinity causing excess ions. Ross and Adam (2013) reported complete submergence of intertidal wetlands. Risk of extinction of species is always high due to melting of glaciers (Cauvy-Fraunié and Dangles, 2019).

Glacier melting has resulted into major flood with huge societal impacts. Kedarnath flood in 2013 in India costs many lives as well as huge loss of socioeconomics (Carrivick and Tweed, 2016). Storm surges and high tides frequency are also increased (Kumar and Pooja, 2020). Slemmons et al. (2013) reported melting of glaciers adversely affects alpine aquatic systems by affecting habitat that is due to change in physical processes such as hydrology, temperature and turbidity, chemical processes such as nutrient cycling and pollution, and ecosystem structure and function. They have also found that streams are to be affected more rather than lakes, and it will disturb the abundance and distribution of aquatic micro- and macrofauna. Also, the nutrient availability is affected negatively, and so the whole food chain is disrupted (Slemmons et al., 2013). All these caused a huge loss of biodiversity. Similarly, Quenta et al. (2016) reported the melting of glaciers has an adverse effect on biodiversity of the high Andean peatlands. Cauvy-Fraunié and Dangles (2019) reported that some of the species will be benefitted by glacier retreat, while some will face major risks of extinction, depending upon their special traits. Body size and trophic position are important traits to determine the sensitivity of the organism toward the climate change (Cauvy-Fraunié and Dangles, 2019).

5. Effects of climate change on nutrient pollution

Nutrient pollution is another challenging environmental problem in the world, which occurs because of excessive nutrient loading mainly nitrogen (N) and phosphorus (P) into the aquatic ecosystem from diverse human activities, causing eutrophication (Shortle and Horan, 2017; Ouyang et al., 2010). N and P are important nutrients for various organisms, living in soil and water ecosystem. However, excess of nutrients added into agricultural soil and water bodies due to anthropogenic activities such as use of excess inorganic fertilizers in agricultural fields, discharge of wastewater into waterways, as well as natural activities such as atmospheric deposition, groundwater discharge, surface loading, and storm water runoff which contains enough N and P, are the principal causes of nutrient pollution (Ouyang, 2012). Climate change stimulates the increasing nutrient loads in the aquatic ecosystem, which helps in eutrophication and gives a favorable environment for the enormous growth of unwanted algae causing algal blooms (O'neil et al., 2012). Large emission of nutrients by anthropogenic activities especially N to air can cause acid rains, which again increase the concentration of N in estuaries, streams, lakes, forests, and grasslands, thus altering their natural properties (Lintern et al., 2018; Van der Weerden and Jarvis, 1997). These concerns have encouraged more studies to assess the possible effects of future climate change on nitrogen emissions and leaching from ecosystems (Gu and Riley, 2010).

Nutrient pollution has enough potential to affect many streams, for instance, coastal waters, bays, rivers, and lakes for the past decades, causing serious environmental and human health issues (Manuel, 2014). Even though nutrient-enriched water bodies promote the growth of harmful algal blooms (HABs) in aquatic and marine ecosystem, it has many direct and indirect negative impacts on other aquatic organisms. In direct effects, nutrient pollution causes impervious surface and turbidity of water bodies, which decrease light availability and dissolve O_2 to submerged vegetation resulting in shading and death of plants (Manuel, 2014). The rapid growth of HABs in nutrients-rich water bodies causes the indirect effect of nutrient pollution on aquatic organisms and ecosystem. These HABs produce bioaccumulative toxins (toxins are accumulated at each trophic level in a food chain), which have been detrimental to the shellfish, small fish, and other organisms (Ferrao-Filho and Kozlowsky-Suzuki, 2011). Furthermore, these are consumed by large fishes and other animals, and thus the toxins come into the food chain. Therefore, the larger impact of HABs can be seen in many big animals such as dolphins, manatees, turtles, birds, and seldom in humans (Berdalet et al., 2016). Sometimes, HABs are not toxic, but still they are deleterious for aquatic and marine life by clogging fish gills and blocking the sunlight (Berdalet et al., 2018). Nutrient pollution can also create hypoxic condition (an area in a water body with low or depleted oxygen) in the littoral zone, which is a threat for aquatic ecosystem functions and services (Jenny et al., 2016). The decomposition process of overgrown HABs consumes most of the dissolved oxygen, which is a major reason for hypoxia in marine and aquatic ecosystem (Manuel, 2014). The primary cause of onset of hypoxia can be climatic (natural hypoxia) or a consequence of human activities. Climate change has the potential to create hypoxic conditions in coastal and marine ecosystems. The anthropogenic P released by urban and cultivated areas plays a major role in triggering hypoxia. However, recent climate change by fluctuations in precipitation and increase of temperatures are not sufficient to trigger hypoxia (Jenny et al., 2016). For example, Europe's perialpine lakes were oxygenated over the last epoch despite very high climatic variability but moved to hypoxic conditions after a minor increase in P beyond a threshold value 8–10 μg P/L (Jenny et al., 2014).

Wang et al. (2018) have reported that total P and nitrate-N content in watershed area of Great Lake have reduced 28%–89% and 17%–28%, respectively, owing to increase in precipitation and stream flow by 4%–21% and 10%–54%, respectively. They have also reported that continuous increases in total P loss at the watershed area are because of changes in extreme storm events, air temperature, and precipitation. Elevated air temperature increases the decomposition rate, nitrification processes, and mineralization, resulting in increase of microbial activities (Verma et al., 2015) and thereafter in increase in solubility of P and production of nitrates into the soil ecosystem. Hence, the surface runoff contains high amount of P and some nitrates, which increase annual nutrient load into waterway (Jennings et al., 2009). The climate change has larger impacts on excess of sediment and nutrient loading into water bodies through the warming climate and higher precipitation frequency than the land-use change (Gu and Riley, 2010). Therefore, it is expected that increase of sediments and magnitude of nutrient delivery is expected under future climate change scenarios (Cousino et al., 2015).

➢ **Possible management strategies**:

Increased sediments and nutrients load can be reduced by implementing some management strategies, but it cannot be fully recovered due to the impact of climate change (Crossman et al., 2013).We should adopt some more appropriate management practices at individual levels for healthier and safer environments. The management plans should be focused on reducing total P loading in wetter environments. Phosphates in soaps and detergents should be banned. Wastewater treatment plants should be improved and increased in number. Point sources should be managed to reduce P loading. Runoff of sediments and nutrients from agricultural lands (nonpoint sources) should be reduced, and a proper remedial action plan should be developed (Cousino et al., 2015).

6. Effects of climate change on thermal pollution

The change in ambient temperature of any natural water bodies by cause of human influence is called thermal pollution. Increase of temperature is the most occurring event in thermal pollution (Kalinowska, 2019). Since the impact of extreme temperature rises is more obvious, relatively small changes in ambient temperature can also be biologically effective. The communities can be altered by a little increase (1–2°C) in temperature of the water bodies as it is lethal for some sensitive species and also can affect the growth and reproduction of other aquatic and amphibious organisms (Dodds and Whiles, 2020). The main reasons of thermal pollution are industries and power plants, withdrawing water (a powerful sink for heat waste) from streams for cooling their generators and machines, and then releasing back into the water bodies, which elevates the temperature of freshwater habitats (Miara et al., 2018). However, among these sources, nuclear power plants are the greatest point source of thermal pollution, which requires 30%–100% additional cool water than the other power plants (Kirillin et al., 2013; EPRI, 2008). Besides the industrial sources, soil erosion, urban runoff, and removing riparian vegetation and forest are also major causes for rising thermal pollution in the water bodies (Herb et al., 2008; Beschta et al., 1987).

During the warm periods, the surface water receives more heat due to decreased water velocity and higher solar penetration in seized water (Dodds and Whiles, 2020). However, artificial cooling of freshwater by releasing the hypolimnetic cold water from large reservoirs to the warm stream has also showed negative impacts on existing aquatic life (Kirillin et al., 2013). The added heat to the streams can adversely affect and decrease the integrity of the ecosystem. The warm temperature diminishes the solubility of O_2 in water and accelerates the rate of decomposition of organic materials, which also consumes oxygen during the process. Therefore, a slight increase in ambient temperature of water may lead to reduce the dissolved oxygen (DO), which could be detrimental for aquatic life (Blumberg and Di Toro, 1990). If warming of water continues, it will create anaerobic conditions. In addition, the surface water algae also grow massively under hot water condition and are likely to decline the dissolved oxygen level of the water. The major impacts on aquatic ecosystems attributable to thermal pollution are (1) loss of biodiversity by massive death of aquatic plants, insects, fish, and amphibians as a consequence of thermal shock, (2) shifting of organisms to a suitable environment due to slight deviation in temperature of water ecosystem, (3) bleaching of corals, as it is very sensitive to temperature, because of increasing water temperature by runoffs and effluents, and (4) accelerating the metabolic rate by increasing enzyme activities of organisms under increased water temperature (Vallero, 2019).

The regional climate warming may enhance the adverse effects of thermal pollution on the environment. The constant warming signal as a result of climate change has now been reported in several lakes (Dokulil et al., 2010) with effects on ecosystems, which were mainly caused by thermal pollution (Parker et al., 1970). The rising of global temperature is certainly real and provides a clear picture of warming world (IPCC, 2001). The temperature was increased between 0.4 and 0.8°C from the past 15 years (EPA, 2001). Therefore, climate change is considered as a nonpoint source of thermal pollution, which affects freshwater habitats extensively. The long-term climate change impacts on aquatic ecosystem are not evident; however, some predictions showed significant shifts in aquatic structure and function mainly in the higher latitudes (White and Visser, 2016). The combined study on global warming and thermal pollution may give an understanding that how aquatic organisms deal with higher temperatures. For instance, the temperature of Lake Victoria has been increased more than 1°C since the 1970s because of climatic change and thermal pollution problem. Another case of lake warming due to nuclear power plant is Stechlin Lake in Germany, a well-studied example of increasing thermal pollution (Koschel and Adams, 2003).

➢ **Preventive measures**

Adaptive measures should be taken for the mitigation of thermal pollution effects. Thermal pollution could be controlled by employing of few efficient scientific techniques such as construction of artificial lakes, cooling ponds, and cooling towers.

(1) Artificial lake: These are human-made water bodies, which gives a potential alternative for cooling of power plants. The discharging of warmed effluents at one end of the lake and withdrawing of cool water from another end of the lake is helpful for water cooling purposes. The heat of artificial lake is finally dissipated through evaporation.
(2) Cooling ponds: This is the simplest and low-cost method for regulating thermal discharges. The maximum dissipation of heat through cooling ponds to the atmosphere can minimize the water area and volume. However, the technique is not so efficient in respect of air–water contact.
(3) Cooling towers: In this technique, a condenser is used for cooling of heated effluents at the time of returning to the water bodies. This is the most efficient technique, and the towers are well designed to control the water temperature. However, the frequent change in local meteorological conditions is a major problem as huge amounts of water vapor come in the atmosphere.

7. Increased risks of climate-related disasters

At present, a global consensus prevails among the scientific communities that anthropogenic activities are the primary cause of climate change around the world (Cook et al., 2013). Climate change has accelerated the risks of occurrences of weather and climate-related events and disasters across the globe, resulting in consequent human casualties and impairing the adaptable ability of the fragile ecosystems and societies (Munang et al., 2013). The economic losses incurred by the climate-related disasters are specifically below the spotlights in several scientific discussion and debates. The number of disasters associated with the natural hazards has continued to grow, projecting a considerable toll on loss of human lives, capitals, and ultimately the economic status. As per the United Nation Office for Disaster Risk Reduction (UNISDR): Global Assessment Report (2009) over 2.2 million people worldwide have led to death due to natural hazard triggered disasters, concomitant with financial damage amounting to $1527.6 billion between 1975 and 2008. Similarly, another report by Bhandari (2015) documented that over the past 10 years (2005–14), human death toll around 403,000 was associated with resultant economic losses of USD 436 billion in Asian countries due to climate-related disaster events. Furthermore, a recent report by National Oceanic and Atmospheric Administration (NOAA, 2019a) documented the occurrences of around 10 weather and climatic extreme events by October 2019, which predicted estimated economic losses over $ 1 billion. Continuous change in the global climatic pattern and the anticipated rise in the rate and severity of extreme climatic occurrences will further augment the risk associated with hurricanes, drought events, flooding, avalanches, and landslides (Intergovernmental Panel on Climate Change/IPCC, 2007), thereby exposing more than a million species of flora and fauna vulnerable to extinction at global level. The report of First (2019) suggested that abatement of emission of GHGs by half within 2030 with an aim of reaching net zero emissions by the middle of this century could help in averting some of the most cataclysmic impacts of global change in climate.

7.1 Climate change aggravates the degradation of ecosystem

Continuous accumulation of stress by the ecological systems due to compulsive jeopardizing of the ecological processes under climate change or other associated forced disturbances (population explosion, pollution, alteration in land-use patterns, misuse of natural resources, etc.) has weakened the resilience of various ecosystems worldwide (Pachauri and Reisinger, 2007). Such disruption in endurance capacity of ecosystem has resulted in unprecedented shift in ecological regimes causing production of some undesirable stable ecosystems. Boreal forest, Mediterranean-type ecosystems, Tundra, coral reefs, sea ice biomes, mangroves, and salt marshes have been listed as most susceptible ecosystems to global climate change (Munang et al., 2013). With reference to context of ecosystem degradation, here we summarize some natural disasters influenced under climate change and their associated risks on the ecosystem functioning.

(a) Extreme waves of heat

The rise in concentration of heat trapping gases in the atmosphere has caused a tremendous elevation of Earth's temperatures with expectation of longer, frequent, and severe prevalence of heat waves. Spiking in temperature may result in mass extinction of key natural engineers of ecosystems affecting both tropical and temperate oceanic regimes. Scientists observed death of several Alaskan salmon species owing to the unprecedented heat waves in 2019 (Berwyn, 2019). As per the report of NOAA's *National State of the Climate* (2019b), the state-wide temperature in Alaska was recorded to be 7.9°F above their normal mean temperature in 2019. The extreme heat created an anoxic condition in the water, causing the fishes to die off under suffocation. Intense heat waves also caused death of macroalgae in marine ecosystems of temperate (Wernberg et al., 2011) and also resulted in destruction of coral reef via bleaching in the tropical system (Baker et al., 2008)

(b) Cyclonic storms and hurricanes

The hurricanes and cyclonic storms are generally energized from the warm water, and climate change is an important factor causing temperatures of ocean to elevate. Rare coastal species are particularly under major threat due to hazardous storms, habitat destruction, and/or alteration in the pattern of food chain. As per the survey report of US Fish and Wildlife Service (USFWS), around 14%–22% of the population of Key deer, which is the smallest deer found in North America, was found dead due to Hurricane Irma that occurred in 2017 (Nobel, 2017). Another disastrous storm named Cyclone Fani hit the coast of Odisha and Tamil Nadu, India, in 2019. The study of

Bharathidasan et al. (2020) reported the mass mortality of painted storks, an aquatic bird commonly found in Indian subcontinent, and listed as "near-threatened" species. Another report by Debata (2019) documented that the severe cyclone resulted in the relative declination of population of sandbar-nesting birds and survival of the chicks, commonly found along Mahanadi River, Odisha, India. Most recently, a supercyclonic storm named Amphan hit West Bengal, Odisha, and Bangladesh and caused widespread destruction in May 2020.

(c) Megafires

Another anticipated impact of global climate change is the rise of incidences of forest fires, resulting in a larger number of record-breaking seasons of fires (Schoennagel et al., 2017). Over the past decades, nearly half of the major wildfires in California have been recorded (Whitcomb, 2018). Vercammen et al. (2018) documented that the Carr Fire, which occurred in Redding, California, exhausted an area >100,000 football grounds. The fire was so massive that a typical weather system was created, causing "fire tornado." The Australian bushfires, which occurred in 2019−20, caused a significant toll of death of both flora and fauna. An ecologist named Chris Dickman reported an estimate that over a billion animals have been killed in such incident, including 8 lakhs in the region of New South Wales (Center for Disaster Philanthropy, 2019). The research of Dickman for the World Wide Fund for Nature estimated that in each hectare around 17.5 mammals, 129.5 reptiles, and 20.7 bird species existed. Moreover, over one-third of the population of koala has been killed in this disaster (Center for Disaster Philanthropy, 2019).

(d) Drought

The frequency and intensity of drought incidences are likely to increase, which exacerbate the vulnerability of several ecosystems (Lucci, 2019). Over the past centuries in Hawaiian Island, the frequency of rainfall has diminished while incidences of wildfires increase have been observed (Helweg and Giardina, 2017). The prevailing dryness associated with drought has increased the likelihood of forest fires, killing native and endemic plant species and causing severe soil erosion and sediment deposition in the near streamlines areas. Drought incidences are usually associated with events like El Niño (Mera et al., 2018), resulting in wetter summer followed by dry winters, inducing the occurrences of such fires by increasing the fuel load. The forest is being damaged by global warming and drought due to enhanced tree mortality and consequent acceleration of invasion of grass species, which simultaneously can degrade the quality cover of forest habitats (Helweg and Giardina, 2017).

7.2 Proper ecosystem management is essential to reduce the risks of weather events

Management of ecosystem can enhance the resilience of ecological systems and communities to the impact of climate-related disasters. Several research studies conducted around the world listed out the potential advantages of ecosystems in retorting the disaster risks (Klein et al., 2019; McVittie et al., 2018; Renaud et al., 2016). It is believed that reduction of disaster risks and adaptation to climate change can be achieved via ecosystem-based approaches. Macroecosystems, such as forests, coastal systems, and wetlands, can act as natural barriers or buffers, thereby safeguarding against the direct exposure to natural hazards and their associated impacts (Barbier, 2006). Restoration of mangroves has been performed in various coastal shores as measures to safeguard against sea level rise and storm surges and can nullify the effect of 70%–90% of energy generated due to strong winds (United Nation Development Program (UNDP), 2004). Restoration of mountain bushes, forests, and meadows could provide protection against the hazards such as flash floods and landslides or due outburst of glacial lakes under accelerated melting of mountain snows or ices (Munang et al., 2013). Furthermore, preservation of coral reefs plays a crucial role in reducing offshore water breaks during the period of tropical storms and sea surges. Such well-managed ecological systems can provide protection naturally against common hazards such as drought, flooding, landslides, avalanches, wildfires storm surges, and so on. Therefore, conservation of such ecosystems plays a pivotal role for sustenance of various ecosystem services and significant for human welfare.

8. Crisis of natural resources

The world is currently witnessing a broadly recognized and prolonged pattern of multiple crises due to global climate change and exponential burst of human population, which is epitomized through uneconomical and unsustainable grabbing of natural resources. Such resulting convolution has afflicted the ecosystem at a transnational level: on one front, the battle of losing important and threatened species along with invasion of exotic species to outcompete the natives, and on the other, the struggle of maintaining environmental sustainability besides the ever-increasing imposing demand on the available natural resources such as land, forests and grasslands, water, fisheries and other marine resources, renewable and nonrenewable energy resources, mineral resources, etc.

8.1 Land degradation

The connection between utilization of land and the climate is complicated. Firstly, land use practices largely shaped the land covers, which have an enormous influence on the concentration of GHGs in the atmosphere. Secondly, although land use patterns influence the global climate change, a sudden change in climatic pattern can also alter the pattern of land use and vegetation cover on a transnational level. For example, natural hazards triggered due to climate change (heavy rainfall, flash flood, drought, or wildfires) can exacerbate the degradation of land, influencing the pattern of land management (Assessment ME, 2005). Such processes may be persuaded by sociopolitical and economic considerations fostering the land use constraints while failing to maintain equilibrium between the societal supply and demand on a sustainable basis (D'Odorico et al., 2013).

Structural alteration in canopy cover of tropical forest and its consequent reduction of biomass production (Miettinen et al., 2014), enhancement of salinization of irrigated lands (Qadir et al., 2014), loss of soil nutrients via erosion (Quinton et al., 2010), and deterioration of soil quality are some of the apparent examples manifesting the land degradation process. Soil quality is the indicator of soil health (UNEP-IRP, 2016). Managing soil health is usually confederated to increased ecological resilience against hazards of climate change. Therefore, alteration in vegetation pattern owing to land degradation may emerge concomitantly with deterioration of soil quality (Bestelmeyer et al., 2015) and may negatively affect the agricultural production processes in responses to the changing climate. Such ecologically and socially sensitive system may impose tremendous pressure on different ecological processes, resulting in utter collapse of ecosystem under induced climate change and anthropogenic stress factors (Marshall et al., 2014).

8.2 Water crisis

Another challenge imposed on the humankind and ecological system under global climate change is the shortage of water or the exceedance of water demand outstripping the supply for sustenance of life. It is evident that climate change is significantly altering the water cycle (Kaushal et al., 2017), resulting in a serious global water crisis scenario, which is likely to be aggravated in the absence of sustainable water management strategies (Wang et al., 2016). Temperature rise has intensified the process of evapotranspiration from ocean, forest and vegetation, land, and different surface water bodies (Middelkoop et al., 2001). Warm climate typically encourages formation of low pressure belt, triggering cyclonic events mostly in the tropical regions and higher precipitations, giving rise to more flooding incidences (Wang et al., 2015; Chen and Xu, 2005). However, occurrences of such incidences are not uniform across the world and may vary on spatial and temporal scales. Therefore, cumulative impact of climate change (temperature rise and variability in precipitation pattern) and population explosion may alter regional to global water supply and demand ratio, subsequently escalating the global scarcity of water supply.

8.3 Loss of biodiversity

The cumulative effect of climate change and anthropogenic interference-inducing habitat fragmentation seems to be a serious concern regulating the fate of global biodiversity. Evidences from the past literature displayed that the multiple elements of climate change have impacted biodiversity at each level: from genetic components over species to the level of biomes (Pires et al., 2018). The overmelting of ice from the polar regions such as Arctic and Antarctic has grabbed much attention of the climate scientists across the world. Such melting has caused a serious impairment in the food chain due to loss of lives surviving under or above the ice. Furthermore, reduction of the sea ice causes temperature variation of seawater and salinity, resulting in alteration of primary productivity and composition of species, affecting the diversity of polar regions. Similarly, losses of native flora and fauna under incidences of wildfires, bleaching of corals reefs in masses due to ocean acidification, invasion of exotic species, and outcompeting the natives all carry the fingerprints of global climate change. The Fifth Assessment Report (AR5) of Pachauri et al. (2014) documented the threat of biodiversity in response to climate change and projected the loss and mass extinction of keystone species of ecological systems if global mean temperature is likely to increase by 1.4°C.

8.4 Marine resources

Tropical oceanic ecosystems such as coral reefs are one of most affected systems due to impacts of climate change (Hoegh-Guldberg and Bruno, 2010), as most of the oceanic species in tropics usually pose limited thermal tolerance range and mostly survive near their upper limit of thermal range (Pörtner and Farrell, 2008). Therefore, the primary organisms dwelling the coral reefs, i.e., scleractinian corals, functioning almost near adjacent to their upper limit of thermal tolerance are exposed to extensive mortality and habitat destruction with a slight increment (1°C) in temperature (Jokiel and Coles, 1990). Mass bleaching of these corals is the most remarkable impact due to increase of temperatures unbalancing the physiological tolerances of a large-scale coral species (Hughes et al., 2003). Furthermore, population reduction of symbiotic zooxanthellae as well as their photosynthetic pigments under thermal stress is causing detrimental impact on the nutrition source for maximum coral species (Hoey et al., 2016). Some recent events of severe and large-scale coral bleaching have evoked considerable coral mortality (Sheppard, 2003), spawning a deep concern regarding the fate of one of the most productive ecosystems of Earth (Cantin et al., 2010).

Besides thermal stress, another cumulative consequence of anthropogenic activities and climate change is on the chemistry of seawater. Altering such chemistry especially associated with the pH reduction along with reduction in availability of carbonate ions is directly pertaining to the anthropogenic increment of atmospheric CO_2 and its consequent increase in ocean (Kleypas et al., 1999), clearly marking the effect of ocean acidification. Experimental studies displayed that an alteration in seawater chemistry will jeopardize the rate of calcification in reef-forming coral species (Comeau et al., 2013, 2014).

Fisheries and aquaculture are one of crucial natural resources, contributing significantly to food security and a way to earn livelihood to millions, mainly to those inhabiting the riverine, coastal, inland, and insular parts of the world. The enhancement in the capacity of fishing vehicles for capturing fishes is one of the major threats to the survival of marine capture fisheries. The population of marine fishes often exposed to human-induced overexploitation, incidental killing of nontargeted species, destruction of habitat, evolutionary demographic shift, increase marine pollution, and alteration in ecosystem structure and functions (Pikitch et al., 2004).

9. National and international meets/conventions on climate change impact and mitigation efforts

The WMO with United Nations Environment Programme (UNEP) set Intergovernmental Panel on Climate Change (IPCC) in 1988 to assess the problems related to climate change and identify solution through research and development programs (IPCC, 2020). The United Nation Framework Convention on Climate Change (UNFCCC) is an intergovernmental agreement addressing the climate change issues around the world, came into effect in 1994. United Nations conferences are one of the largest conferences being organized at the international level. These meetings include conventions related to climate change matters to implement some sustainable goals for future development. Conference of parties (COP) appeared to be the major decision-making sector of UNFCC. The member states of the parties hold meeting every year to discuss and address several challenges underlying the climate change (UN: United Nations). COP meetings are used to occur annually as shown in (Table 6.1) to review the progress and to assess the scheme or actions to reach the new resolutions if needed in future for control the impacts of climate change (IISD RS, 2018).

Stockholm Declaration was the first global environment conference, held in Stockholm, Sweden (5−16 June 1972). Principles and decision had been taken at intergovernmental level on different environmental issues related to pollution, overexploitation of natural resources, and fast developmental programs, which were important from the ecological point of view (Boudes, 2011). For conservation of migratory wild animals, Bonn convention was held in 1979 in Bonn, Germany, with an aim to protect endangered migratory species in their natural migration, which also came to known as Conservation of Migratory Species (JNCC, 2019). World Conservation Strategy (WCS), formed in 1980 by International Union for Conserving Nature and Natural Resources (IUCN), together with UNEP, World Wildlife Fund (WWF), United Nations Educational, Scientific and Cultural Organization (UNESCO), and Food and Agricultural Organization of United Nations (FAO), to preserve biodiversity and maintenance of ecosystem sustainability through conserving natural resources and sustaining ecological processes (IUCN-UNEP-WWF, 1980). With the raising concern of deepening of stratospheric O_3 hole near Antarctic region, conferences for reducing ODSs had been conducted with the aim to reduce ODSs production and consumption. Montreal protocol (1997) came after 2 years of Vienna (Austria) convention (1995) with strong objective of phasing out ODSs till 2030. Although Kyoto Protocol was adopted in 11 December 1987 in Kyoto, Japan, to decrease the emission of GHGs and their sources, which came in force by 2005 with commitments from many industrialized countries (Government of Canada, 2020; UNCC, 2020a; Velders et al., 2007).

Earth summit or Rio summit was one of the biggest conferences related to environment and climate, which was held in Rio de Janeiro, Brazil in 1992. Convention of Biological Diversity (CBD) and Agenda 21 had been major implementations agreed

Table 6.1 COP meeting and UN climate change conferences (CCC) held in different countries to mitigate the climate change problems.

Name of the conferences	Date of conferences	Place of conferences
[a]COP 25/[b]CMP 15/[c]CMA 2	December 2–13, 2019	Madrid, Spain
COP 24/CMP 14/CMA 1-3	December 3–14, 2018	Katowice, Poland
Bonn [d]CCC	April 30–May 10, 2018	Bonn, Germany
Bangkok CCC	September 4–9, 2018	Bangkok, Thailand
Bonn CCC	May 8–18, 2017	Bonn, Germany
COP23/CMP 13/CMA 1-2	November 6–17, 2017	Bonn, Germany
COP22/CMP12/CMA1	November 7–18, 2016	Bab Ighii, Marrakech, Morocco
Bonn CCC	May 16–26, 2016	Bonn, Germany
COP21/CMP 11	November 30–December 11, 2015	Paris, France
Bonn CCC	October 19–23, 2015	Bonn, Germany
Bonn CCC	August 31–September 4, 2015	Bonn, Germany
Geneva CCC	February 8–13	Geneva, Switzerland
Lima CCC	December 1–14	Lima, Peru
Bonn CCC	October 20–25, 2014	Bonn, Germany
Bonn CCC	March 10–14, 2014	Bonn, Germany
COP19/CMP9	November 11–22, 2013	Warsaw, Poland
Bonn CCC	June 3–14, 2013	Bonn, Germany
Bonn CCC	August 29–May 3, 2013	Bonn, Germany
COP 18/CMP 8	November 26–December 7, 2012	Doha, Qatar
Bangkok CCC	August 30–September 5, 2012	Bangkok, Thailand
Bonn CCC	May 14–25, 2012	Bonn, Germany
COP17/CMP 7	November 20–December 9	Durban, South Africa
Bonn CCC	June 6–17, 2011	Bonn, Germany
Bangkok CCC	April 3–18, 2011	Bangkok, Thailand
COP16/CMP 6	November 29–December 10, 2010	Cancun, Mexico

Table 6.1 COP meeting and UN climate change conferences (CCC) held in different countries to mitigate the climate change problems.—cont'd

Name of the conferences	Date of conferences	Place of conferences
Tianjin CCC	October 4–9, 2010	Tianjin, China
Bonn CCC	August 2–6, 2010	Bonn, Germany
Bonn CCC	May 31–June 11, 2010	Bonn, Germany
Bonn CCC	April 9–11, 2010	Bonn, Germany
COP15/CMP 5	December 7–18, 2009	Copenhagen, Denmark
Barcelona CCC	November 2–6, 2009	Barcelona, Spain
Bangkok CCC	April 3–18, 2009	Bangkok, Thailand
Bonn CCC	August 10–14, 2009	Bonn, Germany
Bonn CCC	June 1–12, 2009	Bonn, Germany
Bonn CCC	March 29–April 8, 2009	Bonn, Germany
COP14/CMP 4	December 1–12, 2008	Poznan, Poland
Accra CCC	August 21–27, 2008	Accra, Ghana
Bonn CCC	June 2–13, 2008	Bonn, Germany
Bonn CCC	March 31–April 4, 2008	Bonn, Germany
COP13/CMP 3	December 3–14, 2007	Bali, Indonesia
Vienna CCC	August 27–31, 2007	Vienna, Austria
Bonn CCC	May 7–18, 2007	Bonn, Germany
COP 12/CMP2	November 6–17, 2006	Nairobi, Kenya
Bonn CCC	May 17–26, 2006	Bonn, Germany
COP11/CMP1	November 28–December 9, 2005	Montreal, Canada
Bonn CCC	May 19–27, 2005	Bonn, Germany
COP 10	December 6–17, 2004	Buenos Aires, Argentina
Bonn CCC	June 16–25, 2004	Bonn, Germany
COP 9	December 1–12, 2003	Milan, Italy
Bonn CCC	June 4–13, 2003	Bonn, Germany
COP 8	October 23–November 1, 2002	New Delhi, India
Bonn CCC	June 5–13, 2002	Bonn, Germany
COP 7	October 29–November 9	Marrakech, Morocco

Continued

Table 6.1 COP meeting and UN climate change conferences (CCC) held in different countries to mitigate the climate change problems.—cont'd

Name of the conferences	Date of conferences	Place of conferences
COP 6	November 13–24, 2000	The Hague, Netherlands
Bonn CCC (COP 6-2)	July 16–27, 2001	Bonn, Germany
Lyon CCC	September 11–15, 2000	Lyon, France
Bonn CCC	June 12–16, 2000	Bonn, Germany
COP 5	October 25–November 5, 1999	Bonn, Germany
Bonn CCC	May 31–June 15, 1999	Bonn, Germany
COP 4	November 2–13, 1998	Buenos Aires, Argentina
Bonn CCC	June 2–12, 1998	Bonn, Germany
COP 3	December 1–10, 1997	Kyoto, Japan
Bonn CCC	October 20–31, 1997	Bonn, Germany
Bonn CCC	July 28–August 7, 1997	Bonn, Germany
Bonn CCC	February 25– March 7, 1997	Bonn, Germany
COP 2	July 8–19, 1996	Geneva, Switzerland
COP 1	March 28–Apr 7, 1995	Berlin, Germany

[a]Conference of Parties *(COP);
[b]Conference of Parties serve as meeting of parties to the Kyoto protocol (CMP);
[c]Conference of Parties serve as meeting of parties to the Paris Agreement (CMA).
[d]*CCC*, Climate Change Conference.
Modified from: (UNCC, 2020).

during the Earth Summit to mitigate environmental issues and development through sustainable way. To review and appraise the Agenda 21 and for adopting new sustainable goals and improvement in actions, RIO+5, RIO+10, and RIO+20 summits were also held in New York (1997), Johannesburg (2002), and Rio de Janeiro, Brazil, respectively (Leggett and Carter, 2012; WRM). A conference (COP 21) was held in Paris, France in December 2015 known as Paris Agreement. It was a successful global conference to bring all nations together with an aim to cope with climate crisis with a target to keep global temperature below 1.5-2°C (UNCC, 2020b). Similarly, Global Climate Action Summit was held in 2018, San Francisco (California) having aim to work together against change in climate and reached the climate goals to be successfully implemented at the national to regional and local levels (WWF, 2020). Another meet held in New York, 2019 named United Nation Climate Action Summit, to reduce GHG emissions with target of net zero emission of GHGs by 2050 (Climate Action).

10. Conclusions

This chapter confirms that the capacity of the existing ecosystems to provide essential services for societal benefits is facing severe threat due to population boom. The additional stresses inflicted by anomalies in climatic conditions have altered the framework and functionality of both aquatic and terrestrial ecosystems, which raised the demand for extraordinary measures for adaptation in the present and years ahead. Temperature elevation, global warming, stratospheric O_3 depletion with consequent increase of UV penetration, alteration in seasonality, frequent triggering of high magnitude hazardous events, and subsequent crisis of natural resources, along with increased rates of thermal and nutrient pollution is causing visible change in the ecosystem types. Additionally, the rates of mechanisms underlying the ecological frameworks and association with other ecological systems are negatively affected, often with irreversible changes. Furthermore, loss of system productivity, disaggregation of food chain linkages, disease breakout, and rise in pollutant level accompanied by climatic feedbacks pose critical societal consequences. Such challenges reflect a remarkable threat to life and necessitate the immediate and imperative adaptation plans and mitigation programs by the scientific community to deal with the menace of climate-related anomalies with concomitant anthropic interventions. Although various meets and conventions have been held at international forums with targets to tackle the driving factors of climate change, lack of stringent legislation regarding implementation of such sustainable policies in poor underdeveloped and developing countries has been obstructing the path toward environmental sustainability.

Acknowledgments

The authors are thankful to the Head, Department of Botany and Coordinators, CAS in Botany, and ISLS (DBT) Banaras Hindu University, Varanasi, India, for providing necessary facilities. AG, DSY, and BJ are also grateful to the University Grant Commission, New Delhi, India, and KR is thankful to CAS for extending financial assistance in the form of junior and senior research fellowship.

References

Abram, N.J., Mulvaney, R., Vimeux, F., Phipps, S.J., Turner, J., England, M.H., 2014. Evolution of the southern Annular Mode during the past millennium. Nat. Clim. Change 4 (7), 564–569.

Adedeji, O., 2014. Global climate change. J. Geosci. Environ. Protect. 2 (02), 114.

Al-Aidaroos, A.M., El-Sherbiny, M.M., Satheesh, S., Mantha, G., Agustī, S., Carreja, B., Duarte, C.M., 2015. Strong sensitivity of Red Sea zooplankton to UV-B radiation. Estuar. Coast 38 (3), 846–853.

Albouy, C., Delattre, V., Donati, G., Frölicher, T.L., Albouy-Boyer, S., Rufino, M., Pellissier, L., Mouillot, D., Leprieur, F., 2020. Global vulnerability of marine mammals to global warming. Sci. Rep. 10 (1), 1–12.

Assessment, M.E., 2005. Ecosystems and Human Well-Being, vol. 5. Island press, Washington, DC, p. 563.

Australian Academy of Science, 2020. www.science.org.au/learning/general-audience/science-climate-change/1-what-is-climate-change. (Accessed 4 May 2020).

Bai, K., Chang, N.B., Gao, W., 2016. Quantification of relative contribution of Antarctic ozone depletion to increased austral extratropical precipitation during 1979–2013. J. Geophys. Res. Atmos. 121 (4), 1459–1474.

Bais, A.F., Bernhard, G., McKenzie, R.L., Aucamp, P.J., Young, P.J., Ilyas, M., Jöckel, P., Deushi, M., 2019. Ozone–climate interactions and effects on solar ultraviolet radiation. Photochem. Photobiol. Sci. 18 (3), 602–640.

Bais, A.F., Lucas, R.M., Bornman, J.F., Williamson, C.E., Sulzberger, B., Austin, A.T., Wilson, S.R., Andrady, A.L., Bernhard, G., McKenzie, R.L., Aucamp, P.J., 2018. Environmental effects of ozone depletion, UV radiation and interactions with climate change: UNEP environmental effects assessment panel, update 2017. Photochem. Photobiol. Sci. 17 (2), 127–179.

Baker, A.C., Glynn, P.W., Riegl, B., 2008. Climate change and coral reef bleaching: an ecological assessment of long-term impacts, recovery trends and future outlook. Estuar. Coast Shelf Sci. 80 (4), 435–471.

Ballaré, C.L., Austin, A.T., 2017. UV radiation and terrestrial ecosystems: Emerging perspectives. In: Jordan R, Brian (Ed.), UV-B radiation and plant life: Molecular biology to ecology. CAB Interanational, pp. 23–39.

Barbier, E.B., 2006. Natural barriers to natural disasters: replanting mangroves after the tsunami. Front. Ecol. Environ. 4 (3), 124–131.

Barnes, P.W., Shinkle, J.R., Flint, S.D., Ryel, R.J., 2005. UV-B radiation, photomorphogenesis and plant-plant interactions. In: Progress in Botany. Springer, Berlin, Heidelberg, pp. 313–340.

Barnes, P.W., Throop, H.L., Archer, S.R., Breshears, D.D., McCulley, R.L., Tobler, M.A., 2015. Sunlight and soil–litter mixing: Drivers of litter decomposition in drylands. Progress in Botany (Genetics - Physiology - Systematics - Ecology), 76. Springer, Cham, pp. 273–302.

Berdalet, E., Fleming, L.E., Gowen, R., Davidson, K., Hess, P., Backer, L.C., Moore, S.K., Hoagland, P., Enevoldsen, H., 2016. Marine harmful algal blooms, human health and wellbeing: challenges and opportunities in the 21st century. J. Mar. Biol. Assoc. U. K. 96 (1), 61–91.

Berdalet, E., Kudela, R.M., Banas, N.S., Bresnan, E., Burford, M.A., Davidson, K., Gobler, C.J., Karlson, B., Lim, P.T., Mackenzie, L., Montresor, M., 2018. Global HAB: fostering international coordination on harmful algal bloom research in aquatic systems. In: Global Ecology and Oceanography of Harmful Algal Blooms. Springer, Cham, pp. 425–447.

Berwyn, B., 2019. Alaska Chokes on Wildfires as Heat Waves Dry Out the Arctic. InsideClimate News. https://insideclimatenews.org/news/11072019/arcticwildfires. (Accessed 25 May 2020).

Beschta, R.L., Bilby, R.E., Brown, G.W., Holtby, L.B., Hofstra, T.D., 1987. Stream Temperature and Aquatic Habitat: Fisheries and Forestry Interactions.

Bestelmeyer, B.T., Okin, G.S., Duniway, M.C., Archer, S.R., Sayre, N.F., Williamson, J.C., Herrick, J.E., 2015. Desertification, land use, and the transformation of global drylands. Front. Ecol. Environ. 13 (1), 28–36.

Bhandari, P., 2015. Asia: climate change battleground. Climate 59–61. https://www.climate2020.org.uk/asia-climate-change-battleground/, 2015.

Bharathidasan, M., Vishnugurubaran, D., Gupta, C., Mohanapriya, T., Dharmaceelan, S., Ramprabhu, R., 2020. Clinical management of injured juvenile painted storks (*Mycteria leucocephala*) due to the impact of cyclone Fani. Ind. J. Pure App. Biosci 8 (1), 97–104.

Blumberg, A.F., Di Toro, D.M., 1990. Effects of climate warming on dissolved oxygen concentrations in Lake Erie. Trans. Am. Fish. Soc. 119 (2), 210–223.

Bornman, J.F., Barnes, P.W., Robinson, S.A., Ballare, C.L., Flint, S.D., Caldwell, M.M., 2015. Solar ultraviolet radiation and ozone depletion-driven climate change: effects on terrestrial ecosystems. Photochem. Photobiol. Sci. 14 (1), 88–107.

Bornman, J.F., Barnes, P.W., Robson, T.M., Robinson, S.A., Jansen, M.A., Ballaré, C.L., Flint, S.D., 2019. Linkages between stratospheric ozone, UV radiation and climate change and their implications for terrestrial ecosystems. Photochem. Photobiol. Sci. 18 (3), 681–716.

Boudes, P., 2011. United Nations Conference on the Human Environment.

Bryndum-Buchholz, A., Tittensor, D.P., Blanchard, J.L., Cheung, W.W., Coll, M., Galbraith, E.D., Jennings, S., Maury, O., Lotze, H.K., 2019. Twenty-first-century climate change impacts on marine animal biomass and ecosystem structure across ocean basins. Glob. Change Biol. 25 (2), 459–472.

Cantin, N.E., Cohen, A.L., Karnauskas, K.B., Tarrant, A.M., McCorkle, D.C., 2010. Ocean warming slows coral growth in the central Red Sea. Science 329 (5989), 322–325.

Carrivick, J.L., Tweed, F.S., 2016. A global assessment of the societal impacts of glacier outburst floods. Glob. Planet. Change 144, 1–16.

Cauvy-Fraunié, S., Dangles, O., 2019. A global synthesis of biodiversity responses to glacier retreat. Nat. Ecol. Evolut. 3 (12), 1675–1685.

Center for Disaster Philanthropy, 2019. https://disasterphilanthropy.org/disaster/2019-australian-wildfires/.

Chang, C.W., Ye, H., Miki, T., Deyle, E.R., Souissi, S., Anneville, O., Adrian, R., Chiang, Y.R., Ichise, S., Kumagai, M., Shin-ichiro, S.M., 2020. Long-term warming weakens stabilizing effects of biodiversity in aquatic ecosystems. bioRxiv 1–22. https://doi.org/10.1101/2020.01.06.896746. https://www.biorxiv.org/content/10.1101/2020.01.06.896746v1.article-info.

Chapman, S., 1930. XXXV. On ozone and atomic oxygen in the upper atmosphere. Lond. Edinb. Dublin Philos. Mag. J. Sci. 10 (64), 369–383.

Chattopadhyay, C., Birah, A., Jalali, B.L., 2019. Climate change: impact on biotic stresses afflicting crop plants. In: Natural Resource Management: Ecological Perspectives. Springer, Cham, pp. 133–146.

Chen, Y., Xu, Z., 2005. Plausible impact of global climate change on water resources in the Tarim River Basin. Sci. China Earth Sci. 48 (1), 65–73.

Chipperfield, M.P., Dhomse, S.S., Feng, W., McKenzie, R.L., Velders, G.J., Pyle, J.A., 2015. Quantifying the ozone and ultraviolet benefits already achieved by the Montreal Protocol. Nat. Commun. 6 (1), 1–8.

Climate Action. www.un.org/en/climatechange/un-climate-summit-2019.shtml. (Accessed on dated 14 May 2020).

Climateurope, 2020. Linking Science and Society. www.climateurope.eu/what-is-climate-and-climate-change/. (Accessed 4 May 2020).

Comeau, S., Edmunds, P.J., Spindel, N.B., Carpenter, R.C., 2013. The responses of eight coral reef calcifiers to increasing partial pressure of CO_2 do not exhibit a tipping point. Limnol. Oceanogr. 58 (1), 388–398.

Comeau, S., Edmunds, P.J., Spindel, N.B., Carpenter, R.C., 2014. Fast coral reef calcifiers are more sensitive to ocean acidification in short-term laboratory incubations. Limnol. Oceanogr. 59 (3), 1081–1091.

Cook, J., Nuccitelli, D., Green, S.A., Richardson, M., Winkler, B., Painting, R., Way, R., Jacobs, P., Skuce, A., 2013. Quantifying the consensus on anthropogenic global warming in the scientific literature. Environ. Res. Lett. 8 (2), 024024.

Corell, R.W., 2006. Challenges of climate change: an Arctic perspective. AMBIO A J. Hum. Environ. 35 (4), 148–152.

Cory, R.M., Harrold, K.H., Neilson, B.T., Kling, G.W., 2015. Controls on dissolved organic matter (DOM) degradation in a headwater stream: the influence of photochemical and hydrological conditions in determining light-limitation or substrate-limitation of photodegradation. BioGeoSci. Disc. 12 (13), 9793–9838. https://doi.org/10.5194/bgd-12-9793-2015.

Cousino, L.K., Becker, R.H., Zmijewski, K.A., 2015. Modeling the effects of climate change on water, sediment, and nutrient yields from the Maumee River watershed. J. Hydrol. Reg. Stud. 4, 762–775.

Crossman, J., Futter, M.N., Oni, S.K., Whitehead, P.G., Jin, L., Butterfield, D., Baulch, H.M., Dillon, P.J., 2013. Impacts of climate change on hydrology and water quality: future proofing management strategies in the Lake Simcoe watershed, Canada. J. Great Lake. Res. 39 (1), 19–32.

Cullen, J.J., Neale, P.J., 1994. Ultraviolet radiation, ozone depletion, and marine photosynthesis. Photosynth. Res. 39 (3), 303–320.

D'Odorico, P., Bhattachan, A., Davis, K.F., Ravi, S., Runyan, C.W., 2013. Global desertification: drivers and feedbacks. Adv. Water Resour. 51, 326–344.

Debata, S., 2019. Impact of cyclone Fani on the breeding success of sandbar-nesting birds along the Mahanadi River in Odisha, India. J. Threat. Taxa 11 (14), 14895–14898.

Dodds, W., Whiles, M., 2020. Responses to stress, toxic chemicals, and other pollutants in aquatic ecosystems. Freshw.Ecol. III, 456–581. https://doi.org/10.1016/B978-0-12-813255-5.00016-8.

Dokulil, M.T., Teubner, K., Jagsch, A., Nickus, U., Adrian, R., Straile, D., Jankowski, T., Herzig, A., Padisák, J., 2010. The impact of climate change on lakes in Central Europe. In: The impact of climate change on European lakes. Springer, Dordrecht, pp. 387–409.

Doney, S.C., Ruckelshaus, M., Duffy, J.E., Barry, J.P., Chan, F., English, C.A., Galindo, H.M., Grebmeier, J.M., Hollowed, A.B., Knowlton, N., Polovina, J., 2011. Climate Change Impacts on Marine Ecosystems.

EPA, 2001. Technical Development Document for the Final Regulations Addressing CoolingWater Intake Structures for New Facilities. EPA-821-R-1-36.

EPRI (Electric Power Research Institute), 2008. Water Use for Electric Power Generation. EPRI, Palo Alto, CA, 1014026.

Fabry, C., Buisson, H., 1931. Data on ozone absorption. Compt. Rend. 192, 457.

Fahey, D., Doherty, S., Hibbard, K.A., Romanou, A., Taylor, P., 2017. Physical Drivers of Climate Change.

Farman, J.C., Gardiner, B.G., Shanklin, J.D., 1985. Large losses of total ozone in Antarctica reveal seasonal ClOx/NOx interaction. Nature 315 (6016), 207–210.

Ferrao-Filho, A.D.S., Kozlowsky-Suzuki, B., 2011. Cyanotoxins: bioaccumulation and effects on aquatic animals. Mar. Drugs 9 (12), 2729–2772.

First, P.J., 2019b. IPCC. Global warming of 1.5 C an IPCC special report on the impacts of global warming of 1.5 C above pre-industrial levels and related global greenhouse gas emission pathways. In: The Context of Strengthening the Global Response to the Threat of

Climate Change. Sustainable Development, and Efforts to Eradicate Poverty. https://www.ipcc.ch/sr15/. Accessed, 1.

Folland, C.K., Karl, T.R., Vinnikov, K.Y., 1990. Observed climate variations and change. Clim. Change IPCC Sci. Assess. 195, 238.

Gao, K., Häder, D.P., 2017. Effects of ocean acidification and UV radiation on marine photosynthetic carbon fixation. In: Kumar, M., Ralph, P. (Eds.), Systems Biology of Marine Ecosystems. Springer, Cham, pp. 235–250.

Gérard, M., Vanderplanck, M., Wood, T., Michez, D., 2020. Global warming and plant–pollinator mismatches. Emerg. Top. Life Sci. ETLS20190139.

Giannini, T.C., Costa, W.F., Borges, R.C., Miranda, L., da Costa, C.P.W., Saraiva, A.M., Fonseca, V.L.I., 2020. Climate change in the Eastern Amazon: crop-pollinator and occurrence-restricted bees are potentially more affected. Reg. Environ. Change 20 (1), 9.

Gonzalez, P.L., Polvani, L.M., Seager, R., Correa, G.J., 2014. Stratospheric ozone depletion: a key driver of recent precipitation trends in South Eastern South America. Clim. Dynam. 42 (7–8), 1775–1792.

Government of Canada, 2020. www.canada.ca/en/environment-climate-change/services/air-pollution/issues/ozone-layer/measures-protect/vienna-convention-montreal-protocol.html. (Accessed 15 May 2020).

Grimm, N.B., Chapin III, F.S., Bierwagen, B., Gonzalez, P., Groffman, P.M., Luo, Y., Melton, F., Nadelhoffer, K., Pairis, A., Raymond, P.A., Schimel, J., 2013. The impacts of climate change on ecosystem structure and function. Front. Ecol. Environ. 11 (9), 474–482.

Gu, C., Riley, W.J., 2010. Combined effects of short term rainfall patterns and soil texture on soil nitrogen cycling—a modeling analysis. J. Contam. Hydrol. 112 (1–4), 141–154.

Häder, D.P., Williamson, C.E., Wängberg, S.Å., Rautio, M., Rose, K.C., Gao, K., Helbling, E.W., Sinha, R.P., Worrest, R., 2015. Effects of UV radiation on aquatic ecosystems and interactions with other environmental factors. Photochem. Photobiol. Sci. 14 (1), 108–126.

Hassan, B., Qadri, H., Ali, M.N., Khan, N.A., Yatoo, A.M., 2020. Impact of climate change on freshwater ecosystem and its sustainable management. In: Fresh Water Pollution Dynamics and Remediation. Springer, Singapore, pp. 105–121.

Helweg, D., Giardina, C., 2017. Ecological drought in the Hawaiian Islands: unique tropical systems are vulnerable to drought. Univ. Md. Cent. Environ. Sci. Integr. Appl. Netw. Newsl. 581 (581).

Herb, W.R., Janke, B., Mohseni, O., Stefan, H.G., 2008. Thermal pollution of streams by runoff from paved surfaces. Hydrol. Process. Int. J. 22 (7), 987–999.

History, 2020. Ice Age. www.history.com/topics/pre-history/ice-age. (Accessed 16 May 2020).

Hoegh-Guldberg, O., Bruno, J.F., 2010. The impact of climate change on the world's marine ecosystems. Science 328 (5985), 1523–1528.

Hoey, A.S., Howells, E., Johansen, J.L., Hobbs, J.P.A., Messmer, V., McCowan, D.M., Wilson, S.K., Pratchett, M.S., 2016. Recent advances in understanding the effects of climate change on coral reefs. Diversity 8 (2), 12.

Holz, A., Veblen, T.T., 2011. Variability in the southern Annular Mode determines wildfire activity in patagonia. Geophys. Res. Lett. 38 (14).

Holz, A., Paritsis, J., Mundo, I.A., Veblen, T.T., Kitzberger, T., Williamson, G.J., Aráoz, E., Bustos-Schindler, C., González, M.E., Grau, H.R., Quezada, J.M., 2017. Southern Annular Mode drives multicentury wildfire activity in southern South America. Proc. Natl. Acad. Sci. U.S.A. 114 (36), 9552–9557.

Hughes, T.P., Baird, A.H., Bellwood, D.R., Card, M., Connolly, S.R., Folke, C., Grosberg, R., Hoegh-Guldberg, O., Jackson, J.B., Kleypas, J., Lough, J.M., 2003. Climate change, human impacts, and the resilience of coral reefs. Science 301 (5635), 929–933.
IISD RS, 2018. International Institute of Sustainable Development Reporting Services. In: https://enb.iisd.org/process/climate_atm-fcccintro.html. (Accessed 15 May 2020).
Intergovernmental Panel on Climate Change Working Group II., Adger, W.N., 2007. Climate Change 2007: Impacts, Adaptation and Vulnerability: Working Group II Contribution to the Fourth Assessment Report of the Intergovernmental Panel on Climate Change. Intergovernmental Panel on Climate Change.
International Union for Conservation of Nature, Natural Resources and World Wildlife Fund, 1980. World Conservation Strategy: Living Resource Conservation for Sustainable Development. IUCN, Gland, Switzerland.
IPCC, 2020. Intergovernmental Panel for Climate Change. https://www.ipcc.ch/. (Accessed 6 May 2020).
IPCC, 2001. Climate Change 2001. The Scientific Basis. http://www.ipcc.ch.
Jenkins, G.I., 2009. Signal transduction in responses to UV-B radiation. Annu. Rev. Plant Biol. 60, 407–431.
Jennings, E., Allott, N., Pierson, D.C., Schneiderman, E.M., Lenihan, D., Samuelsson, P., Taylor, D., 2009. Impacts of climate change on phosphorus loading from a grassland catchment: implications for future management. Water Res. 43 (17), 4316–4326.
Jenny, J.P., Arnaud, F., Alric, B., Dorioz, J.M., Sabatier, P., Meybeck, M., Perga, M.E., 2014. Inherited hypoxia: a new challenge for reoligotrophicated lakes under global warming. Glob. Biogeochem. Cycles 28 (12), 1413–1423.
Jenny, J.P., Francus, P., Normandeau, A., Lapointe, F., Perga, M.E., Ojala, A., Schimmelmann, A., Zolitschka, B., 2016. Global spread of hypoxia in freshwater ecosystems during the last three centuries is caused by rising local human pressure. Glob. Change Biol. 22 (4), 1481–1489.
Jeppesen, E., Brucet, S., Naselli-Flores, L., Papastergiadou, E., Stefanidis, K., Noges, T., Noges, P., Attayde, J.L., Zohary, T., Coppens, J., Bucak, T., 2015. Ecological impacts of global warming and water abstraction on lakes and reservoirs due to changes in water level and related changes in salinity. Hydrobiologia 750 (1), 201–227.
JNCC, 2019. https://jncc.gov.uk/our-work/the-convention-on-the-conservation-of-migratory-species-of-wild-animals/. (Accessed 15 May 2020).
Jokiel, P.L., Coles, S.L., 1990. Response of Hawaiian and other Indo-Pacific reef corals to elevated temperature. Coral Reefs 8 (4), 155–162.
Joos, F., Spahni, R., 2008. Rates of change in natural and anthropogenic radiative forcing over the past 20,000 years. Proc. Natl. Acad. Sci. U.S.A. 105 (5), 1425–1430.
Kalinowska, M.B., 2019. Effect of water–air heat transfer on the spread of thermal pollution in rivers. Acta Geophys. 67 (2), 597–619.
Kaushal, S.S., Gold, A.J., Mayer, P.M., 2017. Land Use, Climate, and Water Resources—Global Stages of Interaction.
Kirillin, G., Shatwell, T., Kasprzak, P., 2013. Consequences of thermal pollution from a nuclear plant on lake temperature and mixing regime. J. Hydrol. 496, 47–56.
Kirtman, B., Power, S.B., Adedoyin, J.A., Boer, G.J., Bojariu, R., Camilloni, I., et al., 2013. Near-term climate change: Projections and predictability. In: Stocker, T.F., et al. (Eds.), Climate Change 2013: The Physical Science Basis. Contribution of Working Group I to the Fifth Assessment Report of the Intergovernmental Panel on Climate Change. Cambridge University Press, pp. 953–1028.

Klein, J.A., Tucker, C.M., Steger, C.E., Nolin, A., Reid, R., Hopping, K.A., Yeh, E.T., Pradhan, M.S., Taber, A., Molden, D., Ghate, R., 2019. An integrated community and ecosystem-based approach to disaster risk reduction in mountain systems. Environ. Sci. Pol. 94, 143−152.

Kleypas, J.A., Buddemeier, R.W., Archer, D., Gattuso, J.P., Langdon, C., Opdyke, B.N., 1999. Geochemical consequences of increased atmospheric carbon dioxide on coral reefs. Science 284 (5411), 118−120.

Koschel, R., Adams, D.D., 2003. An approach to understanding a temperate oligotrophic lowland lake (Lake Stechlin, Germany). Archiv fur Hydrobiologie Adv. Limnol. 58, 1−9.

Kumar, P.M., Pooja, R., 2020. Global warming, impacts and mitigation measures: an overview. Disaster Adv. 13 (5), 82−96.

Kumari, R., Singh, S., Agrawal, S.B., 2010. Response of ultraviolet-B induced antioxidant defense system in a medicinal plant, *Acorus calamus*. J. Environ. Biol. 31 (6), 907−911.

Leggett, J.A., Carter, N.T., 2012. June. Rio+ 20: The United Nations Conference on Sustainable Development, June 2012. Library of Congress, Congressional Research Service.

Lenoir, J., Gégout, J.C., Marquet, P.A., De Ruffray, P., Brisse, H., 2008. A significant upward shift in plant species optimum elevation during the 20th century. Science 320 (5884), 1768−1771.

Lindsey, R., 2020. Climate Change: Glacier Mass Balance. NOAA Climate.gov. https://www.climate.gov/news-features/understanding-climate/climate-change-glacier-mass-balance.

Lindsey, R., Dahlman, L., 2020. Climate Change: Global Temperature. NOAA Climate.gov. https://www.climate.gov/news-features/understanding-climate/climate-change-global-temperature.

Lintern, A., Webb, J.A., Ryu, D., Liu, S., Bende-Michl, U., Waters, D., Leahy, P., Wilson, P., Western, A.W., 2018. Key factors influencing differences in stream water quality across space. Wiley Interdiscip. Rev. Water 5 (1), e1260.

Lu, X., Siemann, E., Shao, X., Wei, H., Ding, J., 2013. Climate warming affects biological invasions by shifting interactions of plants and herbivores. Glob. Change Biol. 19 (8), 2339−2347.

Lucci, G.M., 2019. Pastures and drought: a review of processes and implications for nitrogen and phosphorus cycling in grassland systems. Soil Res. 57 (2), 101−112.

Manatsa, D., Morioka, Y., Behera, S.K., Yamagata, T., Matarira, C.H., 2013. Link between Antarctic ozone depletion and summer warming over southern Africa. Nat. Geosci. 6 (11), 934−939.

Manatsa, D., Mudavanhu, C., Mushore, T.D., Mavhura, E., 2016. Linking major shifts in East Africa 'short rains' to the southern Annular Mode. Int. J. Climatol. 36 (4), 1590−1599.

Manuel, J., 2014. Nutrient pollution: a persistent threat to waterways. Environ. Health Perspect. 122, A304−A309.

Mariani, M., Fletcher, M.S., 2016. The Southern Annular Mode determines interannual and centennial-scale fire activity in temperate southwest Tasmania, Australia. Geophys. Res. Lett. 43 (4), 1702−1709.

Marshall, G.J., 2007. Half-century seasonal relationships between the southern Annular Mode and antarctic temperatures. Int. J. Climatol. J. R. Meteorol. Soc. 27 (3), 373−383.

Marshall, G.J., Orr, A., van Lipzig, N.P., King, J.C., 2006. The impact of a changing southern hemisphere Annular Mode on Antarctic Peninsula summer temperatures. J. Clim. 19 (20), 5388−5404.

Marshall, N.A., Stokes, C.J., Webb, N.P., Marshall, P.A., Lankester, A.J., 2014. Social vulnerability to climate change in primary producers: a typology approach. Agric. Ecosyst. Environ. 186, 86–93.

McVittie, A., Cole, L., Wreford, A., Sgobbi, A., Yordi, B., 2018. Ecosystem-based solutions for disaster risk reduction: lessons from European applications of ecosystem-based adaptation measures. Int. J. Disaster Risk Reduct. 32, 42–54.

Melillo, J.M., Callaghan, T.V., Woodward, F.I., Salati, E., Sinha, S., 1990. Effects on Ecosystems. Climate Change: The IPCC Scientific Assessment, pp. 283–310.

Mendelsohn, R., Prentice, I.C., Schmitz, O., Stocker, B., Buchkowski, R., Dawson, B., 2016. The ecosystem impacts of severe warming. Am. Econ. Rev. 106 (5), 612–614.

Mera, Y.E.Z., Vera, J.F.R., Pérez-Martín, M.Á., 2018. Linking El Niño Southern oscillation for early drought detection in tropical climates: the Ecuadorian coast. Sci. Total Environ. 643, 193–207.

Miara, A., Vörösmarty, C.J., Macknick, J.E., Tidwell, V.C., Fekete, B., Corsi, F., Newmark, R., 2018. Thermal pollution impacts on rivers and power supply in the Mississippi River watershed. Environ. Res. Lett. 13 (3), 034033.

Middelkoop, H., Daamen, K., Gellens, D., Grabs, W., Kwadijk, J.C., Lang, H., Parmet, B.W., Schädler, B., Schulla, J., Wilke, K., 2001. Impact of climate change on hydrological regimes and water resources management in the Rhine basin. Clim. Change 49 (1–2), 105–128.

Miettinen, J., Stibig, H.J., Achard, F., 2014. Remote sensing of forest degradation in Southeast Asia—aiming for a regional view through 5–30 m satellite data. Glob. Ecol. Conserv. 2, 24–36.

Molina, M.J., Rowland, F.S., 1974. Stratospheric sink for chlorofluoromethanes: chlorine atom-catalysed destruction of ozone. Nature 249 (5460), 810–812.

Monteith, D.T., Stoddard, J.L., Evans, C.D., de Wit, H., Forsius, M., Høgåsen, T., Wilander, A., Skjelkvåle, B.L., Jeffries, D.S., Vuorenmaa, J., Keller, B., 2007. Rising freshwater dissolved organic carbon driven by changes in atmospheric deposition. Nature 450, 537–540.

Masson-Delmotte, V., Zhai, P., Pörtner, H.O., Roberts, D., Skea, J., Shukla, P.R., Pirani, A., Moufouma-Okia, W., Péan, C., Pidcock, R., Connors, S., 2018. IPCC, 2018: Summary for Policymakers. Global Warming of1.5°C: An IPCC Special Report on the Impacts of Global Warming of, 1.5°C above Pre-industrial Levels and Related Global Greenhouse Gas Emission Pathways, in the Context of Strengthening the Global Response to the Threat of Climate Change, Sustainable Development, and Efforts to Eradicate Poverty (in press).

Munang, R., Thiaw, I., Alverson, K., Liu, J., Han, Z., 2013. The role of ecosystem services in climate change adaptation and disaster risk reduction. Curr. Opin. Environ. Sustain. 5 (1), 47–52.

Muñoz, A.A., González-Reyes, A., Lara, A., Sauchyn, D., Christie, D., Puchi, P., Urrutia-Jalabert, R., Toledo-Guerrero, I., Aguilera-Betti, I., Mundo, I., Sheppard, P.R., 2016. Streamflow variability in the Chilean Temperate-Mediterranean climate transition (35 S–42 S) during the last 400 years inferred from tree-ring records. Clim. Dynam. 47 (12), 4051–4066.

Nagelkerken, I., Connell, S.D., 2015. Global alteration of ocean ecosystem functioning due to increasing human CO_2 emissions. Proc. Natl. Acad. Sci. U.S.A. 112 (43), 13272–13277.

NASA, 2019. https://climate.nasa.gov/news/2865/a-degree-of-concern-why-global-temperatures-matter/Assessed on 10/05/2020.

NASA, 2020a. Climate Kids National. Aeronautics and Space Administration. https://climatekids.nasa.gov/climate-change-meaning/. (Accessed 4 May 2020).

NASA, 2020b. Global Climate Change: Vital Sign of the Planet. National Aeronautics and Space Administration. https://climate.nasa.gov/news/215/just-5-questions-aerosols/. (Accessed 8 May 2020).

NOAA National Centers for Environmental Information (NCEI), 2019a. U.S. Billion- Dollar Weather and Climate Disasters. www.ncdc.noaa.gov/billions/. (Accessed 25 May 2020).

Neale, P.J., Helbling, E.W., Zagarese, H.E., 2003. Modulation of UVR exposure and effects by vertical mixing and advection. In: Helbling, E.W., Zagarese, H. (Eds.), UV Effects in Aquatic Organisms and Ecosystems. Royal Society of Chemistry, UK, pp. 107−134.

NOAA, 2019b. State of the Climate. www.ncdc.noaa.gov/sotc/. (Accessed 23 May 2020).

NOAA. ESRL, Global Monitoring Laboratory. National Oceanic and Atmospheric Administration. www.esrl.noaa.gov/gmd/hats/publictn/elkins/cfcs.html. (Assessed on 8 May 2020).

Nobel, J., 2017. Recent Hurricanes Pushed Rare Island Species Closer to the Brink. www.nationalgeographic.com/news/2017/10/hurricane-irma-hurts-florida-keys-wildlife/. (Accessed 29 May 2020).

Nolan, R.H., Boer, M.M., Collins, L., Resco de Dios, V., Clarke, H., Jenkins, M., Kenny, B., Bradstock, R.A., 2020. Causes and consequences of eastern Australia's 2019-20 season of mega-fires. Glob. Change Biol. 26, 1039−1041.

O'neil, J.M., Davis, T.W., Burford, M.A., Gobler, C.J., 2012. The rise of harmful cyanobacteria blooms: the potential roles of eutrophication and climate change. Harmful Algae 14, 313−334.

Ouyang, W., Skidmore, A.K., Toxopeus, A.G., Hao, F., 2010. Long-term vegetation landscape pattern with non-point source nutrient pollution in upper stream of Yellow River basin. J. Hydrol. 389 (3−4), 373−380.

Ouyang, Y., 2012. Estimation of shallow groundwater discharge and nutrient load into a river. Ecol. Eng. 38 (1), 101−104.

Pachauri, R.K., Reisinger, A., 2007. IPCC, AR4. IPCC Fourth Assessment Report. IPCC, Geneva. https://www.ipcc.ch/report/ar4/syr/.

Pachauri, R.K., Allen, M.R., Barros, V.R., Broome, J., Cramer, W., Christ, R., Church, J.A., Clarke, L., Dahe, Q., Dasgupta, P., Dubash, N.K., 2014. IPCC ARS 5. Climate Change 2014: Synthesis Report. Contribution of Working Groups I, II and III to the Fifth Assessment Report of the Intergovernmental Panel on Climate Change. IPCC, p. 151.

Parker, F.L., Krenkel, P.A., Stevens, D.B., 1970. Physical and engineering aspects of thermal pollution. Crit. Rev. Environ. Sci. Technol. 1 (1−4), 101−192.

Past Interglacials Working Group of PAGES, 2016. Interglacials of the last 800,000 years. Rev. Geophys. 54 (1), 162−219.

Pecl, G.T., Araújo, M.B., Bell, J.D., Blanchard, J., Bonebrake, T.C., Chen, I.C., Clark, T.D., Colwell, R.K., Danielsen, F., Evengård, B., Falconi, L., 2017. Biodiversity redistribution under climate change: impacts on ecosystems and human well-being. Science 355 (6332), eaai9214.

Perin, S., Lean, D.R., 2004. The effects of ultraviolet-B radiation on freshwater ecosystems of the Arctic: influence from stratospheric ozone depletion and climate change. Environ. Rev. 12 (1), 1−70.

Pikitch, E.K., Santora, C., Babcock, E.A., Bakun, A., Bonfil, R., Conover, D.O., Dayton, P., Doukakis, P., Fluharty, D., Heneman, B., Houde, E.D., 2004. Ecosystem-based Fishery Management.

Pires, A.P., Srivastava, D.S., Marino, N.A., MacDonald, A.A.M., Figueiredo-Barros, M.P., Farjalla, V.F., 2018. Interactive effects of climate change and biodiversity loss on ecosystem functioning. Ecology 99 (5), 1203−1213.

Pörtner, H.O., Farrell, A.P., 2008. Physiology and climate change. Science 322 (5902), 690–692.

Qadir, M., Quillérou, E., Nangia, V., Murtaza, G., Singh, M., Thomas, R.J., Drechsel, P., Noble, A.D., 2014. November. Economics of salt-induced land degradation and restoration. Nat. Resour. Forum 38 (4), 282–295.

Quenta, E., Molina-Rodriguez, J., Gonzales, K., Rebaudo, F., Casas, J., Jacobsen, D., Dangles, O., 2016. Direct and indirect effects of glaciers on aquatic biodiversity in high Andean peatlands. Glob. Change Biol. 22 (9), 3196–3205.

Quinton, J.N., Govers, G., Van Oost, K., Bardgett, R.D., 2010. The impact of agricultural soil erosion on biogeochemical cycling. Nat. Geosci. 3 (5), 311–314.

Renaud, F.G., Sudmeier-Rieux, K., Estrella, M., Nehren, U. (Eds.), 2016. Ecosystem-based Disaster Risk Reduction and Adaptation in Practice, vol. 42. Springer, Switzerland.

Robakowski, P., Bielinis, E., Sendall, K., 2018. Light energy partitioning, photosynthetic efficiency and biomass allocation in invasive *Prunus serotina* and native *Quercus petraea* in relation to light environment, competition and allelopathy. J. Plant Res. 131 (3), 505–523.

Robinson, S.A., Erickson III, D.J., 2015. Not just about sunburn–the ozone hole's profound effect on climate has significant implications for Southern Hemisphere ecosystems. Glob. Change Biol. 21 (2), 515–527.

Rogers, B.M., Balch, J.K., Goetz, S.J., Lehmann, C.E., Turetsky, M., 2020. Focus on changing fire regimes: interactions with climate, ecosystems, and society. Environ. Res. Lett. 15 (3), 030201.

Rogers, L.A., Griffin, R., Young, T., Fuller, E., Martin, K.S., Pinsky, M.L., 2019. Shifting habitats expose fishing communities to risk under climate change. Nat. Clim. Change 9 (7), 512–516.

Ross, P.M., Adam, P., 2013. Climate change and intertidal wetlands. Biology 2 (1), 445–480.

Rubin, M.B., 2001. The history of ozone. The Schönbein period, 1839–1868. Bull. Hist. Chem. 26 (1), 40–56.

Scheffers, B.R., De Meester, L., Bridge, T.C., Hoffmann, A.A., Pandolfi, J.M., Corlett, R.T., Butchart, S.H., Pearce-Kelly, P., Kovacs, K.M., Dudgeon, D., Pacifici, M., 2016. The broad footprint of climate change from genes to biomes to people. Science 354 (6313), aaf7671.

Schoennagel, T., Balch, J.K., Brenkert-Smith, H., Dennison, P.E., Harvey, B.J., Krawchuk, M.A., Mietkiewicz, N., Morgan, P., Moritz, M.A., Rasker, R., Turner, M.G., 2017. Adapt to more wildfire in western North American forests as climate changes. Proc. Natl. Acad. Sci. U.S.A. 114 (18), 4582–4590.

Sheppard, C.R., 2003. Predicted recurrences of mass coral mortality in the Indian Ocean. Nature 425 (6955), 294–297.

Sheridan, J.A., Bickford, D., 2011. Shrinking body size as an ecological response to climate change. Nat. Clim. Change 1 (8), 401–406.

Shortle, J., Horan, R.D., 2017. Nutrient pollution: a wicked challenge for economic instruments. Water Econ. Policy 3 (02), 1650033.

Shukla, P.R., Skea, J., CalvoBuendia, E., Masson-Delmotte, V., Pörtner, H.O., Roberts, D.C., Zhai, P., Slade, R., Connors, S., Van Diemen, R., Ferrat, M., IPCC, 2019. Summary for Policymakers. Climate Change and Land: An IPCC Special Report on Climate Change, Desertification, Land Degradation, Sustainable Land Management, Food Security, and Greenhouse Gas Fluxes in Terrestrial Ecosystems (in press).

Slemmons, K.E., Saros, J.E., Simon, K., 2013. The influence of glacial meltwater on alpine aquatic ecosystems: a review. Environ. Sci. J. Integr. Environ. Res. Process. Impacts 15 (10), 1794–1806.

Solomon, S., Ivy, D.J., Kinnison, D., Mills, M.J., Neely, R.R., Schmidt, A., 2016. Emergence of healing in the Antarctic ozone layer. Science 353 (6296), 269–274.
Stohlgren, T.J., Barnett, D., Flather, C., Kartesz, J., Peterjohn, B., 2005. Plant species invasions along the latitudinal gradient in the United States. Ecology 86 (9), 2298–2309.
Stolarski, R.S., Cicerone, R.J., 1974. Stratospheric chlorine: a possible sink for ozone. Can. J. Chem. 52 (8), 1610–1615.
Stott, P.A., Tett, S.F., Jones, G.S., Allen, M.R., Ingram, W.J., Mitchell, J.F., 2001. Attribution of twentieth century temperature change to natural and anthropogenic causes. Clim. Dynam. 17 (1), 1–21.
Sulzberger, B., Austin, A.T., Cory, R.M., Zepp, R.G., Paul, N.D., 2019. Solar UV radiation in a changing world: roles of cryosphere–land–water–atmosphere interfaces in global biogeochemical cycles. Photochem. Photobiol. Sci. 18 (3), 747–774.
Takshak, S., Agrawal, S.B., 2019. Defense potential of secondary metabolites in medicinal plants under UV-B stress. J. Photochem. Photobiol. B Biol. 193, 51–88.
Thuiller, W., 2007. Climate change and the ecologist. Nature 448 (7153), 550–552.
Tripathi, R., Sarkar, A., Pandey Rai, S., Agrawal, S.B., 2011. Supplemental ultraviolet-B and ozone: impact on antioxidants, proteome and genome of linseed (*Linumusitatissimum* L. cv. Padmini). Plant Biol. 13 (1), 93–104.
UNCC, 2020a. United Nations Climate Change. https://unfccc.int/kyoto_protocol. (Accessed 15 May 2020).
UNCC, 2020b. United Nations Climate Change. In: https://unfccc.int/process-and-meetings/the-paris-agreement/the-paris-agreement. (Accessed 16 May 2020).
UNCC, 2020. United Nations Climate Change. In: https://unfccc.int/process-and-meetings/conferences/past-conferences/past-conferences-overview. (Accessed 14 May 2020).
UNDP, 2004. United Nations Development Programme Bureau for Crisis Prevention and Recovery: Reducing Disaster Risk: A Challenge for Development. A Global Report. 2004. http://www.un.org/.
UNEP-IRP (United Nations Environment Programme– International Resource Panel), 2016. Unlocking the Sustainable.
US EPA, 2017. https://19january2017snapshot.epa.gov/climate-impacts/climate-impacts-ecosystems_html.
US EPA, 2018. United State Environment Protection Agency. www.epa.gov/ground-level-ozone-pollution/ground-level-ozone-basics. (Accessed 8 May 2020).
US EPA, 2020. United State Environment Protection Agency. www.epa.gov/climate-indicators/greenhouse-gases. (Accessed 8 May 2020).
UNISDR, G., 2009. Global Assessment Report on Disaster Risk Reduction. United Nations International Strategy for Disaster Reduction Secretariat Geneva, Switzerland.
UN: United Nations. www.un.org/en/sections/issues-depth/climate-change/. (Accessed on 8 May 2020).
Vallero, D.A., 2019. Thermal pollution. In: Waste. Academic Press, pp. 381–404.
Van der Weerden, T.J., Jarvis, S.C., 1997. Ammonia emission factors for N fertilizers applied to two contrasting grassland soils. Environ. Pollut. 95 (2), 205–211.
Van Lipzig, N.P., Marshall, G.J., Orr, A., King, J.C., 2008. The relationship between the southern hemisphere Annular Mode and Antarctic Peninsula summer temperatures: analysis of a high-resolution model climatology. J. Clim. 21 (8), 1649–1668.
Velders, G.J., Andersen, S.O., Daniel, J.S., Fahey, D.W., McFarland, M., 2007. The importance of the Montreal Protocol in protecting climate. Proc. Natl. Acad. Sci. U.S.A. 104 (12), 4814–4819.

Vercammen, P., Vera, A., Chavez, N., 2018. Carr Fire in California Is So Hot It's Creating its Own Weather System. CNN. www.cnn.com/2018/07/30/us/carr-fire-california/index.html. (Accessed 17 May 2020).

Verma, S., Bhattarai, R., Bosch, N.S., Cooke, R.C., Kalita, P.K., Markus, M., 2015. Climate change impacts on flow, sediment and nutrient export in a Great Lakes watershed using SWAT. Clean 43 (11), 1464−1474.

Wang, L., Flanagan, D.C., Wang, Z., Cherkauer, K.A., 2018. Climate change impacts on nutrient losses of two watersheds in the Great Lakes region. Water 10 (4), 442.

Wang, X.J., Zhang, J.Y., Shahid, S., Guan, E.H., Wu, Y.X., Gao, J., He, R.M., 2016. Adaptation to climate change impacts on water demand. Mitig. Adapt. Strateg. Glob. Change 21 (1), 81−99.

Wang, X.J., Zhang, J.Y., Yang, Z.F., Shahid, S., He, R.M., Xia, X.H., Liu, H.W., 2015. Historic water consumptions and future management strategies for Haihe River basin of Northern China. Mitig. Adapt. Strateg. Glob. Change 20 (3), 371−387.

Wernberg, T., Russell, B.D., Thomsen, M.S., Gurgel, C.F.D., Bradshaw, C.J., Poloczanska, E.S., Connell, S.D., 2011. Seaweed communities in retreat from ocean warming. Curr. Biol. 21 (21), 1828−1832.

Whitcomb, D., 2018. Largest Wildfire in California History to Burn for Rest of August. Reuters. https://af.reuters.com/article/worldNews/idAFKBN1KS0PO. (Accessed 17 May 2020).

White, D.A., Visser, J.M., 2016. Water quality change in the Mississippi River, including a warming river, explains decades of wetland plant biomass change within its Balize delta. Aquat. Bot. 132, 5−11.

Williamson, C.E., Neale, P.J., Hylander, S., Rose, K.C., Figueroa, F.L., Robinson, S.A., Häder, D.P., Wängberg, S.Å., Worrest, R.C., 2019. The interactive effects of stratospheric ozone depletion, UV radiation, and climate change on aquatic ecosystems. Photochem. Photobiol. Sci. 18 (3), 717−746.

Wilson, S.R., Madronich, S., Longstreth, J.D., Solomon, K.R., 2019. Interactive effects of changing stratospheric ozone and climate on tropospheric composition and air quality, and the consequences for human and ecosystem health. Photochem. Photobiol. Sci. 18 (3), 775−803.

World Health Organization, 2002. Global Solar UV Index: A Practical Guide: A Joint Recommendation of the World Health Organization. World Meteorological Organization, United Nations Environment Programme and International Commission on Non-Ionizing Radiation Protection.

WRM: World Rainforest Movement. https://wrm.org.uy/browse-by-subject/international-processes-and-actors/rio-92-rio-10-rio-20/. (Accessed on 16 May 2020).

Wu, H., Ismail, M., Ding, J., 2017. Global warming increases the interspecific competitiveness of the invasive plant alligator weed, *Alternanthera philoxeroides*. Sci. Total Environ. 575, 1415−1422.

Wu, Y., Polvani, L.M., 2017. Recent trends in extreme precipitation and temperature over southeastern South America: the dominant role of stratospheric ozone depletion in the CESM Large Ensemble. J. Clim. 30 (16), 6433−6441.

WWF, 2020. World Wildlife Fund. www.worldwildlife.org/pages/global-climate-action-summit-2018. (Accessed 15 May 2020).

Yang, H., Zhao, Z., Qiang, W., An, L., Xu, S., Wang, X., 2004. Effects of enhanced UV-B radiation on the hormonal content of vegetative and reproductive tissues of two tomato cultivars and their relationships with reproductive characteristics. Plant Growth Regul. 43 (3), 251−258.

Yang, Y., Niu, K., Hu, Z., Niklas, K.J., Sun, S., 2018. Linking species performance to community structure as affected by UV-B radiation: an attenuation experiment. J. Plant Ecol. 11 (2), 286–296.

Yang, Y., Yao, Y., Xu, G., Li, C., 2005. Growth and physiological responses to drought and elevated ultraviolet-B in two contrasting populations of Hippophaerhamnoides. Physiol. Plantarum 124 (4), 431–440.

Yu, P., Xu, R., Abramson, M.J., Li, S., Guo, Y., 2020. Bushfires in Australia: a serious health emergency under climate change. Lancet Planet. Health 4 (1), e7–e8.

Climate change: Impact on agricultural production and sustainable mitigation

7

Poonam Yadav[1], *Durgesh Kumar Jaiswal*[1], *Rakesh Kumar Sinha*[2]
[1]Institute of Environment and Sustainable Development, Banaras Hindu University, Varanasi, Uttar Pradesh, India; [2]Department of Civil Engineering, Indian Institute of Technology-Bombay, Mumbai, Maharashtra, India

1. Introduction

Food security is one of the important and sensitive issues in front of the world that is continuously being affected by the climate change. Climate change has affected the food productions and supply to meet the existing and future demand (Lal, 2005; Kang et al., 2009). It has potential to alter the regional and global weather conditions, and thus results in the biophysical effect on agricultural productions. Any variation and uncertainty in the rainfall, temperature, and carbon dioxide (CO_2) rise for long-term have potential to pose threat to agricultural productions, thus making agriculture sector vulnerable. According to the simulations model and projections, agricultural sector can be significantly affected by climate change and result in decreased agricultural productions, increased attack of pest and plant pathogens, affecting water resources, human health and atmospheric nitrogen (N) deposition, changes in the concentration of various gases available in the atmosphere, and changes in soil nutrient availability. Its impact on agriculture has already been reported for many countries of the world, especially in African and hot regions of the world (Magadza, 2000). It is also proposed that due to the CO_2 doubling fertilization effect, overall agricultural production can be increased (as photosynthesis and nitrogen uptake will increase). Water resources will also be affected; thus the overall impact of climate change on agriculture can be moderate to significantly high (Kumar and Gautam, 2014). But coming to any solid conclusion is not possible. According to the IPCC fourth report, climate change is happening at faster rate than expected. As per the IPCC global warming prediction report, global mean temperature increase would range in 1.0–3.5°C by 2100, and in case of CO_2 enrichment, it would range in 1.5–4.5°C. Global mean surface temperature anomalies are also recorded to be increasing consistently with time (NASA GISS). Report data showed that global greenhouse gas (GHG) emissions from 1750 to 2010 abruptly increased with millions of metric tons of carbon released into atmosphere. Increase in temperature affected the crop productivity worldwide, and it has been assumed to be increased due to the rise in mean local temperature from 1 to 3°C. However, an

Global Climate Change. https://doi.org/10.1016/B978-0-12-822928-6.00010-1

increase beyond this range would reduce overall productivity. In contrast, a smaller rise in temperature (1–2°C) in dry tropical areas is expected to significantly reduce crop productivity, leading to an increase in the frequency of mass starvation. Agricultural yield is expected to enhance up to some extent at mid- to high latitudes due to an average increase in temperature of between 1 and 3°C and decrease after that. Rise in mean temperature would facilitate the heavy increase in insect pests' population; thus, damaging effects will be on the natural ecosystems. Crop productivity is also speculated to decrease in arid and semiarid regions of the globe. By the middle of the 21st century, East and Southeast Asia are expected to experience 20% yield enhancement, whereas Central and South Asia are expected to face approximately 20% fall in production. Agricultural land coming under the hot regions of the world is also greatly affected due to climate change as most of the rain water evaporates under high temperature, and tropical storms cause soil erosion and runoff. Rainfall is also highly variable in these hot and dry regions, both within the year and decades. High variability in rainfall is among one of the main causes of loss in agricultural production (Bray et al., 2000). In this regard, increasing population at rapid rate and, thus, increased food demand have come out as a serious issue in parallel to other problems in front of us. It is estimated that approximately 80% of the population that is living in less developed and developing countries will be 90% by 2050 (CGIAR-ISPC, 2010). As, with increasing population, the demand for food will increase and if we fail in keeping the food production rate high, that will result in chronic food shortage and other related issues. Major food crops that meet the global food demand and supply are cereals, and they are highly sensitive to changing climate. Presently, food demand and supplies are being secured under these varying conditions and ongoing climate change. According to the prediction of climate models, it is supposed that, in future, more extreme events and variability in weather will happen with more droughts and number of dry days, heavy rainfall, and storms. These extreme weather events will affect pest attack time and hot spot of diseases, thus posing severe risks to crops. Climate change is an extra threat in front of countries such as India, since increasing population, industrialization, and economic development are already posing pressure on ecological and socioeconomic status. Climate of countries such as India can become warmer under various changing factors such as increased atmospheric carbon dioxide and variations in various environmental conditions. Growing population and pollution are topics of major concern in India. Hence, availability of water resources and increasing population relationship is disturbed. In a projection, a decline in mean per capita annual freshwater availability and population growth for the year 1951–2050 was shown (Kumar and Gautam, 2014; Mall et al., 2006). Hence, an integrated management of all the governing factors controlling the agricultural productions is needed. For the management of these factors, implications and modifications in existing policies are strictly required.

2. Global climate change

With increasing global warming and global mean temperature, an increasing challenge for food security at world and local levels has come as a major challenge in front of the research community. It is observed that global average temperature has increased and variation in the intensity of spatial distribution has been noticed (IPCC, 2018). The Sixth Assessment Report (AR6) of IPCC stated that the climate model projections showed a clear change in global and local climate characteristics (Fig. 7.1). These changes include an increment in average temperature in both ocean and land areas (surface temperature), extreme events mostly in the habitat regions considering high probability of heavy rainfall in many regions, and drought probability in some region with moderate probability. On the other hand, anthropogenic activities caused an increase in temperature by about 1°C in comparison with preindustrial scales with some variations between 0.8 and 1.2°C. The overall global warming is supposed to increase by 1.5°C between 2030 and 2052 (Fig. 7.1), if continuing with the same rate (IPCC, 2018). The five warmest years are recorded in the past four decades after 2010 (https://www.giss.-nasa.gov/research/news/20170118/). The anthropogenic activities responsible for the change in the climate system in comparison with preindustrial time can persist for many decades and will continue with same change in the climate system for the long periods (IPCC, 2018). The increment in temperature and variations in precipitation have significantly hampered the agricultural productions globally. Currently, agricultural production in several regions of the world, particularly in developing countries, is affected and limited because of climate change. The decline in rainfall in these regions is the main factor, leading to hamper agriculture production. The rain-fed agricultural areas in the developing countries are approximately 63% that is climatically suitable for farming, but this percentage changes continuously from one

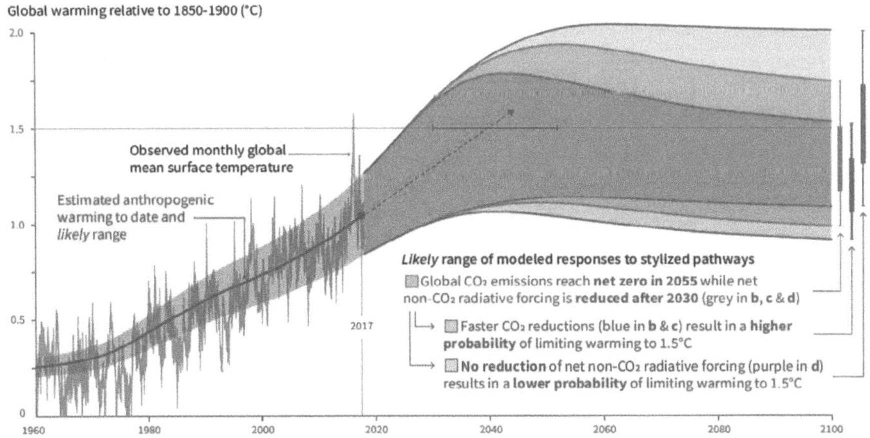

Figure 7.1 Observed global temperature change and modeled responses to stylized anthropogenic emission and forcing pathways.
Source: IPCC AR6 report 2018.

region to another region. This rain-fed region is 85% in South America and 84% in Southeast Asia but is limited to 64% in Central America and 53% in Africa (Parry, 2019) and in China (Chandio et al., 2020). In Southeast Asia approximately, 17% of total areas are covered with mountains and share cool climatic conditions, 65% is dry region, and only 18% is productive land with several climatic limitations (Perry, 2019). Since there are very limited potential rain-fed agricultural areas available, impacts of climate change will surely affect the productivity and food security of that region of the world (Perry, 2019).

3. Impact of climate change on the agricultural sectors

The agricultural sectors can be affected in multiple ways as a consequence of climate change such as food stability, food security, production, water availability, hydrological cycles, and transportation in many countries of the world including India. Also, crops are solely climate dependent, and any variability in climate will have potential impact on agricultural productions on both regional and global levels (Abraha and Savage, 2006). Agricultural production of any region of the world is directly affected by temperature, day—night duration, sunlight and water availability, the major climate factors, and driving force in crop growth. Along with these factors, longer growing seasons and high temperature may also enhance the agricultural production. But, along with these factors, extreme weather events and reduced water availability as a single factor can affect the crop production in many ways. Altogether, these parameters may finally pose negative impact on crop yield. According to an old report of the World Health Organization (WHO), climate change has significantly affected the present agricultural systems, and in future, impact can be even more threatening (World Health Organization, 2009). Temperature rise by 1°C will reduce the wheat yield by 5%—10% (Asseng et al., 2015). It is estimated that crop production in South Asia could decrease by up to 30% by the middle of 21st century (2001). Talking about the tropical and subtropical areas of Africa, its agricultural production is more sensitive and at risk due to global warming than temperate regions (Mendelsohn, 2008). Taking the example of India, there could be major decline in agricultural productions (40%) by the year 2080s under the influence of climate change (IPCC, 2007a,b).

From an economic point of view, the agriculture sector is one of the most important sectors, contributing more than 15% to gross domestic product (GDP) in 2018 (https://pib.gov.in) in India. The contribution of the agriculture sector in the Indian economy is much higher than the world's average contribution (6.4%). Arable land areas (approximately 159 Mha) in India are the second larger producer of agriculture products in the world (after the United States of America), span of tropical and subtropical climates that account for 7.39% of total global agriculture output. The climate change impacts on the crops depend on the local condition of climate and varieties of crops that decide crop yield per unit area. In India, increment in temperature is supposed to be harmful for rain-fed crops and may reduce the net yield. It is observed that global and local climate uncertainty can change both location and capacity of agricultural sector. Generally, the largest water-consuming sector of total water budgets of a country is agriculture sector. The impacts of climate change such as an increase in temperature and other factors including droughts, floods, plant diseases, pests, etc. may affect the net crop production negatively. To manage these climate change barriers, current infrastructure, techniques, crop selection, and cropping systems must be changed considering future demands due to rapidly growing population in India. The reduction in agriculture production due to climate change can be managed by adaptation of suitable strategies such as implementing agroforestry system, shifting in plantation time, change in selection of crop, change in plant–crop varieties, change in irrigation pattern, and so on at regional and local levels.

3.1 Climate change impact on Indian agriculture

The relationship between climate change impacts and the agriculture production is one of the most important topics of debates regarding policy making in developed as well as developing countries such as India. Various studies have concluded that production of a number of crops is dramatically affected by a jump in temperature (Lobell et al., 2012; Aggarwal et al., 2009), variability in rainfall pattern (Prasanna, 2014; Mall et al., 2006), and variability in intensity and frequency of extreme weather events such as temperature extremes, floods, and droughts (Brida and Owiyo, 2013; Singh et al., 2013b). Rahman et al. (2020) found that climate change had significant negative impact on agricultural production in the upper Meghna river basin of India. The impacts of past and future climate change on yield of various cereal crops that too from the different regions worldwide indicated decreasing yield on the basis of simulations and projections studies. Projections showed that there can be approximately 35% reduction in yield of rice, 50% for sorghum, up to 20% for wheat, 13% for barley, and with maximum of 60% for maize. Data were based on location, projection year, and climate scenarios that will be happening in future (Porter et al., 2014).

Growth of crop, water consumption, and yield in the optimum condition depend upon the climate and weather conditions of growing seasons in the local area. Slight change in the normal local weather condition can change the yield of the cereals and can impact negatively. Considering the climatic conditions of India, there are mainly two types of crops growing seasons. The summer (June to September) season is for Kharif crop, which depends on southwest monsoon and winter (October–December) season is for Rabi crop, which depends mostly upon northeast monsoon in India. Kharif crop includes maize, paddy, jute, sugarcane, soybean, cotton, and so on, and depending on the condition of local weather, it can be harvested in November to December. On the other hand, Rabi crops include pulses, mustard, wheat, potato, onion, and so on and are harvested during January and February. As the southwest monsoon plays a sensitive role for Kharif crop, it accounts for >50% of the food grain production and approximately 65% for the oilseeds. Since variability in interannual monsoon rainfall in India results in droughts and floods at a large scale, it leaves major effects on Indian agricultural production (Parthasarathy et al., 1992; Selvaraju, 2003; Kumar et al., 2004) and economy. Therefore, southwest and northeast monsoon both are critical for Kharif and Rabi crop yield in India. Impacts of recent climate change trends in India and other countries on paddy crop, wheat, maize, barley, and pulses are discussed in detail in Section 3.1.2.

3.1.1 Impact on the agricultural ecosystem

Agricultural ecosystems are dynamic and complex systems of climate zones that include various factors such as temperature, precipitation, surrounding conditions influencing the crop growth by either direct or indirect interaction with the crop plants and animals from that area, soil nutrients, and plant growth−promoting microbiota. Changes in any one of the climatic parameter or conditions directly influence all the aspects of a particular agricultural ecosystem and hence affect the system dynamics and cause imbalances in the ecosystem. Additionally, any disturbances to the factors of an agricultural ecosystem may result in frequent occurrence of new types of blights, pest attack, emergence of new pathogens, and damage to plant growth−promoting microbes. These changes result in potential damage to agricultural production, hence impacting the agricultural economy. Damages to wheat, rice, pulses, and other main food and fruit crops including grapes, apples, oranges, peaches, and soybeans, especially by brown-green grasshoppers, are increasing in countries such as Pakistan, India, Ethiopia ((indianexpress), https://indianexpress.com/article/explained/explained-locust-agriculture-crops-india-lockdown-6400755/ on May 9, 2020), and Korea (Kim et al., 2009). *Stripe tenuivirus* of rice, a viral disease of paddy crop, has spread in South Korea and caused damaged to field areas expending up to 14,137 ha (Kim et al., 2009).

3.1.2 Impact on the agricultural production

3.1.2.1 Rice

Rice is one of the important and main staple food and cultivated crop worldwide. It is also the most cultivated crop in arsenic (As)-affected areas of the world especially in Southeast Asia, viz., China, Bangladesh, and India (Poonam et al., 2017; Yadav and Srivastava, 2020). It is estimated that total global rice production is falling behind the demand owing to rapid increase in population and reducing agricultural lands. The land being used for agricultural activities is also suffering several toxic contaminants. Among various toxicants responsible for decline in production of rice, arsenic-contaminated soil of Southeast Asian countries is the leading factor behind decline in rice productivity, rice grain quality, and high grain arsenic. A wide range of approaches have been tested or proposed by scientific community to mitigate this problem. These include phytoremediation, arbuscular mycorrhizal fungi association, bacteria and algal additions, fertilizers/chemicals, use of stress tolerant/low arsenic accumulating rice varieties, and coculture practices (Yadav and Srivastava, 2020). Another serious challenge in achieving high yield and quality rice production is the climate change and its impact. As it is unequivocal that climate change is happening, it may probably be considered as the problem of the century being faced by the whole world whose impacts, however, may vary from region to region. It is mentioned that the growth conditions of rice crop have changed due to the climate change, and rice production would suffer in near future. Several climate model predictions showed that negative impacts of rise in temperature and GHG would be far more devastating in developing countries. Considering methane (CH_4) production from rice fields, about

10%−11% of arable land comes under the rice cultivation areas that act as a major source of methane emission, a potential GHG contributor to global warming. According to the literatures, 20% of methane gas emissions from agricultural field come from paddy field. In the present scenario, rice crop accounts for more than four times higher global warming potential as compared with other crops cultivated worldwide such as wheat, barley, and corn. Methane is produced by methanogenic bacteria present in paddy field, which feeds upon CO_2 for respiration. Hence, increasing CO_2 in atmosphere will favor methanogenic bacteria and rice production. It is also estimated that climate change would affect the rice production in two contrasting ways. It has been shown that increased CO_2 stimulated CH_4 emissions by 42.2% and rice yields by 24.6%, whereas in case of warming, yields were found to be decreased (Van Groenigen et al., 2013). In a metaanalysis of collected data, it is proposed that 4°C warming and CO_2 enrichment together will result in an increase in CH_4 emissions by 58% and decreased rice yields by 33.8% and yield-scaled CH_4 emissions will increase by 105.50% (Van Groenigen et al., 2013). Hence, changes in local climatic conditions would be influencing rice production globally. The rather unexplored scenario is the interaction between rising temperatures, GHGs, soil microbiota, and variability in rainfall and water availability. It is observed that for every 1°C rise in temperature, a reduction of 10% in productivity of rice crops may occurred (IRRI report, 2004). In this support, Agrawal (2011) proposed a significant reduction in rice and is expected reaching up to 6.7% by 2020, 15.1% 2050, and 28.2% by 2080. Paddy crop is very sensitive to high temperature specially at tiller and inflorescence stage (Jagadish et al., 2008). Crop duration gets shorter at temperatures beyond the threshold, and this reduces grain filling duration and promotes the sterility in spikelet resulting in reduced crop yield and low quality of grain (Fitzgerald and Resurreccion, 2009; Van Groenigen et al., 2013). It is also observed that elevated soil temperature increased uptake of arsenic by the rice plants was indeed reported. In this case, the speciation of arsenic in the soil and plants also needs to be quantified, as different arsenic species are known to have variable uptake and transport in plants and differential toxicity to plants and humans. The solution lies in the development of climate-resilient rice varieties, changes in agronomic management practices, and drought-tolerant variety selection that also aim to reduce arsenic in rice grains for the arsenic-affected regions. The challenges are multifaceted and would require targeted research efforts in near future in rather short time.

3.1.2.2 Wheat

Increment in global temperature and depletion in water resources (drought and depletion of ground water table) are among the most concerned abiotic stress posing to world's agriculture system (Zampieri et al., 2017). This can be a big challenge in front of scientific community to face these environmental variabilities and develop adaptive measures (crop improvement and developing high-yielding and stress-tolerant varieties). Wheat production is also reported to be affected by rising temperature that is mainly related to the increasing future CO_2 concentration. Going with the estimations, reports stated 6% decline in wheat production globally for each 1°C increment in global temperature (Asseng et al., 2015) and thus showed a direct risk to global food production and UN development goal for food security. Gahukar (2009) reported that due to increment in mean temperature, wheat yield during 2003 and 2004 faced great loss. Eitzinger et al. (2003) showed that the CO_2 effect helps in increasing wheat crop yield in the experimental area. In the support of above study, similar study was reported by Luo et al. (2005).They also studied the wheat models under all the possible CO_2 levels for Southern Australia projected for 2080s and concluded that wheat yield will increase under CO_2 enrichment.

Strategies adapted by wheat crop in Indo-Gangetic Plains (IGP) to fight climate change for projected years up to 2050s are discussed by Ortiz et al. (2008), and they also showed beneficial side of global warming for wheat crop production in some areas of world like IGP and can pose negative impact in terms of reducing productivity in critical temperature zones. Thus, it is urgently required to develop some high-temperature and drought-tolerant wheat germplasm to fight the impact of climate change. Anwar et al. (2007) simulated the impact of climate change on wheat in southeastern Australia and showed that the future elevated CO_2 level can reduce the wheat yield by $\approx$25%. It can be concluded that wheat productivity and nutritional quality will face a great threat in coming future.

3.1.2.3 Maize

Maize (*Zea mays*) is one of the most growing crops in the world. It is the staple food crop in African countries and growing at large scale in many countries such as Africa, India, Pakistan, China, and America. Luhunga (2017) studied two representative concentration pathways and stated that there would be a high decrease in maize crop production by 9.6% at the end of 21st century. The possible reason behind this reduced production in future is the shorter growing season due to rice in temperature. He further added that there would be expected increment in temperature by 1.84 and 1.43°C in seasonal maximum and minimum temperature under RCP4.5 and by 2.72 and 2.2°C under RCP8.5. Predictions made for Malawi, a Southeastern African country, for mid- to end century period showed that the rain-fed maize crop productivity will decrease up to 14%–33% under the impact of climate change (Msowoya et al., 2016). As per the Lobell et al. (2011), an increment in temperature by 2°C will also affect the maize crop production in Africa as it is a sensitive crop to daytime temperature. Hence, high demand of maize in sub-Saharan Africa region will greatly be affected due to decrease in yield under changing climate. According to the published data worldwide, 10% reduction in maize crop yield is reported under the climate change (IPCC, 2007a,b).

3.1.2.4 Barley

The analysis of crop productivity and yield data of worldwide recorded between 1981 and 2002 showed reduction in yield of barley along with wheat and maize, which contributes to US$5 billion annual loss (Lobell and Field, 2007). Barley grows in arid and semiarid Mediterranean environments and is main crop in that region of the world. It is estimated that barley yield will decrease under the impact of climate change (Samarah, 2005). Relying on the climate projections model, they show that Mediterranean regions will get hotter and thus drier in future due to climate change. A group of projection model showed that there will be an increase in mean season temperature (growing) between 0.9 and 2.16°C, while rainfall will be affected between −24% and +24%. Thus, barley yields will decrease significantly between 25% and 8% in future, as per the climate projections model (Cammarano et al., 2019). According to the study of Tuttolomondo et al. (2009), analysis of complex simulations and model for Italy revealed that with increasing variability in rainfall and temperature, yield variability of barley also increases. Some varieties were showing high yield sensitivity to rainfall variability in comparison with temperature, whereas there was considerable fall in yield under half ($P^1/_2$) reduction in rainfall variability. Thus, barley crop showed both rainfall and temperature dependency under different scenarios. Similar results were proposed by Ko et al. (2019) for South Korea that yield of the barley crop will increase as per the projections of model. As South Korea is a cold country, barley crop is moderate-to-high temperature−dependent crop. In that case, low temperature for barley crop will pose threat and reduce the production (Samarah, 2005). Even drought at some important phase (postanthesis) of barley crop can be damaging and can result in reduced yield with shorter process of grain filling (Samarah, 2005). So a significant increase in temperature favors the crop, thus increasing the growth and yield, whereas rise in CO_2 concentration shows 10%−15% reduction in barley grain (Taub et al., 2008) and reduction in nitrogen content of barley (Manderscheid et al., 1995), rice (Kobayashi et al., 2006), and wheat (Manderscheid et al., 1995; Pleijel and Uddling, 2012).

3.1.2.5 Pulses

Pulses and other crops that are very sensitive to climate change and greatly depend on rainfall will be highly impacted by the climate change and variability in rainfall. Agriculture of Asian and African countries greatly depends on either high temperature range or good rainfall. It is expected that crop productivity will decrease (10%−40%) in Asian countries due to climate variability (Fischer et al., 2005, IPCC). Countries such as India will face threat to cultivation and production of crops such as pulses that depend on precipitation (Basu et al., 2009). According to the projection report of IPCC, in India, mean rainfall is projected to increase by 10%, whereas mean temperature will get increased by 1.7−3.2°C during July−October (Kharif crop) and November−March (Rabi crop), respectively, by 2070 (Gupta, 2011; Bahl, 2015).

We should focus more on high calorific value crops, increasing the agriculture land for their production such as pulses and oilseeds (soybean) to fight the rainfall variability. Pulses play multifaceted role with other food crops, from being protein rich to maintaining soil fertility. Thus, they are important for food nutritional security and sustainability goal that too under the climate change scenarios. It is estimated that demand for the leguminous crop in South Asia will increase by 30% (between 2010 and 2030) with growing population (Aggarwal et al., 2004; Bahl et al., 2015).

3.1.3 Impact on the insect pest and disease development

Crops growing in a given habitat are continuously affected (positively/negatively) via direct and indirect interactions with environmental factors such as soil microorganisms, insect pests, plant pathogens, and herbivores (Bhadouria et al., 2019). According to Bouma (2008), around 10% of the predicted pests are serious threat to agricultural productivity. Changes in global environmental conditions have employed multifold ameliorations in the rate of multiplication, distribution, and survival of several plant pathogens as well as insect pests. Variations in temperature have greatly influenced the behavior distribution and disorder transmission of insects, while plant illness caused by the particular carrier is extensively influenced with the aid of moisture content, temperature, and precipitation (Coakley et al., 1999). Yamamura and Kiritani (1998) stated that every 2°C rise in temperature might also result in speedy completion of the life cycle of the pest. Changes in the temperature of a particular geographical region of the world have a greater influence on insect behavior than plant and animal species. In addition, wet prerequisites can also possibly carry on severe insect and plant pathogen infestation or impact a geographical shift of some pests. Insect profusion may be additionally decided by way of increase in CO_2 concentration, wind-induced dispersal of pests, variations in soil nitrogen content material, and population density (Gatto et al., 2016). According to Weintraub and Beanland (2006), agricultural crops are prone to disease caused by pathogenic fungi, bacteria, nematodes, viruses, and viroids resulting in great loss. But several pathogens, which include viruses and phytoplasmas, are frequently transported via a vector such as insects and pests. Insects and pests respond to climate change in various ways (Koli et al., 2019), such as geographical distribution, variation in temperature throughout, the evolution of species with enhanced virulence, and pest shifts due to changes in food availability and result in phenology changes to influencing community composition and dynamics (Menendez, 2007; Moore and Allard, 2008). However, environmental variables such as moisture content, mild instantaneity, precipitation, and temperature have a remarkable effect on the existence cycle and transfer of pathogenic bacteria, fungus, and virus (Patterson et al., 1999). Ozone layer depletion and varying intensity of UV-B radiation reaching on the earth surface and availability of soil nutrient status exerts a significant impact on disease developments (Thompson et al., 1993; Manning and von Tiedemann, 1995). The trend of temperature, higher to low, more significantly affects the geographic spread of insect pastes during climate change (Menendez, 2007; Boullis et al., 2015). Sharma et al. (2005) and Sharma (2010) reported that greater population shifts in *Helicoverpa armigera* (Hub.) and *Maruca vitrata* (Fab.) to temperate regions due to climate change, whicch will have drastic effects on the hosts of these pests. Kranthi et al. (2009) reported that increased CO_2 levels are responsible for emergence of serious threats of polyphagous pest *Spodoptera litura* (Fab.). Tay et al. (2013) recognized woolly aphid (*Ceratovacuna lanigera*) as an invasive and serious threat in Maharashtra (India) due to climate change. According to Parmesan (2006) and Tripathi et al. (2016), alteration in temperature, moisture regimes, and daylengths might have interrupted diapause activities in some insect species. Substantial economic loss results due to increased pest infestation on crops and significantly increased population of insect pests under climate change.

3.1.4 Impact on the use of agrochemicals

Agrochemicals, especially chemical fertilizers, play a critical role in climate change, as they produce GHGs and reactive nitrogen (Nr) into the environment. GHGs' role in global warming is already well known, whereas Nr circulates easily in atmosphere, air, water, soil, and plants and thus has multiway role in impact, mitigation, and adaptation of climate change (Suddick and Davidson, 2012). As mentioned earlier, up to 24% of the global GHG emission accounted from agricultural sectors. N_2O (nitrous oxide) is the first potent green gas from the agriculture sector that contributes to climate change because of synthetic nitrogen fertilizer as chemical fertilizer. Simultaneously, methane is the second most GHG that contributes to climate change from agricultural field (paddy field), cow, and sheep's digestive tract due to anaerobic fermentation (Hopkins and Del Prado, 2007). With changing climate, behavior, diversity, and population of insect and pest population will change. To control pest population, changes in application pattern and type of agrochemicals (like pesticide) will result in cost variability (Aktar et al., 2009). The change in weather can result in increased geographical distribution of pest and pathogens, increased risk due to invasions by migrant pest, increased resistance of pests to pesticides, and extension of the growing season of crops (Koli et al., 2019), thus providing more time for pest/pathogen buildup, and as a consequence of these events, pesticide use will increase with a 2.4- to 2.7-fold predicted increase by 2050 (Tilman et al., 2001). Due to the increase in diversity of insect pests, diseases, and weeds, performance of pesticides is affected. With changes in climatic condition, the rate of herbicide application exceeded in comparison with insecticides and fungicides due to the adaptive resistance of weed and the declining performance of herbicides (Müller et al., 2010; Ntonifor, 2011; Delcour et al., 2015). Application of high dose of pesticide under changing climate has increased the threat to beneficial insects such as ladybugs, honeybees, and soil microorganisms and risk of environmental contamination (Verma et al., 2014; Jaiswal and Verma, 2016). Yeo et al. (2003) reported the volatilization of agrochemicals (such as pesticide and chemical fertilizers) from the soil system and the surface of vegetables that occurred under increased environmental temperature and caused atmospheric pollution. Additionally, it was also observed that organochlorines showed more volatilization as compared with other groups of pesticide under warmer weather. Thus, warmer climate increases the risk of diffuse release of pesticides into the environment. According to Maksymiv (2015), the diffusion of agrochemicals also results from increase in strength of rainfall that promotes the risk of leaching in soil, and simultaneously, increase in temperature can counteract the leaching of pesticides due to maximum rate of degradation. But the higher temperature poses the risk of drought in some regions and, thus, inhibits the pesticide degradation. Therefore, under continuous change in trends of temperature, intensity of pesticide use is expected to increase in amount, doses, frequencies, and different category/types of applied products. Hence, adapted pesticide use will ultimately influence consumer at the end of the food chain (Delcour et al., 2015).

3.1.5 Impacts on the agricultural economy

The impact of climate change in the agriculture economy might be divided into two broad categories. The first category of agroeconomy is defined as the impact of climate change variation on the yield and production of crops that affect the price of agricultural products and profits and thus will help in consequent implications in national and global economics. The second group of agroeconomic studies focuses on the impact of climate change in developing countries, where agriculture plays an important role in the livelihood of a particular group of population and thus contributes to economy. Overall, these studies focus on the emerging issues of food security, human development, and poverty alleviation in developing countries. The losses or gains of national and global economies are calculated by a combination of computable general equilibrium (or CGE) models. Quantification of losses and gains to GDP is estimated using models coupled with climate models and is relative to some reference models of economic growth. A recent version of these models is integrated assessment model (IAM) referred as fairly simple in comparison with other complicated models. Such type of models has ubiquitous nature in the computation of the impact of climate change on economy and the value migration policies. As these models cover vast and different economic sectors, they create a complexity in terms of uncertainties that derive away economic assumptions from the agronomic and climate modeling. So that, CGE and IAM framework models are open to plenty of criticism (Taylor and Arnim, 2006; Ackerman et al., 2009; Jayaraman, 2011). According to literature, three major methods have been used to study the sensitivity of agriculture to climate variability (FAO, 2000). The first method is focused on the net revenue per hectare across a number of regions with different climate conditions (Sanghi and Mendelsohn, 2008). Variation in net revenue per hectare across farms may also depend on other biophysical variables such as water supply (direct precipitation) and economy (impact of prices and other market effects). According to Reilly (1999), such variables have not yet been included into models based totally on the cross-sectional methods. The second method is to study agronomic—economic approach, based on the crop growth models (Rosenzweig and Parry, 1994). Application of this method into economic models predicts and determines the final economic impact of climate change on agriculture. Agroecological method is the third one, developed by the Food and Agriculture Organization (FAO) (Darwin et al., 1995). This method has been developed for studying the revenue maximization or cost minimization module by using a number of ecophysiological variables under the economic impact of climate change on agriculture. The advanced version of this model agroecological zoning is associated with an applied general equilibrium model to derive more economically relevant estimates (Jayaraman, 2011). However, the analysis of the economic impact of climate change on crop production by using the agroeconomic model creates a problem of underestimation as it does not reflect the indirect impact of climate change, viz., crop conversion and adjustment of input factors for adaptation to climate change (Kim, 2008). The application of the Ricardian model has been proposed as the solution to this problem. This model helped in the estimation of the future economic impact of climate change on agricultural crops by estimating the current value of farmland price as the discounted value of future rent (Sanghi, 1998; Kim, 2008). Sanghi (1998) have estimated the economic impact of climate change on Indian agriculture crops by using the Ricardian model to forecast aggregate change in net revenue for selected climate factors such as temperature and rainfall showing 12.30% losses in net revenue. Kumar and Parikh (2001), reported that the economic impact of climate change might not be uniform throughout the world with variability as loss and gain. So, they used the cross-sectional data on climate, crop, soil, socioeconomic, and geophysical data, from 271 districts of 13 major states of India from 1958 to 1987 in the modified Ricardian model. The result of this study showed that January, April, and July seasons have a negative impact on net revenue and positive effect on October seasons. Regional-level results showed that subtropical states such as Haryana, Punjab, and West Uttar Pradesh were adversely affected along with western coastal regions and southern Indian states. But western regions such as West Bengal, Orissa, and some regions of the Bihar state have beneficial effects as compared with eastern states. Sanghi and Mendelsohn (2008) observed the economic impacts of global warming on the agriculture sector of developing countries such as India and Brazil. The panel data of both countries were incorporated in the Ricardian model and predicted the variation in net farm income with future climate change scenarios. Study also reported that, if temperature increases by ~2°C along with 8% increment in precipitation, the net revenue of agriculture production might go down, up to 12% in India and 20% in Brazil, without carbon fertilization. Simultaneously, they predicted that range of annual damage due to possible climate change varies 4%–26% in India and 1%–36% for Brazil. Thus, the economic effect of climate change on agriculture crops was not uniform in the country and applicable throughout the world with different climatic zone and conditions.

4. Mitigation and adaptation strategies for the agriculture

4.1 Mitigation strategies

In the context of climate change, we are facing two major challenges; one of them is meeting global food demand and another is minimizing the emission of GHGs from the agricultural sector. Smart agriculture practices (SAPs) help in achieving both challenging tasks without any disturbance to the environmental system, as it is the main component of climate change mitigation strategies. Smart agricultural practices include eco-friendly practices that control the rate of land degradation and soil fertility loss due to drought, salinity, waterlogging, and soil erosion. Thus, we observed that agriculture sector plays a critical role in GHGs mitigation by lowering the 10% of N_2O emission, 42% by carbon offset through biofuel production, 32% by absorbing CO_2 emission, and 15% by reducing methane emission under climate change strategies (IPCC, 2007a,b; Arif et al., 2020). Dora (2018) proposed five vital methods to mitigate the impact of climate change on agricultural system that the Indian government should consider:

i. **Smart irrigation system (SIS):** Less than 50% of agricultural land in India is irrigated land, due to increasing water scarcity and depleting groundwater resources. Several states of India, such as Karnataka, Maharashtra, Madhya Pradesh, Rajasthan, Chhattisgarh, and Jharkhand, have extremely poor water irrigation access due to vulnerable climate change. Therefore, effective water management and SIS such as drip or sprinklers method should be incorporated and allocated across state of India facing water scarcity problem.
ii. **Reduce postharvest loss:** Postharvest loss of major agriculture produced is estimated to be US$13 billion. Only 2.2% of vegetables and fruits are consumed in India, and rest are prepared to export; thus, chances of wastage increased. In comparison with India, the United States and China have their own storage and processing facility of vegetables and fruits. Due to lack of centralized large-scale processing and preservation technology, India has faced a big postharvest loss of agriculture products under vulnerable climate change. So there is an overwhelming need to develop advanced technology for local processing, smart packing, and transport facilities to reduce the postharvest loss.
iii. **Environmental data-driven supply chain management:** The agricultural supply chain may be improving by using environmental data generated through new technologies such as sensors, GPS, and satellite imaging. The application of these new technologies can help in collection of significant and meaningful data that make more resilient to Indian agriculture system.
iv. **Farmer-centric crop insurance:** To the protection loss of agricultural production by uncertainty in climate change, an effective crop insurance package is required. Several crop insurance programs are already paying for the loss, but compensation amount is very little as compared with the actual loss. Therefore, the crop insurance package needs to overhaul, so that poor or small agriculture landholder farmers can be protected by long-term insurance cover and low premium.
v. **Evidence-based research:** The agricultural research (breeding, production technology development, resource management innovation, etc.) plays a vital role in increasing yield and also increases the resilience to the climate change problems (such as extreme heat and precipitation, pest, and crop disease). The economically important crops such as pulses and soybean are more vulnerable to climate change. So that advanced agriculture research will directly improve efficiency and boost crop yields along with minimizing the impact of climate change on farming systems.

4.2 Adaptation strategies

According to IPCC (2007a,b), adaptation strategies of climate change are based on initiatives and measures to minimize the uncertain impacts of climate change on environmental system. Various strategies have been implemented to alleviate the effects of climate change. Significant methods adapted for the mitigation of climate change impact are represented in the following section.

4.2.1 Soil management practices

Soil system is negatively affected by an increment in CO_2 level, change in temperature, rainfall, and vegetative cover along with conventional agriculture practices due to impact of climate change. So that some soil practices (conservation tillage, terrace farming, and integrated nutrient management) are found helpful in climate change adaptation (Arif et al., 2020).

4.2.2 Shifting the location of seed production industries

Fischer et al. (2005) projected based on climate change—related models an increment in agricultural area in countries such as Northern Europe, North America, and Russia. As these countries will have favorable conditions for agriculture, rise in temperature will result in more crop-growing seasons, whereas countries such as Africa will face loss of fertile-cultivated land due to deterioration of agricultural land climate change scenario (rise in temperature and moisture stress). Thus, we can conclude that the impact of climate change is geographical locations based. Similarly, Olesen and Bindi (2002) proposed low yield for Mediterranean region and high productivity and expanded farming cover for Europe shifting toward the north of the region. Thus, shifting of agricultural practices with latitude or elevation is important to deal with impact of warming on seed quality and yield (Farrar and Williams, 1991; Bhadouria et al., 2019). Kumar et al. (2009) analyzed the climate change impact on seed production of cabbage in northwestern India between 1981 and 2004. Study reports ~40% reduction in yield of the same cultivar of cabbage seed with increasing minimum and maximum temperature during pod setting. However, a sharp decline in seed yield observed when a mean of maximum temperature exceeded 25°C during the period between 1981 and 2004. Thus, it has been concluded that to enhance the cabbage seed production, shifting toward northward to minimize the effect of increasing temperature is important (Hampton et al., 2016). The seed industries of India, China, North America, northern Europe, southern south American countries, New Zealand, and Australia will possibly move with latitude or elevation, but it will depend on water availability and the availability of land due to increase in the complexity between land use and other agriculture production (Hampton et al., 2012). However, seed industries of lower altitude countries (like Africa) may not have this option, so they are importing more of their seed requirements.

4.2.3 Shifting crop sowing date

The detrimental effect of high temperature on inflorescence stage and seed shed has been reduced by changing the sowing time and recommended as an adaptation strategy to tackle such type of issues (Singh et al., 2013a). do Rio et al. (2016) recommended changing the sowing date in some regions of Brazil as a feasible option to reduce the negative impacts of warming on soybean yields, but only when sowing was delayed in relation to the time presently recommended. Huang et al. (2020) projected that, with rise in temperature, sowing window will extend, hence changing the sowing date and varieties of Maize. Some regions of china can be beneficial in enhancing the yield. Hu et al. (2017) also observed the same effect with this strategy effective for potato crop grown in semiarid regions of China and colder regions of Europe (Olesen et al., 2011). This is an appropriate effective strategy that works with other factors such as soil temperature effects on germination and emergence, daylength responses, water availability, and increased weed, pest, and disease pressure to adapt during seed sowing (Board and Hall, 1984; Parry, 2000; Olesen and Bindi, 2002; Rahman et al., 2006; Hatfield et al., 2011; Hampton et al., 2016).

4.2.4 Plant breeding and focus on developed variety

Over the Past 50 years, plant breeding has been a success story in improving the yields and developing cultivars that can help the crop in adapting drought and temperature stress resistance, resistance to pests and diseases, and the ability to cope with salinity and waterlogging (Jaggard et al., 2010; Ceccarelli et al., 2010; Singh et al., 2013a). The aforementioned resilient trait cultivars against environmental stress can be achieved by using a combination of conventional breeding methodologies, molecular technologies (such as marker–trait associations and genome-wide selection), and transgenic approaches (Ortiz et al., 2008; Varshney et al., 2011; Cairns et al., 2013). The traditional and contemporary molecular breeding methodology and transgenic approaches are a potential genetic strategy to develop plant heat-stress tolerance and drought-stress tolerance cultivars (Reif et al., 2005; Wahid et al., 2007; Habash et al., 2009; Bita and Gerats, 2013).

5. Policy implications

To project and implement an effective policy for management of the future climate change impacts, it is very important to project about the sensitive sectors that will be affected more, and in what ways agriculture will be affected. It is the future agricultural system that will experience future climate change at greater extent and will be greatly affected. So, changes in techniques, modifications, and changes in existing policy must be considered.

The priorities for adaptation policy measures to climate change are determined through the EEFA models, which includes effectivity (E), effectiveness (E), feasibility(F), and acceptance (A) as evaluation policy criteria. The efficiency (E) confirms the cost and benefit analysis of adaptation policy; the effectiveness (E) is to check the extent of achievement of the adaptation covering the objective; feasibility(F) is to check whether the policy-making authorities can really put in force the adaptation policy; and acceptance (A) is to investigate whether or not the rural population would comfortably accept the adaptation policy. Thus, there are various options for adaptation measure to climate change impact on agricultural sector that includes research and development, infrastructure management, legal and institutional improvement, economic means, manpower training and education, monitoring, and technology and management (Kim, 2008). Thus, policy implications for adaptation management of climate change impacts on agriculture systems are multidisciplinary and include possible adaptations to address the various issues (http://www.indiaenvironmentportal.org.in/files/Defra-india-cc-agri.pdf) as discussed in the following:

i. **Food security policy:** To account for changing crop yields properly transferring boundaries for crops, and the impact that this can have on food supply.
ii. **Trade policy:** Changes in certain crops can affect imports/exports, depending on the crops (like cash crops such as chillies, tomato, and onion).
iii. **Livelihoods:** Agriculture contributes significantly to gross national product; it is quintessential that policy addresses issues of loss of livelihood with changes in crops, as well as the need to shift some regions to new crops, and the associated abilities training required.
iv. **Water policy:** Because impacts vary significantly according to whether crops are rain fed or irrigated, water policy will want to reflect on consideration on the implications for water demand of agricultural change due to climate change.
v. **Adaptive measures:** Along with the policy making, it is also very important to consider adaptive measures to cope with altering agricultural patterns. These measures might also include choices for selection of crops, changes in cropping patterns, and promotion of advanced and traditional techniques for water conservation and irrigation.

6. Conclusion and future prospective

In this chapter, climate change impact on agricultural systems and mitigation approaches have been discussed. Various factors as a consequence of climate change have been described, including crop yields, CO_2 enrichments, and rising temperature. Crop yield, water productivity, and available soil water balances are integral parts of a productive agricultural system. It is widely discussed that variation in temperature and rainfall will be (happening now) major consequence of global warming, which will greatly affect the level of ground water and moisture status in soil. So, with rising global temperature, variability in rainfall, and decrease in ground water level, crop productivity will decrease. According to projections and experiments, it is estimated that climate change would dramatically affect the agricultural production by 2020, 2040, and 2080, if adaptive strategies will not be considered. Most visible risk of climate change being experienced by tea cultivation in Kenya is supporting example. Keeping the main challenges and issues of climate change in mind, we will have to move toward the adaptability with reference to selection of crop and changes in agricultural practices.

Climate change and its challenges are multifaceted and dynamic; thus, targeted research efforts are required. So the complex interaction between climate change impacts and varying environmental factors, irrigation techniques, regional factors, and variations in crop types and the combined impacts of all these factors on agricultural production and water cycle need to be investigated further with deep understanding. Hence, the following recommendations may be helpful for further future research:

- Precision and accuracy in climate change prediction models that can be achieved through integration of experimental and simulated data and linking of these with agricultural production system for suitable options for sustaining agricultural production.
- Evaluation of the impacts of climate change in selected locations and development of models for pest population dynamics.
- Preparation of agricultural productivity database under climate change.

References

Abraha, M.G., Savage, M.J., 2006. Potential impacts of climate change on the grain yield of maize for the midlands of KwaZulu-Natal, South Africa. Agric. Ecosyst. Environ. 115 (1−4), 150−160.

Ackerman, F., DeCanio, S.J., Howarth, R.B., Sheeran, K., 2009. Limitations of integrated assessment models of climate change. Clim. Change 95 (3−4), 297−315.

Aggarwal, P.K., Joshi, P.K., Ingram, J.S., Gupta, R.K., 2004. Adapting food systems of the Indo-Gangetic plains to global environmental change: key information needs to improve policy formulation. Environ. Sci. Pol. 7 (6), 487−498.

Aggarwal, P.K., Singh, A.K., Samra, J.S., Singh, G., Gogoi, A.K., Rao, G.G.S.N., Ramakrishna, Y.S., 2009. Introduction. In: Global Climate Change and Indian Agriculture. Indian Council of Agricultural Research, New Delhi, India.

Aktar, W., Sengupta, D., Chowdhury, A., 2009. Impact of pesticides use in agriculture: their benefits and hazards. Interdiscipl. Toxicol. 2 (1), 1−12.

Anwar, M.R., O'Leary, G., McNeil, D., Hossain, H., Nelson, R., 2007. Climate change impact on rainfed wheat in south-eastern Australia. Field Crops Res. 104 (1−3), 139−147.

Arif, M., Jan, T., Munir, H., Rasul, F., Riaz, M., Fahad, S., Mian, I.A., 2020. Climate-smart agriculture: assessment and adaptation strategies in changing climate. In: Global Climate Change and Environmental Policy. Springer, Singapore, pp. 351−377.

Asseng, S., Ewert, F., Martre, P., Rötter, R.P., Lobell, D.B., Cammarano, D., et al., 2015. Rising temperatures reduce global wheat production. Nat. Clim. Change 5 (2), 143.

Bahl, P.N., 2015. Climate change and pulses: approaches to combat its impact. Agric. Res. 4, 103−108.

Basu, P.S., Ali, M., Chaturvedi, S.K., 2009. Terminal heat stress adversely affects chickpea productivity in northern India-strategies to improve thermotolerance in the crop under climate change. In: ISPRS Archives XXXVIII-8/W3 Workshop Proceedings: Impact of Climate Change on Agriculture, New Delhi, pp. 189−193.

Bhadouria, R., Singh, R., Singh, V.K., Borthakur, A., Ahamad, A., Kumar, G., Singh, P., 2019. Agriculture in the era of climate change: consequences and effects. In: Climate Change and Agricultural Ecosystems. Woodhead Publishing, pp. 1−23.

Bita, C., Gerats, T., 2013. Plant tolerance to high temperature in a changing environment: scientific fundamentals and production of heat stress-tolerant crops. Front. Plant Sci. 4, 273.

Board, J.E., Hall, W., 1984. Premature flowering in soybean yield reductions at nonoptimal planting dates as influenced by temperature and photoperiod 1. Agron. J. 76 (4), 700−704.

Boullis, A., Francis, F., Verheggen, F.J., 2015. Climate change and tritrophic interactions: will modifications to greenhouse gas emissions increase the vulnerability of herbivorous insects to natural enemies? Environ. Entomol. 44 (2), 277−286.

Bouma, E., 2008. Weather and Climate Change in Relation to Crop Protection. www.knpv.org/./Pests%20and%20climate%20change/ABSTRACTS_Pests_and_Climate_Change.pdf.

Bray, E.A., Bailey-Serres, J., Weretilnyk, E., 2000. Responses to abiotic stress. In: Gruissem, W., Buchannan, B., Jones, R. (Eds.), Biochem. Mol. Biol. Plants. American Society of Plant Physiologists, Rockville, pp. 1158−1249.

Brida, A.B., Owiyo, T., 2013. Loss and damage from the double blow of flood and drought in Mozambique. Int. J. Glob. Warming 5 (4), 514−531.

Cairns, J.E., Hellin, J., Sonder, K., Araus, J.L., MacRobert, J.F., Thierfelder, C., Prasanna, B.M., 2013. Adapting maize production to climate change in sub-Saharan Africa. Food Security 5 (3), 345−360.

Cammarano, D., Ceccarelli, S., Grando, S., Romagosa, I., Benbelkacem, A., Akar, T., Al-Yassin, A., Pecchioni, N., Francia, E., Ronga, D., 2019. The impact of climate change on barley yield in the Mediterranean basin. Eur. J. Agron. 106, 1−11.

Ceccarelli, S., Grando, S., Maatougui, M., Michael, M., Slash, M., Haghparast, R., Labdi, M., 2010. Plant breeding and climate changes. J. Agric. Sci. 148 (6), 627−637.

CGIAR-ISPC, May 2010. Product Stewardship and Liability in the Context of IPR: Report of a Study. ISPC Secretariat, Rome, Italy, p. 110.

Chandio, A.A., Jiang, Y., Rehman, A., Rauf, A., 2020. Short and long-run impacts of climate change on agriculture: an empirical evidence from China. Int. J. Clim. Change Strateg. Manag. 12 (12), 201−221. https://doi.org/10.1108/IJCCSM-05-2019-0026.

Coakley, S.M., Scherm, H., Chakraborty, S., 1999. Climate change and plant disease management. Annu. Rev. Phytopathol. 37 (1), 399−426.

Darwin, R., Tigras, M., Lewandrowski, J., and Raneses, A., 1995. (World).

Delcour, I., Spanoghe, P., Uyttendaele, M., 2015. Literature review: impact of climate change on pesticide use. Food Res. Int. 68, 7−15.

do Rio, A., Sentelhas, P.C., Farias, J.R.B., Sibaldelli, R.N., Ferreira, R.C., 2016. Alternative sowing dates as a mitigation measure to reduce climate change impacts on soybean yields in southern Brazil. Int. J. Climatol. 36 (11), 3664−3672.

Dora, M., 2018. Five Ways India Must Help its Farmers Face the Threat of Climate Change. https://theconversation.com/five-ways-india-must-help-its-farmers-face-the-threat-of-climate-change-91587.

Eitzinger, J., Štastná, M., Žalud, Z., Dubrovský, M., 2003. A simulation study of the effect of soil water balance and water stress on winter wheat production under different climate change scenarios. Agricultural Water Management 61 (3), 195−217.

Farrar, J.F., Williams, M.L., 1991. The effects of increased atmospheric carbon dioxide and temperature on carbon partitioning, source-sink relations and respiration. Plant Cell Environ. 14 (8), 819−830.

Fischer, G., Shah, M., Tubiello, F.N., Van Velhuizen, H., 2005. Socio-economic and climate change impacts on agriculture: an integrated assessment, 1990−2080. Phil. Trans. Biol. Sci. 360 (1463), 2067−2083.

Fitzgerald, M.A., Resurreccion, A.P., 2009. Maintaining the yield of edible rice in a warming world. Funct. Plant Biol. 36, 1037−1045.

Food and Agriculture Organization (FAO), 2000. Two Essays on Climate Change and Agriculture: A Developing Country Perspective. FAO Social and Economic Development Paper No. 145. Food and Agricultural Organization, Rome. Available from: http://www.fao.org/docrep/003/x8044e/x8044e04.htm#TopOfPage. Viewed on September 25, 2010.

Gahukar, R.T., 2009. Food security: the challenges of climate change and bio-energy. Curr. Sci. 96 (1), 26−28.

Gatto, M.P., Cabella, R., Gherardi, M., 2016. Climate change: the potential impact on occupational exposure to pesticides. Ann. Ist. Super. Sanita 52 (3), 374−385.

https://www.giss.nasa.gov/research/news/20170118/.

Gupta, P.K., 2011. Herbicides and fungicides. In: Gupta, R.C. (Ed.), Reproductive and Developmental Toxicology. Academic Press.

Habash, D.Z., Kehel, Z., Nachit, M., 2009. Genomic approaches for designing durum wheat ready for climate change with a focus on drought. J. Exp. Bot. 60 (10), 2805−2815.

Hampton, J.G., Rolston, M.P., Pyke, N.B., Green, W., 2012. Ensuring the long-term viability of the New Zealand seed industry. Agron. Soc. N.Z. 42, 129−140.

Hampton, J.G., Conner, A.J., Boelt, B., Chastain, T.G., Rolston, P., 2016. Climate change: seed production and options for adaptation. Agriculture 6 (3), 33.

Hatfield, J.L., Boote, K.J., Kimball, B.A., Ziska, L.H., Izaurralde, R.C., Ort, D., et al., 2011. Climate impacts on agriculture: implications for crop production. Agron. J. 103 (2), 351−370.

Hopkins, A., Del Prado, A., 2007. Implications of climate change for grassland in Europe: impacts, adaptations and mitigation options: a review. Grass Forage Sci. 62 (2), 118−126.

Hu, Q., Yang, N., Pan, F., Pan, X., Wang, X., Yang, P., 2017. Adjusting sowing dates improved potato adaptation to climate change in semiarid region, China. Sustainability 9 (4), 615.

Huang, M., Wang, J., Wang, B., Li Liu, D., Yu, Q., He, D., et al., 2020. Optimizing sowing window and cultivar choice can boost China's maize yield under 1.5 C and 2 C global warming. ERL 15 (2), 024015.

https://indianexpress.com/article/explained/explained-locust-agriculture-crops-india-lockdown-6400755/.

IPCC, 2007a. Climate Change: Synthesis Report. Fourth Assessment Report. Intergovernmental Panel on Climate Change, IPCC, Valencia.

IPCC, 2007b. Summary for Policy-Makers, Climate Change 2007: Mitigation. Contribution of Working Group III to the Fourth Assessment Report of the IPCC. Cambridge University Press, Cambridge, United.

IPCC, 2018. Summary for policymakers. In: Masson-Delmotte, V.P., Zhai, H.-O., Pörtner, Roberts, D., Skea, J., Shukla, P.R., Pirani, A., Moufouma-Okia, W., Péan, C., Pidcock, R., Connors, S., Matthews, J.B.R., Chen, Y., Zhou, X., Gomis, M.I., Lonnoy, E., Maycock, T., Tignor, M., Waterfield, T. (Eds.), Global Warming of 1.5°C. An IPCC Special Report on the Impacts of Global Warming of 1.5°C above Pre-industrial Levels and Related Global Greenhouse Gas Emission Pathways, in the Context of Strengthening the Global Response to the Threat of Climate Change, Sustainable Development, and Efforts to Eradicate Poverty. World Meteorological Organization, Geneva, Switzerland, p. 32.

Jagadish, S.V., Craufurd, P.Q., Wheeler, T.R., 2008. Phenotyping parents of mappingpopulations of rice for heat tolerance during anthesis. Crop Sci. 48, 1140−1146.

Jaggard, K.W., Qi, A., Ober, E.S., 2010. Possible changes to arable crop yields by 2050. Phil. Trans. Biol. Sci. 365 (1554), 2835−2851.

Jaiswal, D.K., Verma, J.P., 2016. The significance of plant-associated microbial rhizosphere for the degradation of xenobiotic compounds. In: Plant Responses to Xenobiotics. Springer, Singapore, pp. 307−316.

Jayaraman, T., 2011. Climate change and agriculture: a review article with special reference to India. Rev. Agrarian Stud. 1 (2), 16−78.

Kang, Y., Khan, S., Ma, X., 2009. Climate change impacts on crop yield, crop water productivity and food security−A review. Prog. Nat. Sci. 19 (12), 1665−1674.

Kim, C.G., 2008. The Impact of Climate Change on the Agricultural Sector: Implications of the Agro-Industry for Low Carbon, Green Growth Strategy and Roadmap for the East Asian Region. Korea Rural Economic Institute.

Kim, C.-G., et al., 2009. Impacts and countermeasures of climate change in Korean agriculture (in Korean). Research Report No. 593. Korea Rural Economic Institute.

Ko, J., Ng, C.T., Jeong, S., Kim, J.H., Lee, B., Kim, H.Y., 2019. Impacts of regional climate change on barley yield and its geographical variation in South Korea. Int. Agrophys. 33 (1), 81−96.

Kobayashi, K., Okada, M., Kim, H.Y., Lieffering, M., Miura, S., Hasegawa, T., 2006. Paddy rice responses to free-air [CO_2] enrichment. In: Nösberger, J., Long, S.P., Norby, R.J., Stitt, M., Hendrey, G.R., Blum, H. (Eds.), Managed Ecosystems and CO_2: Case Studies, Processes, and Perspectives. Springer-Verlag, Berlin, pp. 87−104.

Koli, P., Bhardwaj, N.R., Mahawer, S.K., 2019. Agrochemicals: harmful and beneficial effects of climate changing scenarios. In: Climate Change and Agricultural Ecosystems. Woodhead Publishing, pp. 65–94.

Kranthi, K.R., Kranthi, S., Gopalakrishnan, N., Asokan, R., Mayee, C.D., January 2009. Bt resistance-its management and prospects in the current context of climate change. In: Proc Natn Symp IPM Strategies to Combat Emerging Pests in the Current Scenario of Climate Change, pp. 237–261.

Kumar, R., Gautam, H.R., 2014. Climate change and its impact on agricultural productivity in India. J. Climatol. Weather Forecast. 2 (1). https://www.longdom.org/open-access/climate-change-and-its-impact-on-agricultural-productivity-in-india-2332-2594.1000109.pdf. https://doi.org/10.4172/2332-2594.1000109.

Kumar, K., Parikh, J., 2001. Socio-economic impacts of climate change on Indian agriculture. Int. Rev. Environ. Strat. 2 (2).

Kumar, K.K., Kumar, K.R., Ashrit, R.G., Deshpande, N.R., Hansen, J.W., 2004. Climate impacts on Indian agriculture. Int. J. Climatol. 24 (11), 1375–1393.

Kumar, P.R., Yadav, S.K., Sharma, S.R., Lal, S.K., Jha, D.N., 2009. Impact of climate change on seed production of cabbage in North Western Himalayas. World J. Agric. Sci. 5 (1), 18–26.

Lal, R., 2005. Climate change, soil carbon dynamics, and global food security. In: Lal, R., Stewart, B., Uphoff, N., et al. (Eds.), Climate Change and Global Food Security. CRC Press, Boca Raton (FL), pp. 113–143.

Lobell, D.B., Field, C.B., 2007. Global scale climate-crop yield relationships and the impact of recent warming. Environ. Res. Lett. 2, 1–7.

Lobell, D.B., Bänziger, M., Magorokosho, C., Vivek, B., 2011. Nonlinear heat effects on African maize as evidenced by historical yield trials. Nat. Clim. Change 1 (1), 42–45.

Lobell, D., Sibley, A., Ortiz-Monasterio, J.I., 2012. Extreme Heat Effects on Wheat Senescence.

Luhunga, P.M., 2017. Assessment of the impacts of climate change on maize production in the southern and western highlands sub-agro ecological zones of Tanzania. Front. Environment. Sci. 5, 51.

Luo, Q., Bellotti, W., Williams, M., Bryan, B., 2005. Potential impact of climate change on wheat yield in South Australia. Agricultural and Forest Meteorology 132, 273–285.

Magadza, C.H.D., 2000. Climate change impacts and human settlements in Africa: prospects for adaptation. Environ. Monit. Assess. 61, 193–205.

Maksymiv, I., 2015. Pesticides: benefits and hazards. J. Vasyl Stefanyk Precarpathian Natl. Univ. 2 (1), 70–76.

Mall, R.K., Gupta, A., Singh, R., Singh, R.S., Rathore, L.S., 2006. Water resources and climate change: an Indian perspective. Curr. Sci. 1610–1626.

Manderscheid, R., Bender, J.J., Jäger, H.-J., Weigel, H.J., 1995. Effects of season long CO_2 enrichment on cereals. II. Nutrient concentrations and grain quality. Agric. Ecosyst. Environ. 54, 17–185.

Manning, W.J., Tiedemann, A.V., 1995. Climate change: potential effects of increased atmospheric carbon dioxide (CO_2), ozone (O_3), and ultraviolet-B (UV-B) radiation on plant diseases. Environ. Pollut. 88 (2), 219–245.

Mendelsohn, R., 2008. The impact of climate change on agriculture in developing countries. J. Nat. Resour. Pol. Res. 1, 5–19.

Menéndez, R., 2007. How are insects responding to global warming? Tijdschr Entomol. 150 (2), 355.

Moore, B.A., Allard, G.B., 2008. Climate Change Impacts on Forest Health. Forest Health & Biosecurity Working Papers, Forestry Department Food and Agriculture Organization of the United Nations. Working Paper FBS/9E. FAO, Rome, p. 35.

Msowoya, K., Madani, K., Davtalab, R., et al., 2016. Climate change impacts on maize production in the warm Heart of Africa. Water Resour. Manag. 30, 5299–5312.

Müller, R., Bandow, C., Seeland, A., Fennel, D., Coors, A., Ebke, P.K., et al., 2010. BiK F AdaMus: a novel research project studying the response and adaptive potential of single species and communities to climate change in combination with other stressors. J. Soils Sediments 10 (4), 718–721.

Ntonifor, N.N., 2011. Potentials of tropical African spices as sources of reduced-risk pesticides. J. Entomol. 8 (1), 16–26.

Olesen, J.E., Bindi, M., 2002. Consequences of climate change for European agricultural productivity, land use and policy. Eur. J. Agron. 16 (4), 239–262.

Olesen, J.E., Trnka, M., Kersebaum, K.C., Skjelvåg, A.O., Seguin, B., Peltonen-Sainio, P., Micale, F., 2011. Impacts and adaptation of European crop production systems to climate change. Eur. J. Agron. 34 (2), 96–112.

Ortiz, R., Sayre, K.D., Govaerts, B., Gupta, R., Subbarao, G.V., Ban, T., et al., 2008. Climate change: can wheat beat the heat? Agric. Ecosyst. Environ. 126 (1–2), 46–58.

Parry, M.L., 2019. Climate Change and World Agriculture. Routledge.

Parmesan, C., 2006. Ecological and evolutionary responses to recent climate change. Annu. Rev. Ecol. Evol. Syst. 37 (1), 637–669.

Parry, M.L. (Ed.), 2000. Assessment of Potential Effects and Adaptations for Climate Change in Europe; The Europe ACACIA Project. Jackson Environment Institute, University of East Anglia, Norwich, UK.

Parthasarathy, B., Rupa Kumar, K., Munot, A.A., 1992. Forecast of rainy season food grain production based on monsoon rainfall. Indian J. Agric. Sci. 62, 1–8.

Patterson, D.T., Westbrook, J.K., Joyce, R.J.V., Lingren, P.D., Rogasik, J., 1999. Weeds, insects, and diseases. Clim. Change 43 (4), 711–727.

Pleijel, H., Uddling, J., 2012. Yield vs. quality trade-offs for wheat in response to carbon dioxide and ozone. Global Change Biol. 18, 596–605.

Poonam, Srivastava, S., Pathare, V., Suprasanna, P., 2017. Physiological and molecular insights into rice-arbuscular mycorrhizal interactions under arsenic stress. Plant Gene 11, 232–237.

Porter, J.R., Xie, L., Challinor, A.J., Cochrane, K., Howden, S.M., Iqbal, M.M., Lobell, D.B., Travasso, M.I., 2014. Food security and food production systems. In: Climate Change 2014: Impacts, Adaptation, and Vulnerability. Part A: Global and Sectoral Aspects. Contribution of Working Group II to the Fifth Assessment Report of the Intergovernmental Panel on Climate Change. Cambridge University Press, Cambridge, UK and New York, USA, pp. 485–533.

Prasanna, V., 2014. Impact of monsoon rainfall on the total food grain yield over India. J. Earth Syst. Sci. 123 (5), 1129–1145.

Rahman, M.M., Hampton, J.G., Hill, M.J., 2006. Soybean seed yield as affected by time of sowing in a cool temperate environment. J. N. Seeds 7 (4), 1–15.

Rahman, M.M., Thompson, J.R., Flower, R.J., 2020. Hydrological impacts of climate change on rice cultivated riparian wetlands in the Upper Meghna River Basin (Bangladesh and India). Hydrol. Sci. J. 65 (1), 33–56.

Reif, J.C., Zhang, P., Dreisigacker, S., Warburton, M.L., van Ginkel, M., Hoisington, D., Melchinger, A.E., 2005. Wheat genetic diversity trends during domestication and breeding. Theor. Appl. Genet. 110 (5), 859–864.

Reilly, J., 1999. What does climate change mean for agriculture in developing countries? A comment on Mendelsohn and Dinar. World Bank Res. Obs. 14 (2), 295–305.

Rosenzweig, C., Parry, M.L., 1994. Potential impact of climate change on world food supply. Nature 367 (6459), 133–138.

Samarah, N.H., 2005. Effects of drought stress on growth and yield of barley. Agron. Sustain. Dev. 25, 145–149.

Sanghi, A., 1998. Global Warming and Climate Sensitivity: Brazilian and Indian Agriculture (Unpublished Ph. D. thesis). University of Chicago, USA.

Sanghi, A., Mendelsohn, R., 2008. The impacts of global warming on farmers in Brazil and India. Global Environ. Change 18 (4), 655–665.

Selvaraju, R., 2003. Impact of El Nino- southern oscillation on Indian food grain production. Int. J. Climatol. 23, 187–206.

Sharma, H.C., 2010. Effect of climate change on IPM in grain legumes. In: 5th International Food Legumes Research Conference (IFLRC V), and the 7th European Conference on Grain Legumes (AEP VII), pp. 26–30.

Sharma, H.C., Stevenson, P.C., Gowda, C.L.L., 2005. Heliothis/Helicoverpa Management: Emerging Trends and Prospects for Future Research, p. 453.

Singh, G., Mishra, D., Singh, K., Parmar, R., 2013a. Effect of rainwater harvesting on plant growth, soil water dynamics and herbaceous biomass during rehabilitation of degraded hills in Rajasthan, India. For. Ecol. Manag. 310, 612–622.

Singh, R.P., Prasad, P.V., Reddy, K.R., 2013b. Impacts of changing climate and climate variability on seed production and seed industry. Adv. Agron. 118, 49–110 (Academic Press).

Suddick, E.C., Davidson, E.A., 2012. The role of nitrogen in climate change and the impacts of nitrogen-climate interactions on terrestrial and aquatic ecosystems, agriculture, and human health in the United States: a technical report submitted to the US National Climate Assessment, 149. North American Nitrogen Center of the International Nitrogen Initiative (NANC-INI), Woods Hole Research Center, p. 1644, 02540.

Taub, D.R., Miller, B., Allen, H., 2008. Effects of elevated CO_2 on the protein concentration of food crops: a meta-analysis. Global Change Biol. 14 (3), 565–575.

Tay, W.T., Soria, M.F., Walsh, T., Thomazoni, D., Silvie, P., Behere, G.T., et al., 2013. A brave new world for an old world pest: *Helicoverpa armigera* (Lepidoptera: noctuidae) in Brazil. PLoS One 8 (11).

Taylor, L., Arnim, R.von, July 2006. Modeling the impact of trade liberalisation: a critique of computable general equilibrium models. Oxfam Int. Res. Rep. Available from: http://www.oxfam.org.uk/resources/policy/trade/research_trade_.

Thompson, G.B., Brown, J.K.M., Woodward, F.I., 1993. The effects of host carbon dioxide, nitrogen and water supply on the infection of wheat by powdery mildew and aphids. Plant Cell Environ. 16 (6), 687–694.

Tilman, D., Fargione, J., Wolff, B., D'Antonio, C., Dobson, A., Howarth, R., Swackhamer, D., 2001. Forecasting agriculturally driven global environmental change. Science 292 (5515), 281–284.

Tripathi, A., Tripathi, D.K., Chauhan, D.K., Kumar, N., Singh, G.S., 2016. Paradigms of climate change impacts on some major food sources of the world: a review on current knowledge and future prospects. Agric. Ecosyst. Environ. 216, 356–373.

Tuttolomondo, T., La Bella, S., Lecardane, G., Leto, C., June 2009. Simulation of the effects of climate change on barley yields in rural Italy. In: Second Meeting on Statistics on Rural Development and Agriculture Household Income, pp. 11–12.

Van Groenigen, K.J., Van Kessel, C., Hungate, B.A., 2013. Increased greenhouse-gas intensity of rice production under future atmospheric conditions. Nat. Clim. Change 3 (3), 288–291. https://doi.org/10.1038/nclimate1712.

Varshney, R.K., Bansal, K.C., Aggarwal, P.K., Datta, S.K., Craufurd, P.Q., 2011. Agricultural biotechnology for crop improvement in a variable climate: hope or hype? Trends Plant Sci. 16 (7), 363–371.

Verma, J.P., Jaiswal, D.K., Sagar, R., 2014. Pesticide relevance and their microbial degradation: a-state-of-art. Rev. Environ. Sci. Biotechnol. 13 (4), 429–466.

Wahid, A., Gelani, S., Ashraf, M., Foolad, M.R., 2007. Heat tolerance in plants: an overview. Environ. Exp. Bot. 61 (3), 199–223.

Weintraub, P.G., Beanland, L., 2006. Insect vectors of phytoplasmas. Annu. Rev. Entomol. 51, 91–111.

WHO (World Health Organization), 2009. Protecting Health from Climate Change: Global Research Priorities. WHO, Geneva.

Yadav, P., Srivastava, S., 2020. Co-culturing *Hydrilla verticillata* with rice (*Oryza sativa*) plants ameliorates arsenic toxicity and reduces arsenic accumulation in rice. Environ. Technol. Innov. 100722.

Yamamura, K., Kiritani, K., 1998. A simple method to estimate the potential increase in the number of generations under global warming in temperate zones. Appl. Entomol. Zool 33 (2), 289–298.

Yeo, H.G., Choi, M., Chun, M.Y., Sunwoo, Y., 2003. Concentration distribution of polychlorinated biphenyls and organochlorine pesticides and their relationship with temperature in rural air of Korea. Atmos. Environ. 37 (27), 3831–3839.

Zampieri, M., Ceglar, A., Dentener, F., Toreti, A., 2017. Wheat yield loss attributable to heat waves, drought and water excess at the global, national and subnational scales. Environ. Res. Lett. 12 (6), 064008.

Geological records of climate change

8

Vijay Kumar[1], *Khushaboo Verma*[2]
[1]Department of Applied Sciences and Humanities, Faculty of Engineering and Technology, Rama University, Kanpur, Uttar Pradesh, India; [2]Faculty of Education, Banaras Hindu University, Varanasi, Uttar Pradesh, India

1. Introduction

The temperature, pressure, humidity, density, and other atmospheric properties change from place to place and time to time. The hour-to-hour and day-to-day changes in atmospheric properties called weather. The statistical average of the weather taken over for a long time, typically 30 years, is called climate (Change et al., 2006; Turner, 2018). The climate is never static; it has been changing since the origin of the atmosphere by 3.5 billion years ago (Pittock, 2017). There have been many shifts in the climate from an oxygen-deficient environment to an oxygen-rich environment, from glaciated to deglaciated period, warm to cold, and wet to dry. For the past 50 million years ago, the climate of the earth was gradually cooling. When the earth started to cool, the global average temperature was 6–7°C warmer than today (Zachos et al., 2001). It has been reported that about 34 million years ago, a continental ice sheet on Antarctica was formed by the merge of ice caps. At the time of global cooling (2.6 million years ago), the local ice caps and mountain glaciers coalesce to form large ice sheets (Maslin et al., 1998). At this time, the earth's climate was colder than today, which was known as "Ice Age." This ice age was a series of glacial episodes separated by a short interglacial period of 10,000–30,000 years (Lisiecki and Raymo, 2005). The earth is currently going through one of the interglacial periods, which is the warming period (Holocene). This warming period was established by 11,500 years ago. But since the year 1901, the warming of the earth has been increased due to the anthropogenic activity and resulted in climate change.

Now geologists have also confirmed climate change through the study of the geological history of the earth. From such climatic changes, the most noticeable was the Pleistocene glaciation that has occurred 2 million years. This glaciation indicated that a similar phenomenon was involved in Gondwana's Carboniferous-Permian glaciation (Chaloner and McElwain, 1997; McElwain and Chaloner, 1996; McElwain, 1998).

The evidence of climate change is preserved in several geological formations such as oceanic sediments, oxygen isotope ratios, ices sheets, fossils, pollen grains, tree rings, and so on. Numerous studies are based on field observation, laboratory techniques, and numerical modeling, which advocate climate change in the past. For example, the oceanic sediments corroborate the records of the organism of several

Global Climate Change. https://doi.org/10.1016/B978-0-12-822928-6.00015-0

geological periods reaching millions of years ago (Kemp, 2003; Lane et al., 2019), oxygen isotope analysis reveals the temperature records of the past, and the air bubbles trapped in the ice sheets represent the atmospheric composition of the past (Stern et al., 1997; Clarke and Jenkyns, 1999). Tree ring shows important climatic records (Briffa et al., 1996); the study of the buried pollen in an area gives an idea about the types of vegetation in the past (Muller et al., 2003; Zhang et al., 2020a,b; Barreda et al., 2019). The hard calcium carbonate skeletons of the corals are used to determine the temperature of the past water bodies and variations in atmospheric composition. The drilling of the cores through the ice sheets produces a record of polar temperatures and atmospheric composition ranging back to 120,000 years in Greenland and 800,000 years in Antarctica (https://www.geolsoc.org.uk/climaterecord).

2. Geological evidence of climate change

2.1 Oceanic sediments

The surface-dwelling organism in the sea changes as the climate changes. The shells of such get settled at the ocean floor and undergo the formation of sedimentary structures. The remains of these organisms are beneficial in studying the records of climate change. For example, the shells of pelagic Foraminifers are very sensitive to the variation in oceanic water temperature, and the deposition of such remains could help study climate change (Fig. 8.1). Heroy et al. (2008) have investigated the sediment analysis of the Bransfield Basin of the Antarctic Peninsula that had extended through Holocene. They found that the detailed analysis of sediment revealed three

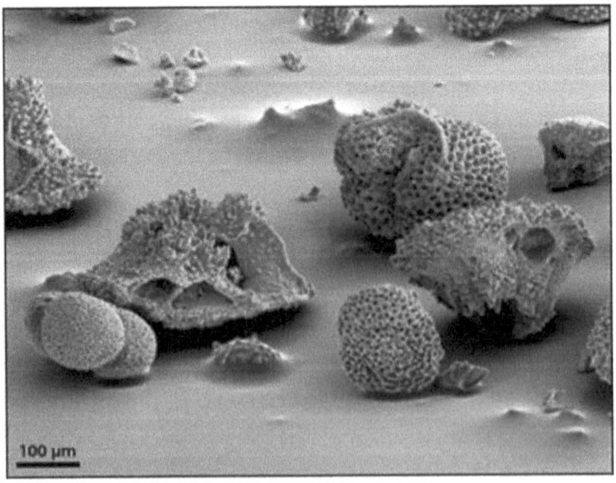

Figure 8.1 Electron optic image of the Foraminifera skeleton present in cores of sediments. Such evidence contains information about ocean chemistry, temperature, currents, and surface winds. https://earthobservatory.nasa.gov/features/Paleoclimatology_SedimentCores/paleoclimatology_sediment_cores_2.php.

stratigraphic units in PC-61, associated with three different sedimentary environments. They had identified two different volcanic ash layers from Deception Island. Based on the analysis of diatom accumulation, they have determined five climatic regimes that emphasized a little Mid-Holocene period. Furthermore, they have added that the Mid-Holocene period was abundant with *Eucampia antarctica* and *Fragilariopsis curta*. This period was marked by a vast abundance of diatom and the highest sediment accumulation rate (Heroy et al., 2008). Charles and Fairbanks have found the importance of North Atlantic Deep Water in glacial and interglacial climate change (Charles and Fairbanks, 1992). Kemp has reported that the laminated marine sediments place great importance as evidence of climate change. The marine sediments are enriched with plenty of records of several variations, which are caused by varying precipitation patterns, runoff through the land, changes in biological productivity, and change in the water column. Such sediments can capture the changes occurring from interannual to decadal. Kemp found that the laminated sediments are distributed intermittently in the sediment column, which holds any abrupt change in climate on the onset and cessation of laminae, such as the oxygenation rate of bottom water. Hence, these records play a significant role in revealing the change in the climate. Such records help to understand the mechanism of climate change. It has been discovered that the Equatorial and Southern Ocean are rich in laminated diatom sediments. These sediments have put significant evidence against climate change (Kemp, 2003). A study made by Fricke et al. (1998) has reported rapid warming with the help of the Cenozoic oxygen isotope record. This rapid warming event is commonly called as latest Paleocene thermal maximum (LPTM). They analyzed the oxygen isotope ratio of carbonate and phosphate of hydroxyapatite present in the tooth enamel of mammal and body scales of river fish.

A comparative study of carbonate and phosphate isotope present in modern and fossils has indicated diagenetic changes that have occurred in fossils. Mammals recorded a higher shift in oxygen isotope ratio and fishes via carbonate and phosphate at the same time and of the same duration as LPTM. Such transformation revealed a significant change that has occurred in the isotopic structure of local precipitation. Thus, these become the first evidence against the continental climate change during LPTM (Fricke et al., 1998). Similarly, Xiao et al. (2005) have isolated 2000 years' old mud from the East China Sea's inner shelf. They found that at least nine abrupt increases in grain size in the recent 2000 years corresponded to China's historical low-temperature events. They concluded that this low temperature might be due to the periodical strengthening of the East Asian Winter Monsoon (Xiao et al., 2005).

2.2 Oxygen isotope ratio

The oxygen isotope ratio is the ratio of lighter isotope ^{16}O to the heavier isotope ^{18}O. The formation of a water molecule occurs from both the isotopes. The lighter isotope (^{16}O) evaporates readily than the heavier isotope (^{18}O) from the ocean. Therefore, the amount of the lighter isotope (^{16}O) in the precipitation is higher, which is why the glacial ice also contains a higher concentration of ^{16}O. Due to this, the higher concentration of the heavier isotope (^{18}O) is left behind in the ocean water. Thus, during the extensive glacial periods, the lighter isotope ^{16}O is tied up in the form of ice, and the concentration of heavier isotope (^{18}O) in seawater increases. But during the warmer period, the glacial ice melts and releases a huge amount of ^{16}O to the sea, which decreases the relative proportion of ^{18}O to the ^{16}O. These records of the ratio of two isotopes ($^{18}O/^{16}O$) could help determine climate change by identifying the glacial and interglacial periods.

We do have these records in the form of ocean sediments. There are some microorganisms that secrete calcium carbonate shells where the prevailing dominant oxygen isotope is utilized. If that microorganism belongs to the glacial period, it uses more ^{18}O, and if it belongs to the warmer period, it utilizes ^{16}O. This $^{18}O/^{16}O$ can be identified from the composition of the shells made up of calcium carbonate because when the shells of such organisms settle down to the floor, they form the sedimentary layers. Therefore, the oxygen isotope ratio present in the shells of the organism can be useful in determining glacial periods and, hence, climate change.

The $^{18}O/^{16}O$ ratio also changes with the change in temperature. When the temperature is higher, more ^{18}O is evaporated, and when the temperature is lower, its evaporation is less. Thus, the higher temperature increases the heavier isotope concentration, making it abundant in the precipitation of warm eras. Conversely, in the colder periods, the heavier isotope is less productive. The above principle is used to study the ice and snow layers deposited in the glaciers. These depositions represent a record of changes in past temperatures. Stern et al. (1997) have studied the oxygen isotope analysis of smectite and kaolinite present in the stratigraphic sequence of paleosols of Siwalik Himalayan. They reported that the O^{18} of smectite was increased by 3–4%, which was synchronous with the O^{18} of pedogenic calcite (3.5%). The covariation of O^{18} of smectite and calcite suggested a pedogenic origin of smectite. Such analysis revealed that O^{18} values of clay mineral in paleosols are a significant indicator of paleoclimate. They stated that a 3.5‰ increase in oxygen isotope ratio of smectite and calcite from 8.6 to 6.5 Ma indicated that this increase was due to climate change (Stern et al., 1997).

Similarly, Terranes and MacKenzie have studied lacustrine oxygen isotope record of 108 years with the annual laminae of calcite in a small lake in central Switzerland (Teranes and MacKenzie, 2001). Based on data obtained from 163 drip sites and 39 caves on five continents, Baker et al. (2019) have presented a global analysis. They found that the $\delta^{18}O$ of drip water was most similar to the $\delta^{18}O$ of amount-weighted precipitation, where the mean annual temperature was less than 10°C, whereas during seasonal climate, when the mean annual temperature was greater than 10°C and less than 16°C, the drip water $\delta^{18}O$ has recorded the recharge weighted $\delta^{18}O$. It indicated that $\delta^{18}O$ of the speleothems reflected the meteoric precipitation only in a cold climate (Baker et al., 2019).

2.3 Tree rings

The tree logs have several concentric rings, which tend to continue larger from the center to outward. Every year a growing tree undergoes secondary growth and produces a new layer of wood under the bark. The size, density, and so on of the tree rings are the consequent of the prevailing environmental conditions during the ring's growth (Fig. 8.2).

Figure 8.2 The growing tree produces a layer of new cells beneath the bark per year. It can be easily seen by examining the trunk of the tree. Since the thickness of the tree ring depends on precipitation and temperature, these are useful records of past climate. https://www.thoughtco.com/dendrochronology-tree-rings-170704.

These tree rings show characteristics pattern of growth during different environmental conditions. The tree produces a wide ring during favorable environmental conditions, whereas a narrow ring is formed under unfavorable conditions. If the tree's cutting year is known, one can easily determine the tree's age and the year by counting the tree ring from outside since, each year, a single round is added. Such a study of annual tree rings and dating is called dendrochronology.

The study of the tree rings is not only done with the trees that have been cut down. It can also be done through living trees by taking small nondestructive core samples. The extended pattern of a tree ring is called ring chronologies, which are developed by comparing rings among various trees in that area. These tree ring chronologies extending back to thousands of years have been established for some regions which are used as a reference for matching the tree ring of unknown samples. Thus the tree ring chronologies establish a special record of environmental conditions that can reconstruct the climate variation within a region for spans of thousands of years before human historical records. Zhang et al. (2020a,b) have studied the width of the tree rings and $\delta^{13}C$ chronologies based on samples of 99 tree rings of Schrenk spruce in Kazakhstan. The analysis of tree ring chronologies and the climate data revealed that the summer means temperature was the most potent climate signal recorded by tree ring $\delta^{13}C$. They have also reconstructed a temperature change since 1850 C.E. with the help of tree ring δ^{13}Ccorr chronology. They found that for the past 166 years, the temperature of Southern Kazakhstan has risen by the rate of 0.27°C per decade. But for the past 30 years, the temperature has been increased by 0.44°C per decade. The temperature and precipitation records revealed that the climate has alternated between warm–dry and cold–humid periods for the past 166 years (Zhang et al., 2020a,b). Chen has studied six stem sections of Chinese Torreya trees, which were 60–90 years old. These tree sections were analyzed with regional climatic data. The study revealed that the stem's radial growth increased linearly even up to 90 years, where the average radial increase of each tree ranged from 1.9 to 5.1 mm per year. They found that the variances of the basal area increment increased with time. They correlated this increase with the variances of precipitation in growing seasons. The analysis of the data revealed a triangle-shaped relationship between precipitation, growing periods, and annual radial increment. The standard precipitation index data, maximum monthly temperature, and minimum monthly air temperature also showed similar patterns. Such annual increases were correlated with the local climate (Chen, 2020).

Montwe et al. (2018) studied the cold adaptation of Western America's coniferous species for informing the seed transfer limit. They used tree ring signatures of cold damage from garden trials designed to study genetic population differentiation. They identified the different geographic clines for spring frost and fall frost damage. They also identified that the northern region's provenances were sensitive to spring frosts, whereas the provenances that had greater productivity were more susceptible to fall frosts. Montwe et al. have concluded that while implementing seed transfer to avoid the harmful effects of climate change, the cold adaptation should remain an important consideration (Montwé et al., 2018).

2.4 Fossils pollen

The distribution of different vegetation types in every part of the earth results from types of climatic conditions present over there. Therefore, the plant community reflects the then climatic condition. The different plants show different phases of the life cycle. Pollen and spores are part of that very life cycle having resistant walls. Therefore, the pollen and spores remain preserved in the sediments for thousands of years (Fig. 8.3).

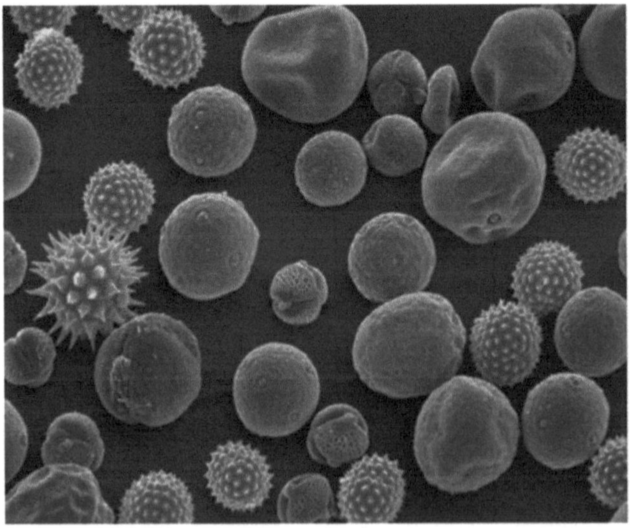

Figure 8.3 A false-color image of pollen grains using an electron microscope. The model depicted how the pollen grains of one species differ in size, shape, and surface characteristics from one another one. Analysis of the types and abundance of pollen in lake sediments and peat deposits provides information on how climate has changed over time.
Lutgens, F.K., Tarbuck, E.J., Tusa, D. The atmosphere, 12th ed. Englewood Cliffs, NJ: Prentice-Hall.

The analysis of such pollen and spores from accurately dated sediments can help develop the records of changing vegetation in an area that can help reconstruct the past climate. Naidina and Richards have analyzed the pollen and spore deposited in Akchagylian–Apsheronian in the northwestern Caspian Sea to investigate past vegetation and climate change from late Pliocene to early Pleistocene. The climatic variations, and the related changes in vegetation from 3.6 to 0.8 Ma, can be identified from the pollen accumulated in sediment cores and outcrops from the North Caucasus. The lower Akchagylian is an open landscape with steppe vegetation. They recognized that the transgression of the palaeo-Caspian spread present in the middle of the lower Akchagylian and the treeless landscapes of the earliest Akchagylian were replaced by forest with thermophilic relicts. It was observed that during the middle–upper Akchagylian and Apshernian periods, the forest of the North Caucasus was replaced with steppe and semiarid vegetation. These changes were obvious in the dendroflora. Thus, the North Caucasus' vegetation showed changes that were consistent with the warming of the climate at around 3.2 Ma. Such events coincided with a warming period in the Mediterranean and represented the warm period "mid-Pliocene" (Naidina and Richards, 2016). In 1993, Jarvis collected the pollen from lake Shayema, Manning County, for investigating the postglacial vegetation and changes in southwestern Sichuan China's climate. They stated that the sclerophyllous nature of the current warm temperate vegetation represents extreme drought occurring in spring before the onset of the Asian monsoons. According to the Javis study, pollen records represented various vegetation changes for the past 11,000 year. The cold-tolerant species, such *as Abies*, *Betula*, and deciduous oaks, were dominant from 11,000–9100 year B.P. But from 9100 to 7800, these cold-tolerant species decreased and replaced with evergreen oaks. This change indicated that the climate has warmed with increased precipitation (Jarvis, 1993). Shulmeister and Lees have studied the pollen evidence for reexamining the Holocene climatic history of tropical Northern Australia. For this study, they have used the recently published pollen record from Groot Eylandt. Their research was dedicated to the pollen evidence related to environmental changes at c. 4000 BP that mark a break amid a constantly increasing effective precipitation climate. But this climate was slightly deviated in the earlier Holocene and highly variable in later Holocene. According to them, such break represented the first evidence of the onset of "modern" ENSO-dominated ocean–atmosphere interactions in the Holocene from the monsoonal lowlands of northern Australia (Shulmeister and Lees, 1995). Wick et al. (2003) have studied the evidence of late-glacial and Holocene climate change with high-resolution pollen records from the laminated sediments present in Lake Van in Turkey. They found that during 6200 and 4000 year BP, the proxy data available indicated optimum climatic conditions, lower salinity, and high lake level. But after 4000 year BP, the modern climatic condition was established with the increased aridity (Wick et al., 2003). Guo et al. (2020) have evaluated the Paleoclimatic trend from the terrestrial Northern Qaidam Basin, NW China. They found that the carbon-isotopic, petrological, and floral records of 20-m-thick coal represented Bajocian records of terrestrial climate change. ^{13}C values from 88 samples indicated that during the Middle Jurassic period, C_3 plants were the main coal-forming vegetation (Guo et al., 2020). Duller et al. (year) have studied the aggregate temporal response of sediment routing systems (SRS) in Spain for investigating climate warming at the Paleocene–Eocene thermal maximum (PETM). They observed that the terrestrial sites in northern Spain had recorded 16.5 ± 7.5 k.y. of the time lag between the onset of the PETM (Duller et al., 2019).

2.5 Coral reefs

Corals are the rich biodiversity in the ocean, which forms reef when the colonies of coral are living in warm and shallow water. They form atop the hard material left behind by the past corals. The skeleton of the corals is made up of calcium carbonate ($CaCO_3$). The oxygen isotope ratio ($^{18}O/^{16}O$) present in the carbonate can give an idea of the temperature of the water. The density of the skeleton formed in winter is different than summer due to the temperature and the related environmental factors that affect the growth rate (Fig. 8.4).

Therefore, seasonal growth bands are formed, which are very similar to that of tree rings. When the recent instrumental records were compared with the coral records for the same period, it was noticed that the climatic data obtained from corals were more accurate and reliable. The oxygen isotope analysis of the coral growth rings formed by the deposition of the skeleton can also be used for the measurement of precipitation, especially in the area of the large variation of rainfall (Tarbuck et al., 2015). It has been reported that since the year 1900, the ocean temperature has increased by 1°C due to climate change. Hoegh-Guldberg has reported that the ocean water's increasing temperature has affected the coral and zooxanthellae, its photosynthetic symbionts. In the recent past 20 years, the coral bleaching event has occurred at mass levels, which led to the significant loss of corals at the global level (Hoegh-Guldberg, 1999). Toth et al. have stated that the geological records could be important evidence for the long-term controls on reef development. From the geological records, the present coral crisis could be solved. They analyzed the cores of new and existing coral reefs present throughout the Florida Keys reef tract (FKRT). They observed that the reef development was spatially and temporarily variable during Holocene. The maximum

Figure 8.4 The figure represented the dying of coral due to heatwaves.
https://www.dw.com/en/coral-reefs-rapidly-die-from-marine-heat-waves-study/a49960345.

Holocene reef thickness was in Dry Tortugas. The Holocene reef of Florida had relatively thin accumulations. They found that when the reef was developing actively, the accretion rate was 3 m/k.y. throughout the FKRT. The thickness of the reef was spatially variable because of differences existed in the duration of reef development. It has been investigated that there was a significant decline in the accretion of the reef from 6000 years ago to the present. Most of the FKRT was geologically senescent by 3000 years ago. They stated that the variability of subregional hydrography modulated the timing of reef senescence. They concluded that climatic cooling was the ultimate cause of the reef shutdown (Toth et al., 2018).

3. Conclusions

The climate is changing for millions of years, which is being stored in several geological records. These geological records can be used as evidence of climate change because of the storage of information of that particular period. For example, the precipitation and temperature of a given period define the tree ring of vegetation. Hence, the precipitation and temperature records can be stored in the tree rings. Similarly, various geological evidences such as fossils, pollen, oxygen isotope ratios, oceanic sediments, and ice sheets, etc. also records potential data of climate change. This chapter defines the different geological evidence of climate change in detail.

References

Baker, A., Hartmann, A., Duan, W., Hankin, S., Comas-Bru, L., Cuthbert, M.O., et al., 2019. Global analysis reveals climatic controls on the oxygen isotope composition of cave drip water. Nat. Commun. 10 (1), 1–7.

Barreda, V.D., Palazzesi, L., Olivero, E.B., 2019. When flowering plants ruled Antarctica: evidence from Cretaceous pollen grains. New Phytol. 223 (2), 1023–1030.

Briffa, K.R., Jones, P.D., Schweingruber, F.H., Karlén, W., Shiyatov, S.G., 1996. Tree-ring variables as proxy-climate indicators: problems with low-frequency signals. In: Climatic Variations and Forcing Mechanisms of the Last 2000 Years. Springer, Berlin, Heidelberg, pp. 9–41.

Chaloner, W.G., McElwain, J., 1997. The fossil plant record and global climatic change. Rev. Palaeobot. Palynol. 95 (1–4), 73–82.

Change, A.D.C., Blair, T., Pachauri, R., 2006. Avoiding Dangerous Climate Change. Cambridge University Press.

Charles, C.D., Fairbanks, R.G., 1992. Evidence from Southern Ocean sediments for the effect of North Atlantic deep-water flux on climate. Nature.

Chen, X., 2020. Historical radial growth of Chinese Torreya trees and adaptation to climate change. Atmosphere 11 (7), 691.

Clarke, L.J., Jenkyns, H.C., 1999. New oxygen isotope evidence for long-term Cretaceous climatic change in the Southern Hemisphere. Geology 27 (8), 699–702.

Duller, R.A., Armitage, J.J., Manners, H.R., Grimes, S., Jones, T.D., 2019. Delayed sedimentary response to abrupt climate change at the Paleocene-Eocene boundary, northern Spain. Geology 47 (2), 159–162.

Fricke, H.C., Clyde, W.C., O'Neil, J.R., Gingerich, P.D., 1998. Evidence for rapid climate change in North America during the latest Paleocene thermal maximum: oxygen isotope compositions of biogenic phosphate from the Bighorn Basin (Wyoming). Earth Planet Sci. Lett. 160 (1–2), 193–208.

Guo, B., Eley, Y., Hilton, J., Zou, M., Wang, Q., 2020. Carbon-isotope, petrological and floral record in coals: implication for Bajocian (Middle Jurassic) climate change. Int. J. Coal Geol. 220, 103417.

Heroy, D.C., Sjunneskog, C., Anderson, J.B., 2008. Holocene climate change in the Bransfield Basin, Antarctic Peninsula: evidence from sediment and diatom analysis. Antarct. Sci. 20 (1), 69.

Hoegh-Guldberg, O., 1999. Climate change, coral bleaching and the future of the world's coral reefs. Mar. Freshw. Res. 50 (8), 839–866.

https://www.geolsoc.org.uk/climaterecord.

Jarvis, D.I., 1993. Pollen evidence of changing Holocene monsoon climate in Sichuan Province, China. Quat. Res. 39 (3), 325–337.

Kemp, A.E., 2003. Evidence for abrupt climate changes in annually laminated marine sediments. Philos. Trans. R. Soc. Lond. A Math. Phys. Eng. Sci. 361 (1810), 1851–1870.

Lane, S.N., Bakker, M., Costa, A., Girardclos, S., Loizeau, J.L., Molnar, P., Silva, T., Stutenbeckar, L., Schlunegger, F., 2019. Making stratigraphy in the Anthropocene: climate change impacts and economic conditions controlling the supply of sediment to Lake Geneva. Sci. Rep. 9 (1), 1–11.

Lisiecki, L.E., Raymo, M.E., 2005. A Pliocene-Pleistocene stack of 57 globally distributed benthic δ18O records. Paleoceanography 20 (1).

Maslin, M.A., Li, X.S., Loutre, M.F., Berger, A., 1998. The contribution of orbital forcing to the progressive intensification of Northern Hemisphere glaciation. Quat. Sci. Rev. 17 (4), 411–426.

McElwain, J.C., 1998. Do fossil plants signal palaeoatmospheric carbon dioxide concentration in the geological past? Phil. Trans. Roy. Soc. Lond. B Biol. Sci. 353 (1365), 83–96.

McElwain, J.C., Chaloner, W.G., 1996. The fossil cuticle as a skeletal record of environmental change. Palaios 11 (4), 376–388.

Montwé, D., Isaac-Renton, M., Hamann, A., Spiecker, H., 2018. Cold adaptation recorded in tree rings highlights risks associated with climate change and assisted migration. Nat. Commun. 9 (1), 1–7.

Muller, U.C., Pross, J., Bibus, E., 2003. Vegetation response to rapid climate change in Central Europe during the past 140,000 yr based on evidence from the Furamoos pollen record. Quat. Res. 59 (2), 235–245.

Naidina, O.D., Richards, K., 2016. Pollen evidence for late pliocene–early Pleistocene vegetation and climate change in the North Caucasus, North-Western Caspian region. Quat. Int. 409, 50–60.

Pittock, A.B., 2017. Climate Change: Turning up the Heat. Routledge.

Shulmeister, J., Lees, B.G., 1995. Pollen evidence from tropical Australia for the onset of an ENSO-dominated climate at c. 4000 BP. Holocene 5 (1), 10–18.

Stern, L.A., Chamberlain, C.P., Reynolds, R.C., Johnson, G.D., 1997. Oxygen isotope evidence of climate change from pedogenic clay minerals in the Himalayan molasse. Geochem. Cosmochim. Acta 61 (4), 731–744.

Tarbuck, E.J., Lutgens, F.K., Tasa, D.G., 2015. *Earth Science*. Pearson Higher Ed.

Teranes, J.L., McKenzie, J.A., 2001. Lacustrine oxygen isotope record of 20th-century climate change in central EuropeL evaluation of climatic controls on oxygen isotopes in precipitation. J. Paleolimnol. 26 (2), 131–146.

Toth, L.T., Kuffner, I.B., Stathakopoulos, A., Shinn, E.A., 2018. A 3,000-year lag between the geological and ecological shutdown of Florida's coral reefs. Glob. Chang. Biol. 24 (11), 5471–5483.

Turner, C., 2018. Climate Change and Biodiversity. Scientific e-Resources.

Wick, L., Lemcke, G., Sturm, M., 2003. Evidence of Lateglacial and Holocene climatic change and human impact in eastern Anatolia: high-resolution pollen, charcoal, isotopic and geochemical records from the laminated sediments of Lake Van, Turkey. Holocene 13 (5), 665–675.

Xiao, S., Li, A., Jiang, F., Li, T., Huang, P., Xu, Z., 2005. Recent 2000-year geological records of mud in the inner shelf of the East China Sea and their climatic implications. Chin. Sci. Bull. 50 (5), 466.

Zachos, J., Pagani, M., Sloan, L., Thomas, E., Billups, K., 2001. Trends, rhythms, and aberrations in global climate 65 Ma to present. Science 292 (5517), 686–693.

Zhang, Z., Li, Y., Li, C., Xu, Q., Zhang, R., Ge, Y., Li, B., Deng, C., Li, Z., Zhang, L., 2020a. Pollen evidence for the environmental context of the early Pleistocene Xiashagou fauna of the Nihewan Basin, north China. Quat. Sci. Rev. 236, 106298.

Zhang, R., Qin, L., Shang, H., Yu, S., Gou, X., Mambetov, B.T., et al., 2020b. Climatic change in southern Kazakhstan since 1850 CE inferred from tree rings. Int. J. Biometeorol. 1–11.

Global climate change: the loop between cause and impact

9

Vijay Kumar[1], *Deeksha Ranjan*[1], *Khushaboo Verma*[2]
[1]Department of Applied Sciences and Humanities, Faculty of Engineering and Technology, Rama University, Kanpur, Uttar Pradesh, India; [2]Faculty of Education, Banaras Hindu University, Varanasi, Uttar Pradesh, India

1. Introduction

The term climate is a long term shift in the weather pattern. It is an average of at least 30 years of weather condition of a particular place (Kumar, 2015; Nwankwoala, 2015; The Royal Society, 2014; Climate Change, 2018). (The Royal Society, 2014) The average weather condition is taken via a systematic recording and processing of various climate segments such as atmospheric pressure, temperature, sunshine, humidity, wind speed, clouds, and precipitation (Tranter, 2017). Based on the collected data, maps, and charts are prepared by which one can quickly notice the cumulative changes that may have occurred in all the segments over a period of time.

The climate does not vary only place to place; but it is variable over time also. The history of the Earth reveals that the climate has experienced many shifts. This shift was observed from warmer to colder and wet to dry (Dyurgerov and Meier, 2000; Haeberli and Beniston, 1998; Dowdeswell et al., 1997). Every part of the planet earth has experienced a significant change in the climate, for example, from the ice age to desert dunes. These changes have been proved by several proxy data recorded by nature, such as seafloor spreading, pollen grains, glacial ice, and tree growth rings (Royal Society, 2020).

Since 1900, Earth's surface air temperature has increased by 1°C, where over half of the temperature has increased since mid-1970. Despite this, the extent of ice and snow in the Northern Hemisphere has reduced, the Greenland ice sheet has shrunk, the sea level has increased, and the temperature-sensitive species of fish, mammals, and insects have started to shift toward the pole (Keisling, 2020). These incidents advocate that the climate has changed (Ogunbode et al., 2020).

Climate change is caused by both natural and anthropogenic causes (Haman, 2008). Naturally, the climate change is caused due to the variation of Earth's orbit, altering the solar energy distribution. Solar variability due to increasing or decreasing the number of a sunspot in 11-year cycle, plate tectonic which affect the energy balance of the Earth by redistributing the landmass (Cliver et al., 1998; Schurer et al., 2014; Feynman, 2007; Beer, 2005; Hay, 1996; Ruddiman, 2013; Whittaker, 2012; Seybold et al., 2020). It is also caused due to the albedo and volcanic eruption which emits a huge amount of gases into the atmosphere (Potter et al., 2020; Hall and Xu, 2006; Fletcher et al., 2009;

Global Climate Change. https://doi.org/10.1016/B978-0-12-822928-6.00002-2

Self and Rampino, 1988). The anthropogenic sources release greenhouse gases in a tremendous amount from deforestation, aerosol, industrialization, etc (Rosenzweig et al., 2008; Irving et al., 2019).

According to the United Nations Framework Convention on Climate Change (UNFCCC), climate change is human-induced change while the term "Variability" for the changes caused by the natural phenomenon (Solomon, 2007).

The climate change is not an isolated rather a complex process which has a very serious effects on a range of other global phenomenon such as stratospheric ozone depletion, rise in sea level, and ocean acidification (Kirk-Davidoff et al., 1999; Hartmann et al., 2000; Larin, 2020; Ward, 2020; Barnes, 2019). Climate change may result in extreme weather events, such as drought and desertification (Burrell et al., 2020). The desertification converts the fertile land into unfertile land, which affects agricultural productivity directly. The declining agricultural productivity creates food insecurity and becomes a threat to the human and cattle also. The drying up of the water resources such as pond, river, and stream are also severely affected. The global increase in temperature leads to forest fire like the recent forest and bush fire of Australia and Amazon, which caused the massive loss of biodiversity. Ocean acidification is also one of the significant effects of climate change. The ocean's acidification affects biodiversity and nutrient cycling (Lovejoy, 2006; Grandon, 2020; Frauendorf et al., 2020). Several other impacts are evident from climate change, such as the shrinking of Arctic sea ice because of its increasing heat content. The melting glaciers and ice caused to rise in sea level. According to a record available, the global sea level has been increased on an average of 16 cm since 1901 (Royal Society, 2020). The unexpected warming and unequal precipitation affect flora and fauna worldwide via drought and devastating flood (Whetton et al., 1993).

2. Natural causes of climate change

2.1 Orbital variations and climate change

Earth's climate is affected by several solar and celestial factors, which occur in cycles. The global climate is affected by several factors dealing with Earth as a whole and its position in Sun's space (Jochmann, 1993). The factors of the Earth include its obliquity (Earth's axial tilt), the eccentricity (how circular/elliptical Earth's orbit is) and the precession of the equinoxes and solstices (with different Earth-Sun distances during any given season) (Jochmann and Greiner-Mai, 1996). These factors are even now not well understood, e.g., the periodicity of the ice age, which is characterized by the climate oscillations, might be linked to the changes in Earth's orbit.

Almost a century ago, Yugoslavian astronomer Milutin Milankovitch proposed a hypothesis that relates Earth's motion and climate variations. He explained the long-term and collective effect of variation in the incoming solar radiation on the Earth's climate. Thus, they are responsible for periodicity and climate fluctuations of Ice ages (glaciation periods). Milankovitch formulated a comprehensive mathematical model based on three types of Earth's orbit movements (i.e., eccentricity, obliquity, and precession). These cyclic orbital movements are known as Milankovitch cycles (Crowley et al., 1993). The Milankovitch cycle are explained as follows:

2.1.1 Orbital eccentricity

As it is well known that the revolution of Earth around the Sun is not perfectly circular but slightly elliptical because of Jupiter and Saturn's gravity pull. Eccentricity is the measure of how much the shape of Earth's orbit deviates from being perfectly circular, which affects its distance from the Sun (Fig. 9.1A). These variations between Earth and Sun have played a significant role in producing and understanding the climate changes over hundreds and thousands of years, although a little substantial in understanding current seasonal variations. The reasons behind the summers in the Northern Hemisphere are about 4.5 days longer than winters and springs about 3 days longer than autumns is eccentricity. The different lengths of seasons can be explained with eccentricity with the decrease in eccentricity; the seasons become almost of the same length (Climate, 2018).

The distance between the perihelion (Earth's closest approach to the Sun) and aphelion (Earth's farthest approach to the Sun) is 5.1 million km. This small variation in distance means that each January, Earth receives about 6% more insolation than in July.

Earth's orbit changes in a cyclic period that spans about 90,000 to 100,000 years, with orbit stretching to a longer elliptical shape and then returning to a more circular shape (Shackleton, 2000). The Earth receives about 20%–30% more insolation at perihelion than at aphelion. It would undoubtedly result in a substantially changed climate than the current (Lutgens et al., 2013).

2.1.2 Earth's obliquity

It is the inclination of Earth's axis from its orbit's plane and is the most significant reason for Earth's seasonal temperature changes (Williams, 1993). The current tilt is about 23.5° but this tilt changes and with it, changes the severity of seasons. During the cyclic period of around 41,000 years, the axis's inclination varies between 22.1 and 24.5° (Fig. 9.1B). The smaller the inclination angle, the smaller the difference between the temperature of summer and winter (Levrard and Laskar, 2003). It could lead to more ice sheets, i.e., glaciation at high altitudes as it could result in milder seasons, i.e., warmer winters and cooler summers. The warmer winters could result in more snowfall as the moisture-holding capacity of air would increase, and in cooler summers, there would be less snow melting. This increase in ice covering would reflect more solar radiation into space, leading to a further drop in temperature (Lutgens et al., 2013).

2.1.3 Precession

Like a slightly run-down top, the Earth wobbles slowly during spinning on its axis. Wobbling happens because the Earth is bulged at its equator due to the tidal forces caused by Sun and Moon's gravitational effect, which affects its rotation (Fig. 9.1C). As a result of this axial wobbling, the Earth's axis points toward different sky spots during a cyclic period of around 26,000 years. Currently, Earth's axis is pointing toward Polaris (known as the North Star). However, about the year A.D.

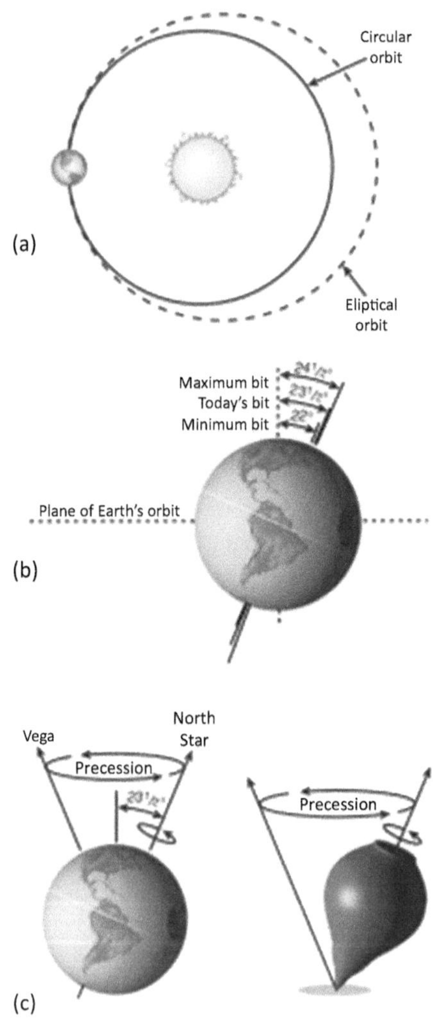

Figure 9.1 Earth's Orbital variations. (a) The shape of Earth's orbit changes gradually from circular to elliptical which return to its original shape again during a cycle of about 100,000 years. (b) showing how the axis of rotation changes from 21.5 to 24.5° (current 23.5°) to the plane of Earth's orbit during a cycle of 41,000 years, (c) showing how the axis of the earth wobbles like a spinning top. As a result, the axis of the earth points to different spots in the sky during a cycle of about 26,000 years.
Source: Lutgens, F.K., Tarbuck, E.J., Tusa, D., 2013. The Atmosphere, 12th ed. Prentice-Hall, Englewood Cliffs, NJ.

14,000, Earth's North Star would be Vega as the axis will be pointing toward it, and by the year 27,000, Polaris will again become the North Star. Thousands of years ago, Kochab and Pherkad were the North Star of Earth (Huybers and Wunsch, 2003; Pokras and Mix, 1987).

This cyclic axial wobbling causes the climate to change significantly, making the seasonal contrast more extreme in one hemisphere than the other. In about 12,000 years, when the axis would be inclined toward Vega, the orbital position at which the winter and summer solstices (when the Sun is at its most significant distance from the celestial equator) occur will be flipped. Consequently, in total contrast with the current position, the Northern hemisphere will receive more extreme solar radiation and experience winter near aphelion and summer near perihelion. The southern hemisphere, however, will experience moderate climatic variations. This will cause winter to be colder and summers to be warmer than at present, resulting in a more excellent seasonal contrast. Axial wobbling also changes the timing of the seasons gradually, causing them to begin earlier over time (Climate, 2018).

2.2 Solar variability and climate change

Sun is a variable star; the energy is given out by it also varies with timescale. It forms a basis for the hypothesis of climate change. An increase in solar output would result in a warmer atmosphere, and a decrease would cause a colder atmosphere, which could be used to explain climate changes of any time length and intensity. However, it was only with the launch of specialized instruments through satellites around the 1980s that the accurate measurement of variations in the Sun's energy output became possible (Haigh, 1996).

Various aspects of solar activity that can cause climate changes are variations in the solar magnetic field, which affects the number of cosmic rays coming from deep space and entering the Earth's atmosphere, and also fluctuations in UV rays from the Sun cause photochemical changes in upper Earth's atmosphere. The best-known aspect of solar activity variation, which has an impact on climate, relates to regular fluctuation in the number of sunspots (Sunspot cycles), which are conspicuous dark blemishes on the Sun's surface. It develops as a massive magnetic storm, where an intense magnetic field weakens the gas flow transporting heat energy deep from Sun's interior (Fig. 9.2). It appears dark because it has a lower temperature than the surrounding area.

The number of sunspots varies regularly. It was observed that a variation of about 1.5 W/m^2 (equivalent to about 0.1% average insolation i.e., 1370 W/m^2) in solar output is received by Earth over cyclic period of 11 years. That means the number of sunspots changes from maximum to minimum number in this cycle (Solanki, 2002; Beer, 2005; Crowley and Kim, 1996). There is also a 22-year cycle, which is based on the fact that the magnetic polarities of sunspot clusters reverse every successive 11 years.

There are two examples that are debated usually to find correlations on time scales ranging from days to tens of thousands of years for the possibility of Sun and climate change.

2.2.1 Sunspots and temperature

According to studies, there were periods of no or reduced sunspots, called the Little Ice Age. It is a time span between about 1300 and 1870, corresponding to colder winters in Europe and North America than during the 20th century. The temperature was as much

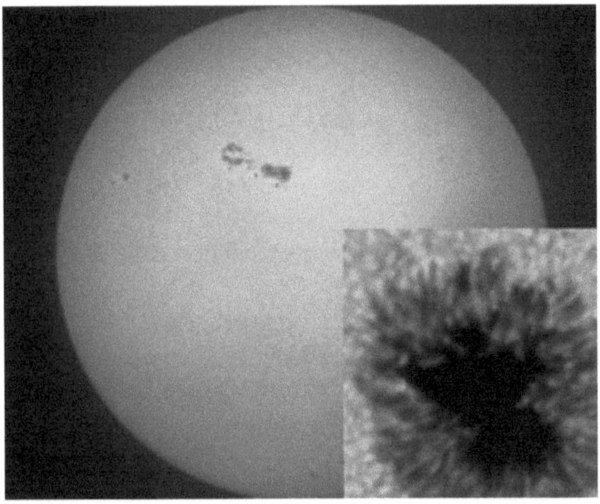

Figure 9.2 Showing the group of large sunspot on the solar disk (b) Sunspots having visible umbra (dark central area) and penumbra (the lighter area surrounding umbra).
Source: Lutgens, F.K., Tarbuck, E.J., Tusa, D., 2013. The Atmosphere, 12th ed. Prentice-Hall, Englewood Cliffs, NJ.

as 2°C lower than present average winter temperatures. In contrast to this, the time span when there were more sunspots coincided with warmer periods in these regions.

It is believed that during the period of a maximum number of sunspots, the Sun emits slightly more radiation, i.e., more solar output. As the sunspots are suppressing heat, the heat flows to the surrounding areas, which becomes brighter than usual (called faculae) and thus radiating more heat. This means periods of more sunspots may contribute to warmer climate, and fewer sunspots may be associated with a colder climate. Overall, the heating effect of the faculae outweighs the cooling effect of the sunspots.

With these matches, between the variation of Sun's activity and temperature change, in mind, some scientists are of the belief that Sun's activity is essential to cause climate change. While other scientists, based on many investigations using climate records from different parts of the globe, failed to find any significant correlation between solar activity variation and temperature and thus climatic change.

2.2.2 Sunspots and drought

There is also a possibility of a correlation between the Sun's activity and variations in precipitation. A detailed study of tree rings revealed the recurrence period of 22 years in droughts in the western United States. This periodicity coincided well with the 22-year cycle of the Sun.

The National Research Council pointed out that there is no effective mechanism that might connect so subtle a feature of the Sun to drought patterns in limited regions has yet appeared. Moreover, the cyclic pattern of droughts found in tree rings is itself a subtle feature that shifts from place to place within the broad region of study.

To properly understand the correlation between solar variability and climate change, the research focused on identifying any links between the Sun and lower atmosphere (Lutgens et al., 2013).

2.3 Plate tectonics and climate change

Earth's orbital variations explain alternating glaciation and interglaciation periods, but it could not explain why glaciers have been absent throughout most of the Earth's history? The Plate-tectonics theory provided the answer to this, along with many other climatic changes over the years.

The theory explained that the outer Earth's surface is made up of rigid crustal slabs called plates. These plates keep on shifting to one another over a weak layer of rocks below it. This movement is prolonged (hardly a few centimetres a year). Examples of the few most massive plates, i.e., landmasses with the seafloor, are North America, South America, and Africa. As these tectonic plates move ponderously over a geological timescale, the landmasses, including continents, also shift to different latitudes and positions (Hay, 1996).

An example of this is that the coal mines formed in tropical areas over millions of years ago are found at higher latitudes today. This shifting of tectonic plates allows geologists and climate scientists to understand its effect on the global circulation of air and ocean water and thus continent's climatic changes. It is now well understood that the shifting of plates caused an oceanic current shift, generating more heat (Berger et al., 2008). The variation in the oceanic circulation changed the transport of heat and moisture and, thus, the climate.

The shifting also resulted in volcanic eruptions generating more CO_2 and SO_2 in the atmosphere, leading to temperature rise (Ruddiman, 2013; Sanchez et al., 2010; Zhang et al., 2012; Stadler et al., 2020). The temperature levels record a fall; if on the other hand, the shift is insignificant, the overall leading to climate change over a period of time, although the rate is slow.

With this theory's help, scientists could also explain why the presently warm climate of Africa, Australia, South America and India once had been freezing like Greenland and Antarctica. There is evidence of glacial activity in these places that these regions experienced ice age about 250 million years ago. This is because these areas in their ancient glaciation period were joined as a single landmass, i.e., 'supercontinent' and was located toward the South Pole. As the tectonic plates moved and spread apart, the portions of landmass, each on a different plate, slowly moved to different positions and migrated to their present positions. Thus, the large glaciated land portions became widely scattered in subtropical areas (Lutgens et al., 2013).

This is now well understood that the really slow tectonic plates movement causes the shifting of land masses over a large geological timespan, leading to too gradual climatic changes that actually happen on a time scale of millions of years. Thus, this theory cannot explain the climate changes occurring over a short period like tens, hundreds, and thousands of years.

2.4 Albedo and climate change

Albedo is the measurement of reflectivity of sunlight by any object (Hall and Qu, 2006). Objects reflecting more sunlight have high albedo while those reflecting less have a low albedo. It is measured on a scale of 0−1 or given in percentage with 1 being a perfect reflector (100%) and 0 (0%) absorbing all incoming light. The albedo of clouds ranges from 40% to 90%, whereas the albedo of fresh snow range from 75% to 90%. The albedo of forests varies from 5% to 15%, and that of water ranged is from 2% to >99% (Bonfils et al., 2001; Ridgwell et al., 2009).

Objects with higher albedo are cooler in temperature and while those with lower albedo are hotter in temperature. White objects, like snow-covered hills, have high albedo (most sunlight hitting the surface bounces back toward space) and dark objects, like pavement, have a low albedo (Otterman, 1977; Angstrom, 1962). The albedo of a given region depends on soil composition and moisture, vegetation types, level of urbanization, etc. Albedo differs significantly between surfaces, and it also changes with time. Albedo changes with change in cloud cover and surface cover (Qu and Hall, 2006; Hadley and Kirchstetter, 2012).

The balance between the amount of solar radiation reflected and absorbed by the Earth's surface, which is determined by albedo, plays an essential role in regulating global temperature and climate. Earth's overall albedo measured to be 0.30 significantly affects Earth's equilibrium temperature as it varies the amount of solar energy reflected by Earth, not how much is absorbed. This changes the Earth's energy budget balance and, thus, Earth's heat balance (Stephens et al., 2015).

The troposphere and Earth's surface absorbs the solar energy that is not reflected, raises Earth's temperature by various processes. Evaporation of water present on Earth's surface requires energy, provided by the absorbed Sun's radiation. This energy remains stored in water vapor as latent heat and is released back into the atmospheric air with vapor condensation. The total IR rays that are reradiated from Earth's surface, about 83%, radiate back to Earth and do not go back to the atmosphere directly. These back radiations have almost double the amount of energy than those which are absorbed by the Earth's surface directly from the Sun. This energy adds to the atmospheric energy warming the Earth to current temperature (Twomey, 1974).

The Earth's albedo shifts as it warms, resulting in global warming, which causes the melting of ice. This melting of ice causes a decrease in Earth's white surface area, leading to less solar energy to be reflected (lower albedo) than absorbed. This whole process warms the Earth even more. Arctic ice melting is a major cause of concern as it causes positive feedback (Fritz, 1949).

In addition to this, the conditions arising because of the greenhouse effect also shifts Earth's albedo. This effect allows more trapping of IR rays, as the concentration of greenhouse gases increases in the atmosphere. The more the IR rays are trapped, the lower Earth's albedo by allowing less energy to reflect from the atmosphere. In simple words, if the concentration of greenhouse gases does not increase in the atmosphere means remains constant, the temperature of Earth also remains in equilibrium. But, since the greenhouse effect can shift the albedo of Earth, which results in variations in Earth's equilibrium temperature, a change in albedo can contribute to overall climate change.

2.5 El Nino-Southern oscillation (ENSO) cycle

ENSO is an important climatic phenomenon occurring on Earth that can change the global atmospheric circulations, leading to variation in temperature and precipitation. It consists of three phases; the opposite phases "El Niño" and "La Nina" and "Neutral". El Niño is the warm phase of ENSO, while La Niña is the cold phase of ENSO and Neutral is between the continuums (Gergis and Fowler, 2009).

ENSO cycle describes the irregular, periodic variations between the sea surface temperatures (SST) and atmosphere in the east-central Equatorial pacific. These deviations from the normal temperatures can cause a large scale effect on climate in most of the tropics and subtropics.

The El Nino events occur at irregular intervals of two-to-seven years. Sir Gilbert Walker in the 1930s discovered the "Southern Oscillations," which is a large-scale change in atmospheric pressure over Indonesia and the tropical Pacific Ocean (Trenberth and Hoar, 1997). In the 1960s, Jacob Bjerknes and other climatologists determined that El Niño and Southern Oscillations are linked and occur simultaneously when the sea surface temperature increases in the eastern tropical Pacific. This is how the hybrid term ENSO was born (Collins, 2005).

El Nino brings severe climatic changes. During El Nino, rainfall increases drastically in Ecuador and northern Peru, which generally have relatively arid climates. Coastal flooding and erosion are common El Nino events. It brings rain to South America and droughts to Indonesia and Australia. It is likely to develop over North America during the upcoming winter season, which includes warmer temperatures than normal over western and central Canada and western and northern United States. Wetter-than-average conditions are likely to develop over US Gulf Coast and Florida, while drier than average conditions can be expected in the Ohio Valley and the Pacific Northwest. El Nino episodes of 1982–83 and 1997–98 were the strongest of the 20th century. During 1982–83 the eastern tropical Pacific temperature became many degrees above normal, causing Australia to experience harsh droughts, and Chile experienced heavy rainfall and flooding. While during 1997–98, Indonesia, Malaysia, and the Philippines had drought conditions, and Peru experienced severe rain and flooding. California experienced heavy winter rainfall, and Midwest was hit by record-breaking warm temperatures (the period known as 'the year without a winter'). Thus, the presence of El Nino has a significant impact on weather patterns, ocean conditions, and marine fisheries across large portions of the globe for an extended period (Aladenola et al., 2016; Supriatin, 2016).

2.6 Volcanic activity and climate change

Volcanism is essential and long been implicated as a possible natural cause of weather and global climate change on many timescales. Explosive volcanic eruptions release massive quantities of gases, ash, and fine-grained debris into the atmosphere, which may last only a few days, but can influence climatic patterns for years. The powerful eruptions are sufficient to throw these materials high into the stratosphere, circling the globe for months to years. The Sulfur gases convert to Sulfate aerosols, sub-micron droplets containing about 75% sulfuric acid (Bethke et al., 2017). These aerosols can linger for as long as three to 4 years in the stratosphere. The suspended volcanic material spewed into the atmosphere forms a blanket around the globe. The particles absorb sunlight and locally heat the stratosphere while filtering out and thus reducing the amount of solar radiation reaching the Earth, which in turn lowers the temperature in the troposphere and thus influencing the atmospheric circulation pattern (Stenchikov, 2016; Grant, 1996).

The most conspicuous cold period was linked to the 1815 Mount Tambora eruption in Sumbawa Island, Indonesia producing the "year without a summer" in 1816, leading to crop failures and famine, killing more than 60,000 people. Similarly, after the 1883 Krakatau eruption and 1980 Mount St. Helens eruption, the temporary extensive atmospheric loading reduced the amplitude of the diurnal cycle of surface air temperature in the tropospheric region cloud. The effects, however, disappeared as soon as the particles settled down to the ground.

The three recent major volcanic eruptions, Mount ST. Helens (Washington) in 1980, El Chichon (Mexico) in 1982, and Mount Pinatubo (Philippines) in 1991 provided the opportunity to make more detailed measurements and to study the atmospheric impact of volcanic eruptions with the aid of Satellite images and remote-sensing instruments (Lutgens et al., 2013).

The eruption of Mount St. Helens speculated any possible effect on climate as a slight cooling, probably less than 0.1°C occurred, that it could not be differentiated from other natural temperature variations. After 2 years of monitoring and studying the eruption of El Chicon it was noticed that its cooling effect on global mean temperature was more significant than Mount St. Helens and was around 0.3–0.5°C, although being less explosive than Mount Helens. This is because St. Helens mostly ejected fine ash, which settled out soon. However, El Chichon released a more considerable amount of Sulfur dioxide, combined with water vapors in the stratosphere, forming a dense cloud of sulfuric acid particles called aerosols, which did not settle down soon and lasted in the atmosphere for many years. The cloud reflected the solar radiation into space and produced a colder effect in the lower troposphere (Andreae, 1995).

The event of Mount Pinatubo explosion released the enormous amount of sulfur aerosol particles into the stratosphere, which produced a vast dust cloud surrounding the globe by forming a veil and creating quite a substantial cooling of the globe (about 0.2°C) and troposphere (perhaps 0.4°C) from late 1991 to 1994 (Hansen et al., 1992; Soden et al., 2002).

3. Anthropogenic cause

The Sun is the primary controlling factor of the climate. The incoming solar radiation from the Sun travels in the form of electromagnetic radiation, which contains several rays of different wavelengths varying from 0.0001 nm to 100 m. The radiation with a wavelength in the range of 0.0001 nm–4 mm is called shortwave radiation, and the radiation with a wavelength greater than 4 mm is called longwave radiation. 95% of the solar radiation reaching the Earth's surface range from 0.1 to 2.5 mm are shortwave radiations.

These shortwave radiations reach the Earth's surface without hindering and absorbed by the atmosphere and earth surface. 90% of the solar radiation absorbed by the Earth is re-emitted as longwave radiation in the form of heat, which is absorbed and reradiated by few naturally occurring atmospheric gases, i.e., greenhouse gases (GHGs). The GHGs are carbon dioxide, water vapor, methane, nitrous oxide, etc. which absorb and emit the longwave radiation and prevent their escape into outer space. In this way, they maintain the Earth's global temperature and lower atmosphere, as shown in Fig. 9.1. The Earth's global temperature is maintained at 15°C (Ehhalt et al., 2001).Thus the phenomenon of absorbing and reradiating heat by the GHGs is called the greenhouse effect.

If all the radiation emitted by the Earth's surface escaped to outer space, the Earth's average temperature would be 10°C less than the present. If all the greenhouse gases are removed from the atmosphere, the Earth's average temperature would be −18°C.

At present, the concentration of the GHGs is increasing due to several anthropogenic activities such as industrialization and fossils fuel burning. The increased greenhouse gases in the atmosphere will absorb more infrared radiation (longwave radiation) and enhance the greenhouse effect. The continued enhanced greenhouse effect gradually increases the global mean temperature called global warming. The long-term impact of global warming on a global scale is called climate change (Xu and Cui, 2020).

3.1 Greenhouse gases

The example of GHGs are, water vapor (H_2O) carbon dioxide (CO_2), methane (CH_4), nitrous oxide (N_2O), chlorofluoro carbon (CFC), sulfur hexafluoride (SF_6) and tropospheric ozone (O_3). Among these, H_2O vapor and CO_2 are significant GHGs. These GHGs absorb the infrared radiation emitted by the Earth's surface and return it, responsible for warming the Earth's surface as shown in Fig. 9.3 (Watson et al., 1990). The trend of increasing the concentration of GHGs are given in Table 9.1.

The GHGs are released from transportation (29%), electricity (28%), industrial (22%), commercial and residential (12%), and agricultural sectors (9%). The concentration of GHGs is increasing due to enhanced anthropogenic activity such as rapid industrialization, fossils fuel burning, deforestation, land use pattern, and intensive agricultural activities. The changes in land use patterns add regional impact on climate change.

For example, the conversion of forest into cropland is a local activity. In contrast, soil organic matter loss adds a global impact on nutrient cycling, especially carbon cycling (Ledley et al., 1999; Hockstad and Hanel, 2018).

3.1.1 Carbon dioxide (CO_2)

CO_2 is released both from natural and anthropogenic sources. But at present, the concentration of CO_2 is increasing because of anthropogenic activities. The primary anthropogenic source of CO_2 in the atmosphere is the burning of fossil fuel, deforestation, and change in land use (Ramanathan and Feng, 2009). Among these, fossil fuel

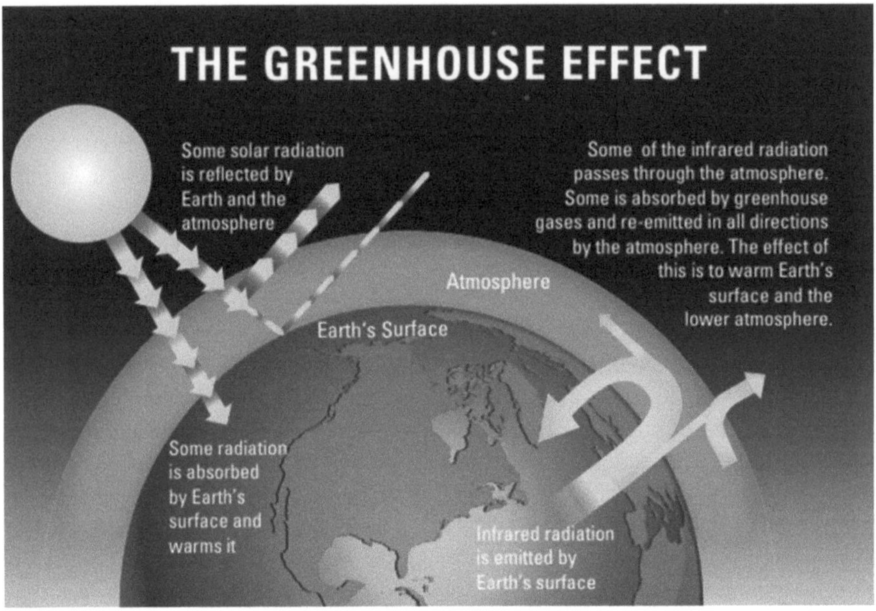

Figure 9.3 Absorption and emission of heat energy by GHGs, maintain the temperature of the earth surface and lower atmosphere. The introduction of more GHGs into the atmosphere will enhance the warming process.
Source: Royal Society, 2020. Climate change, evidence and causes.

Table 9.1 Trend of increasing concentration of GHGs.

Greenhouse gases	Pre-industrial concentration (in ppm) 1750 AD	Present concentration(in ppm)	% Increase (since 1750)	Atmospheric life time (in years)
CO_2	280 ppm	412 ppm	40%	5–200
CH_4	700 ppb	1875 ppb	168%	12
N_2O	270 ppb	334 ppb	23.7%	114
Tropospheric ozone	237	337 ppb	42.2	na
CFC-11	0	232 ppt	–	55
CFC-12	0	516 ppt	–	140
CFC-13	0	72 ppt	–	85
HCFC-22	0	233 ppt	–	12
HFC-134a	0	84 ppt	–	13.4
CCl_4	0	82 ppt	–	26
SF_6	0	8.6	–	3200

combustion, such as coal and petroleum, has been the primary source that releases 30 billion tonnes of CO_2 in the atmosphere per annum. It is the largest contributor to global warming (Watson et al., 1992; Mintzer, 1990).

The average concentration of CO_2 is measured from Mauna Loa Observatory in Hawaii. At present, the concentration of CO_2 is 412 ppm. The records showed that the present concentration of CO_2 is 147% greater than preindustrial time (280 ppm) (Fig. 9.4). More than 50% of the total increase in CO_2 concentration has been recorded since 1970 (Denisov et al., 2019).

Out of total emission, the developed countries releases about 60% of CO_2. The developing nations are also releasing a massive amount of CO_2, which is likely to increase more than developed nations. The most significant CO_2 is emitted by China (28%), which is followed by United States (15%), India (7%), and Russian Federation (5%) (Peters et al., 2020).

The analysis of air trapped in Antarctica revealed that the concentration of CO_2 extending back 10,000 years was in the range of 260–280 ppm, which began to increase significantly in the 19th century (Kellogg, 2019). The scientists have also examined the air trapped in ice core extending back 80,00,00 years, which revealed that CO_2 concentrations remained within the range of 170–300 ppm throughout many "ice age" cycles. They didn't find any CO_2 concentration records above 300 ppm until the past 200 years (Dilmore and Zhang, 2018; Barbera et al., 2019).

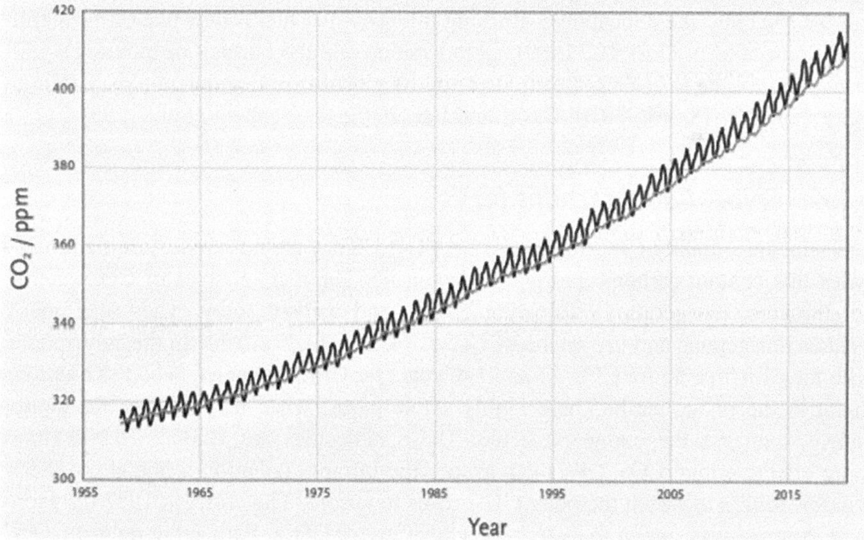

Figure 9.4 Graphical representation of atmospheric CO_2 measurement since 1958 from the Mauna Loa Observatory in Hawaii (*black up and down lines*) and the South Pole (*red line: gray line in printed version*). The *lines* are showing an increase in atmospheric CO_2 concentration at a steady rate each year.

Source: Royal Society, 2020. Climate change, evidence and causes.

3.1.2 Methane

Methane is also one of the potential GHGs. Its global warming potential is 25 times greater than CO_2. Methane is also released from both natural and anthropogenic sources. At present, its concentration level has reached more than double (1875 ppm) than that of preindustrial time (700 ppm) (Saunois et al., 2016).

The major source of methane emission is biological, accounting for more than 80% of total atmospheric methane. These biological sources are anthropogenic in origins, such as landfill, livestock management (enteric fermentation by ruminants), anaerobic respiration, rice field, and biomass burning. The emission of methane from a freshwater body occurs due to the incomplete decomposition of organic matter. Termites also produce methane while digesting cellulose. Several methanogens of rice field, bogs, tundra, wetlands, and lakes also add methane into the atmosphere. The production and use of fossil fuel also responsible for the increased emission of methane. Coal mining is also one source of methane production (Tiwari et al., 2020; Archer, 2007; Dean, 2020).

3.1.3 Nitrous oxide (N_2O)

The important sources of N_2O are agricultural practices, burning of biomass, and industrial activities. For the last few decades, its concentration has increased mainly because of nitrogen-based fertilizer and land-use changes. Nitrate contaminated groundwater also releases N_2O. The industrial activities, such as nylon production, release N_2O into the atmosphere. Burning nitrogen-rich fuel is also one of the significant sources of N_2O (NRC, 1989). Deforestation and the burning of biomass significantly contribute to N_2O to the atmosphere. Its global warming potential is 310 times more than CO_2 (Smith, 2010; Ussiri and Lal, 2013; Rennenberg et al., 1995).

3.1.4 Chlorofluorocarbons (CFCs)

CFCs include, mainly CFC-11, CFC-12, CFC-13. They are non-toxic, non-flammable gases that contain carbon, chlorine, and fluorine. The main sources of CFCs are air conditioners, refrigeration, insulating foams, and aerosol spray. They are mainly used as refrigerants and fire retardants. The CFCs are highly stable in the atmosphere, with their lifetime 55 for CFC-11 and 140 years for CFC-12, which becomes a serious threat to the planet earth. These highly stable gases, when they reaches the stratosphere, damages the ozone molecule. These molecules are 12,000–15,000 times more effective than CO_2. CFCs can absorb the infrared radiation in the atmospheric window region to which the other GHGs fails to absorb. Therefore, in 1987 the Montreal Protocol was signed to restrict the global use of CFCs. It aimed to reduce CFCs' production by 50% before the year 2000 to prevent the ozone layer (Montzka et al., 2011; Say et al., 2019; Vollmer et al., 2018).

3.1.5 Water vapor (H_2O) and aerosol

It is the largest contributor to the greenhouse effect. About 60% of warming is caused by H_2O vapor. It absorbs the radiation of the same wavelength as CO_2. H_2O vapor does not control the temperature of the Earth; rather, it is controlled by temperature. The higher the temperature, the more the H_2O vapor will be added to the atmosphere by evaporation. The amount of H_2O vapor, cloud cover, an albedo of the cloud, and rainfall duration and intensity affect the climatic processes. The natural H_2O vapor content in the atmosphere varies from 1% to 5%.

Aerosols are very fine, microscopic, liquid and solid particles suspended in the air. These are mainly present in dry air. These are composed of different materials such as soil, salts, smoke, and sulfuric acid. The anthropogenic aerosols are generated from the burning of agricultural residue, burning of fossils fuel. Aerosol play an important role in climate change. It affects the climate directly by absorbing and reflecting the incoming solar radiation or indirectly by modifying the cloud (Mitchell et al., 1995). The primary anthropogenic sources of aerosols are sulfur dioxide emitted by the burning of vegetation and the combustion of fossil fuel (Kiehl and Briegleb, 1993). Most of the aerosols are regionally distributed and show a high degree of spatial and temporal variation. These have very short residence time (about a week) in the atmosphere. The vegetation fire releases fine organic particles with varying amounts of black carbon capable of absorbing solar radiation. The forest fire emits mainly smoke, whereas the burning of grasses emits a large quantity of black carbon. The aerosols from the ocean contain the salt and sulfate particles.

4. Effect of climate change

4.1 Change in sea level

Change on sea level is one of the significant consequences of changing climate. At present, the global environment has changed because of enhanced global warming. The increase in sea level has resulted due the meltwater from mountain glaciers, Greenland ice sheets, Antarctica ice sheets, and thermal expansion of seawater due to warming. According to Lindsey, the global sea level has reached up to 23 cm of rising since 1880 (Lindsey, 2017). In 2019, the global mean sea level rose 8.7 cm above the 1993 level, which was the highest level until the present. The rate of increasing sea level has also increased from 1.5 mm (during the 19 century) to 3.6 mm per year (2006–15). It has been predicted that the rise in sea level will be continued even in the following centuries of 2100. Rao et al. (2011) have studied the impact of climate change on the sea-level rise along the Andhra Pradesh coast. It has been identified that thermal expansion and addition of ice met water were the cause of stearic and eustatic sea-level rise (Allen and Komar, 2006; Meehl et al., 2005). In 2007, IPCC estimated the global sea level of 59 cm, but according to a new estimate, the global sea level would rise more than 1 m by 2100 AD (IPCC, 2007). Rao et al. have suggested an urgent need to initiate a mitigation plan against the possible impact of predicted sea-level rise based on the geomatics model (Rao et al., 2011).

According to Alley et al., the mass balance of ice sheets, ice caps, and regional glaciers directly and indirectly consequences on sea-level positions (Alley et al., 2005). During the Pleistocene glacial period, there was huge formation of ice sheets in the Northern Hemisphere. Such formation had subtracted the water from the ocean. Thus the sea level was dropped, which indicated the positive mass balances. Conversely, during the deglaciation period, the negative mass balance released a huge amount of water to the ocean. Thus the sea level was increased.

Parker and Boyer have studies that the 10% wetland of North America's west coast has lost due to the rise in sea level (Parker and Boyer, 2019).The increased sea level results in increased intensity and frequency of storm surges.

The increase in sea level also results in the change in shoreline, which leads to floods. It can also inundate the sewage infrastructure, which leads to increased sewer leak and pathogen in the coastal region. It can also cause the salinization of surface and groundwater (Nriagu, 2019).

4.2 Ocean acidification

The anthropogenic emission of CO_2 has also a great effect on the ocean chemistry and marine biota. The absorption of anthropogenic CO_2 decreases the pH of the ocean. It also lowers the calcium carbonate level (Doney et al., 2009; Feely et al., 2004). It is well-established that when CO_2 dissolves in water, it forms a weak acid i.e., carbonic acid (H_2CO_3) (Honisch et al., 2012). The ocean has absorbed about a third of CO_2 released by anthropogenic activities. The continuous increase absorption of human-caused CO_2 leads to a decrease in the ocean's pH and disturbs the equilibrium of chemicals present naturally in seawater. This equilibrium shift is expected to be more in the next century. Since preindustrial time, the ocean's pH has decreased by 0.1 due to the absorption of huge CO_2 (Raven et al., 2005). The scientist has estimated that if the current trend of CO_2 increases in the atmosphere will continue, the ocean's pH will be decreased by 0.3 by the year 2100 (Orr et al., 2005). This shift of pH representing a significant change in seawater chemistry, which has not been observed for millions of years. The change toward acidity will affect ocean chemistry to a small extent. It can pose a significant threat to some shell-forming creatures ranging from phytoplanktons to benthic, molluscs, echinoderms, and corals, as building their hard using calcium carbonate. The decline in pH also affects the nutrients cycling and natural marine ecosystem (Fabry et al., 2008). Spero et al., have reported the 8%−14% reduction in shell mass in Orbulina universa (Spero et al., 1997). Gazeau et al. have found that the calcification rate decreased by 25% in *Mytilus edulis* (Gazeau et al., 2007).

4.3 Ozone depletion

Global climate change is caused by the increased concentration of GHGs and natural carbon sinks reduction, which induces the greenhouse effect. The greenhouse effect is the natural phenomenon of maintaining global temperature. The depletion of the ozone layer is related to climate change in several ways. First, the GHGs like CFC, HCFC, SF_6 are directly involved in ozone depletion. Second, it has been projected that climate change has induced the global temperature to increase in the range of 0.6−4°C by 2050. This increased temperature of the troposphere becomes the cause of the cooling of stratosphere. The cooling of stratosphere causes the enhanced formation of free radical and ozone depletion. The depletion of ozone itself leads to cooling of stratosphere which further increases the depletion of ozone layer. Austin et al. have predicted that the increasing concentration of CO_2 causes the cooling of the stratosphere. This stratosphere cooling induces the formation of the polar stratospheric cloud, which activates the ozone-depleting species and causes Arctic ozone hole (Austin et al., 1992). Shindell has also made an investigation where he found that the GHGs are responsible for the depletion of the ozone layer (Shindell et al., 1998). Mackezie et al. have reported that climate change is affected by ozone depletion and ozone depletion affects climate change. Hartmann et al. have also reported that climate change has led to rapid cooling and ozone depletion in the lower polar stratospheric region (Hartmann et al., 2000). They also studied that GHGs are responsible for the depletion of stratospheric ozone (McKenzie et al., 2011). Thus there is a strong interaction between the climate change ozone depletion (Diffey, 2003).

4.4 Melting of polar ice and glaciers

The mountain glaciers and polar ice caps are extremely sensitive to change in temperature. The climate change induced by global warming due to the GHGs is diminishing the polar ice. Bajracharya et al. have reported that the increased rate of retreating of Himalayan glaciers ranges from 10 to 60 m per annum. Several small glaciers have already disappeared. They have studied that in the last 50 years, the glaciers have shifted more than 100 m vertically. The retreating of glaciers has led to the development of several lakes in number and in size, also in the Himalayan region (Bajracharya et al., 2008). Zhang et al. have studied the effect of climate change on the melting of glaciers. For this, they have chosen Niyang River Basin in the Southeast Qinghai−Tibet Plateau in China. They found significantly increased streamflow due to glaciers' melting (Zhang et al., 2011). Wang et al. have reported that climate change causes glaciers melting (Wang et al., 2019).

Since late 1970, the Arctic sea ice has decreased dramatically, especially in summer and autumn. The satellite data has reported that Arctic sea ice has reduced by 40%. The extent of declining sea ice content may be unprecedented in at least the past 1450 years (The Royal Society, 2014).

4.5 Enhanced extreme weather events

Due to the enhanced human-induced GHG emission, the lower atmosphere of the Earth is becoming warmer. The warming of the Earth's atmosphere provides potential energy for developing storms and other extreme weather events. The warming of the climate intensifies, leading to hot days and nights, which amplifies heat waves, due to the warming, evaporation increases, promoting the drought and wildfire. It also contributes to heavy rainfall and snowfall due to the high moisture content of the atmosphere. The El Nino and La Nina events become more extreme in the warm climate. The El Nino, which promotes drought conditions in tropical and subtropical land areas, becomes more extreme. Similarly, La Nina also becomes wetter in many places.

The warming climate also causes the ocean to be warm, which produces massive water vapor to the atmosphere. The warming and excess water vapor promote the development of strong hurricanes with intense rainfall. In addition to this, the seawater content also increases due to the melting of the glaciers. During the coastal storm, the increased seawater and simultaneous rainfall led to the development of destructive storm surge and flooding (The Royal Society, 2014). Menezes et al. have reported the changing climate is the reason behind extreme weather events occurring in Brazil (Menezes et al., 2020). Choobari et al. have analyzed Iran's meteorological records from 1951 to 2013 to investigate climate change and its effect on extreme weather events. They have concluded that the temperature of Iran has increased by 1.3°C from 1951 to 2013. This raises the temperature and increases the heat, and decreases precipitation (Alizadeh-Choobari and Najafi, 2018). Wuebbles et al. have also concluded that the changing climate has increased the intensity and duration of the extreme weather events (Wuebbles et al., 2014).

4.6 Food security

The global population is increasing day by day. The current human population is 7.6 billion, which is expected to be raised by 8.6 billion during 2030 and 9.2 billion during 2050 (Foresight, 2011). While during 2100, it is believed that the global population will increase by 11.2 billion. This increasing human population pattern needs more than 60% extra food than produce today to meet out their food demand. The changing climate induces extreme weather events such as drought, flood, storm surge, hurricane, and tornado. Due to such extreme weather phenomenon, about 10 million peoples are facing hunger, and more than 100 million could die by 2030 because of the unavailability of food (Koebler, 2012; Romm, 2011). Beddington et al. have reported that pest infestation and water scarcity have also put agricultural production under great stress (Beddington et al., 2012). According to Gillis et al., the adverse effect of climate change on food security is related to soil, water, and crops. The changing climate affects the soil quality by depleting the soil organic carbon pool, available water capacity. The changing climate affects the soil by crusting, compacting, and accelerating erosion and salinization. Similarly, climate change may also affect water availability by altering the hydrological cycle (Gillis, 2013).

5. Conclusions

Since 1900, the anthropogenic emission of GHGs has posed a serious threat via increasing the temperature of the Earth by 1°C. The warming of the planet earth has reduced the extent of ice and snow in Northern Hemisphere. The melting of the glaciers, ice, and snow has increased the sea level. The temperature increase on the ocean water has also forced to migrate the temperature-sensitive species of fish, mammals, and insects. Such incidents have advocated that the climate has changed. Both natural and anthropogenic processes cause a change in the climate. The natural causes include a variation of Earth's orbit, solar variability, plate tectonics, and volcanic eruption. Climate change has resulted in extreme weather events such as drought, desertification, forest fire, loss of biodiversity, ocean acidification, shrinking of Arctic sea ice, and sea-level rise.

References

Aladenola, O., Cashman, A., Brown, D., 2016. Impact of El Niño and climate change on rainwater harvesting in a Caribbean state. Water Resour. Manag. 30 (10), 3459–3473.

Alizadeh-Choobari, O., Najafi, M.S., 2018. Extreme weather events in Iran under a changing climate. Clim. Dynam. 50 (1–2), 249–260.

Allen, J.C., Komar, P.D., 2006. Climate controls on US west coast erosion processes. J. Coast Res. 22, 511–529.

Alley, R.B., Clark, P.U., Huybrechts, P., Joughin, I., 2005. Ice-sheet and sea-level changes. Science 310 (5747), 456–460.

Andreae, M.O., 1995. Climatic effects of changing atmospheric aerosol levels. World Survey Climatol. 16 (06), 347–398.

Angstrom, A., 1962. Atmospheric turbidity, global illumination and planetary albedo of the earth. Tellus 14 (4), 435–450.

Archer, D., 2007. Methane Hydrate Stability and Anthropogenic Climate Change.

Austin, J., Butchart, N., Shine, K.P., 1992. Possibility of an Arctic ozone hole in a doubled-CO_2 climate. Nature 360 (6401), 221–225.

Bajracharya, S.R., Mool, P.K., Shrestha, B.R., 2008. Global climate change and melting of Himalayan glaciers. In: Melting Glaciers and Rising Sea Levels: Impacts and Implications, pp. 28–46.

Barbera, A.C., Vymazal, J., Maucieri, C., 2019. Greenhouse gases formation and emission. In: Fath, B. (Ed.), Encyclopedia of Ecology, second ed., pp. 329–333.

Barnes, P.W., Williamson, C.E., Lucas, R.M., Robinson, S.A., Madronich, S., Paul, N.D., et al., 2019. Ozone depletion, ultraviolet radiation, climate change and prospects for a sustainable future. Nat. Sustain. 2 (7), 569–579.

Beddington, J.R., Asaduzzaman, M., Clark, M.E., Bremauntz, A.F., Guillou, M.D., Howlett, D.J.B., et al., 2012. What next for agriculture after Durban? Science 335 (6066), 289–290.

Beer, J., 2005. Solar variability and climate change. Memorie-Soc. Astron. Ital. 76 (4), 751.

Berger, A.L., Gulick, S.P., Spotila, J.A., Upton, P., Jaeger, J.M., Chapman, J.B., et al., 2008. Quaternary tectonic response to intensified glacial erosion in an orogenic wedge. Nat. Geosci. 1 (11), 793–799.

Bethke, I., Outten, S., Otterå, O.H., Hawkins, E., Wagner, S., Sigl, M., Thorne, P., 2017. Potential volcanic impacts on future climate variability. Nat. Clim. Change 7 (11), 799–805.

Bonfils, C., de Noblet-Ducoudré, N., Braconnot, P., Joussaume, S., 2001. Hot desert albedo and climate change: Mid-Holocene monsoon in North Africa. J. Clim. 14 (17), 3724–3737.

Burrell, A.L., Evans, J.P., De Kauwe, M.G., 2020. Anthropogenic climate change has driven over 5 million km^2 of drylands towards desertification. Nat. Commun. 11 (1), 1–11.

Climate, 2018. www.climate.nasa.gov/news.

Climate Change, August 1, 2018. Global Sea Level. www.NOAA climate.gov.

Cliver, E.W., Boriakoff, V., Feynman, J., 1998. Solar variability and climate change: geomagnetic aa index and global surface temperature. Geophys. Res. Lett. 25 (7), 1035–1038.

Collins, M., 2005. El Niño-or La Niña-like climate change? Clim. Dynam. 24 (1), 89–104.

Crowley, T.J., Kim, K.Y., 1996. Comparison of proxy records of climate change and solar forcing. Geophys. Res. Lett. 23 (4), 359–362.

Crowley, T.J., Yip, K.J.J., Baum, S.K., 1993. Milankovitch cycles and Carboniferous climate. Geophys. Res. Lett. 20 (12), 1175–1178.

Dean, J.F., 2020. Old methane and modern climate change. Science 367 (6480), 846–848.

Denisov, S.N., Eliseev, A.V., Mokhov, I.I., September 2019. Contribution of natural and anthropogenic emissions of CO_2 and CH_4 to the atmosphere from the territory of Russia to global climate changes in the twenty-first century. Dokl. Earth Sci. 488 (1), 1066–1071 (Pleiades Publishing).

Diffey, B., 2003. Climate change, ozone depletion and the impact on ultraviolet exposure of human skin. Phys. Med. Biol. 49 (1), R1.

Dilmore, R., Zhang, L., 2018. Greenhouse gases and their role in climate change. In: Greenhouse Gases and Clay Minerals. Springer, Cham, pp. 15–32.

Doney, S.C., Fabry, V.J., Feely, R.A., Kleypas, J.A., 2009. Ocean Acidification: The Other CO_2 Problem.

Dowdeswell, J.A., Hagen, J.O., Björnsson, H., Glazovsky, A.F., Harrison, W.D., Holmlund, P., Jania, J., et al., 1997. The mass balance of circum-Arctic glaciers and recent climate change. Quat. Res. 48 (1), 1–14.

Dyurgerov, M.B., Meier, M.F., 2000. Twentieth century climate change: evidence from small glaciers. Proc. Natl. Acad. Sci. U.S.A. 97 (4), 1406–1411.

Ehhalt, D., Prather, M., Dentener, F., Derwent, R., Dlugokencky, E.J., Holland, E., et al., 2001. Atmospheric Chemistry and Greenhouse Gases (No. PNNL-SA-39647). Pacific Northwest National Lab. (PNNL), Richland, WA (United States).

Fabry, V.J., Seibel, B.A., Feely, R.A., Orr, J.C., 2008. Impacts of ocean acidification on marine fauna and ecosystem processes. ICES J. Mar. Sci. 65 (3), 414–432.

Feely, R.A., Sabine, C.L., Lee, K., Berelson, W., Kleypas, J., Fabry, V.J., Millero, F.J., 2004. Impact of anthropogenic CO_2 on the $CaCO_3$ system in the oceans. Science 305 (5682), 362–366.

Feynman, J., 2007. Has solar variability caused climate change that affected human culture? Adv. Space Res. 40 (7), 1173–1180.

Fletcher, C.G., Kushner, P.J., Hall, A., Qu, X., 2009. Circulation responses to snow albedo feedback in climate change. Geophys. Res. Lett. 36 (9).

Foresight, U.K., 2011. The Future of Food and Farming. Final Project Report. The Government Office for Science, London.

Frauendorf, T.C., MacKenzie, R.A., Tingley III, R.W., Infante, D.M., El-Sabaawi, R.W., 2020. Using a space-for-time substitution approach to predict the effects of climate change on nutrient cycling in tropical island stream ecosystems. Limnol. Oceanogr. https://doi.org/10.1002/lno.11577.

Fritz, S., 1949. The albedo of the planet earth and of clouds. J. Meteorol. 6 (4), 277–282.

Gazeau, F., Quiblier, C., Jansen, J.M., Gattuso, J.P., Middelburg, J.J., Heip, C.H., 2007. Impact of elevated CO_2 on shellfish calcification. Geophys. Res. Lett. 34 (7).

Gergis, J.L., Fowler, A.M., 2009. A history of ENSO events since AD 1525: implications for future climate change. Clim. Change 92 (3–4), 343–387.

Gillis, J., 2013. Not Even Close: 2012 was Hottest Ever in US. New York Times.

Grandon, N., 2020. Effects of Ocean Acidification on Species Composition and Biodiversity in the Region of Kongsfjord (Svalbard).

Grant, W.B., 1996. Tropical stratospheric ozone changes following the eruption of Mount Pinatubo. In: The Mount Pinatubo Eruption. Springer, Berlin, Heidelberg, pp. 161–175.

Hadley, O.L., Kirchstetter, T.W., 2012. Black-carbon reduction of snow albedo. Nat. Clim. Change 2 (6), 437–440.

Haeberli, W., Beniston, M., 1998. Climate change and its impacts on glaciers and permafrost in the Alps. Ambio 258–265.

Haigh, J.D., 1996. The impact of solar variability on climate. Science 272 (5264), 981–984.

Hall, A., Qu, X., 2006. Using the current seasonal cycle to constrain snow albedo feedback in future climate change. Geophys. Res. Lett. 33 (3).

Haman, K., 2008. Natural and Anthropogenic Causes of Climate Changes. Nauka (1).

Hansen, J., Lacis, A., Ruedy, R., Sato, M., 1992. Potential climate impact of Mount Pinatubo eruption. Geophys. Res. Lett. 19 (2), 215–218.

Hartmann, D.L., Wallace, J.M., Limpasuvan, V., Thompson, D.W., Holton, J.R., 2000. Can ozone depletion and global warming interact to produce rapid climate change? Proc. Natl. Acad. Sci. 97 (4), 1412–1417.

Hay, W.W., 1996. Tectonics and climate. Geol. Rundsch. 85 (3), 409–437.

Hockstad, L., Hanel, L., 2018. Inventory of US Greenhouse Gas Emissions and Sinks (No. Cdiac: EPA-EMISSIONS). Environmental System Science Data Infrastructure for a Virtual Ecosystem.

Hönisch, B., Ridgwell, A., Schmidt, D.N., Thomas, E., Gibbs, S.J., Sluijs, A., et al., 2012. The geological record of ocean acidification. Science 335 (6072), 1058–1063.

Huybers, P., Wunsch, C., 2003. Rectification and precession signals in the climate system. Geophys. Res. Lett. 30 (19).

IPCC, 2007. Summary for policymakers. In: Soloman, S.D., Manning, Q.M., Chen, Z., Miller, H.L. (Eds.), Climate Change: The Physical Science Basis. Contribution of Working Group I to the Fourth Assessment Report of the Intergovernmental Panel on Climate Change. Cambridge University Press, Cambridge, pp. 1–18.

Irving, D.B., Wijffels, S., Church, J.A., 2019. Anthropogenic aerosols, greenhouse gases, and the uptake, transport, and storage of excess heat in the climate system. Geophys. Res. Lett. 46 (9), 4894–4903.

Jochmann, H., 1993. Earth rotation and global change. Adv. Space Res. 13 (11), 271–280.

Jochmann, H., Greiner-Mai, H., 1996. Climate variations and the earth's rotation. J. Geodyn. 21 (2), 161–176.

Keisling, B.A., 2020. Modeling the Pleistocene History of the Greenland Ice Sheet.

Kellogg, W.W., 2019. Climate Change and Society: Consequences of Increasing Atmospheric Carbon Dioxide. Routledge.

Kiehl, J.T., Briegleb, B.P., 1993. The relative roles of sulfate aerosols and greenhouse gases in climate forcing. Science 260 (5106), 311–314.

Kirk-Davidoff, D.B., Hintsa, E.J., Anderson, J.G., Keith, D.W., 1999. The effect of climate change on ozone depletion through changes in stratospheric water vapour. Nature 402 (6760), 399–401.

Koebler, J., 2012. Report: 100 Million Could Die from Climate Change by 2030.

Kumar, S., 2015. Climate change and its causes and effects: a review. IJRSI 106–108.

Larin, I.K., 2020. Effect of global warming on the stratospheric ozone depletion rate in catalytic cycles. Russ. J. Phys. Chem. B 14, 344–350.

Ledley, T.S., Sundquist, E.T., Schwartz, S.E., Hall, D.K., Fellows, J.D., Killeen, T.L., 1999. Climate change and greenhouse gases. Eos Trans. Am. Geophys. Union 80 (39), 453–458.

Levrard, B., Laskar, J., 2003. Climate friction and the Earth's obliquity. Geophys. J. Int. 154 (3), 970–990.

Lindsey, R., 2017. Climate News, Stories, Images, & Video (Climate Watch Magazine).

Lovejoy, T.E., 2006. Climate Change and Biodiversity. The Energy and Resources Institute (TERI).

Lutgens, F.K., Tarbuck, E.J., Tusa, D., 2013. The Atmosphere. Prentice-Hall, Englewood Cliffs, NJ.

McKenzie, R.L., Aucamp, P.J., Bais, A.F., Björn, L.O., Ilyas, M., Madronich, S., 2011. Ozone depletion and climate change: impacts on UV radiation. Photochem. Photobiol. Sci. 10 (2), 182–198.

Meehl, G.A., Washington, W.M., Collins, W.D., Arblaster, J.M., Hu, A., Buja, L.E., et al., 2005. How much more global warming and sea level rise? Science 307, 1769–1772.

Menezes, J.A., dos Santos, R.B., Vommaro, F., Confalonieri, U., de Lima Barata, M.M., Margonari, C., 2020. Extreme weather events under a changing climate: a brief context for Brazil and the role of the health sector. In: Extreme Weather Events and Human Health. Springer, Cham, pp. 293–314.

Mintzer, I.M., 1990. Energy, greenhouse gases, and climate change. Annu. Rev. Energy 15 (1), 513–550.

Mitchell, J.F.B., Johns, T.C., Gregory, J.M., Tett, S.F.B., 1995. Climate response to increasing levels of greenhouse gases and sulphate aerosols. Nature 376 (6540), 501–504.

Montzka, S.A., Dlugokencky, E.J., Butler, J.H., 2011. Non-CO_2 greenhouse gases and climate change. Nature 476 (7358), 43–50.

National Research Council, 1989. Ozone Depletion, Greenhouse Gases, and Climate Change. National Academies Press.

Nriagu, J.O., 2019. Encyclopedia of Environmental Health. Elsevier.

Nwankwoala, H.N.L., 2015. Causes of climate and environmental changes: the need for environmental-friendly education policy in Nigeria. J. Educ. Pract. 6 (30), 224–234.

Ogunbode, C.A., Doran, R., Böhm, G., 2020. Exposure to the IPCC special report on 1.5°C global warming is linked to perceived threat and increased concern about climate change. Climatic Change 158 (3), 361–375.

Orr, J.C., Fabry, V.J., Aumont, O., Bopp, L., Doney, S.C., Feely, R.A., et al., 2005. Anthropogenic ocean acidification over the twenty-first century and its impact on calcifying organisms. Nature 437 (7059), 681–686.

Otterman, J., 1977. Anthropogenic impact on the albedo of the earth. Clim. Change 1 (2), 137–155.

Parker, V.T., Boyer, K.E., 2019. Sea-level rise and climate change impacts on an urbanized Pacific Coast estuary. Wetlands 39 (6), 1219–1232.

Peters, G.P., Andrew, R.M., Canadell, J.G., Friedlingstein, P., Jackson, R.B., Korsbakken, J.I., et al., 2020. Carbon dioxide emissions continue to grow amidst slowly emerging climate policies. Nat. Clim. Change 10 (1), 3–6.

Pokras, E.M., Mix, A.C., 1987. Earth's precession cycle and Quaternary climatic change in tropical Africa. Nature 326 (6112), 486–487.

Potter, S., Solvik, K., Erb, A., Goetz, S.J., Johnstone, J.F., Mack, M.C., et al., 2020. Climate change decreases the cooling effect from postfire albedo in boreal North America. Global Change Biol. 26 (3), 1592–1607.

Qu, X., Hall, A., 2006. Assessing snow albedo feedback in simulated climate change. J. Clim. 19 (11), 2617–2630.

Ramanathan, V., Feng, Y., 2009. Air pollution, greenhouse gases and climate change: global and regional perspectives. Atmos. Environ. 43 (1), 37–50.

Rao, K.N., Subraelu, P., Kumar, K.C.V.N., Demudu, G., Malini, B.H., Ratheesh, R., Rajawat, A.S., 2011. Climate change and sea-level rise: impact on agriculture along Andhra Pradesh coast—a geomatics analysis. J. Indian Soc. Remote Sens. 39 (3), 415–422.

Raven, J., Caldeira, K., Elderfield, H., Hoegh-Guldberg, O., Liss, P., Riebesell, U., et al., 2005. Ocean Acidification Due to Increasing Atmospheric Carbon Dioxide. The Royal Society.

Rennenberg, H., Wassmann, R., Papen, H., Seiler, W., 1995. Role of methane and nitrous oxide in climate change. Clim. Change Rice 60–68.

Ridgwell, A., Singarayer, J.S., Hetherington, A.M., Valdes, P.J., 2009. Tackling regional climate change by leaf albedo bio-geoengineering. Curr. Biol. 19 (2), 146–150.

Romm, J., November 2011. Oxfam: Extreme Weather Has Helped Push Tens of Millions into Hunger and Poverty in Grim Foretaste of Warmed World. Think Progress.

Rosenzweig, C., Karoly, D., Vicarelli, M., Neofotis, P., Wu, Q., Casassa, G., et al., 2008. Attributing physical and biological impacts to anthropogenic climate change. Nature 453 (7193), 353–357.

Royal Society, 2020. Climate change, evidence and causes.

Ruddiman, W.F. (Ed.), 2013. Tectonic Uplift and Climate Change. Springer Science & Business Media.

Sanchez, G., Rolland, Y., Corsini, M., Braucher, R., Bourlès, D., Arnold, M., Aumaître, G., 2010. Relationships between tectonics, slope instability and climate change: cosmic ray exposure dating of active faults, landslides and glacial surfaces in the SW Alps. Geomorphology 117 (1–2), 1–13.

Saunois, M., Jackson, R.B., Bousquet, P., Poulter, B., Canadell, J.G., 2016. The growing role of methane in anthropogenic climate change. Environ. Res. Lett. 11 (12), 120207.

Say, D., Ganesan, A.L., Lunt, M.F., Rigby, M., O'Doherty, S., Harth, C., ... Bauguitte, S., 2019. Emissions of halocarbons from India inferred through atmospheric measurements. Atmos. Chem. Phys. 19, 9865–9885.

Schurer, A.P., Tett, S.F., Hegerl, G.C., 2014. Small influence of solar variability on climate over the past millennium. Nat. Geosci. 7 (2), 104–108.

Self, S., Rampino, M.R., 1988. The relationship between volcanic eruptions and climate change: still a conundrum? Eos Trans. Am. Geophys. Union 69 (6), 74–86.

Seybold, H., Berghuijs, W., Prancevic, J.P., Kirchner, J., 2020. Global Dominance of Tectonics over Climate in Shaping River Longitudinal Profiles.

Shackleton, N.J., 2000. The 100,000-year ice-age cycle identified and found to lag temperature, carbon dioxide, and orbital eccentricity. Science 289 (5486), 1897–1902.

Shindell, D.T., Rind, D., Lonergan, P., 1998. Increased polar stratospheric ozone losses and delayed eventual recovery owing to increasing greenhouse-gas concentrations. Nature 392 (6676), 589–592.

Smith, K.A. (Ed.), 2010. Nitrous Oxide and Climate Change. Earthscan.

Soden, B.J., Wetherald, R.T., Stenchikov, G.L., Robock, A., 2002. Global cooling after the eruption of Mount Pinatubo: a test of climate feedback by water vapour. Science 296 (5568), 727–730.

Solanki, S.K., 2002. Solar variability and climate change: is there a link? Astron. Geophys. 43 (5), 5–9.

Solomon, S., 2007. IPCC (2007): Climate Change the Physical Science Basis. AGUFM, 2007, U43D-01.

Spero, H.J., Bijma, J., Lea, D.W., Bemis, B.E., 1997. Effect of seawater carbonate concentration on foraminiferal carbon and oxygen isotopes. Nature 390 (6659), 497–500.

Stalder, N.F., Herman, F., Fellin, G.M., Coutand, I., Aguilar, G., Reiners, P.W., Fox, M., 2020. The relationships between tectonics, climate and exhumation in the Central Andes (18–36°S): evidence from low-temperature thermochronology. Earth Sci. Rev. 103276.

Stenchikov, G., 2016. The role of volcanic activity in climate and global change. In: Climate Change. Elsevier, pp. 419–447.

Stephens, G.L., O'Brien, D., Webster, P.J., Pilewski, P., Kato, S., Li, J.L., 2015. The albedo of Earth. Rev. Geophys. 53 (1), 141–163.

Supriatin, L.S., Martono, M., December 2016. Impacts of climate change (El Nino, La Nina, and sea level) on the coastal area of Cilacap Regency. Forum Geografi 30 (2), 106–111.

The Royal Society, 2014. National academy of sciences. Science. National Academies Press, p. 36.

Tiwari, S., Singh, C., Singh, J.S., 2020. Wetlands: a major natural source responsible for methane emission. In: Restoration of Wetland Ecosystem: A Trajectory towards a Sustainable Environment. Springer, Singapore, pp. 59–74.

Tranter, B., 2017. It's only natural: conservatives and climate change in Australia. Environ. Sociol. 3 (3), 274–285.

Trenberth, K.E., Hoar, T.J., 1997. El Niño and climate change. Geophys. Res. Lett. 24 (23), 3057–3060.

Twomey, S., 1974. Pollution and the planetary albedo. Atmos. Environ. 8 (12), 1251–1256.

Ussiri, D., Lal, R., 2013. The role of nitrous oxide on climate change. In: Soil Emission of Nitrous Oxide and its Mitigation. Springer, Dordrecht, pp. 1–28.

Vollmer, M.K., Young, D., Trudinger, C.M., Mühle, J., Henne, S., Rigby, M., et al., 2018. Atmospheric Histories and Emissions of Chlorofluorocarbons CFC-13 ($CClF_3$), ΣCFC-114 ($C_2Cl_2F_4$), and CFC-115 (C_2ClF_5).

Wang, N.Y., Wang, L.K., Wang, M.H.S., Wong, J.O., September 2019. Scientific studies of global warming, climate change, glacier melting and salmon protection. Evol. Prog. Sci. Technol. Eng. Arts Math. 1 (9), 1–110.

Ward, P.L., 2020. The Ozone Depletion Theory of Global Warming.

Watson, R.T., Rodhe, H., Oeschger, H., Siegenthaler, U., 1990. Greenhouse gases and aerosols. Clim. Change IPCC Sci. Assess. 1, 17.

Watson, R.T., Meira Filho, L.G., Sanhueza, E., Janetos, A., 1992. Greenhouse gases: sources and sinks. Clim. Change 92, 25–46.

Whetton, P.H., Fowler, A.M., Haylock, M.R., Pittock, A.B., 1993. Implications of climate change due to the enhanced greenhouse effect on floods and droughts in Australia. Clim. Change 25 (3–4), 289–317.

Whittaker, A.C., 2012. How do landscapes record tectonics and climate? Lithosphere 4 (2), 160–164.

Williams, G.E., 1993. History of the Earth's obliquity. Earth-Sci. Rev. 34 (1), 1–45.

Wuebbles, D.J., Kunkel, K., Wehner, M., Zobel, Z., 2014. Severe weather in United States under a changing climate. Eos Trans. Am. Geophys. Union 95 (18), 149–150.

Xu, Y., Cui, G., 2020. Influence of spectral characteristics of the Earth's surface radiation on the greenhouse effect: principles and mechanisms. Atmos. Environ. 117908.

Zhang, M., Ren, Q., Wei, X., Wang, J., Yang, X., Jiang, Z., 2011. Climate change, glacier melting and streamflow in the Niyang River basin, Southeast Tibet, China. Ecohydrology 4 (2), 288–298.

Zhang, W., He, M., Li, Y., Cui, Z., Wang, Z., Yu, Y., 2012. Quaternary glacier development and the relationship between the climate change and tectonic uplift in the Helan Mountain. Chin. Sci. Bull. 57 (34), 4491–4504.

Development of abiotic stress–tolerant mustard genotype through induced mutagenesis

10

Archana Rai, Shankar Bhujbal, Sanjay J. Jambhulkar
Nuclear Agriculture and Biotechnology Division, Bhabha Atomic Research Centre, Mumbai, Maharashtra, India

1. Introduction

Climate change is one of the greatest threats to agricultural sustainability and food security (Arora, 2019). Increased frequency of abrupt environmental conditions either alone or with simultaneous occurrence leads to negative impacts on crop production, which may lead up to 70% reduction in global crop production (Boyer, 1982). Abiotic stresses include water scarcity, water logging, salinity, rise in temperature, heavy metal toxicity, and rising atmospheric CO_2 concentration. Among these drought, salt, and temperature stresses are the major stresses affecting the crop geographical distribution and productivity. Changes in the rain patterns and increase in evapotranspiration due to global warming have increased the frequency and severity of drought stress (Dai, 2011), and approximately 40% of the global land area is affected by drought (Trenberth et al., 2014). Over 800 million hectare of land is estimated to be affected by salinity worldwide (FAO-ITPS-GSP, 2015), and it has been increased by 37% during the past two decades in irrigated lands (1990–2013) (Ghassemi et al., 1995; Qadir et al., 2014). Global average temperature is likely to rise from 2.0 to 4.9°C by 2100 (Raftery et al., 2017). Zhao et al. (2017) predicted that increase in mean global temperature by each degree celsius is likely to reduce global yields of wheat, rice, maize, and soybean by 6.0%, 3.2%, 7.4%, and 3.1% respectively. In Indian scenario, there is projection of temperature increase of 0.5–1.2°C by 2020, 0.88–3.16°C by 2050, and 1.56–5.44°C by 2080 (IPCC, 2007) and 10%–40% yield reduction (Fischer et al., 2007; IPCC, 2007). The impact of these abiotic stresses on plants varies with factors such as phonological stage of crop plant, stress duration, and the intensity of stress.

The worldwide food market is composed of mainly cereals, oil crops, and legumes. Among oil crops, *Brassicaceae* play a prominent role in oilseed production (Raymer, 2002). *Brassica* spp. (oilseed rape/rapeseed/canola) rank second with 41 billion US dollars of annual value (USDA FAS, 2018). Like other crops, these crops are also facing the effects of climate change. Seed yield and oil are determined by temperature and

Global Climate Change. https://doi.org/10.1016/B978-0-12-822928-6.00004-6

precipitation (Vernon and Van Gool 2006; Namazkar et al., 2016). High temperature, fog, cold temperature, and uncertainty in rains during mustard crop establishment affect the crop adversely and cause significant yield losses either through physiological disorder alone or along with aphid pest attack, white rust, downy mildew, and stem rot diseases (Kumar, 2005). InfoCrop model–based simulation study in mustard showed that seed yield is likely to reduce in irrigated and rain-fed conditions and increasing atmospheric CO_2 and temperature change. Results further revealed simulated mustard yield increase with elevated CO_2 concentration, whereas this positive effect of increased CO_2 will be nullified by rise in temperature (Boomiraj et al., 2010). Upon exposure to various abiotic stresses, plants use different mechanisms for stress adaptation, and hence, studying these mechanism(s) will help to generate elite crop species with enhanced abiotic stress tolerance.

To meet future oilseed demand, climate-resilient crops varieties are needed, as these have adaptive ability of maintaining its productivity even when these are being affected by a disturbance in optimum growth conditions (Singh, 2016). Generation and utilization of desirable variability are utmost important factors to develop a variety with stress tolerance. Mustard crop improvement through mutation breeding for development of high-yielding varieties has been critically analyzed (Jambhulkar 2007, 2015). Generation of gene-specific mutations and its characterization are the precise approaches to strengthen the trait-specific variability. TILLING and CRISPR technologies are the advances in mutation breeding to induce gene-specific mutations and are being exploited to tailor designer crops. Thus, resilient crop varieties development is a compelling agenda among researchers.

Abiotic stress response in *Brassica* spp. (drought, salinity, high temperature): Physiological, biochemical, and molecular processes are involved in the adaptation of the plant to various abiotic stresses. Thus, it is essential to understand the plant responses under the stress regime. Genus *Brassica* consists of different species, and their response against the stress also varies. Among various species, it was found more tolerant to water deficient, followed by *Brassica rapa* and *Brassica juncea*, whereas *Brassica carinata* was found to be comparatively susceptible (Zhu et al., 2011; Hasanuzzaman et al., 2017).

In response to salt stress, amphidiploid species of *Brassica* were found less sensitive as compared with diploid species *B. rapa* (Kumar, 1995). It indicated that source of salt tolerance is from the A (*Brassica campestris*) and C (*Brassica oleracea* L.) genomes. Contrasting result has been reported regarding the salinity stress response of these species at different developmental stages. However, most of the study suggests that salt tolerance consistently maintains their degree throughout the plant ontogeny (Ashraf and McNeilly, 2004).

The ideal temperature for the germination of *Brassica* ranges from 25 to 28°C, and plant development and growth starts to reduce when temperature rises above 28°C (Sharma et al., 2007). Less flower number were reported under the high heat stress during vegetative growth stage in *B. rapa*, *Brassica napus*, and *B. juncea* (Zhang et al., 2015a,b). At seedling stage, the most thermotolerant species, *B. carinata* was reported followed *by B. campestris* and *Brassica tournefortii*. These three species were found comparatively more tolerant than *Brassica alba*, *B. juncea*, and *B. napus*. *B. napus* was found to be most susceptible to high temperature (Sharma et al., 2007).

Genetics of abiotic stress tolerance: Abiotic stress tolerance is a complex trait, and exploration of the heritable potential of these traits within the existing germplasm helps breeder to identify parents to be used in a breeding program. For drought tolerance traits, germination rate, leaf chlorophyll stability, and proline accumulation were found genetically control factors and identified as selection criteria in *Brassica* spp. (Richards and Thurling, 1979). Broadsense heritability for chlorophyll stability and proline accumulation in *B. napus* -were found 40% and 64%, respectively, whereas for germination rate, broad-sense heritability was 55% (Richards and Thurling, 1979) in *B. rapa*. To determine the gene action governing yield and its components under drought conditions, generation means in *B. napus* was calculated (Cheema and Sadaqat, 2004). Results suggested additive and nonadditive gene action involvement in majority of the traits. Most of the traits are governed by duplicate/complementary type of epistasis. Plant height, days to flowering, and maturity were found to be governed by more number of genes, which acts in a composite manner (Cheema and Sadaqat, 2004).

For salinity stress seedling emergence, number of siliquae on main shoot, secondary branch number per plant, and seed yield showed high broad-sense heritability in *B. juncea* (Sinha and Yadav, 1991; Kumar, 1993). Under nonsaline and saline conditions, combining ability analysis for seed germination, rate of germination, and seedling vigor in *B. juncea* showed both additive and nonadditive gene effects (Thakral and Prakash, 1998). High ratio of narrow-sense heritability estimates was observed for stress tolerance index and the additive effect in *B. napus* (Rezai and Saeidi, 2005). Qiu and Li (2009) suggested dominant genes with an additive effect for salt tolerance in *B. rapa*. Long et al. (2013) found additive dominance effects of three major genes with in *B. napus* for salt tolerance.

To investigate the genetics of thermotolerant traits, half diallel crosses of *B. juncea* genotypes along with their F1s were evaluated during terminal stage for canopy temperature and chlorophyll fluorescence at 40 DAS, 50 DAS, 60 DAS, and 70 DAS along with percentage seedling mortality and seed yield per plant. Heterosis was observed for most of the traits. Additive gene effects or additive × additive gene were observed based on overall appraisal of GCA (Singh et al., 2017). The physiological traits were investigated through combining ability and their gene action by Ram et al. (2018). Analysis displayed significant general and specific combining effects for physiological traits such as population survival after 10 DAS, population survival after 25 DAS, membrane stability index, excised-leaf water loss, relative water content, leaf water retention capacity (within 24 h), and seed yield per plant. The additive genetic effects were predominant for all the traits as general combining ability (GCA) was more than the specific combining ability (SCA) (Ram et al., 2018).

Physiobiochemical alterations during stress conditions: Abiotic stresses alter major biological processes in plants such as photosynthesis, rate of transpiration, protein biosynthesis, and metabolite accumulation and consequently limit plant growth (Zhu, 2016; Elferjani and Soolanayakanahally, 2018). Relative water content is an important physiological measure of plant cell hydration status and found to be reduced under drought (Sabagh et al., 2019) and salinity (Yang et al., 2012) in *B. napus* and *B. juncea* under salinity (Ghuge et al., 2011). Reduced chlorophyll content because of pigments loss and thylakoid membrane damage has been reported under drought

(Nasri et al., 2008; Ahmar et al., 2019) and causes significant effect on net photosynthesis (A), stomatal conductance (gs) intercellular CO_2 concentration (Ci), and leaf area at both vegetative and flowering stages of *B. napus* (Naderikharaji et al., 2008). Reduction in photosynthetic pigment has also been observed under salinity (Hayat et al., 2011; Ahmad et al., 2012) and heat (Hayat et al., 2009; Soengas et al., 2018) in *Brassica* spp. In plant species, nitrogen (N) uptake and its absorption process are affected by abiotic stresses. Various enzymes such as nitrite reductase, glutamine synthetase, glutamate dehydrogenase, and asparagine synthetase were found to be changed by stresses in *B. juncea* (Goel and Singh, 2015). Reduced uptake of essential nutrients such as nitrogen and potassium has been reported in *B. napus* under salt stress condition (Ahmar et al., 2019). Various stresses generate the ROS (reactive oxygen species) in plants, which leads to oxidative stress in plant. To combat these stresses, there is activation of scavenging machinery and functioning through different antioxidant enzymes (Kusvuran et al., 2016). Drought stress enhances the activity of antioxidant enzymes (catalase and peroxidase) in canola genotypes, but the enzyme activity increase varies among genotypes (Hosseini and Hassibi, 2011). Salt (Ghuge et al., 2011) and heat (Rai et al., 2020) stresses also lead to increased antioxidant enzyme activities in *B. juncea*. Osmotic adjustment has been found as a valuable component of osmotic stress tolerance and osmoprotectants (proline, glycine–betaine, gamma-aminobutyric acid and sugars); accumulation is a vital response observed in different plant systems (Suprasanna et al., 2016). *Brassica* species showed induction of proline under drought stress, whereas application of trehalose reduced the drought effect (Alam et al., 2014). Significant accumulation of glycine–betaine was also observed in *B. juncea* under salt stress (Ghuge et al., 2011). Reduction in linolenic acid was reported in *B. juncea* under rain-fed conditions (Enjalbert et al., 2013).

Impact on growth and yield: Abiotic stresses severely affect the germination, plant growth, flowering, yield, and seed quality. Flowering is the most susceptible stage for drought injury, resulting in an extreme loss in seed as well as oil yields by 29.5% and 31.7%, respectively (Shekari et al., 2015). Drought also reduces seed oil content and seed yield at a rate of 2.6% and 25%, respectively, in *B. napus* (Ali et al., 2017). Drought stress at flowering stage causes decline in pods number, seed yield, and productivity, but seed number per pod is not much affected in canola (Sinaki et al., 2007). Drought stress causes 35% and 18% yield loss at flowering and grain filling stages, respectively, in canola (Sinaki et al., 2007), respectively. Water stress increase also causes reduction in saturated fatty acid and oleic acid, however increases linolenic and linoleic acid (Aslam et al., 2009). The yield reduction under moisture stress in *B. juncea* ranged from 17% to 94% due to considerable effects on yield components (Chauhan et al., 2007). Crops growing in salt-affected soils suffer from physiological drought stress and ion toxicity, which then lead to reduced growth and productivity. The detrimental effect of salinity on *Brassica* has been observed for the reduction of plant height, yield contributing characters along with product quality deterioration (Zamani et al., 2010). Plant height reduction, increase in the days to the first flowering, and decreasing trend in yield performance were observed as salinity levels increase in *Brassica* (Bybordi, 2010). Grain yield significantly decreased with increasing salinity levels in *B. napus* (Tahmasebpour et al., 2018).

For *Brassica* species, the threshold temperature during flowering is 29.5°C. Flower number reduction, seeds produced per flower, and ultimately seed yield loss were observed due to high mean temperature during vegetative development (Morrison and Stewart, 2002). Pollen viability, anthesis time, fertilization process, and grain-filling stages are the sensitive stages in *Brassica* spp. for high temperature (Rieu et al., 2017). High temperatures even for a short period of heat stress can lead to yield reduction of 15%—20% (Rahaman et al., 2018).

2. Indicator for different stresses

The efficiency of any breeding programs for development of climate-resilient crop can be significantly improved by understanding the associations between seed yield and different selection criteria used for stress tolerance estimation in genotypes (Collard and Mackill, 2008; Xu, 2016). In many crops, main criteria for evaluating tolerance to various environmental stresses are yield performance and yield stability under different growth conditions (Clarke et al., 1992). Based upon these features, stress tolerance (TOL) was defined as the difference between yield under stress (Ys) and nonstress (Yp) environments and mean productivity (MP) of Ys and Yp (Rosielle and Hamblin, 1981), whereas stress tolerance index (STI) was defined by Fernandez (1992) which can identify genotypes that produce high yield under stress as well as nonstress conditions and geometric mean (GMP). Stress susceptibility index (SSI) was proposed by Fischer and Maurer (1978).

Among the stress tolerance indicators, higher values of TOL and STI represent comparatively more sensitive to stress, and selection based on these two indices favors genotypes with low yield potential under nonstress and high yield under stress condition. However, STI- and GMP-based selection favors genotypes with higher stress tolerance and yield potential (Fernandez, 1992). Based upon specific stress condition, different index has been calculated to identify tolerant genotypes such as drought stability index (DSI) (Chauhan et al., 2007), heat susceptibility index (HIS) (Chauhan et al., 2009), and stress intensity (Singh et al., 2014) in *Brassica*. Besides yield parameters, several physiological and biochemical parameters have also been used to screen the stress-tolerant genotypes in *Brassica* species, which is listed in Table 10.1.

Identification of molecular determinants and quantitative trait loci (QTLs) for abiotic stresses: For molecular breeding of climate-resilient mustard genotypes, it is important to identify stress tolerance—specific genic component which is mostly polygenic in nature. When abiotic stress is perceived by the plant, gene expression reprogramming takes place to bring ionic and water homeostasis and cellular stability under stress conditions (Zhu, 2016). Technologies such as microarray, RNA-seq, and bioinformatics software tools have generated massive data sets, which have been successfully employed to understand the abiotic stress responses of *Brassica* species.

In *B. rapa* leaves, genome-wide gene expression analysis under drought stress identified 1092 drought-responsive genes, which consist of the genes for transcription factors (37), signal transduction (28), and water and osmosensing-responsive pathways (61)

Table 10.1 Indices used to assess the abiotic stress impact on *Brassica* spp.

Method or indicator	Species	Developmental stage	Stress	References
Biomass and water potential	*Brassica napus* *Brassica juncea* *Brassica campestris* *Brassica carinata*	Seedling	Drought	Ashraf and Mehmood (1990)
DSI	*B. juncea*	Mature	Drought	Chauhan et al. (2007)
LS, PSS, RVIb	*B. napus*	Germination	Drought	Yang et al. (2007)
Leaf wilting index	*B. napus*	Seedling	Drought	Li et al. (2012)
TDM, LAI, RGR, CGRd	*B. napus*	Whole stages	Drought	Moaveni et al. (2010)
Photosynthetic rate	*B. napus; Raphanus sativus* L.	20 days' stage 6 weeks' old	Salinity	Ulfat et al. (2007), Munir et al. (2013)
Osmolyte accumulation (proline, glycine–betaine, and total soluble sugars); proline	*B. napus* *B. juncea*	6 weeks' old; germinating seeds	Salinity	Ulfat et al. (2007), Pandey and Penna (2017)
Relative cell membrane permeability and antioxidant enzyme	*B. napus*	30-day-old plant	Salinity	Ashraf and Ali (2008)
Transcript accumulation pattern for various salt overly sensitive member genes	*Brassica* spp.	Seedling stage	Salinity	Kumar et al. (2009)
Shoot Na+; ion discrimination	*B. napus*	Vegetative stage; 6 weeks' old	Salinity	Ulfat et al. (2007), Khayat et al. (2010)

Table 10.1 Indices used to assess the abiotic stress impact on *Brassica* spp.—cont'd

Method or indicator	Species	Developmental stage	Stress	References
Mitochondrial respiration via alternative oxidase	*B. juncea*	Germinating seeds	Salinity	Pandey and Penna (2017)
Heat susceptibility index; stress intensity	*B. juncea*	60 days after sowing	Terminal heat stress	Chauhan et al. (2009), Sharma and Sardana (2013) Singh et al. (2014)
Electrolyte leakage	*B. juncea*	Terminal stage	Heat stress	Kumar et al. (2013a,b), Kavita and Panday (2017)
Membrane stability index	*B. juncea*	25 days after sowing	Heat stress	Ram (2014)
Melondialdehyde content	*B. juncea*	Seedling stage	Early heat stress	Rai et al. (2020)

CGR, crop growth rate; *DSI*, drought susceptibility index; *LAI*, leaf area index; *RGR*, relative growth rate; *RVI*, relative vigor index; *TDM*, total dry matter.

(Yu et al., 2012). In *Sinapis alba*, differentially expressed genes involved in cell division, catalytic, and metabolic processes were detected under drought stress (Dong et al., 2011).

Combining the RNA-seq-detected differentially expressed genes (DEGs) with genome-wide association study found that 16 loci significantly associated loci were identified, which showed significant association with water stress response in *B. napus* (Zhang et al., 2015a,b). RNA-seq technique has also been used for identification of genes involved in salinity tolerance in *Brassica* spp. Yong et al. (2014) identified transcription factor and transporter genes from *B. napus* root and leaves, which are salinity responsive. Detoxification of ROS, sulfur assimilation, and calcium signaling pathway genes were detected in *B. juncea* seedling (Sharma et al., 2015). Bhardwaj et al. (2015) identified the heat stress—responsive differential transcripts of transcription factor and heat shock proteins using RNA-seq. Other abiotic stress tolerance—related transcriptomic studies in *Brassica* species have been summarized in Table 10.2.

Table 10.2 Transcriptomic studies in *Brassica* spp. for identification of abiotic stress–related molecular determinants.

Species	Tissue	Stress condition	Platform	Responsive biological processes	References
Brassica napus	Seed	PEG- or ABA analog PBI429-inhibited germination	Microarray	Late seed development, carbohydrate metabolism, cell wall loosening, ROS scavenging, lipolysis	Li et al. (2005)
B. napus	Seed (during ripening stage)	Natural desiccation during seed-ripening stage	Microarray	Signal transductions, protein synthesis	Fei et al. (2007)
B. napus	Whole plant	Drought (air-dried)	Microarray	Transcription factors	Lee et al. (2008)
B. napus	Seed (during ripening stage)	Natural desiccation	cDNA chip	Fatty acid biosynthesis, auxin, and jasmonate signaling	Niu et al. (2009)
B. napus	Seedling root	Drought (mannitol simulation)	Microarray	Metabolism, transcription, signal transduction, hormone and abiotic stress responses, growth and development	Chen et al. (2010)
Brassica juncea	Seedling	Drought (mannitol simulation)	RNA-seq	Stress/defense responses, metabolism, phosphorylation, signal transduction, transcription and translation, cell growth, cell structure, membrane transport, circadian rhythm, catalytic activity	Bhardwaj et al. (2015)
B. napus	Leaf	Drought (mannitol simulation)	Expressed sequence tag	Transcription factors, kinases, phosphatase, microRNAs	Shamloo-Dashtpagerdi et al. (2015)

B. napus	Seedlings	PEG-6000 (200 g/L)	RNA-seq	Osmoprotectant (LEA, proline)	Zhang et al. (2015a,b)
B. napus	Seedlings	PEG-6000 (200 g/L)	RNA-seq	Hormone signal transduction, ABC transporters, and oxidative phosphorylation pathways	Wang et al. (2017)
Brassica rapa	Whole plant	NaCl (250 mM)	Microarray	Transcription factor, stress responsive genes	Lee et al. (2008)
B. napus	Root, leaf	NaCl (200 mM)	RNA-seq	Transcription factor, transporter genes	Yong et al. (2014)
B. juncea	Seedling	300 mM mannitol	RNA-seq	Signaling proteins, LEA protein	Bhardwaj et al. (2015)
B. napus	Germinating seed	NaCl solution (1.25%)	RNA-seq	Proline metabolism, inositol metabolism, carbohydrate metabolic processes, and oxidation–reduction processes	Long et al. (2015)
B. juncea	Seedling	salinity stress (200 mM)	RNA-seq	ROS detoxification, sulfur assimilation and calcium signaling pathways	Sharma et al. (2015)
B. napus	Siliques at the seed-filling stage	High temperature	Microarray	Heat shock transcription factors and heat shock proteins	Yu et al. (2014)
B. juncea	Seedling	High temperature (42°C)	RNA-seq	Transcription factor, Heat shock proteins	Bhardwaj et al. (2015)
B. rapa	Seedling	High temperature (37°C)	RNA-seq	Transcription factors, kinases/phosphatases, genes related to photosynthesis	Wang et al. (2016)
Brassica alboglabra	Seedling	High temperature (42°C)	RNA-seq	GS biosynthesis genes, heat shock transcription factors, and heat shock proteins	Guo et al. (2019)

GS, glutamine synthetase; *ROS*, reactive oxygen species.

Most of the abiotic stress–related traits are quantitative and governed by a number of genetic loci. A QTL is defined as a genomic region responsible for the variation of a quantitative trait. QTLs mapping identifies the genomic region along with their positions in the genome. It compares gene action and pleiotropic and phenotypic effects, along with epistasis interactions with other QTLs (Xiao et al., 1996; Zhu et al., 1999). QTLs mapping in *B. oleracea* for water use efficiency and photosynthetic traits identified 30 significant QTLs (Hall et al., 2005). Uptmoor et al. (2009) used the combination of QTL analysis and ecophysiological modeling and identified 14 QTLs on the parameters describing preflowering leaf area development under drought. Fletcher et al. (2014) conducted a QTL analysis of flowering time and root mass traits in *B. napus* under drought conditions. Zhang et al. (2015a,b) reported 16 loci, 79 candidate genes, and 8 putative candidate genes significantly associated with water stress response in *B. napus*. Using genome-side association study technique, Yong et al. (2015) identified a gene controlling variation in salt tolerance in *B. napus*. In *B. napus*, heat susceptibility indices for all three traits were calculated using association mapping with single nucleotide polymorphisms (37,539 SNPs). A total of 5, 7, and 8 QTLs were detected, which showed associated with pollen sterility, number of pods on main raceme, and sterile/aborted pods, respectively. Together, they explained, respectively, 46.3%, 60.6%, and 60.5% of phenotypic variation (Rahaman et al., 2018). Furthermore, a list of QTL analysis, which is used to identify the abiotic stress–related traits, has been given in Table 10.3.

Genetic improvement of *B. juncea* (Indian mustard) for abiotic stress tolerance through induced mutagenesis: Mutation breeding is an important breakthrough in crop breeding, which utilizes the induced genetic variability for various morphological, biochemical, physiological, higher yield, nutritional value, disease resistance, and climate-resilient factors (Chopra, 2005; Bhatia et al., 1999; Maluszynski et al., 1995; Oladosu et al., 2016). This technology is being used to develop 3302 mutant varieties in various plant species (MVD IAEA). Among them, 245 varieties have been reported to have abiotic stress tolerance such as drought, salinity, and high temperature tolerance (Fig. 10.1A) in different crop species. Induced mutagenesis has been engaged for genetic improvement of traits (qualitative and quantitative) in oleiferous *Brassica* species (Robbelen, 1990; Bhatia et al., 1999). Mutation breeding technique has been successfully employed for development of varieties with high yield (Jambhulkar 2009) and climate-resilient mustard crop (Jambhulkar 2017). Isolation of desirable mutant for specific traits, their molecular characterization, and utilization for mustard crop improvement have been critically analyzed (Jambhulkar 2007, 2015). Till today, 41varieties in genus *Brassica* have been released for cultivation through mutation breeding; among them, four varieties are abiotic stress tolerant. Their speciation has been shown in Fig. 10.1B. This suggests that mutation breeding technology has tremendous potential to develop the climate-resilient *Brassica* crops (Table 10.4). The Bhabha Atomic Research Centre, Mumbai, has developed yellow seed coat early maturing variety TPM1 thought mutagenesis. It has additional character of seedling-stage thermotolerance (Rai et al., 2020) and tolerance against arsenate and arsenite (Srivastava et al., 2010). Mutation breeding derived line Trombay Mustard 179 (TM179) has also been identified as drought tolerant. *B. juncea* line CS 614-4-1-4 developed through irradiation showed 100% germination up to EC 22 dS/m (Sharma et al., 2008).

Table 10.3 Studies in *Brassica* spp. for identification of abiotic stress–related quantitative trait loci (QTL).

Species	Stress condition	QTL features	References
Brassica oleracea	Drought	WUE; photosynthetic traits	Hall et al. (2005)
B. oleracea	Drought	Leaf development	Uptmoor et al. (2009)
Brassica napus	Drought	Root morphology, flowering time	Fletcher et al. (2014)
B. napus	Drought	Seedling traits (root fresh weight, root length, and seedling vigor)	Khanzada et al. (2020)
B. napus	Drought	Drought tolerance index for seed yield	Raman et al. (2020)
B. napus	Salinity	Salt tolerance, shoot biomass, and ion homeostasis–related traits	Yong et al. (2015)
B. napus	Salinity	Chlorophyll; Salt-tolerant gene	Lang et al. (2017)
B. napus	Salinity	Tolerance traits at germination stage (FW: fresh weight, RL: root length, GR: growth rate)	Wan et al. (2018)
Brassica juncea	Salinity	Yield attributes and physiological traits	Patel (2017)
B. napus	Heat	Pollen sterility, sterile/aborted pods, and number of pods on main raceme	Rahaman et al. (2018)
B. oleracea	Heat	Floral development and/or heat response pathways	Branham et al. (2017)

WUE, water use efficiency.

Genome editing using CRISPR/Cas9 technology has shown potential to generate new alleles. This technology has been used in *B. napus* for editing the genes ALCATRAZ, GA1-3, FRUITFULL, DA1, DA2, CLAVATA, and SPL3 associated with pod/plant, BnFAD2 for fatty acid synthesis, and BnWRKY11and -70 related to biotic stress response (Lohani et al., 2020). Though its application in editing genes related to abiotic stress in rice has been demonstrated (Kumar et al., 2020; Alfatih et al., 2020), it is not yet reported in *Brassica* spp.

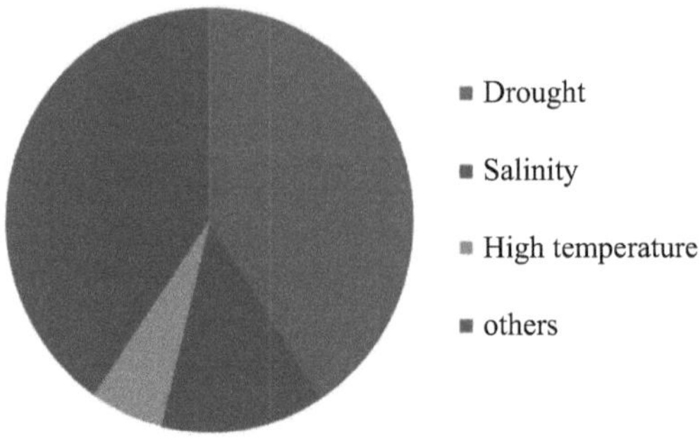

Figure 10.1A Abiotic stress–tolerant varieties developed through mutation breeding.

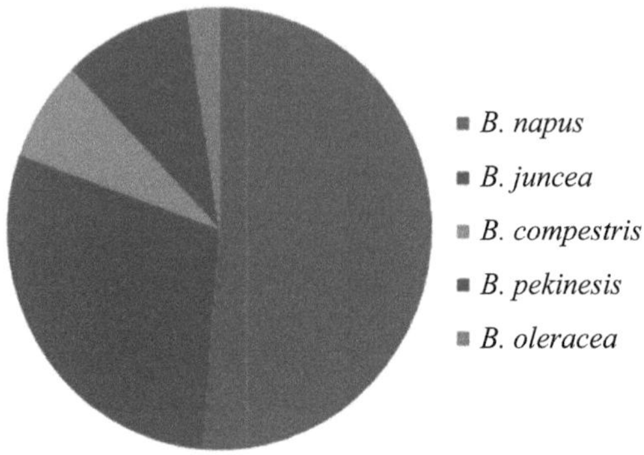

Figure 10.1B *Brassica* varieties developed through mutation breeding.

Mutant characterization: Targeting-induced local lesions in genomes (TILLING) is one of the powerful reverse genetics approach and enables high-throughput analysis of large number of mutants (Mccallum et al., 2000). TILLING has an application in crop improvement as the mutations identified by TILLING, which inherited stably, can be utilized in crop improvement (Kumar et al., 2013a,b). TILLING screening can also identify heterozygous recessive mutations in the individuals which are generally missed in phenotyping, and the target gene polymorphism can further be used in marker-assisted selection during varieties development (Wang et al., 2012). The detailed methodology and its application have been summarized in various reviews (Wang et al., 2012; Chen et al., 2014; Irshad et al., 2020). TILLING has been widely used to study gene function in various plant species including *B. napus* (Harloff et al., 2012; Wang et al., 2008) and *B. rapa* (Stephenson et al., 2010) and has been utilized to explore the new genetic diversity in *B. napus* mutant population (Gilchrist et al., 2013). Using TILLING, three new variants of CAX1 Ca^{2+} transporter were identified in *B. rapa*, and these mutants were further evaluated for the tolerance to saline conditions (Graham et al., 2014; Navarro-León et al., 2020). Besides TILLING, other nest-generation sequencing–based mutation mapping approaches such as MutMap, MutMap-Gap, mutant chromosome sequencing (Mut Chrom Seq), exon capture, and Mut-Ren-Seq can be explored to identify the mutant loci (Kumawat et al., 2019).

Table 10.4 Released varieties of rapeseed–mustard using mutation breeding for abiotic stress tolerance.

S. no.	Variety	Country and year of release	Mutagen and mutant in cross	Improved characters
1.	Binasarisha-5	Bangladesh, 2002	γ-Rays	Salt-tolerant variety (tolerance level up to 13 dS/m)
2.	Binasarisha-6	Bangladesh, 2002	γ-Rays	Salt-tolerant variety (tolerance level up to 13 dS/m)
3.	Huyou 4	China, 1970	γ-Rays, 600 Gy	Tolerance to low temperature
4.	Ganyou 5	China, 1984	γ-Rays, 1400 Gy	Tolerance to low temperature
5.	Xiuyou 1	China, 1979	γ-Rays, 80 Gy, hybrid seeds of F1 generation	Cold resistance
6.	RLM 185	India, 1983	γ-Rays	Tolerant to drought

From Kumar, P.R., Chauhan, J.S., Singh, A. K., Yadav, S.K., 2000. Rapeseed-Mustard Varieties in India, NRCRM, Sewar, Bharatpur, India, pp. 151–167, https://mvd.iaea.org/.

References

Ahmad, P., Kumar, A., Ashraf, M., Akram, N.A., 2012. Salt-induced changes in photosynthetic activity and oxidative defense system of three cultivars of mustard (*Brassica juncea* L.). Afr. J. Biotechnol. 11 (11), 2694–2703.

Ahmar, S., Liaqat, N., Hussain, M., Salim, M.A., Shabbir, M.A., Ali, M.Y., et al., 2019. Effect of abiotic stresses on *Brassica* species and role of transgenic breeding for adaptation. Asian J. Crop Sci. 3 (1), 1–10.

Alam, M.M., Nahar, K., Hasanuzzaman, M., Fujita, M., 2014. Trehalose-induced drought stress tolerance: a comparative study among different *Brassica* species. Plant Omics 7, 271–283.

Alfatih, A., Wu, J., Jan, S.U., Zhang, Z.-S., Xia, J.-Q., Xiang, C.-B., 2020. Loss of rice PARAQUAT TOLERANCE 3 confers enhanced resistance to abiotic stresses and increases grain yield in field. bioRxiv 2020, 02.22.961151.

Ali, G., Bazrafshan, F., Zare, M., Faraji, H., Reza, A., Langeroodi, S., 2017. Studying the effect of drought stress on yield and physiological characteristics in genotypes of canola (*Brassica napus* L.) Helix. Int. J. 8, 1250–1258.

Arora, N.K., 2019. Impact of climate change on agriculture production and its sustainable solutions. Environ. Sustain. 2, 95–96.

Ashraf, M., Ali, Q., 2008. Relative membrane permeability and activities of some antioxidant enzymes as the key determinants of salt tolerance in canola (*Brassica napus* L.). Environ. Exp. Bot. 63, 266–273.

Ashraf, M., McNeilly, T., 2004. Salinity tolerance in *Brassica* oilseeds. Crit. Rev. Plant Sci. 23 (2), 157–174.

Ashraf, M., Mehmood, S., 1990. Response of four *Brassica* species to drought stress. Environ. Exp. Bot. 30, 93–100.

Aslam, M.N., Kailism, G.S., Bayliss, L.K., Speijers, J., Cowling, A.W., 2009. Canola oil increases in polyunsaturated fatty acids and decreases in oleic acid in drought-stressed Mediterranean-type environments. Plant Breed. 128, 348–355.

Bhardwaj, A.R., Joshi, G., Kukreja, B., et al., 2015. Global insights into high temperature and drought stress regulated genes by RNA-Seq in economically important oilseed crop *Brassica juncea*. BMC Plant Biol. 15, 9.

Bhatia, C.R., Nichterlein, K., Maluszynski, M., 1999. Oilseed cultivars development from induced mutations and mutation altering fatty acid composition. Mutat. Breed. Rev. 11, 1–36.

Boomiraj, K., Chakrabarti, B., Aggarwal, P.K., Choudhary, R., Chander, S., 2010. Assessing the vulnerability of Indian mustard to climate change. Agric. Ecosyst. Environ. 138, 265–273.

Boyer, J.S., 1982. Plant productivity and environment. Science 218, 443–448.

Branham, S.E., Stansell, Z.J., Couillard, D.M., Farnham, M.W., 2017. Quantitative trait loci mapping of heat tolerance in broccoli (*Brassica oleracea* var. italica) using genotyping-by-sequencing. Theor. Appl. Genet. 130 (3), 529–538.

Bybordi, A., 2010. Effects of salinity on yield and component charactersin canola (*Brassica napus* L.) cultivars. Not. Sci. Biol. 2 (1), 81–83.

Chauhan, J., Meena, M., Saini, M., Singh, M., Meena, S., Singh, K., 2009. Heat stress effects on morpho-physiological characters of Indian mustard (*Brassica juncea* L.). In: Conference Paper 16thAustralian Research Assembly on *Brassicas*, Ballarat Victoria.

Chauhan, J.S., Tyagi, M.K., Kumar, A., Nashaat, N.I., Singh, M., Singh, N.B., Jakhar, et al., 2007. Drought effects on yield and its components in Indian mustard (*Brassica juncea* L.). Plant Breed. 126, 399–402.

Cheema, K.L., Sadaqat, H.A., 2004. Potential and genetic basis of drought tolerance in canola (*Brassica napus*): I. Generation mean analysis for some phenological and yield components. Int. J. Agric. Biol. 6, 1560–8530.

Chen, L., Chen, L., Ren, F., Zhong, H., Feng, Y., Jiang, W., et al., 2010. Identification and expression analysis of genes in response to high-salinity and drought stresses in *Brassica napus*. Acta Biochim. Biophys. Sin. 42 (2), 154–164.

Chen, L., Hao, L., Parry, M.A.J., Phillips, A.L., Hu, Y.G., 2014. Progress in TILLING as a tool for functional genomics and improvement of crops. J. Integr. Plant Biol. 56 (5), 425–443.

Chopra, V.L., 2005. Mutagenesis: investigating the process and processing the outcome for crop improvement. Curr. Sci. 89, 353–359.

Clarke, J.M., De Pauw, R.M., Townley-Smith, T.M., 1992. Evaluation of methods for quantification of drought tolerance in wheat. Crop Sci. 32, 728–732.

Collard, B.C.Y., Mackill, D.J., 2008. Marker-assisted selection: an approach for precision plant breeding in the twenty-first century. Philos. Trans. R. Soc. London, Ser. B 363, 557–572.

Dai, A., 2011. Drought under global warming: a review Wiley Interdiscip. Rev. Clim. Change 2, 45–65.

Dong, C.-H., Li, C., Yan, X.-H., Shunmou, H., Huang, J.-Y., Wang, L.-J., et al., 2011. Gene expression profiling of Sinapis alba leaves under drought stress and rewatering growth conditions with Illumina deep sequencing. Mol. Biol. Rep. 39, 5851–5857.

Elferjani, R., Soolanayakanahally, R., 2018. Canola responses to drought, heat, and combined stress: shared and specific effects on carbon assimilation, seed yield, and oil composition. Front. Plant Sci. 9 https://doi.org/10.3389/fpls.2018.01224.

Enjalbert, J.N., Zheng, S., Johnson, J.J., Mullen, J.L., Byrne, P.F., McKay, J.K., 2013. *Brassicaceae germplasm* diversity for agronomic and seed quality traits under drought stress. Ind. Crop. Prod. 47, 176–185.

FAO-ITPS-GSP, 2015. Status of the World's Soil Resources. FAO-ITPS-GSP Main Report. Food and Agriculture Organization of the United Nations, Rome, Italy, pp. 125–127.

Fei, H., Tsang, E., Cutler, A., 2007. Gene expression during seed maturation in *Brassica napus* in relation to the induction of secondary dormancy. Genomics 89, 419–428.

Fernandez, G.C.J., 1992. Effective selection criteria for assessing plant stress tolerance. In: Kuo, C.G. (Ed.), Adaptation of Food Crops to Temperature and Water Stress. Asian Vegetable Research and Development Center, Shanhua, Taiwan, pp. 257–270.

Fischer, G., Tubiello, F.N., Velthuizen, H.V., Wiberg, D.A., 2007. Climate change impacts on irrigation water requirements: effects of mitigation, 1990–2080. Technol. Forecast. Soc. Change 74, 1083–1107.

Fischer, R.A., Maurer, R., 1978. Drought resistance in spring wheat cultivars. I. Grain yield responses. Aust. J. Agric. Res. 29, 897–912.

Fletcher, R., Mullen, J., Heiliger, A., Mckay, J., 2014. QTL analysis of root morphology, flowering time, and yield reveals trade-offs in response to drought in *Brassica napus*. J. Exp. Bot. 66 https://doi.org/10.1093/jxb/eru423.

Goel, P., Singh, A.K., 2015. Abiotic stresses down regulate key genes involved in nitrogen uptake and assimilation in *Brassica juncea* L. PLoS One 10 (11), e0143645.

Ghassemi, F., Jakeman, A.J., Nix, H.A., 1995. Salinisation of Land and Water Resources: Human Causes, Extent, Management and Case Studies. Sydney, Australia, and CAB International. UNSW Press, Wallingford, UK.

Ghuge, S.A., Rai, A.N., Khandagale, B.G., Penna, S., 2011. Salt-induced stress responses of *Brassica* (*Brassica juncea* L.) genotypes. Arch. Agron Soil Sci. 57 (2), 127–136.

Gilchrist, E.J., Sidebottom, C.H.D., Koh, C.S., MacInnes, T., Sharpe, A.G., Haughn, G.W., 2013. A mutant *Brassica napus* (canola) population for the identification of new genetic diversity via TILLING and next generation sequencing. PLoS One 8 (12), e84303.

Graham, N.S., Hammond, J.P., Lysenko, A., et al., 2014. Genetical and comparative genomics of Brassica under altered Ca supply identifies Arabidopsis Ca-transporter orthologs. Plant Cell 26, 2818–2830.

Guo, R., Wang, X., Han, X., et al., 2019. Comparative transcriptome analyses revealed different heat stress responses in high- and low-GS *Brassica alboglabra* sprouts. BMC Genom. 20, 269.

Hall, N.M., Griffiths, H., Corlett, J.A., Jones, H.G., Lynn, J., King, G.J., 2005. Relationships between water use traits and photosynthesis in *Brassica oleracea* resolved by quantitative genetic analysis. Plant Breed. 124, 557–564.

Harloff, H., Lemcke, S., Mittasch, J., Frolov, A., Wu, J., Dreyer, F., et al., 2012. A mutation screeing platform for rapeseed (*Brassica napus* L.) and the detection of sinapine biosynthesis mutants. Theor. Appl. Genet. 124, 957–969.

Hasanuzzaman, M., Nahar, K., Hossain, M.S., Mahmud, J.A., Rahman, A., Inafuku, M., Fujita, M., 2017. Coordinated actions of glyoxalase and antioxidant defense systems in conferring abiotic stress tolerance in plants. Int. J. Mol. Sci. 18 (1), 200–228.

Hayat, S., Masood, A., Yusuf, M., Fariduddin, Q., Ahmad, A., 2009. Growth of Indian mustard (*Brassica juncea* L.) in response to salicylic acid under high-temperature stress. Braz. J. Plant Physiol. 21 (3), 187–195.

Hayat, S., Mir, B.A., Wani, A.S., Hasan, S.A., Irfan, M., Ahmad, A., 2011. Screening of salt-tolerant genotypes of *Brassica juncea* based on photosynthetic attributes. J. Plant Interact. 6 (1), 53–60.

Hosseini, M., Hassibi, P., 2011. Effects of water deficit stress on several quantitative and qualitative characteristics of canola (*Brassica napus* L.) cultivars. Not. Sci. Biol. 3 (3), 120.

IAEA. Mutant Variety Database. Available online: http://mvd.iaea.org/. (Accessed on 10 August 2020).

IPCC, 2007. In: Parry, M.L., Canziani, O.F., Paultikof, J.P., van der Linden, P.J., Hanon, C.E. (Eds.), Climate Change – Impacts, Adaptation and Vulnerability Technical Summary of Working Group II to Fourth Assessment Report of Intergovernmental Panel on Climate Change. Cambridge University Press, Cambridge, UK, pp. 23–78.

Irshad, A., Guo, H., Zhang, S., Liu, L., 2020. TILLING in cereal crops for allele expansion and mutation detection by using modern sequencing technologies. Agron 10 (3), 405.

Jambhulkar, S.J., 2007. Mutagenesis: generation and evaluation of induced mutations. In: Gupta, P.C. (Ed.), Advances in Botanical Research (Rapeseed Breeding), vol. 45. Academic Press: Elsevier Publication Ltd., USA, pp. 418–428.

Jambhulkar, S.J., 2009. Mutation induction, evaluation and utilization for development of high yielding varieties in Indian mustard and sunflower: an overview of BARC work. In: D'Souza, S.F., Sharma, A. (Eds.), Thematic Volume: Food and Agriculture of International Conference on Peaceful Uses of Atomic Energy Sept.29- Oct.1, 2009, New Delhi, pp. 49–54.

Jambhulkar, S.J., 2015. Induced mutagenesis and allele mining. In: Kumar, A., Banga, S.S., Meena, P.D., Kumar, P.R. (Eds.), *Brassica* Oilseeds Breeding and Management. CAB International, Wallingford UK, pp. 53–67. https://doi.org/10.1079/9781780644837.0053.

Jambhulkar, S.J., 2017. Induced mutagenesis for breaking the yield barriers in *Brassica juncea* resilient to climate change. In: Souvenir of 3rd National *Brassica* Conference-2017, pp. 51–54.

Kavita, K., Pandey, A., 2017. Physiological attributes for screening of Indian mustard (*Brassica juncea* L. Czern and Coss) genotypes during terminal heat stress. Int. J. Curr. Microbiol. App. Sci. 6 (9), 2908–2913.

Khanzada, H., Wassan, G.M., He, H., et al., 2020. Differentially evolved drought stress indices determine the genetic variation of *Brassica napus* at seedling traits by genome-wide association mapping. J. Adv. Res. 24, 447–461.

Khayat, P.N., Jamaati-e-Somarin, S., Zabihi-e- Mahmoodabad, R., Yari, A., Khayatnezhad, M., Gholamin, R., 2010. Screening of salt tolerance canola cultivars (*Brassica napus* L.) using physiological markers. World Appl. Sci. J. 10, 817–820.

Kumar, A., 2005. Rapeseed-mustard in India: current status and future prospects. In: Winter School on Advances in Rapeseed-Mustard Research Technology for Sustainable Production of Oilseeds, National Centre on Rapeseed-Mustard, Sewar, Bharatpur, December 15 to January 04, 2005, pp. 278–288.

Kumar, A.P., Boualem, A., Bhattacharya, A., Parikh, S., Desai, N., Zambelli, A., Leon, A., Chatterjee, M., Bendahmane, A., 2013a. SMART—sunflower mutant population and reverse genetic tool for crop improvement. BMC Plant Biol. 13, 38.

Kumar, D., 1993. Variability studies in Indian mustard on normal and saline soils. Ann. Arid Zone 32, 25–28.

Kumar, D., 1995. Salt tolerance in oilseed *Brassicas*-present status and future prospects. Plant Breed. 65 (10), 1439–1477.

Kumar, G., Purty, R.S., Sharma, M.P., Singla-Pareek, S.L., Pareek, A., 2009. Physiological responses among *Brassica* species under salinity stress show strong correlation with transcript abundance for SOS pathway-related genes. J. Plant Physiol. 166, 507–520.

Kumar, P.R., Chauhan, J.S., Singh, A.K., Yadav, S.K., 2000. Rapeseed-Mustard Varieties in India. NRCRM, Sewar, Bharatpur, India, pp. 151–167.

Kumar, S., Sairam, R.K., Prabhu, K.V., 2013b. Physiological traits for high temperature stress tolerance in *Brassica juncea*. Indian J. Plant Physiol. 18, 89–93.

Kumar, S., Verma, V.V., Yadav, R.K., et al., 2020. CRISPR-Cas9 mediated genome editing of drought and salt tolerance (OsDST) gene in indica mega rice cultivar MTU1010. Physiol. Mol. Biol. Plants 26, 1099–1110.

Kumawat, S., Rana, N., Bansal, R., Vishwakarma, G., Mehetre, S.T., Das, B.K., et al., 2019. Expanding avenue of fast neutron mediated mutagenesis for crop improvement. Plants 8, 164.

Kusvuran, S., Kiran, S., Ellialtioglu, S.S., 2016. Antioxidant enzyme activities and abiotic stress tolerance relationship in vegetable crops. In: Abiotic and Biotic Stress in Plants–Recent Advances and Future Perspectives. IntechOpen, London, UK. https://doi.org/10.5772/62235.

Lang, L., Xu, A., Ding, J., Zhang, Y., Zhao, N., Tian, Z., et al., 2017. Quantitative trait locus mapping of salt tolerance and identification of salt-tolerant genes in *Brassica napus* L. Front. Plant Sci. 8 https://doi.org/10.3389/fpls.2017.01000.

Lee, S.C., Lim, M.H., Kim, J.A., et al., 2008. Transcriptome analysis in *Brassica rapa* under the abiotic stresses using *Brassica* 24K oligo microarray. Mol. Cell 26 (6), 595–605.

Li, F., Wu, X., Tsang, E., Cutler, A., 2005. Transcriptional profiling of imbibed *Brassica napus* seed. Genomics 86, 718–730.

Li, Z., Wu, B.J., Lu, G.Y., Chen, Y., Zou, C.S., Zhang, X.K., 2012. Differences in physiological responses of *Brassica napus* genotypes under water stress during seedling stage. Chin. J. Oil Crop Sci. 34, 033–039.

Lohani, N., Jain, D., Singh, M.B., Bhalla, P.L., 2020. Engineering multiple abiotic stress tolerance in canola, *Brassica napus*. Front. Plant Sci. 11 (3) https://doi.org/10.3389/fpls.2020.00003.

Long, W., Zou, X., Zhang, X., 2015. Transcriptome analysis of canola (*Brassica napus*)under salt stress at the germination stage. PLoS One 10 (2), e0116217.

Long, W.H., Pu, H.M., Zhang, J.F., Qi, C.K., Zhang, X.K., 2013. Screening of *Brassica napus* for salinity tolerance at germination stage. Chin. J. Oil Crop Sci. 35, 271–275.

Maluszynski, M., Ahloowalia, B.S., Sigurbjornsson, B., 1995. Application of in vivo and in vitro mutation techniques for crop improvement. Euphytica 85, 303–315.

Mccallum, C.M., Comai, L., Greene, E.A., Henikoff, S., 2000. Targeting induced local lesions in genomes (TILLING) for plant functional genomics. Plant. Physiol. 123, 439.

Moaveni, P., Ebrahimi, A., Farahani, H.A., 2010. Physiological growth indices in winter rapeseed (*Brassica napus* L.) cultivars as affected by drought stress at Iran. J. Cereals Oilseeds 1, 11–16.

Morrison, M.J., Stewart, D.W., 2002. Heat stress during flowering in summer *Brassica*. Crop Sci. 42 (3), 797–803.

Munir, S., Hussain, E., Bhatti, K., Nawaz, K., Hussain, K., Rasheed, R., et al., 2013. Assessment of inter-cultivar variations for salinity tolerance in winter radish (*Raphanus sativus* L.) using photosynthetic attributes as effective selection criteria. World Appl. Sci. J. 21, 384–388.

Naderikharaji, R., Pakniyat, H., Biabani, A.R., 2008. Effect of drought stress on photosynthetic rate of four rapeseed (*Brassica napus*) cultivars. J. Appl. Sci. 8 (23), 4460–4463.

Namazkar, S., Stockmarr, A., Frenck, G., Egsgaard, H., Terkelsen, T., Mikkelsen, T., et al., 2016. Concurrent elevation of CO_2, O_3 and temperature severely affects oil quality and quantity in rapeseed. J. Exp. Bot. 67 (14), 4117–4125.

Nasri, M., Khalatbari, M., Zahedi, H., Paknejad, F., Moghadam, H.R., 2008. Evaluation of micro and macro elements in drought stress condition in cultivars of rapeseed (*Brassica napus* L.). Am. J. Agric. Biol. Sci. 3 (3), 579–583.

Navarro-León, E., López-Moreno, F.J., Torre-González, A.d.l., Ruiz, J.M., Esposito, S., Begoña, B., 2020. Study of salt-stress tolerance and defensive mechanisms in *Brassica rapa* CAX1a TILLING mutants. Environ. Exp. Bot. 175, 104061.

Niu, Y., Wu, G.-Z., Ye, R., Lin, W.-H., Shi, Q.-M., Xue, L.-J., et al., 2009. Global analysis of gene expression profiles in *Brassica napus* developing seeds reveals a conserved lipid metabolism regulation with *Arabidopsis thaliana*. Mol. Plant 2, 1107−1122.

Oladosu, Y., Rafii, M.Y., Abdullah, N., Hussin, G., Ramli, A., Rahim, H.A., et al., 2016. Principle and application of plant mutagenesis in crop improvement: a review. Biotechnol. Biotechnol. Equip. 30, 1−16.

Pandey, M., Penna, S., 2017. Time course of physiological, biochemical, and gene expression changes under short-term salt stress in *Brassica juncea* L. Crop J. 5 (3), 219−230.

Patel, R., 2017. Molecular Marker Analysis for QTL Identification in *Indian Mustard* (*Brassica Juncea* L. Czern & Coss.) for Salinity Tolerance. Ph.D thesis.

Qadir, M., Quillérou, E., Nangia, V., Murtaza, G., Singh, M., Thomas, R.J., et al., 2014. Economics of salt-induced land degradation and restoration. In: Natural Resources Forum. Wiley Online Library, United Nations, pp. 282−295.

Qiu, Y., Li, X.X., 2009. Genetic analysis of salinity tolerance in *Brassica campestris* L. China Vegetab 1, 21−25.

Raftery, A.E., Zimmer, A., Frierson, D.M.W., Startz, R., Liu, P., 2017. Less than 2 °C warming by 2100 unlikely. Nat. Clim. Change 7, 637-641.

Rahaman, M., Mamidib, S., Rahmana, M., 2018. Genome-wide association study of heat stress-tolerance traits in spring-type *Brassica napus* L. under controlled conditions. Crop J. 6 (2), 115−125.

Rai, A.N., Saini, N., Yadav, R., Suprasanna, P., 2020. A potential seedling-stage evaluation method for heat tolerance in Indian mustard (*Brassica juncea* L. Czern and Coss). 3 Biotech 10 (3), 114.

Ram, B., 2014. High temperature stress tolerance in Indian mustard (*Brassica juncea*) germplasm as evaluated by membrane stability index and excised-leaf water loss techniques. J. Oilseed *Brassica* 5 (2), 149−157.

Ram, B., Priyamedha, Meena, H.S., Kumar, A., Singh, B.K., Rani, R., et al., 2018. Studies on combining ability and gene action for heat stress tolerance traits in Indian mustard (*Brassica juncea* L.). J. Oilseed *Brassica* 9 (2), 139−145.

Raman, H., Raman, R., Mathews, K., Diffey, S., Salisbury, P., 2020. QTL mapping reveals genomic regions for yield based on incremental tolerance index to drought stress and related agronomic traits in canola. Crop Pasture Sci. 71 (6), 562−577.

Raymer, P.L., 2002. Canola: an emerging oilseed crop. Trends New Crops New Uses 1, 122−126.

Rezai, A.M., Saeidi, G., 2005. Genetic analysis of salt tolerance in early growth stages of rapeseed (*Brassica napus* L.) genotypes. Indian J. Genet. Plant Breed. 65, 269−273.

Richards, R.A., Thurling, N., 1979. Genetic analysis of drought stress response in rapeseed (*Brassica campestris* and *B. napus*). III. physiological characters. Euphytica 28, 755−759.

Rieu, I., Twell, D., Firon, N., 2017. Pollen development at high temperature: from acclimation to collapse. Plant Physiol. 173, 1967−1976.

Robbelen, G., 1990. Mutation breeding for quality improvement a case study for oilseed crops. Mutat. Breed. Rev. 6, 1−43.

Rosielle, A.A., Hamblin, J., 1981. Theoretical aspects of selection for yield in stress and non-stress environments. Crop Sci. 21, 943−946.

Sabagh, A.E.L., Hossain, A., Celaleddin, B., Ozgul, G., Zahoor, A., Saddam, H., et al., 2019. Effects of drought stress on the quality of major oilseed crops: implications and possible mitigation strategies - a review. Appl. Ecol. Environ. Res. 17, 4019—4043.

Shamloo-Dashtpagerdi, R., Razi, H., Ebrahimie, E., 2015. Mining expressed sequence tags of rapeseed (*Brassica napus* L.) to predict the drought responsive regulatory network. Physiol. Mol. Biol. Plants 21 (3), 329—340.

Sharma, P., Sardana, V., 2013. Screening of Indian mustard (B. juncea) for thermo tolerance at seedling and terminal stage. J. Oilseed *Brassica* 4, 61—67.

Sharma, P.C., Prashat, R., Singh, G.D., Pareek, A., 2008. Improving salt tolerance and seed yield in Indian mustard (*Brassica Juncea* L.) through radiation induced mutagenesis. Report no. IAEA-CN—167. INIS 40 (1), 54.

Sharma, R., Chhabra, M.L., Dhawan, A., Dhawan, K., Singh, D., 2007. Thermo tolerance in *Brassica*: refined rapid screening method to identify thermo tolerant genotypes in *Brassica*. Proc. Genet. Breed. Breed. Stress Resist. 430—432.

Sharma, R., Mishra, M., Gupta, B., Parsania, C., Singla-Pareek, S.L., Pareek, A., 2015. De novo assembly and characterization of stress transcriptome in a salinity-tolerant variety CS52 of *Brassica juncea*. PLoS One 10 (5), e0126783.

Shekari, F., Soltaniband, V., Javanmard, A., Abbasi, A., 2015. The impact of drought stress at different stages of development on water relations, stomatal density and quality changes of rapeseed (*Brassica napus* L.). Iran Agric. Res. 34 (2), 81—90.

Sinaki, J.M., Heravan, E.M., Shirani Rad, H.A., Nour Mohammadi, G., Zarei, H., 2007. The effects of water deficit during growth stages of canola (*B. napus* L.). Am.-Eurasian J. Agric. Environ. Sci. 2, 417—422.

Singh, B., Thakral, N.K., Munjal, R., Boken, G., 2017. Combining ability analysis: physiological traits for high temperature stress tolerance in Indian mustard [*Brassica juncea* (L.) Czern & Coss.]. Int. J. Pure App. Biosci. 5 (5), 725—735.

Singh, M., Rathore, S., Raja, P., 2014. Physiological and stress studies of different rapeseed-mustard genotypes under terminal heat stress. Int. J. Genet. Eng. Biotechnol. 5, 133—142.

Singh, R., 2016. Yield increase in major oil seeds in India through replacement by new climate resilient varieties. Clim. Change Environ. Sustain. 4, 211.

Sinha, T.S., Yadav, R.S., 1991. Genetic variability, heritability of quantitative traits and path analysis of Indian mustard in normal, saline and alkali soil. In: Golden Jubilee Symposium on Genetic Resources and Education: Current Trends and Next Fifty Years, New Delhi, 12—15 February, 1991.

Soengas, P., Rodríguez, V.M., Velasco, P., Cartea, M.E., 2018. Effect of temperature stress on antioxidant defenses in *Brassica oleracea*. ACS Omega 3 (5), 5237—5243.

Srivastava, S., Srivastava, A.K., Suprasanna, P., et al., 2010. Comparative antioxidant profiling of tolerant and sensitive varieties of *Brassica juncea* L. To arsenate and arsenite exposure. Bull. Environ. Contam. Toxicol. 84, 342—346.

Stephenson, P., Baker, D., Girin, T., Perez, A., Amoah, S., King, J., et al., 2010. A rich TILLING resource for studying gene function in *Brassica rapa*. BMC Plant Biol. 10, 62.

Suprasanna, P., Nikalje, G.C., Rai, A.N., 2016. Osmolyte accumulation and implications in plant abiotic stress tolerance pages 1-12. In: Osmolytes and Plants Acclimation to Changing Environment: Emerging Omics Technologies. Sringer, ISBN 978-81-322-2615-4.

Tahmasebpour, B., Nojadeh, M.S., Esmaeilpour, M., 2018. Salt stress tolerance of spring canola (*Brassica napus* L.)Cultivars. Int. J. Plant Biol. Res. 6 (4), 1098.

Thakral, N.K., Prakash, K., 1998. Combining ability for vigour characters under normal and saline environments in Indian mustard (*Brassica juncea* (L.) Czern.& Coss.). J. Oilseeds Res. 15, 234—237.

Trenberth, K., Dai, A., van der Schrier, G., Jones, P.D., Barichivich, J., Briffa, K.R., et al., 2014. Global warming and changes in drought. Nat. Clim. Change 4, 17–22.

Ulfat, M., Athar, H.R., Ashraf, M., Akram, N.A., Jamil, A., 2007. Appraisal of physiological and biochemical selection criteria for evaluation of salt tolerance in canola (*Brassica napus* L.). Pakistan J. Bot. 39, 1593–1608.

Uptmoor, R., Osei-Kwarteng, M., Gürtler, S., Stützel, H., 2009. Modeling the effects of drought stress on leaf development in a *Brassica oleracea* doubled haploid population using two-phase linear functions. J. Am. Soc. Hort. Sci. 134 (5), 543–552.

USDA FAS, 2018. Oilseeds: World Markets and Trade. Foreign Agricultural Service/USDA. https://apps.fas.usda.gov/psdonline/circulars/oilseeds.pdf.

Vernon, L., Van Gool, D., 2006. Potential Impacts of Climate Change on Agricultural Land Use Suitability: Canola. Department of Agriculture and Food, Western Australia. Report 303.

Wan, H., Yinke, W., Qian, J., Gao, Y., Wen, J., Yi, B., et al., 2018. Association mapping of salt tolerance traits at germination stage of rapeseed (*Brassica napus* L.). Euphytica Dordrecht 214 (10), 1–16.

Wang, A., Hu, Huang, J., Li, X., Zhou, G., Yan, Z., 2016. Comparative transcriptome analysis reveals heat-responsive genes in Chinese cabbage (*Brassica rapa* ssp. chinensis). Front. Plant Sci. 7 https://doi.org/10.3389/fpls.2016.00939.

Wang, N., Wang, Y., Tian, F., King, G., Zhang, C., Long, Y., et al., 2008. A functional genomics resource for *Brassica napus*: development of an EMS mutagenized population and discovery of FAE1 point mutations by TILLING. New Phytol. 180 (4), 751–765.

Wang, P., Yang, C., Chen, H., Song, C., Zhang, X., Wang, D., 2017. Transcriptomic basis for drought-resistance in *Brassica napus* L. Sci. Rep. 7, 40532.

Wang, T.L., Uauy, C., Robson, F., Till, B., 2012. TILLING in extremis. Plant Biotechnol. J. 10, 761–772.

Xiao, J., Li, J., Yuan, L., Tanksley, S.D., 1996. Identification of QTLs affecting traits of agronomic importance in a recombinant inbred population derived from a sub specific rice cross. Theor. Appl. Genet. 92, 230–244.

Xu, Y., 2016. Envirotyping for deciphering environmental impacts on crop plants. Theor. Appl. Genet. 129, 653–673.

Yang, C.J., Zhang, X.K., Zou, C.S., Cheng, Y., Zheng, P.Y., Li, G.Y., 2007. Effects of drought simulated by PEG-6000 on germination and seedling growth of rapeseed (*Brassica napus* L.). Chin. J. Oil Crop Sci. 29, 425–430.

Yang, Y., Zheng, Q., Liu, M., Long, X., Liu, Z., Shen, Q., et al., 2012. Difference in sodium spatial distribution in the shoot of two canola cultivars under saline stress. Plant Cell Physiol. 53 (6), 1083–1092.

Yong, H.Y., Wang, C.L., Bancroft, I., Li, F., Wu, X.M., Kitashiba, H., et al., 2015. Identification of a gene controlling variation in the salt tolerance of rapeseed (*Brassica napus* L.). Planta 242, 313–326.

Yong, H.-Y., Zou, Z., Kok, E.-P., Kwan, B.-H., Chow, K., Nasu, S., et al., 2014. Comparative transcriptome analysis of leaves and roots in response to sudden increase in salinity in *Brassica napus* by RNA-seq. BioMed Res. Int. 467395.

Yu, E., Fan, C., Yang, Q., Li, X., Wan, B., Dong, Y., et al., 2014. Identification of heat responsive genes in *Brassica napus* siliques at the seed-filling stage through transcriptional profiling. PLoS One 9 (7), e101914.

Yu, S., Zhang, F., Yu, Y., et al., 2012. Transcriptome profiling of dehydration stress in the Chinese cabbage (*Brassica rapa* L. ssp. pekinensis) by tag sequencing. Plant Mol. Biol. Rep. 30, 17–28.

Zamani, Z., Nezami, M.T., Habibi, D., Khorshidi, M.B., 2010. Effect of quantitative and qualitative performance of four canola cultivars (*Brassica napus* L.) to salinity conditions. Adv. Environ. Biol. 4 (3), 422–427.

Zhang, J., Jiang, F., Yang, P., Li, J., Yan, G., Hu, L., 2015a. Responses of canola (*Brassica napus* L.) cultivars under contrasting temperature regimes during early seedling growth stage as revealed by multiple physiological criteria. Acta Physiol. Plant. 37 (2), 1–10.

Zhang, J., Mason, A.S., Wu, J., Liu, S., Zhang, X., Luo, T., et al., 2015b. Identification of putative candidate genes for water stress tolerance in canola (*Brassica napus*). Front. Plant Sci. 6, 1058.

Zhao, C., Liu, B., Piao, S., et al., 2017. Temperature increase reduces global yields. Proc. Natl. Acad. Sci. U.S.A. 114 (35), 9326–9331.

Zhu, J.-K., 2016. Abiotic stress signaling and responses in plants. Cell 167, 313–324.

Zhu, H., Bricero, G., Dovel, R., Hayes, P.M., Liu, B.H., Ullrich, S.E., 1999. Molecular breeding for grain yield in barley: an evaluation of QTL effects in a spring barley cross. Theor. Appl. Genet. 98, 772–779.

Zhu, Z.H., Zheng, W.Y., Zhang, X.K., 2011. Principal component analysis and comprehensive evaluation on morphological and agronomic traits of drought tolerance in rapeseed (*Brassica napus* L.). Sci. Agric. Sin. 44, 1775–1787.

Impact of tropospheric ozone pollution on wheat production in Southeast Asia: an update

11

Manisha Sarkar[1], *Divya Pandey*[2], *Randeep Rakwal*[3,4,5], *Ganesh Kumar Agrawal*[4,5], *Abhijit Sarkar*[1]
[1]Laboratory of Applied Stress Biology, Department of Botany, University of Gour Banga, Malda, West Bengal, India; [2]Leibniz Centre for Agricultural Landscape Research (ZALF), Müncheberg, Germany; [3]Faculty of Health and Sport Sciences, University of Tsukuba, Tsukuba, Ibaraki, Japan; [4]Research Laboratory for Biotechnology and Biochemistry (RLABB), Kathmandu, Nepal; [5]Global Research Arch for Developing Education (GRADE) Academy Pvt. Ltd., Birgunj, Parsa, Nepal

1. Introduction

Ground-level ozone (O_3) is the atmospheric pollutant and greenhouse gas (third) most likely to intimidate universal food production due to its extreme pernicious effect toward arable crops and pervasiveness over important agricultural regions along with ecosystem, productivity, and human health (Review of evidence on health aspects of air pollution, 2013; United States Environmental Protection Agency, 2013; Monks et al., 2015; CLRTAP, 2017). O_3 is the main stressor in regions with warm climates where high temperatures and large emission of precursors (nitrogen oxides and hydrocarbons) from anthropogenic sources lead to the formation of this secondary pollutant. However, O_3 having the background levels of 25−55 ppb is known to occur crucially in the northern hemisphere, which is eventually thought to shift from North America and Europe to Asia (Granier et al., 2011; Wild et al., 2012; Cooper et al., 2014; Zhang et al., 2016; Mills et al., 2018b). The study is based on Eastern and Southern Asia with their Mongolian origin containing several ethnic groups as per the countries are concerned along with their food habits, as well as their way of living, are also found to be related with their origin, culture along with the environmental factors. Wheat has been considered as the basic staple food for the major civilizations of Europe as well as in Asia since 8000 years, and now, it is regarded to be the second major domestic food crop in the world. Wheat is served as the primary diet for two-thirds of the global population along with its contribution toward one-third of grain production in India (Sarkar and Agrawal., 2010a,b). As per Development Research by the International Maize and Wheat Improvement Center (CIMMYT) during the past two decades, it has been shown that the production of wheat in much warmer areas is technologically attainable; thus, the food crop conveniently covers the largest share in global crop area along with global food trade. Additionally, wheat is rudimentary to human civilization

Global Climate Change. https://doi.org/10.1016/B978-0-12-822928-6.00008-3

and has played a major role in feeding a hungry world and revamped global food security as the crop contributes about 20% of the total dietary calories and proteins worldwide (Shiferaw et al., 2013). It is claimed as the "miracle crop"of the past century as it helped to overcome famines and gave lives to so many people through green revolution in the developing countries of Asian continent. In contrary, it has been widely reported that wheat is one of the most O_3-sensitive crops and the mean global wheat yield is remarkably suppressed by negative influence of current level of O_3 in northern hemisphere by 7.1%–9.9% per year along with all the regions of the world (Biswas et al., 2008; Broberg et al., 2015; Feng et al., 2016; Mills et al., 2018a,c). The global yields of key domesticated crops have declined by 2%–15% in exposure to present-day ozone concentration that has been verified by recent studies (Feng and Kobayashi, 2009; Van Dingenen et al., 2009; Fishman et al., 2010; Avnery et al., 2011a). According to Van Dingenen et al. (2009), the global wheat yield loss was reported to be in the range of 7%–12%; however, the relative regional wheat yield loss in developing countries of East Asia (India, Pakistan, and Bangladesh) was estimated as 8%–27%, whereas another report reveals that global losses would increase to 25.8% by 2030, with 44.4% and 25.7% losses particularly in South and East Asia, respectively (Avnery et al., 2011b). The intervarietal responses may also differ from one another, which have been projected by several studies on wheat (Barnes et al., 1990; Pleijel et al., 2006; Biswas et al., 2008; Feng et al., 2011; Saintanis et al., 2014).

Tropospheric O_3, one of the severe causes of global climate change, is predicted to be altered by the increased tropospheric temperature and water vapor (Lu et al., 2019) The photooxidation of precursor gases such as carbon monoxide (CO), methane (CH_4), and volatile oxygen compounds (VOCs) in the availability of nitrogen oxide (NOx) is considered to be the responsible cause behind the generation and significant increment in concentration of particular secondary air pollutant ozone in ground level since the past few decades (Tai and Martin, 2017). Apart from being two largest wheat producers of Asian wheat production, India and China are considered to be the most affected wheat-cultivated areas (Van Dingenen et al., 2009; Avnery et al., 2011a). If the global O_3 precursor outrush continues to increase in this way, then the O_3-sensitive crops are predicted to see a further 10% decline in their yields by the year 2030. Though the ancillary future yield depletion could be prevented by the efficient mitigation of O_3 precursors, indeed, with hostile emission controls, global year 2030 deprivation could remain substantial, especially for O_3-sensitive crops (for wheat up to 17% globally along with considerable regional variability; Van Dingenen et al., 2009; Avnery et al., 2011b).

2. Modern bread wheat: the second most important cash crop

2.1 *Origin and evolution of modern wheat*

Wheat is one of the most dominant food crops of Eastern and Southern Asia being used for human food and livestock, which readily provides essential amino acids, vitamins and minerals, and beneficial phytochemicals along with dietary fiber components to the human diet particularly enriched in whole grain products. The genetic and genomic structure of various group of wheat species(genera *Aegilops*, *Amblyopyrum*, and *Triticum*) and their virtual relationship among the wild wheat varieties along in between them and domesticated wheat species led scientists to perform extensive studies on cytogenetic, molecular, and evolutionary aspects of wheat cultivars. As a part of the "Neolithic Revolution" among the cereals, wheat was the first domesticated food crop more than 10,000 years ago in the Middle East. The relationship between the earliest cultivated forms diploid (genome AA) (einkorn) and tetraploid (genome AABB) (emmer) revealed that they are mainly originated from the southeastern part of Turkey (Heun et al., 1997; Nesbitt, 1998; Dubcovsky and Dvorak, 2007); since 9000 years, the first hexaploid wheat made its first appearance (Feldman, 2001). Earliest cultivated forms of wheat were selected for their higher yield and other important characteristics, but due to domestication, the newer cultivars got separated from their wild relatives with special mention firstly about the loss of shattering of spike at maturity ensuring seed dispersal by seed loss, whereas the nonshattering traits acknowledged by mutations at the Br (brittle rachis) locus (Nalam et al., 2006). Another important trait is the alteration from hulled forms arose by Q locus involved in modification of the impact upon the recessive mutants at the Tg (tenacious glume) locus (Jantasuriyarat et al., 2004; Simons et al., 2006; Dubcovsky and Dvorak, 2007).

The modern traits of tetraploid and hexaploid wheat are free-threshing, whereas the early domesticated forms (einkorn, emmer, and spelt) are hulled like the cultivated form diploid, tetraploid, and hexaploid wheat, which have the tough rachis without spelt form of bread wheat.

The hybridization between cultivated emmer with that of the unrelated wild grass *Triticum tauschii* (*Aegilops tauschii* and *Aegilops squarosa*), leading to the generation of einkorn and emmer occurred independently with novel hexaploid (AABBDD) several times popular for its superior characteristics (Shewry, 2009).

The significance of the genetic changes during domestication can be epitomized when the modern wheat traits being impotent to endure wild against the better adapted species. In the year 1880, John Bennet Lawes efficiently demonstrated that event by granting part of the famous long-term Broadbalk experiment at Rothamsted to get back to its natural state (Dyke, 1993). In 1882, for the said purpose, some portions of the unharvested wheat were allowed to grow and kept on observations for the rest successive years, which resulted into a good crop harvest in 1883, but by the year 1885, weeds dominated and few of the wheat plants were monitored (Shewry, 2009).

Extensive studies reveal the polyploidy traits of wheat to be served as the classical example of genomic evolution via allopolyploids (Sax, 1921, 1927; Sears, 1948, 1969; Kihara et al., 1959; Morris and Sears, 1967; Feldman and Levy, 2012). According to the studies, tetraploid wheat contains one genome A, which was derived from the diploid wheat species, and another genome B, which was derived from unknown wheat species, whereas hexaploid wheat comprised of two tetraploid wheat genomes A and B and newer addition D from other traits. Genome A of wild and cultivated einkorn clearly is related to genomes of tetraploid and hexaploid varieties of wheat, whereas D genome is derived from the hexaploid wheat *T. tauschii* and the development being so recent that D genome divergence is considered to be very little between the tetraploid and hexaploid wheat. On contrary, B genome is considered to be derived from the S genome appeared in the Sitopsis section of *Aegilops*, along with *Aegilopsspeltoides* found to be the proximate extant species. Apart from that, the S genome of *A. speltoides* and G genome of *Triticum timopheevii* (tetraploid species with A and G genomes) are also closest (Feldman, 2001).

According to the summarized spread of wheat across the globe after Feldman (2001), it can be mentioned that the main path of wheat inside Europe was through Anatolia to Greece (8000 BP), after that northward via Balkans to the Danube (7000 BP), from there across Italy, France, and Spain (7000 BP) and by about 5000 BP finally reached the United Kingdom and Scandinavia. Likewise, wheat reached China by about 3000 BP via Iran into Central Asia and through Egypt initially arrived in Africa. Wheat was introduced to Mexico and Australia by the Spaniards in 1529 and 1788, respectively.

2.2 Wheat as a major player in green revolution

As the world makes its journey toward achieving Zero Hunger, the crop of huge importance has been selected as wheat, which is considered to be the most widely grown cereal crop across the globe. A series of researches, development, and technology transfer initiatives occurring between the 1940s and late 1970s that increased agricultural production worldwide is called as "Green Revolution." The initiative acquired by the diligent leadership of Norman E Borlaug, the "Father of Green Revolution," credited with saving over billions of people from starvation by the active involvement of the Rockefeller and Ford foundation. The main reason behind the success of Green Revolution was the development of modern wheat varieties by the CIMMYT. The varieties soon after their development highly accepted in large scale by developing countries and apparently increased the yield by many folds owing to their short stature along with high responsiveness toward fertilizers. The implementation of biotechnology in accordance with the use of genetically modified crops is of great importance in view of future perspective (Plantbreeding.coe.uga.edu). A major breakthrough occurred in the attempt to provide food for ever-increasing population, which was achieved by the utilization of newly developed wheat varieties by CIMMYT in South Asia. The remarkable increase in the wheat production was observed in developing countries, which was termed Green Revolution by William Gaud of US Agency for International Development(USAID) in 1968 (Murphy, 2007; Swaminathan, 2006). The successiveness of this event accomplished by the adoption of high-yielding varieties (HYVs) along with use of fertilizers, pesticides, and irrigation observing its impact in India and Pakistan during 1960–70 and in China after 1980 (Borlaug, 2000; Davies, 2003).

The success of Green Revolution was achieved by the semidwarf varieties of wheat, which were exceptional by their huge adaptive nature, strong stature, and higher disease-resistant capacity. The semidwarf variety of wheat was considered to be developed from the semidwarf wheat variety Norin 10 originated by a Japanese scientist by introducing the Daruma as the donor of semidwarfing trait, and the responsible recessive genes were rht1 and rht2 (Rajaram and van Ginkel, 1996). This wheat has various advantages over the normal wheat cultivars as it is short and contains stronger stalk that would not fall over on ground and also take much less time to grow to the height needed for grain production and also become less labor intensive. Along with these, the variety is rust-resistant and can handle in higher fertilizer (reported in crop life International, 2016 on semidwarf wheat: the game changer) with special mention to wheat cultivar "Sonalika," which played a major role in Green Revolution in India. The CIMMYT wheat program also gave effort to create breed against rust by utilizing variety Hope, which had stem rust resistance and Fronatna, which had durable resistance to leaf rust. The resistance is conferred by minor genes, which contain an additive interaction relationship (Rajaram, 2005).

Due to the impact of Green Revolution, the world cereal production of wheat increased by 2.53 times in the year 1961–2006 (FAO, 2007); along with that, 2.7 times higher yield was obtained at that period of time in the developing countries compared with 2.3 times in developed countries. The irrigated land area was increased by 210 million hectare (ha) and 271 million ha in 1980 and 2000, respectively, than that of 139 million ha in 1961. Area planted for HYVs of wheat in 1978–79 was 72.4% of the total area under wheat in Asia (Dalrymple, 1978). In many of the irrigated areas, the older HYVs get replaced by the second- and third-generation modern varieties (Evenson and Gollin, 2003), and eventually, the total cereal production in India has increased by 186 million tons from 70 million tons (in the year 1961) during the year 1961–99, whereas the cereal production in China has inclined from 91 million tons to 390 million tons at the same time (FAO, 2000).

3. Tropospheric ozone: the most toxic secondary air pollutant and climate factor

3.1 Good ozone and bad ozone

Ozone, the triatomic molecule of three oxygen atoms, first published by a German-Swiss chemist, Christian Friedrich Schönbein in 1840 which can occur both in the Earth's upper atmosphere and at ground level. Depending upon the occurrence of ozone, the effectiveness or the harmfulness is determined as only about 10% of total ozone is situated in the troposphere, especially near ground level typically called as tropospheric ozone, whereas rest of the 90% of atmospheric ozone is located in the stratosphere, particularly between 19 and 30 km above the Earth's surface (Lagzi et al., 2013).

"Good ozone"—Generally, ozone occurs in the Earth's upper atmosphere 6–30 miles above the Earth's surface where it forms a protective layer that plays an extremely pivotal and beneficial contribution in sheltering the lower layers of the atmosphere and the surface of the Earth from the Sun's harmful ultraviolet radiations; thus, the stratospheric ozone is renounced as the "good ozone," but due to manmade chemicals such as ozone-depleting substances (ODSs), chlorofluorocarbons (CFCs), hydrochlorofluorocarbons (HCFCs), halons, methyl bromide, carbon tetrachloride, and chloroform, the beneficial ozone layer is getting depleted, resulting into the formation of "ozone hole." Once released into the air, these ozone-depleting substances degrade very slowly.

"Bad ozone"—In the Earth's lower atmosphere (tropospheric zone) near the ground level, harmful layer of O_3 is prevalent that is known as ground-level ozone or "bad" ozone, which is not emitted directly into the air but is created by chemical reactions between NOx and VOCs, discharged from industrial facilities, electric utilities, motor vehicle exhaust, gasoline vapors, power plants, industrial boilers, refineries and chemical plants, etc. Although the amount of tropospheric ozone is only about 10% of the total amount of ozone present in a vertical column of the atmosphere, this relatively small amount of tropospheric ozone has a huge importance because it has inevitable role in the generation of photochemical air pollution and oxidizing impact of the gas near surface layer of the globe (Krupa and Manning, 1988). The concentration of tropospheric O_3 varies with the places; it is abundant in the troposphere typically varying from less than 10 ppb over remote tropical oceans up to about 100 ppb in the upper troposphere, and sometimes exceeding 100 ppb downwind of polluted metropolitan regions (IPCC, 2001). The adverse effect of tropospheric O_3 was first observed in grape vines (Richards et al., 1958); from then till now, it has already remarked as the most important rural air pollutant affecting human health, vegetation, or material that is susceptible to oxidation (Ashmore, 2005; Fuhrer and Booker, 2003; Karnosky et al., 2007; Karlsson et al., 2003; Laurence and Andersen, 2003; Matyssek and Sandermann, 2003).

3.2 The genesis of ozone

Being a secondary air pollutant, O_3 does not get emitted directly in the atmosphere. So the efficient source of tropospheric ozone production would be the stratospheric ozone through some unavoidable reactions, and the tropopause (upper portion of the troposphere) is the most active cyclonic region of biosphere where the tropospheric and stratospheric ozone exchange takes place. Another major source of O_3 formation is the chemical reaction between the primary air pollutants, NOx and VOCs.

Ozone production in the troposphere: Firstly O_3 can be generated by the recombination of an oxygen molecule and an oxygen atom, same as in the stratosphere where ozone is produced from a single recombination reaction of an atomic oxygen and a molecular oxygen with the presence of a third body M (M is usually O_2 or N_2) required to maintain a natural ozone cycle.

$$O + O_2 + M \rightarrow O_3 + M \qquad (11.1)$$

In the troposphere, nitrogen dioxide is the only known compound that can produce O atom during its photodissociation at available radiation:

$$NO_2 + h\nu\ (\lambda < 420\ \text{nm}) \rightarrow O + NO \qquad (11.2)$$

With the presence of NO, the molecular ozone (from reaction 11.1) gets depleted followed by the generation of NO_2:

$$O_3 + NO \rightarrow O_2 + NO_2 \qquad (11.3)$$

These reactions only maintain the threshold level of NO_2 and O_3 in the ambient by recycling. The utmost net ozone production can take place in the presence of other primary pollutants or precursors, such as CO, CH_4, nonmethane hydrocarbons (NMHCs), or some VOCs by the stimulation of single reaction scheme via oxidation of CO in the presence of NO:

$$CO + OH \rightarrow H + CO_2$$

$$H + O_2 + M \rightarrow HO_2 + M$$

$$HO_2 + NO \rightarrow OH + NO_2$$

$$NO_2 + h\nu\ (\lambda < 420\ \text{nm}) \rightarrow O + NO$$

$$O + O_2 + M \rightarrow O_3 + M$$

$$\text{Net reaction: } CO + 2O_2 \rightarrow CO_2 + O_3 \qquad (11.4)$$

Similar chain reactions occur with the oxidation of methane in NO-rich environment when O_3 and formaldehyde (CH_2O) are formed rapidly:

$$CH_4 + OH \rightarrow CH_3 + H_2O$$

$$CH_3 + O_2 + M \rightarrow CH_3O_2 + M$$

$$CH_3O_2 + NO \rightarrow CH_3O + NO_2$$

$$CH_3O + O_2 \rightarrow CH_2O + HO_2$$

$$HO_2 + NO \rightarrow OH + NO_2$$

$$NO_2 + h\nu\ (\lambda < 420\ \text{nm}) \rightarrow O + NO\ (2x)$$

$$O + O_2 + M \rightarrow O_3 + M\ (2x)$$

$$\text{Net reaction: } CH_4 + 4O_2 + 2h\nu \rightarrow CH_2O + H_2O + 2O_3$$

Later on, the further O_3 production is carried out by formaldehyde photodissociation to generate other radicals, which can participate in forward reaction. Apart from methane, other organic compounds (NMHC, VOC) are involved in tropospheric O_3 generation.

Along with the aforementioned strategies, long-range transport of O_3 from distant pollution sources (Kondratyev and Varotsos, 2001a,b; Varotsos et al., 2004) is also responsible for the increment of ozone concentration at some places (Lagzi et al., 2013).

3.3 Absorption in plants and general consequences

O_3, being a potent oxidant responsible for the photooxidation of living organisms along with plants due to its high oxidative capacity and ozone level, is also considered to be most detrimental as well as damaging air pollutant to which plants are exposed and produces substantive plant biomass and yield reductions (Thompson, 1992; Agrawal et al., 2005; Manning, 2005; Hassan and Tewfik, 2006; Singh et al., 2009a, 2014; Wahid, 2006a,b; Sarkar and Agrawal, 2010a,b; Tripathi and Agrawal, 2013). O_3 transports to the plant surface through the wind turbulence, although O_3 uptake is restricted by the cuticular surface of leaf (Kerstein and Lendzian, 1989). One of the efficient ways for O_3 to enter inside the plant system is the stomata; however, the stomatal conductance gets declined due to the O_3 exposure (Mansfield and Pearson, 1996; Castagna et al., 2001; Guidi et al., 2001; Pasqualini et al., 2002; Singh et al., 2009a). The actual cause behind this physiological action is debatable (Paoletti and Grulke, 2005) though it is considered that O_3 acts on altering the abscisic acid signaling or generating an active oxygen species (AOS) burst straight inside the stomatal guard cells (Kangasjärvi et al., 2005).

After invading through stomata, the fate of O_3 is to rest inside the substomatal cavity and also in the air spaces of the leaf. The O_3 is unable to persist inside the apoplastic zone for longer period of time as it rapidly degrades to form various reactive oxygen species (ROS) and eventually gets involved into the reaction with the biomolecules present in cell wall, apoplastic fluid, or plasma membrane (Laisk et al., 1989; Mishra et al., 2013b). The H_2O_2 accumulation in extracellular area is detected among the earliest responses of O_3 exposure along with the manifestation of leaf lesion (Pellinen et al., 1999; Wohlgemuth et al., 2002). As a consequence of ROS (H_2O_2) generation, NAD(P)H-dependent oxidase gets activated, and the photosynthesis becomes disrupted, leading to the accumulation of ROS (Park et al., 1998; Rao and Davis, 1999). Not only superoxide radicals are responsible for leaf injury under O_3 exposure (Alscher and Hess, 1993), but also the hydroxyl radicals, being one of the most reactive molecules, cause cellular damage by altering the normal characteristics of proteins, lipids, and DNA (Iqbal et al., 1996). ROS after generation leads to a series of chain reactions, which cause significant alteration in cellular metabolism of plants.

3.4 Consequences

ROS, which is generated by O_3, causes deleterious damage to the normal functioning of plant cell. To combat against the cellular damage, plants have developed several enzymatic (SOD, various peroxidases, catalases and glutathione reductase, etc.) and nonenzymatic antioxidants (ascorbic acid, carotenoids, glutathione, and tocopherol) in various cellular compartments (Das and Roychoudhury, 2014). To maintain the chlorophyll from the photooxidative damage, nonphotosynthetic pigments and carotenoids play the major role, with special mention to glutathione (GSH), which is known to be abundant thiol in plant antioxidants, which help in scavenging cellular ROS (Larson, 1988). Ascorbic acid is considered to be the main ROS detoxifying compound and mentioned to be the most abundant low-molecular-weight antioxidant synthesized to create first line of defense against O_3 exposure by donating electrons to various enzymatic and nonenzymatic reactions (Castillo and Greppin, 1988; Smirnoff, 2000; Polle et al., 1995). Apart from that, α-tocopherol deserves special mention as, being an oxidant–antioxidant, it readily prevents the chain reaction during lipid peroxidation due to O_3 exposure (Serbinova and Packer, 1994). Both kinds of antioxidant systems have their potentiality to combat against ROS, highlighting their crucial role in stress tolerance and cellular survival (Das and Roychoudhury, 2014).

4. Wheat production under ozone pollution: the world scenario

The conventional secondary air pollutant, i.e., tropospheric O_3 is considered to be one of the most phytotoxic gases known to cause substantial damage to vegetation specifically agricultural crops worldwide (Ashmore, 2005; Emberson et al., 2009; Fowler et al., 2009; Teixeira et al., 2011; Avnery et al., 2011b) produced via the photooxidation of precursor gases such as carbon CO, CH_4, and VOCs with the availability of NOx in ground level (Tai and Martin, 2017). However, in the past few decades, although the thickness of O_3 column of stratosphere is getting declined, in contrast tropospheric O_3 column on the other side reveals a continuous increasing trend as per the Intergovernmental Panel on Climate Change (IPCC) Assessment Report Five (AR5), which declared that continuation of this reckless emission trend causes elevation of O_3 concentration by 40%−60% possibly by the end of this century (Tiwari et al., 2008; IPCC, 2013; Cooper et al., 2014; Lin et al., 2014). Among the highest O_3-polluted regions, Asia precisely East and South Asia have emerged as the hot spot area for O_3 formation, as these parts have reported to be experiencing mean 7 h O_3 concentration up to 50−90 ppb (Emberson et al., 2009; Tiwari and Agrawal, 2018). Being the second most important food crop globally, the major concern relies over the fact that wheat is one of the most O_3-sensitive C_3 crop plants and reported to be negatively influenced by tropospheric O_3 throughout the world (Biswas et al., 2008; Broberg et al., 2015; Feng et al., 2016), including 8%−27% of yield loss in developing countries of East Asia (i.e., India, Pakistan, and Bangladesh) (Avnery et al., 2011b).

Ozone is noxious to plants (Iriti and Faoro, 2008) as it produces ROS after entering into the substomatal space of the leaves. ROS-mediated oxidative stress conclusively remolds the plant cell ultrastructure, physiological as well as metabolical activities resulting into alterations in plant growth, carbon allocation patterns, antioxidative systems, and reduction in plant yield (Ainsworth, 2008; Puckette et al., 2007; Singh et al., 2010, 2014). To be very precise, not only the sensitivity and manifestation of O_3 pollution on plants differ among species, but also cultivars of the same species differ in their sensitivity toward O_3 stress in case of wheat, maize, and mung reportedly (Biswas et al., 2008; Chaudhary and Agrawal, 2015; Singh et al., 2014). The large-scale cultivar-specific responses were evaluated over world's second most popular crop plant wheat. Not only in Bangladesh (Saintanis et al., 2014) with 10 cultivars but also in China (Biswas et al., 2008) involving 20 wheat cultivars and in India (Singh et al., 2018) having 14 wheat cultivars against elevated levels of ozone and the path analysis approach made the variables and ultimately plant yield influenced in a cultivar-specific way recognizing the cultivars that are either sensitive or tolerant against the O_3 stress.

Unlike any other crop species, the foremost concern of wheat is its yield, which gets negatively influenced by the effects of elevated O_3 (EO) along with various other related variables such as the number of grains per plant and the weight of grains per plant and gets immensely reduced in case of the most sensitive cultivars and minimum in tolerant cultivars (Sarkar and Agrawal, 2010a,b; Mishra et al., 2013a; Feng et al., 2008; Singh et al., 2018). The CSRI (Cumulative Stress Response Index) of wheat cultivars generally shows the same trend with that of the plant yield responses such as highest reduction in growth and its indices, photosynthesis, and stomatal conductance at separate stages of sampling remarked in sensitive cultivars subsequent by intermediately sensitive and least in tolerant cultivars (Singh et al., 2018). Interesting fact is that the earlier supplied wheat cultivars (before year 2000) were reported to be much tolerant than that of the modern cultivars (Biswas et al., 2008; Singh et al., 2018) supported by the earlier studies, which also support positive relationship between the cultivar sensitivity and year of their release (Pleijel et al., 2006; Biswas et al., 2008; Biswas and Jiang, 2011) where greater stomatal conductance and thus higher O_3 flux containing modern wheat cultivars were found to be more ozone sensitive (Pleijel et al., 2006). Therefore, the determining factors of intraspecific sensitivity and tolerance of plants toward O_3 stress are dependent on the response of cultivars against EO (ambient+30 ppb) that alter the growth, biomass, resource allocation, and physiology of that plant with respect to CSRI and path analysis O_3 (Singh et al., 2018). While developing new varieties, breeders make sure about the maximum photosynthetic rate of that particular crop so that highest yield can be achieved. As a consequence, those plants absorb more amount of O_3 along with CO_2 through the open stomatal pore for much longer time and damage the cellular component by oxidative stress, which could be an explanation of higher ozone sensitiveness of recent wheat cultivars against the older cultivars.

Varietal response and stomatal conductance: The study of Biswas et al., (2008) depicted similar effect of O_3 on older and newer varieties as year of release is positively correlated with stomatal conductance. The recently released cultivars possess higher adaption capacity of O_3 tolerance but at the same time had higher O_3 uptake through stomata due to higher stomatal conductance; on contrary, the previously released cultivars possess lower O_3 uptake capacity yet higher chance to repair due to lower stomatal conductance (Biswas et al., 2008).

Apparently in much studies, antioxidants (ascorbic acid content) get declined greatly in case of newly released cultivars accredited to insufficient reducing power (NADPH) resulting into the increased limitation of ascorbic acid recycling caused by greater loss in assimilation rate or dark respiration by O_3(Foyer and Harbinson, 1994; Rao et al., 1995; Biswas et al., 2008), while decrease in POD level of newer cultivar might be due to the insufficient scavenge of O_3-generated ROS in cellular level to protect photosynthetic apparatus. Thus, higher oxidative damage to the cellular membrane and enzymes relatively leads to a negative alteration of photosynthetic rate and chlorophyll content (Anderson et al., 2003; Calatayud et al., 2003; Biswas et al., 2008).

Besides morphological aspects, the biochemical and molecular approaches in determining plant sensitiveness or resistance toward O_3 damage have become another matter of concern that is proved by Fatima et al. (2018a,b) in their report where among the three selected cultivars, the sensitive (late released cultivar) one suffered the most oxidative stress counteracted by the antioxidant enzymes exposed to EO (ambient+ 30 ppb), which was not enough to control oxidative damage and led to decline in final productivity (Fatima et al., 2018a).

Antithetically, another report mentioned independence of O_3 sensitivity of wheat cultivars with the year of cultivar release as per the results of the principal component analysis (PCA) biplot analysis, which showed that the O_3-sensitive cultivars did not follow the year of release trend suggesting the ozone sensitivity being independent of the factor cultivar breeding (Saintanis et al., 2014). Most importantly, the traits that provide tolerance to wheat varieties against high O_3 could be potentially useful for developing future O_3-tolerant wheat varieties, and this concept will prompt more breeders to select O_3 resistance as a desirable trait for wheat varieties.

5. Case studies: Southeast Asia

5.1 *Pakistan*

From Pakistan, we have found four reports regarding the effect of O_3 on wheat cultivars (Pak-81, Chakwal-86, Inqilab-91, Punjab-96, and Pasban-90). In the year 1994 (Wahid et al., 1995), two winter wheat varieties were experimentalized using open-top chambers (OTCs) ventilated with ambient or charcoal-filtered air (FA) nearby Lahore, Pakistan. The application of 6-h daily mean O_3 concentration (25–45 nL/L) resulted into the higher yield reduction for Pak-81 (sensitive with reduction by 46.7%) and Chakwal-86 (less sensitive with reduction by34.8%) (Wahid et al., 1995). In the experiment by Maggs et al. (1995), the analyzed data project significant outcome in terms of the maintenance of agricultural yield as pollution emissions rise in South and Southeast Asia. In the year 1995, Maggs and his colleagues selected two wheat cultivars (Pak-81 and Chakwal-86) along with two rice cultivars (Basmati-385 and IRRI-6) to the study the yield response against the O_3 pollution using OTCs ventilated with ambient or charcoal-FA nearby the site in Lahore, Pakistan. At this location, 6-h mean O_3 concentration reaches 60 ppb in certain months, and as a result of that, significant yield reduces in two consecutive seasons, which varies from 33% to 46% in wheat and from 37% to 51% in rice. In the year 2006, Wahid in his experiment selected three wheat varieties (Inqilab-91, Punjab-96, and Pasban-90) and started the investigation using OTCs with charcoal-FA, unfiltered air (UFA), and unchambered field plots (AA) during 2003–04 season at a semiurban site in Lahore, Pakistan, by the application of 8-h daily mean O_3, NO_2, and SO_2 concentrations for 72, 28, and 15 ppb, respectively, which by the end of the season caused grain yield reduction in UFA with 43% for Pasban 90, 39% for Punjab-91, and 18% for Inqilab-91 compared with FA control plants (Wahid, 2006, Table 11.1).

Table 11.1 Different wheat cultivars and their responses against tropospheric ozone.

Country	Test cultivars	O_3 concentration	Observed results	References
Bangladesh	Sufi, Bijoy	60 and 100 ppb	11.5% and 45% in Sufi and 33% and 45% in Bijoy for both the O_3 exposure	Akhtar et al. (2010)
	Akbar, Bari 25, Bari 26, Bijoy, Gourab, Kanchan, Prodip, Shatabdi, Sourav, and Sufi	50, 60, 80, 100, 120, 135, 150, and 200 ppb	Shoot biomass, total chlorophyll, leaf greenness, and carotenoid content were reduced	Saintanis et al. (2014)
China	Jingdong 6	200, 100, 50 ppb	80.4%, 58.6%, and 10.5% decrease in grain yield	Feng et al. (2003)
	Yangmei 185	Frequently exceeded 40 ppb	3.4% and 12.7% increment in yield for 150 and 300 ppm EDU treatment, respectively	Wang et al. (2007)
	Jia 403	CF + O_3-105 (mm^3/m^3)	1000-grain mass and grain yield of wheat exposed to CF + O_3 treatment were reduced by 20.6% and 36.0%, respectively	Feng et al. (2007)
	Aegilops tauschii; *Triticum boeoticum* Bioss; *Triticosecale wittmack*; *Triticum monococcum*; *Triticum dicoccum*; *Triticum timopheevii* Zhuk.; *Triticum polonicum* L.; *Triticum turgidum* ssp. duruma; *Triticum aestivum* L. cv. Nongda 3214; *T. aestivum* L. cv. Xinong 2611; *T. aestivum* L. cv. Hannong 2; *T. aestivum* L. cv. Xiaoyan 22	100 ppb	54%, 37%, and 7% reduction in biomass, photosynthetic rate, and photochemical efficiency, respectively	Biswas et al. (2008)

	M7	AOT 40 increased by 6–9 ppm h (predicted)	RYL (relative yield loss)was estimated to be 7.1%–8.4% based on simulated O_3 dosages	Liu et al. (2009)
	Yangmai16 (Y16) and Yangfumai 2 (Y2)	Elevated level of ozone (EO), 56.4 ppb	Pigment amount, photosynthetic rate decreased; leaf senescence increased	Feng et al. (2011)
		Seasonal EO, 10.9 and 17.5 ppm h	The FACE-O_3 thus provides a reliable means to study the impacts of [O_3] elevation on crops without any alterations to the canopy microclimate	Tang et al. (2011)
	Yannong 19 (Y19), Yannong 16 (Y16), Yannong 15 (Y15), Yannong 2 (Y2)	AOT 40 14.7, 16.4, and 15.1 ppm h for three consecutive years, respectively	Average 20% yield loss for four cultivars	Zhu et al. (2011)
	JM22	EO, 75–85 ppb	20%, 33%, 33%, and 29% decrease in net photosynthetic rate, stomatal conductance, chlorophyll content, and total biomass, respectively	Li et al. (2011)
	Yannong 19 (Y19), Yangmai 16 (Y16), Yangmai 15 (Y15), and Yangfumai 2 (Y2)	50% higher than ambient	Ozone uptake threshold of 12 nmol/m^2 s was judged most reasonable for the wheat flux response relationship	Feng et al. (2012)
	Yangmai 185, Jia 002	70–100 ppb for the O_3-1 treatment and 150–200 ppb for the O_3-2 treatment	Yields decreased by 8.5%–58% and 40%–73% compared with CF control for O_3-1 and O_3-2 treatments	Wang et al. (2012)

Continued

Table 11.1 Different wheat cultivars and their responses against tropospheric ozone.—cont'd

Country	Test cultivars	O_3 concentration	Observed results	References
	Jia 002	AOT 40 of 2007—1.58–9.17; 2008— 15.32–27.67 ppm h	Yield reduced by elevated ozone exposure	Zheng et al. (2013)
	Yangmai 16 (Y16), Yangmai 19 (Y19)	1.5 × ambient O_3 (AO)	Higher SOD, POD, APX level in Y19 cultivar than Y16	Wang et al. (2014a)
	Yangmai 16 (Y16), Yangmai 19 (Y19)	1.5 × AO	12.96% increase in total phenolic content in Y16	Wang et al. (2014b)
		Minimum (16.1 ppbV) and maximum (53.3 ppbV)	12.9% on average wheat yield decrease	Zhu et al. (2015)
	Yangmai 16 (Y16), Yangmai 15 (Y15), Yangfumai 2(Y2), Yannong 19 (Y19), and Jiaxing 002 (J2)	EO at 1.5 times over AO concentration	EO-induced loss reached statistically significant levels in Y19 (58%), Y2 (78%), and Y15 (62%), but not in J2 (21%) or Y16 (16%)	Feng et al. (2016)
		AOT 40, 4.33–15.82 ppm h	9.86%–36.05% relative yield loss	Zhao et al. (2018)
	HN 6049, HN 6425, JD22, JM22, JN7, JM58, L639, SN22, SX616, SX 633, SX 733, SX 828, SY 20, TM8, ZM 411	AOT 40, 17.4 ppmh	Decreased yield, weight of 1000 grains, and harvest index by 20.3%, 15.1%, and 14.2%, respectively	Feng et al. (2018)
	Liangxing99	Elevated O_3 (80 ppb + O_3)	22.6% yield loss without SNP treatments	Li et al. (2018)
	Yannong 19, Yangmai 16	60 ppb over ambient conditions (40 ppb)	Soil NEA decreased by 35.42% and 51.90% under EO with Y16 and Y19, respectively	Chen et al. (2018)

India	Malviya 234		Yield decreased in case of (O_3 + UV radiation) combined treatment than that of the single treatments	Ambasht and Agrawal (2003)
	Malviya 533, Malviya 234	40 ppb	Yield increased by 18.8%, 19.1%, and 20.5% in M 533 and 24.8%, 66.9%, and 66.8% in M 234 as compared with non−EDU-treated plants	Tiwari et al. (2005)
	HUW 234	O_3 concentration, 40.1 ppb in nonfiltered chambers (NFCs); 90.4 in filtered chambers (FCs)	Weight and number of grains/plant significantly increased by 20.7% and 14.5%, respectively, in FCs as compared with NFCs	Rai et al. (2007)
	HUW 234, HUW 468, HUW 510, PBW 343, Sonalika	Mean monthly O_3 concentrations ranged between 35.3 and 54.2 ppb	25.6%, 24%, 20.4%, 8.6%, and 1.9% increased yield in EDU-treated cultivars than the non−EDU-treated one	Singh and Agrawal (2009)
	HUW 234, HUW 468, HUW 510, PBW 343, Sonalika	34.2−54.2 ppb O_3 concentration	Yield was higher in cultivars treated with EDU as compared with respective non−EDU-treated ones, with increments of 25.8%, 20.5%, 11.2%, 10.2%, and 1.9% in HUW 468, HUW 510, HUW 234, Sonalika, and PBW 343, respectively	Singh et al. (2009a,b)
	Sonalika, HUW 510	Mean AO concentration was 47.3 ppb	Decreases in RuBisCO subunits and lipoxygenase(s) and induced accumulation of APX, SOD, and thaumatin-like protein	Sarkar et al. (2010)

Continued

Table 11.1 Different wheat cultivars and their responses against tropospheric ozone.—cont'd

Country	Test cultivars	O_3 concentration	Observed results	References
	Sonalika, HUW 510	45.3 and 47.3 ppb O_3 in two successive years	Total number of grains per plant decreased by 18%, 26.5%, and 33% in Sonalika and 12%, 25%, and 28% in HUW 510 at NFCs, NFCLOs, and NFCHOs, respectively, as compared with FCs	Sarkar and Agrawal (2010b)
	HUW 468	27.7–59.1 ppb O_3 concentration from December 2007–April 2008	400 ppm EDU application increased yield, i.e., 11.7% increment in 100 seed weight was observed	Singh and Agrawal (2010)
	M 533	45.7 ppb during wheat growth and 50.2 ppb after flag leaf development	Reductions in photosynthetic rate (PN), stomatal conductance (gs), Fv/Fm ratio, photosynthetic pigments, and starch, and increases in intercellular CO_2 concentration (Ci), F0, and soluble sugars in NFCs compared with FCs	Rai et al. (2011)
	PBW 343, M 533	Open-top chambers with mean O_3 concentrations were 50.2 and 53.2 ppb, and AOT 40 values were 9 and 12 ppm h	6% yield reduction in two wheat cultivars	Rai and Agrawal (2014)
	HUW 37, K 9107	Plants were exposed to elevated CO_2 (700 ppm) and elevated O_3 (ambient + 10 ppb, i.e., nearly 25% of ambient)	Number and weight of grains per plant increased by 12.8% and 34.8% in HUW 37, and 19% and 37.5% in K 9107, respectively, as compared with NFCs	Mishra et al. (2013a)

	Kundan, PBW 343	60 ppb AO concentration	50% and 25% increment in proteome level at vegetative stage in Kundan and PBW 343, respectively	Gupta et al. (2018)
	HP 1209, HUW 234	70 (T1) and 100 (T2) ppb O_3 for 4 h daily	15% yield loss at a 100 ppb O_3 for 4 h	Singh et al. (2017)
	HD 2967, HD 3059, HD 2987, Kharchiya 65, PBW 502, PBW 550, DBW50, DBW 77, NIAW 34, Kundan, HUW 12, HUW 55, HUW 213, and HUW 251	Seasonal mean O_3 concentration for the experimental period was 47.2 ppb and ambient AOT 40 value was 8000 ppb/h	Compared with other cultivars, reduction in weight of grains per plant and number of grains per plant were highest in HD28 and least for cultivar KH65	Singh et al. (2018)
	HD 2987, Kharchiya 65	47.7 ppb in December to 56.7 ppb in March, whereas seasonal mean O_3 concentration was 49.6 ppb in ambient condition	Study also showed that rhizosphere of sensitive cultivar HD 2987 was affected more under EO as compared with tolerant Kharchiya 65	Fatima et al. (2018a)
	HD 2967, Sonalika	52.4 (AO) and 59.6 ppb (EO) (daytime average based on hourly values between 09.00 and 17.00 h)	Sonalika showed significant reduction in grain weight per plant by 18% and 19% under EO at RN and HN treatments, respectively, and HD 2967, grain weight per plant was decreased by 10.2% (insignificantly), whereas increase in the husk weight/plant under EO at RN treatment	Pandey et al. (2018)
	HD 2987, PBW 502, and Kharchiya 65	Seasonal mean O_3 concentration for the experimental period was 49.6 ppb	Yield decreased by 42.2% in HD 2987, 27.1% in PBW 502, and only 12.9% in Kharchiya 65	Fatima et al. (2018b)

Continued

Table 11.1 Different wheat cultivars and their responses against tropospheric ozone.—cont'd

Country	Test cultivars	O_3 concentration	Observed results	References
Korea	Keumkang	120 ppm for 15, 30, 45, and 60 min	The water absorption index, peak viscosity, and final viscosity of flour increased by ozone treatment	Lee et al. (2017)
Pakistan	Pak-81, Chakwal-86	25–45 nL/L	Yield loss of cultivar Pak-81 and Chakwal-86 are 46.7% and 34.8%, respectively	Wahid et al. (1995)
	Pak-81, Chakwal-86	60 ppb	Yield loss of 33%–46% in wheat and from 37% to 51% in rice	Maggs et al. (1995)
	Inqilab-91, Punjab-96, Pasban-90	72 ppb	Grain yield loss in UFA with 43% for Pasban-90, 39% for Punjab-91, and 18% for Inqilab-91 compared with FA control plan	Wahid. (2006)
Russia	Novosibirskaya	At ambient level	Significant change in morphological characteristics not in germination ability	Lazukin et al. (2018)

EDU, ethylene diurea; *NFCHOs*, nonfiltered air with additional higher (+20 ppb) O_3; *NFCLOs*, nonfiltered air with additional lower (+10 ppb) O_3; *UFA*, unfiltered air.

5.1.1 India

From India, several researchers reported several experiments regarding the effects on wheat plants' yield due to O_3. Wheat cultivars Malviya 533 and Malviya 234 were selected by Tiwari et al. (2005), in the year 2005 for experimentalizing the impact of ground level ambient O_3 (AO) on growth and productivity with the help of ethylene diurea (EDU) (0, 150, 300, and 450 ppm as soil drenches at 10 days interval) in the suburban site near Varanasi, India. As a result of the experiment, 150 ppm of EDU application caused improvement for M 533, whereas 300 and 450 ppm of EDU application improved the yield of M 234, thus showing that EDU treatment affected plant growth varying with cultivar, age, and EDU concentration (Tiwari et al., 2005). Rai and Agrawal (2014) in their experiment reported the intraspecific variability of two wheat cultivars (PBW 343 and M 533) at different developmental stages using OTCs with mean O_3 amount of 50.2 and 53.2 ppb, and accumulated O_3 over a threshold concentration of 40 ppb (AOT 40) values were 9 and 12 ppm h, respectively, for 2008–09 and 2009–10 successive years. In this study, the yield of two wheat cultivars gets reduced by 6% in exposure of critical levels of a 3-month AOT 40 of 3 ppm h for two consecutive years. Though the yield reductions were same in both cultivars under AO, the approach of obtaining the resistance is different between the cultivars (Rai and Agrawal, 2014). In the experiment conducted by Singh et al. (2018), 14 wheat cultivars were widely grown in India and exhibited to 30 ppb EO above ambient level using OTPs to assess their regulate against O_3 stress by evaluating their cumulative stress response index; this resulted into variable yield reduction of test cultivars. The main findings of the study were to detect the differential sensitivity and tolerance of wheat cultivars and developing resistance against EO (Singh et al., 2018). An experiment performed by Rai et al. (2007) assessed the effects of ambient gaseous air pollution on wheat (*Triticum aestivum* L. var. HUW 234) grown in a suburban area situated along the eastern Gangetic plain of India, using OTCs, i.e., filtered chambers (FCs), nonfiltered chambers (NFCs), and open plots (OPs) inside which 8-hourly air monitoring was supervised to determine ambient concentrations of O_3, NO_2, and SO_2. As per the result, the plants grown in NFCs and OPs revealed much lower photosynthetic rate, stomatal conductance, chlorophyll content, and Fv/Fm ratio in comparison with the plants in FC, thus showing that the yield of plants does result the same as the morphological parameters did that helps in remarking O_3 and NO_2 to be the major pollutant of suburban area responsible for yield loss of tropical wheat (Rai et al., 2007). Singh et al. (2017) in their study observed the response of two wheat cultivars (HUW 234 and HP 1209) at constant levels of EO and eventually compared the results of the present OTCs with free air controlled enrichment (FACE) condition after the implementation of 70 and 100 ppb O_3 for 4 h on a daily basis from germination to maturity. Both of the cultivar showed significant variation in quantity as well as quality of grains especially higher yield reduction in cultivar HP 1209 (Singh et al., 2017).

In the study (Singh et al., 2009a,b), different concentrations of EDU (0 and 400 ppm as soil drench for 12 days) were applied to five wheat cultivars (Sonalika, PBW 343, HUW 468, HUW 510, and HUW 234) at the study site of Varanasi, Uttar Pradesh, in the presence of 34.2–54.2 ppb concentration of O_3 during growth period of wheat. As a result of the experiment, EDU application refurbished the mechanism of grain yield significantly by maintaining higher levels of antioxidants, metabolites, and enzymes in cultivars HUW 468 and HUW 510. However, cultivars Sonalika and PBW 343 exhibited greater resistance to O_3 (Singh et al., 2009a,b). Mishra et al. (2013a) conducted the experiment to assess the impact of elevated levels of CO_2 and O_3 singly and in combination of two wheat cultivars HUW 37 and K 9107 on their growth, yield attributes and grain quality in OTCs, i.e., EO, elevated CO_2 (EC), and elevated O_3+CO_2 (ECO). In ECO exposure, EC fully protected wheat cultivars against negative effects of O_3 that resulted into yield parameters, which showed significant increase in EC followed by ECO and in EO. Both the cultivars were subjected to significant yield reduction; however, cultivar-specific responses were clearly observed when cultivar K 9107 was proved to be much tolerant as compared with cultivar HUW 37 against EO or AO (Mishra et al., 2013a). In the year 2018, Fatima et al. (2018a) selected two wheat cultivars HD 2987 (O_3 sensitive) and Kharchiya 65 (O_3 tolerant) for their experiment. The study was conducted in Varanasi, Uttar Pradesh (UP) under AO and elevated (ambient + 30 ppb) levels of O_3 (EO) using OTPs. After investigation, soil enzymatic activities and microbial biomass carbon and nitrogen were found to be reduced more in sensitive cultivars by 11.1% and 21.2%, respectively, due to ozone exposure. Overall, the study revealed that rhizosphere of tolerant Kharchiya 65 remained less affected than the sensitive cultivar HD 2987 under EO. Pandey et al. (2018) in their experiment grew two wheat cultivars (HD 2967 and Sonalika) in OTC under AO and EO (ambient + 20 ppb O_3), and they applied (1) recommended nitrogen (RN), (b) 1.5 times the recommended N (HN) to observe any intervarietal change in plants against O_3 stress, which showed positive result in the sense that nitrogen application found to decrease O_3 damage to some extent as one of the two cultivars, HD 2967, manifested higher biomass, efficient resource partitioning, and intensified antioxidant defense system with less subtraction in economic yield against O_3 at higher N dose (Pandey et al., 2018). Rai et al. (2011) performed the experiment by selecting a modern variety M533 at a rural site in an agricultural area situated 20 km south of Varanasi city in the presence of 45.7 ppb during wheat growth and 50.2 ppb after flag leaf development having six OTCs with three treatments such as OTCs ventilated with ambient nonfiltered air (NFCs), OTCs ventilated with activated charcoal filters (FCs), and OPs without chambers, respectively, including the OP just to observe the chamber effect, and the results showed reductions in photosynthetic rate (PN), stomatal conductance (gs), Fv/Fm ratio, photosynthetic pigments and starch, and increases in intercellular CO_2 concentration (Ci), F0, and soluble sugars in NFCs compared with FCs. As per the study, significant alteration in wheat variety possesses negative influence in yield by the higher concentration of O_3 (Rai et al., 2011). Ambasht and Agrawal (2003) conducted an experiment at the botanical garden of BHU, Varanasi, under the natural field condition having 12 plots divided into four treatments, i.e., control (C), UV-B treated (UV-B), O_3 treated (O_3), combination of UV-B and O_3 (UV-B + O_3) with the wheat cultivar Malviya 234 to assess the biochemical, physiological, and growth response of the wheat plant to supplemental UV radiation and also enhanced O_3 concentration. The experiment resulted that though UV- and O_3-treated plants' sample showed reduction in chlorophyll content, photosynthetic rate, stomatal conductance, carotenoid content, anthocyanin content, biomass, and yield, but for most of the cases, the combined effect of O_3 and UV radiation made the highest amount of reduction in those selected parameters (Ambasht and Agrawal, 2003). The present investigation was exhibited with two cultivars of Indian wheat (Sonalika and HUW 510) projecting into AO and EO levels. Daytime AO amount throughout the first year (2007–08) was 45.3 ppb in average with a range of 4.0–125 ppb, and during the second year (2008–09), O_3 was 47.3 ppb in average with an extent of 2.0–137 ppb by using OTCs, i.e., FCs, nonfiltered air NFCs, nonfiltered air with additional lower (+10 ppb) O_3 (NFCLOs), and nonfiltered air with additional higher (+20 ppb) O_3 (NFCHOs). Both the cultivars indicated toward different effects; however, the pair of cultivars reported negative impact of O_3 on various growth, reproductive, and yield parameters. The total number of grains per plant decreased by 18%, 26.5%, and 33% in Sonalika and 12%, 25%, and 28% in HUW 510 at NFCs, NFCLOs, and NFCHOs, respectively, as compared with FCs (Sarkar and Agrawal, 2010b). In the study by Gupta et al. (2018), the experiment was performed under AO concentration of 60 ppb by selecting Kundan and PBW 343, a pair of cultivars to study the effect of EDU on plant proteome, growth, and yield against AO phytotoxicity. After applying EDU as foliar spray (200 and 300 ppm), both of the cultivars exhibited less lipid peroxidation along with increased chlorophyll content, biomass, and yield due to EDU-mediated increment in antioxidant level. EDU treatment provides 50% and 15% protein increment in the vegetative phases of Kundan and PBW 343, respectively, thus providing much protection to the Kundan and less protection for PBW 343 against O_3 toxicity (Gupta et al., 2018). Singh and Agrawal conducted an experiment under natural field conditions including the use of OTCs at a suburban site of Varanasi, India, in the year 2010 exerting 27.7–59.1 ppb O_3 concentration from December 2007–April 2008 to detect the optimal EDU dosage in different stages of wheat plant cultivar HUW 468. The result showed higher yield in 400 ppm of EDU dose as it showed a higher yield than any other dose as 1000 grain weight was higher (11.7%) in 400 ppm EDU treatment than any other treatment (Singh and Agrawal, 2010). Sarkar et al. (2010) conducted the study at Agricultural Research Farm, Banaras Hindu University, Varanasi, using OTCs, i.e., FCs, NFCs, NFCLOs, and NFCHOs with the mean ambient daytime. The O_3 concentration for the complete wheat growth period was 47.3 ppb (4.9 ppb in FCs, 47.3 ppb in NFCs, 50.4 ppb in NFCLOs, and 54.9 ppb in NFCHOs, respectively). Two wheat cultivars (Sonalika and HUW 510) were selected to evaluate the impact of elevated concentrations of O_3 on phenotypical, physiological, and biochemical traits followed by leaf proteome analysis. As a result of the experiment, O_3 stress significantly decreases various physiological and biochemical characteristics of the plant including drastic decline occurred in RuBisCO large and small subunits, ATP synthase aldolase, and phosphoglycerate kinase, whereas some stress defense–related proteins got induced by O_3 stress analyzed through one-dimensional gel electrophoresis (1-DGE) and two-dimensional gel electrophoresis (2-DGE) (Sarkar et al., 2010). Field experiment was conducted by Singh and Agrawal (2009) on five local wheat cultivars (HUW 234, HUW 468, HUW 510, PBW 343, and Sonalika) to analyze the impact of O_3 on their growth and yield characteristics under natural condition (AO) in a suburban area of

Varanasi during December 2006 to March 2007 with the mean monthly O_3 concentrations ranged between 35.3 and 54.2 ppb. As a result, EDU treatment positively affects the plant growth and yield, i.e., 25.6%, 24%, 20.4%, 8.6%, and 1.9% increased yield in EDU-treated cultivars, HUW468, Sonalika, HUW 510, HUW 234, and PBW 343, respectively than non-EDU-treated ones Singh and Agrawal (2009). In this study, Fatima et al. (2018b) selected three wheat cultivars (HD 2987—O_3 sensitive; PBW 502—intermediately sensitive, Kharchiya 65—O_3tolerant) under elevated level (ambient + 30 ppb) levels of O_3 at Botanical garden, BHU, during the winter season (December 2015 to March 2016) with the mean seasonal O_3 concentration of 49.6 ppb to investigate the effect of EO on histochemical study, native PAGE, antioxidative enzyme, phenylpropanoid pathway, altered yield losses, and also related change of reproductive structure. As per the result of the experiment, higher induction in antioxidative enzyme was reported from the modern cultivar HD 2987 than that of the Kharchiya 65 indicating more susceptibility of modern wheat cultivar toward O_3. Furthermore, the yield (test weight) was also decreased by 42.2% in HD 2987, 27.1% in PBW 502, and only 12.9% in Kharchiya 65 under EO compared with AO substantiating the same trend (Fatima et al., 2018b, Table 11.1).

5.2 Bangladesh

From Bangladesh, two experimental studies regarding the effect on wheat plant yield due to O_3 were reported. Akhtar et al. (2010) in their experiment selected two wheat cultivars of Bangladesh (Sufi and Bijoy) to analyze the effect of O_3 on crop plant cultivation and grown them inside plastic boxes containing Andisol provided with O_3 exposure of 60 and 100 ppb for March 13 to June 4, 2008. As per the result is concerned, the plant growth and yield was reduced (11.5% and 45% in Sufi and 33% and 45% in Bijoy) under both the O_3 exposures i.e., O_3 pollution and AO, respectively, thereby affecting the wheat production in Bangladesh adversely (Akhtar et al., 2010). In another experiment under controlled environmental condition of the chamber, Saintanis et al. (2014) tested the O_3 sensitivity of 10 selected wheat cultivars by exposing plants to 50, 60, 80, 100, 120, 135, 150, and 200 ppb of O_3 exposure for 14, 11, 8, 6, 5, 4, 3, and 1 days, respectively, for 8 h/day and thus reported reduction of shoot biomass, total chlorophyll, leaf greenness, and carotenoid content, but the effects were mainly dependent upon O_3 exposure regimes and wheat cultivar. Additionally, as per the results of the PCA biplot analysis, the O_3-sensitive cultivars did not follow the year of release trend, suggesting the O_3 sensitivity being independent of the factor cultivar breeding (Table 11.1).

5.3 China

From China, several experimental studies have been reported by the scientists regarding the O_3 effect on the wheat cultivars of China. In the study by Feng et al. (2011), two phonologically similar winter wheat cultivars (Yangfumai 2 (Y2) and Yangmai 16 (Y16)) were exposed to EO treatment (27% of the AO concentration) at Xiaoji town, Jiangdu county, Jiangsu province, China, to determine the impact on plant physiological characteristics related to photosynthesis in fully open-air field condition. EO significantly declined the pigment amounts and photosynthetic rates, whereas it accelerated the leaf senescence due to high amount of lipid peroxidation in the cultivar Y2, but in case of cultivar Y16, contrast effects were observed at much last stage of flag leaf aging suggesting Y2 as much more sensitive toward O_3 than Y16 (Feng et al., 2011). Zhu et al. (2011) performed an experiment under EO concentration in fully open-field condition at Xiaoji town, Jiangdu county, Jiangsu province, China, for three consecutive growing seasons from 2006 to 09 by selecting four wheat cultivars (Y19, Y16, Y15, and Y2). As a result, EO significantly reduced the grain yield from 0.78 to 0.62 kgm^{-2} (20.2%) when an average was made across four cultivars and three growing seasons. The season-based response suggested that the grain yield was highest for the cultivar Y15; however, it was lowest for cultivar Y19 (Zhu et al., 2011).

Wang et al. (2014a) conducted an experiment at Xiaoji Town, Jiangdu County, Jiangsu Province, China, selecting two O_3-sensitive wheat cultivars (Y16 and Y19) to determine the different apoplastic antioxidant mechanisms of O_3-treated cultivars under O_3-FACE condition. Analysis of the apoplast of cultivar Y19 projected the pronounced amount of superoxide anion, hydroxyl radicals, hydrogen peroxide, and lipid peroxidation under elevated $1.5 \times$ AO than those of Y16. But in later stage, O_3 stress induced a decrease in antioxidant level of Y19 cultivar and thus appeared to be sensitive, but Y16 cultivar—projected O_3-induced higher antioxidative response implicated higher O_3 tolerance level and thus indicated that there is varietal difference in wheat apoplastic antioxidant enzyme responses toward elevated O_3 (Wang et al., 2014a).

Wang et al. (2014b) in the other experiment under same condition as described earlier determined the different metabolic mechanisms of phenolic compounds, secondary metabolism-related enzymes activities, lipid peroxidation extent, and ROS content of two wheat cultivars under O_3-FACE, and as per the result, Y19 variety revealed more increment in lipid peroxidation and ROS content and thus confirmed the Y19 variety to be more sensitive than the other cultivar (Y16) against O_3 stress (Wang et al., 2014b).

Feng et al. (2018) performed an experiment at Xiaotangshan town, Changping district, Northwest of Beijing, China, by selecting 15 wheat cultivars exposed to 17.4 ppm h AOT 40 (accumulated O_3 over an hourly concentration threshold of 40 ppb) over the growing season to detect the intraspecific variation to the sensitivity of wheat cultivar by the application of antiozonant tool EDU. The plants treated with EDU significantly increased photosynthetic rate, photosynthetic pigments (chlorophyll and carotenoids contents) along with antioxidant capacity and reduced the malondialdehyde and ascorbate contents. Non—EDU-treated plants showed reduction in yield, weight of 1000 grains, and harvest index by 20.3%, 15.1%, and 14.2%, respectively, across all the cultivars. The EDU interaction with plant sensitivity supposed to be cultivar specific and was ranked by highly sensitive to less sensitive by comparing EDU-treated and nontreated plants average grain yield (Feng et al., 2018, Table 11.1).

Acknowledgments

AS acknowledges the receipt of financial assistance in the form of a research project (Reference No.F.30-393/2017 (BSR), dated September 27, 2018) funded by the University Grants Commission, MHRD, Government of India. The research project helped to develop the idea and outcome of this chapter.

References

Akhtar, N., Yamaguchi, M., Inada, H., Hoshino, D., Kondo, T., 2010. Effects of ozone on growth, yield and leaf gas exchange rates of two Bangladeshi cultivars of wheat (*Triticum aestivum* L.). Environ. Pollut. 158, 1763–1767.

Agrawal, S.B., Singh, A., Rathore, D., 2005. Role of ethylene diurea (EDU) in assessing impact of ozone on *Vigna radiata* L. plants in suburban areas of Allahabad (India). Chemosphere 61, 218–228.

Ainsworth, E.A., 2008. Rice production in a changing climate: a meta-analysis of responses to elevated carbon dioxide and elevated ozone concentration. Glob. Change Biol. 14, 1642–1650.

Alscher, R.G., Hess, J.L., 1993. Antioxidants in Higher Plants. CRC, Boca Raton.

Ambasht, N.K., Agrawal, M., 2003. Effects of enhanced UV-B radiation and tropospheric ozone on physiological and biochemical characteristics of field grown wheat. Biol. Plant. (Prague) 47 (4), 625–628.

Anderson, P.D., Palmer, B., Houpis, J.L.J., Smith, M.K., Pushnik, J.C., 2003. Chloroplastic responses of ponderosa pine (*Pinus ponderosa*) seedling to ozone exposure. Environ. Int. 29, 407–413.

Ashmore, M.R., 2005. Assessing the future global impacts of ozone on vegetation. Plant Cell Environ. 28, 949–964.

Avnery, S., Mauzerall, D.L., Liu, J., Horowitz, L.W., 2011a. Global crop yield reductions due to surface ozone exposure: 1 Year 2000 crop production losses and economic damage. Atmos. Environ. 45, 2284–2296.

Avnery, S., Mauzerall, D.L., Liu, J., Horowitz, L.W., 2011b. Global crop yield reductions due to surface ozone exposure: 2. Year 2030 potential crop production losses and economic damage under two scenarios of O_3 pollution. Atmos. Environ. 45, 2297–2309.

Barnes, P.W., Beyschlag, W., Rayel, R., Flint, S.D., Caldwell, M.M., 1990. Plant competition for light analyzed with a multispecies canopy model. III. In fluence of canopy structure in mixtures and monocultures of wheat and wild oat. Oecologia 82 (3), 374–380. https://doi.org/10.1007/BF00317486.

Biswas, D.K., Xu, H., Li, M.Z., Liu, Y.G., Chen, Y.H., Sun, J.Z.J.Z., Jiang, G.M., 2008. Assessing the genetic relatedness of higher ozone sensitivity of modern wheat to its wild and cultivated progenitors/relatives. J. Exp. Bot. 59, 951–963.

Biswas, D.K., Jiang, G.M., 2011. Differential drought-induced modulation of ozone tolerance in winter wheat species. J. Exp. 62 (12), 4153–4162.

Borlaug, N. 2000. The Green Revolution revisited and the road ahead. Anniversary Nobel Lecture. Oslo, Norway. Available from http://www.nobelprize.org/nobel_prizes/peace/laureates/1970/borlaug-lecture.pdf [Accessed 8 May 2012].

Broberg, M.C., Feng, Z., Xin, Y., Pleijel, H., 2015. Ozone effects on wheat grain quality –a summary. Environ. Pollut. 197, 203–213.

Calatayud, A., Iglesias, D.J., Talon, M., Barreno, E., 2003. Effect of 2 month ozone exposure in spinach leaves on photosynthesis, antioxidant systems and lipid peroxidation. Plant Physiol. Biochem. 41, 839–845.

Castagna, A., Nali, C., Ciompi, S., Lorenzini, G., Soldatini, G.F., Ranieri, A., 2001. Ozone exposure affects photosynthesis of pumpkin (*Cucurbita pepo*) plants. New Phytol. 152, 223–229.

Castillo, F.J., Greppin, H., 1988. Extracellular ascorbic acid and enzyme activities related to ascorbic acid metabolism in *Sedum album* L. leaves after ozone exposure. Environ. Exp. Bot. 28, 231–238.

Chaudhary, N., Agrawal, S.B., 2015. The role of elevated ozone on growth, yield and seed quality amongst six cultivars of mung bean. Ecotoxicol. Environ. Saf. 111, 286–294. https://doi.org/10.1016/j.ecoenv.2014.09.018.

Chen, W., Hou, H., Sun, C., Wu, Z., Zhang, L., Yu, W., 2018. The effect of elevated ozone concentration on enzymes increases NO_3 content in the soil at the jointing stage of wheat field. Eur. J. Soil Biol. 89, 14–19.

CLRTAP, 2017. Mapping critical levels for vegetation, Chapter III of manual on methodologies and criteria for modelling and mapping critical loads and levels and air pollution effects, risks and trends. UNECE Convention on Long-range Transboundary Air Pollution. http://icpvegetation.ceh.ac.uk/ [Accessed 18 October 2017].

Cooper, O.R., Parrish, D.D., Ziemke, J., Balashov, N.V., Cupeiro, M., 2014. Global distribution and trends of tropospheric ozone: an observation-based review. Elementa. https://doi.org/10.12952/journal.elementa.000029.

Dalrymple, D.G., 1978. Development and Spread of High Yielding Varieties of Wheat and Rice in Less Developed Nations. Foreign agricultural economic report. USDA.

Das, K., Roychoudhury, A., 2014. Reactive oxygen species (ROS) and response of antioxidants as ROS-scavengers during environmental stress in plants. Front. Environ. Sci. 2, 53. https://doi.org/10.3389/fenvs.2014.00053.

Davies, W.P., 2003. An historical perspective from the green revolution to the gene revolution. Nutr. Rev. 61 (6), S124–S134. https://doi.org/10.1301/nr.2003.jun.S124-S134.

Dubcovsky, J., Dvorak, J., 2007. Genome plasticity a key factor in the success of polyploidy wheat under domestication. Science 316, 1862–1866.

Dyke, G.V., 1993. John Lawes of Rothamsted. Pioneer of Science Farming and Industry. Hoos Press, Harpenden, UK.

Emberson, L.D., Buker, P., Ashmore, M.R., Mill, G., Jackson, L.S., Agrawal, M., Atikuzzaman, M.D., Cinderby, S., Engardt, M., Jamir, C., Kobayashi, K., Oanh, N.T.K., Quadir, Q.F., Wahid, A., 2009. A comparison of North American and Asian exposure data for ozone effects on crop yields. Atmos. Environ. 43, 1945–1953.

EPA US, 2013. Integrated Science Assessment (ISA) of Ozone and Related Photochemical Oxidants (Washington, D.C).

Evenson, R.E., Gollin, D., 2003. Assessing the impact of the green revolution, 1960 to 2000. Science 300, 758.

FAO, 2000. FAOSTAT 2000.

FAO, 2007. Cereal Production in Drylands. FAO.

Fatima, A., Singh, A.A., Agrawal, M., Agrawal, S.B., 2018a. Effect of elevated O_3 on rhizospheric enzymatic activities of ozone sensitive and tolerant wheat cultivars. Arch. Agron Soil Sci. 64 (12), 1768–1776. https://doi.org/10.1080/03650340.2018.1448929.

Fatima, A., Singh, A.A., Mukherjee, A., Agrawal, M., Agrawal, S.B., 2018b. Variability in defence mechanism operating in three wheat cultivars having different levels of sensitivity against elevated ozone. Environ. Exp. Bot. 115, 66–78. https://doi.org/10.1016/j.envexpbot.2018.06.015.

Feldman, M., 2001. Origin of cultivated wheat. In: Bonjean, A.P., Angus, W.J. (Eds.), The World Wheat Book: A History of Wheat Breeding. Lavoisier Publishing, Paris, France, pp. 3–56.

Feldman, M., Levy, A.A., 2012. Genome evolution due to allopolyploidization in wheat, 192 (13), 763–774.

Feng, Z., Kobayashi, K., 2009. Assessing the impacts of current and future concentrations of surface on crop yield with meta-analysis ozone. Atmos. Environ. 43, 1510–1519.

Feng, Z., Jin, F., Zhang, F., Huang, 2003. Effect of ground level ozone (O_3) pollution on yields of rice and winter wheat in the Yangtze River Delta. J. Environ. Sci. 15 (3), 360–362.

Feng, Z., Fao, F., Chen, Z., Wang, K., Zheng, W., Feng, W., 2007. Response of gas exchange and yield components of field-grown *Triticum aestivum* L. to elevated ozone in China. Photosynthetica 45 (3), 441–446.

Feng, Z., Kobayashi, K., Ainsworth, E.A., 2008. Impact of elevated ozone concentration on growth, physiology and yield of what (*Triticum aestivum L.*): a meta analysis. Glob. Change Biol. 14, 2696–2708. https://doi.org/10.1111/j.1365-2486.2008.01673.x.

Feng, Z., Pang, J., Kobayashi, K., Zhu, J., Ort, D.R., 2011. Differential responses in two varieties of winter wheat to elevated ozone concentration under fully open-air field conditions. Glob. Change Biol. 17, 580–591.

Feng, Z., Tang, H., Uddling, J., Pleijel, H., Kobayashi, K., Zhu, J., Oue, H., Guo, W., 2012. A stomatal ozone flux response relationship to assess ozone-induced yield loss of winter wheat in subtropical China. Environ. Pollut. 164, 16–23.

Feng, Z., Wang, L., Pleijel, H., Zhu, J., Kobayashi, K., 2016. Differential effects of ozone on photosynthesis of winter wheat among cultivars depend on antioxidative enzymes rather than stomatal conductance. Sci. Total Environ. 572, 404–411.

Feng, Z., Jiang, L., Calatayud, V., Dai, L., Paoletti, E., 2018. Intraspecific variation in sensitivity of winter wheat (*Triticum aestivum L.*) to ambient ozone in Northern China as assessed by ethylene diurea (EDU). Environ. Sci. Pollut. Res. Int. 25 (29), 29208–29218. https://doi.org/10.1007/s11356-018-2782-8.

Fishman, J., Creilson, J.K., Parker, P.A., Ainsworth, E.A., Vining, G.G., et al., 2010. An investigation of widespread ozone damage to the soybean crop in the upper midwest determined from ground-based and satellite measurements. Atmos. Environ. 44, 2248–2256. https://doi.org/10.1016/j.atmosenv.2010.01.015.

Fowler, D., Pilegaard, K., Sutton, M.A., et al., 2009. Atmospheric composition change: ecosystems-atmosphere interactions. Atmos. Environ. 43, 5193–5267. https://doi.org/10.1016/j.atmosenv.2009.07.068.

Foyer, C.H., Harbinson, J., 1994. Oxygen metabolism and the regulation of photosynthetic electron transport. In: Foyer, C.H., Mullineaux, P. (Eds.), Causes of Photooxidative Stress and Amelioration of Defense Systems in Plants. CRC Press Inc, Boca Raton, FL, pp. 1–42.

Fuhrer, J., Booker, F., 2003. Ecological issues related to ozone: agricultural issues. Environ. Sci. Pol. 29, 141–154.

Granier, C., Bessagnet, B., Bond, T., D'Angiola, A., Denier van der Gon, H., et al., 2011. Evolution of anthropogenic and biomass burning emissions of air pollutants at global and regional scales during the 1980–2010 period. Clim. Change 109 (1–2), 163–190.

Guidi, L., Nali, C., Lorenzini, G., Filippi, F., Soldatini, G.F., 2001. Effect of chronic ozone fumigation on the photosynthesis process of poplar clones showing different sensitivity. Environ. Pollut. 113, 245–254.

Gupta, S.K., Sharma, M., Majumder, B., Maurya, V.K., Lohani, M., Deeba, F., Pandey, V., 2018. Impact of Ethylene diurea (EDU) on growth, yield and proteome of two winter wheat varieties under high ambient ozone phytotoxicity, 196, 161–173. https://doi.org/10.1016/j.chemosphere.2017.12.150.

Hassan, I.A., Tewfik, I., 2006. CO_2 photoassimilation, chlorophyll fluorescence, lipid peroxidation and yield in cotton (*Gossypium hirsutum* L. cv Giza 65) in response to O_3. World Rev. Sci. Technol. Sustain. Dev. 3, 70–78.

Heun, M., Schäfer-Pregl, R., Klawan, D., Castagna, R., Accerbi, M., Borghi, B., Salamini, F., 1997. Site of einkorn wheat domestication identified by DNA fingerprinting. Science 278, 1312–1314.

IPCC, 2001. Climate Change 2001: The Scientific Basis. Contribution of Working Group I to the Third Assessment Report of the Intergovernmental Panel on Climate Change. Cambridge University Press, Cambridge and New York. https://doi.org/10.1002/joc.763.

IPCC, 2013. Climate Change 2013: The Physical Science Basis. Contribution of Working Group I to the Fifth Assessment Report of the Intergovern-Mental Panel on Climate Change. Cambridge University Press, Cambridgeand New York.

Iqbal, M., Abdin, M., Mahmooduzzafar, Z., Yunus, M., Agrawal, M., 1996. Resistance mechanisms in plants against air pollution. In: Iqbal, M., Yunus, M. (Eds.), Plant Response to Air Pollution. Wiley, New York, pp. 195–240.

Iriti, M., Faoro, F., 2008. Oxidative stress, the paradigm of ozone toxicity in plants and animals. Water Air Soil Pollut. 187, 285–301.

Jantasuriyarat, C., Vales, M.I., Watson, C.J.W., Riera-Lizarazu, O., 2004. Identification and mapping of genetic loci affecting the free threshing habit and spike compactness in wheat (*Triticum aestivum* L.). Theor. Appl. Genet. 108, 261–273.

Kangasjärvi, J., Jaspers, P., Kollist, H., 2005. Signalling and cell death in ozone-exposed plants. Plant Cell Environ. 28, 1021–1036.

Karlsson, P.E., Uddling, J., Skarby, L., Wallin, G., Sellden, G., 2003. Impact of ozone on the growth of birch (*Betula pendula*) seedlings. Environ. Pollut. 124, 485–495.

Karnosky, D., Skelly, J., Percy, K., Chappelka, A., 2007. Prospectives regarding 50 years of research on effects of tropospheric ozone air pollution on US forests. Environ. Pollut. 147, 489–506.

Kerstein, G., Lendzian, K.J., 1989. Interaction between ozone and plant cuticles.1. Ozone deposition and permeability. New Phytol. 112, 1989–2004.

Kihara, H., Yamashita, H., Tanaka, M., 1959. Genomes of 6x species of *Aegilopsis*. Wheat Inf. Serv. 8, 3–5.

Kondratyev, K.Y., Varotsos, C.A., 2001a. Global tropospheric ozone dynamics: part I: tropospheric ozone precursors. Environ. Sci. Pollut. Res. 8 (1), 57–62. https://doi.org/10.1007/BF02987295.

Kondratyev, K.Y., Varotsos, C.A., 2001b. Global tropospheric ozone dynamics: part II: numerical modelling of tropospheric ozone variability. Environ. Sci. Pol. 8 (1), 113–119.

Krupa, S.V., Manning, W.J., 1988. Atmospheric ozone: formation and effects on vegetation. Environ. Pollut. 50, 101–137.

Lagzi, I., Meszaros, R., Gelybo, G., Leelossy, A., 2013. Atmospheric Chemistry. Eötvös Loránd University.

Laisk, A., Kull, O., Moldau, H., 1989. Ozone concentration in leaf intercellular air spaces is close to zero. Plant Physiol. 90, 1163–1167.

Larson, R.A., 1988. The antioxidants of higher plants. Phytochemistry 27, 969–978.
Laurence, J.A., Andersen, C.P., 2003. Ozone and natural systems: understanding exposure, response, and risk. Environ. Int. 29, 155–160.
Lazukin, A.V., Pinchuk, M., Stepanova, O., 2018. Treatment of spring wheat seeds by ozone generated from humid air and dry oxygen. Res. Agric. Eng. 64, 34–40. https://doi.org/10.17221/106/2016-RAE.
Lee, M.J., Kim, M.J., Kwak, H.S., Lim, S., Kim, S.S., 2017. Effects of ozone treatment on physicochemical properties of Korean wheat flour. Food Sci. Biotechnol. 26 (2), 435–440. https://doi.org/10.1007/s10068-017-0059-5.
Li, Y., Li, C., Zheng, Y., Wu, G., Wuyun, T., Xu, H., Jiang, G., 2011. Cadmium pollution enhanced ozone damage to winter wheat: biochemical and physiological evidences. J. Environ. Sci. (China) 23 (2), 255–265.
Li, C., Song, Y., Guo, L., Gu, X., Muminov, M.A., Wang, T., 2018. Nitric oxide alleviates wheat yield reduction by protecting photosynthetic system from oxidation of ozone pollution. Environ. Pollut. 236, 296–303.
Lin, M., Horowitz, L.W., Oltmans, S.J., Fiore, A.M., Fan, S., 2014. Tropospheric ozone trends at Mauna Loa Observatory tied to decadal climate variability. Nat. Geosci. 7, 136–143. https://doi.org/10.1038/ngeo2066.
Liu, F., Wang, X., Zhu, Y., 2009. Assessing current and future ozone-induced yield reductions for rice and winter wheat in Chongqing and the Yangtze River Delta of China. Environ. Pollut. 157, 707–709.
Lu, X., Zhang, L., Shen, L., 2019. Meteorology and climate influences on tropospheric ozone: a review of natural sources, chemistry, and transport patterns. Curr. Pollut. Rep. 5, 238–260. https://doi.org/10.1007/s40726-019-00118-3.
Maggs, R., Wahid, A., Shamsi, S.R.A., Ashmore, M.R., 1995. Effects of ambient air pollution on wheat and rice yield in Pakistan. Water Air Soil Pollut. 85, 1311–1316.
Manning, W.J., 2005. Establishing a cause and effect relationship for ambient ozone exposure and tree growth in the forest: progress and an experimental approach. Environ. Pollut. 137, 443–454.
Mansfield, A.T., Pearson, M., 1996. Disturbance in stomatal behaviour in plants exposed to air pollution. In: Iqbal, M., Yunus, M. (Eds.), Plant Response to Air Pollution. Wiley, Chichester, pp. 178–193.
Matyssek, R., Sandermann, H., 2003. Impact of ozone on trees: an ecophysiological perspective. In: Progress in Botany, vol. 64. Springer-Verlag, Heidelberg, Germany, p. 349.
Mills, G., Sharps, K., Simpson, D., Pleijel, H., Broberg, M., Uddling, J., Jaramillo, F., Davies, W.J., Dentener, F., Van den Berg, M., et al., 2018a. Ozone pollution will compromise efforts to increase global wheat production. Glob. Change Biol. 24, 3560–3574.
Mills, G., Pleijel, H., Malley, C.S., Sinha, B., Cooper, O.R., Schultz, M.G., Neufeld, H.S., Simpson, D., Sharps, K., Feng, Z., Gerosa, G., Harmens, H., Kobayashi, K., Saxena, P., Paoletti, E., Sinha, V., Xu, X., 2018b. Tropospheric ozone assessment report: present-day tropospheric ozone distribution and trends relevant to vegetation. Elem. Sci. Anth. 6 (1), 47. https://doi.org/10.1525/elementa.302.
Mills, G., Sharps, K., Simpson, D., Pleijel, H., Frei, M., Burkey, K., Emberson, L., Uddling, J., Broberg, M., Feng, Z., et al., 2018c. Closing the global ozone yield gap: quantification and cobenefits for multistress tolerance. Glob. Change Biol. 24, 1–26.
Mishra, A.K., Rai, R., Agrawal, S.B., 2013a. Differential response of dwarf and tall tropical wheat cultivars to elevated ozone with and without carbon dioxide enrichment: growth, yield and grain quality. Field Crop. Res. 145, 21–32.

Mishra, A.K., Rai, R., Agrawal, S.B., 2013b. Individual and interactive effects of elevated carbon dioxide and ozone on tropical wheat (*Triticum aestivum* L.) cultivars with special emphasis on ROS generation and activation of antioxidant defence system. Ind. J. Biochem. Bio. 50, 139–149.

Monks, P.S., Archibald, A.T., Colette, A., Cooper, O., Coyle, M., Derwent, R., Fowler, D., Granier, C., Law, K.S., Mills, G.E., Stevenson, D.S., Tarasova, O., Thouret, V., von Schneidemesser, E., Sommariva, R., Wild, O., Williams, M.L., 2015. Tropospheric ozone and its precursors from the urban to the global scale from air quality to short-lived climate forcer. Atmos. Chem. Phys. 15, 8889–8973.

Morris, R., Sears, E.R., 1967. The cytogenetics of wheat and relatives. In: Quisenberry, K.S., Reitz, L.P. (Eds.), Wheat and Wheat Improvement. The American Society of Agronomy, Madison. USA, pp. 19–87.

Murphy, D., 2007. Plant Breeding and Biotechnology: Societal Context and the Future of Agriculture, first ed. Cambridge Univ. Press., U.K.

Nalam, V.J., Vales, M.I., Watson, C.J.W., Kianian, S.F., Riera-Lizarazu, O., 2006. Map-based analysis of genes affecting the brittle rachis character in tetraploid wheat (*Triticum turgidum* L.). Theor. Appl. Genet. 112, 373–381.

Nesbitt, M., 1998. Where was einkorn wheat domesticated? Trends Plant Sci. 3, 1360–1385.

Pandey, A.K., Ghosh, A., Agrawal, M., Agrawal, M., Agrawal, S.B., 2018. Effect of elevated ozone and varying levels of soil nitrogen in two wheat (*Triticum aestivum* L.) cultivars: growth, gas-exchange, antioxidant status, grain yield and quality. Ecotoxicol. Environ. Saf. 158, 59–68.

Paoletti, E., Grulke, N.E., 2005. Does living in elevated CO_2 ameliorate tree response to ozone? A review on stomatal responses. Environ. Pollut. 137, 483–493.

Park, J.I., Grant, C.M., Davies, M.J., Dawes, I.W., 1998. The cytoplasmic Cu, Zn superoxide dismutase of *Saccharomyces cerevisiae* is required for resistance to freeze-thaw stress generation of free radicals during freezing and thawing. J. Biol. Chem. 273, 22921–22928.

Pasqualini, S., Antonielli, H., Ederli, L., Piccioni, C., Loreto, F., 2002. Ozone uptake and its effect on photosynthetic parameters of two tobacco cultivars with contrasting ozone sensitivity. Plant Physiol. Biochem. 40, 599–603.

Pellinen, R., Palva, T., Kangasjarvi, J., 1999. Subcellular localization of ozone-induced hydrogen peroxide production in birch (*Betula pendula*) leaf cells. Plant J. 20, 349–356.

Pleijel, H., Eriksen, A.B., Danielsson, H., Bondesson, N., Sellden, G., 2006. Differential ozone sensitivity in an old and a modern Swedish wheat cultivar grain yield and quality, leaf chlorophyll and stomatal conductance. Environ. Exp. Bot. 56, 63–71.

Polle, A., Wieser, G., Havranek, W.M., 1995. Quantification of ozone influx and apoplastic ascorbate content in needles of Norway spruce trees (*Picea abies* L., *Karst.*) at high altitude. Plant Cell Environ. 18, 681–688.

Puckette, M.C., Weng, H., Mahalingam, R., 2007. Physiological and biochemical responses to acute ozone-induced oxidative stress in *Medicago truncatula*. Plant Physiol. Biochem. 45, 70–79.

Rai, R., Agrawal, M., 2014. Assessment of competitive ability of two Indian wheat cultivars under ambient O_3 at different developmental stages. Environ. Sci. Pol. 21, 1039–1053.

Rai, R., Agrawal, M., Agrawal, S.B., 2007. Assessment of yield losses in tropical wheat using open top chambers. Atmos. Environ. 41, 9543–9554.

Rai, R., Agrawal, M., Agrawal, S.B., 2011. Effects of ambient O_3 on wheat during reproductive development: gas exchange, photosynthetic pigments, chlorophyll fluorescence, and carbohydrates. Photosynthetica 49 (2), 285–294.

Rajaram, S., 2005. Role of conventional plant breeding and biotechnology in future wheat production. Turk. J. Agric. For. 29 (2), 105–111.

Rajaram, S., van Ginkel, M., 1996. In: Reynolds, M.P., Rajaram, S., McNab, A. (Eds.), Increasing the Yield Potential in Wheat: Breaking the Barriers. CIMMYT, Mexico D.F., pp. 11–18

Rao, M.V., Davis, K.R., 1999. Ozone-induced cell death occurs via two distinct mechanisms in *Arabidopsis*: the role of salicylic acid. Plant J. 17, 603–614.

Rao, M.V., Halle, B.A., Ormrod, D.P., 1995. Amelioration of ozone-induced oxidative damage in wheat plants grown under high carbon dioxide. Plant Physiol. 109, 421–432.

REVIHAAP, 2013. Review of Evidence on Health Aspects of Air Pollution – REVIHAAP Project. WHO European Centre for Environment and Health, WHO Regional Office for Europe, Bonn. Available at: http://www.euro.who.int/data/assets/pdf_file/0004/193108/REVIHAAP-Final-technical-report-final-version.pdf.

Richards, B.L., Middleton, J.T., Hewitt, W.B., 1958. Air pollution with relation to agronomic crops. V. Oxidant stipple of grape. Agron. J. 50, 559–561.

Saintanis, C.J., Bari, S.M., Burkey, K.O., Stamatelopoulos, D., 2014. Screening of Bangladeshi winter wheat (*Triticum aestivum* L.) cultivars for sensitivity to ozone. Environ. Sci. Pol. 21 (23), 13560–13571.

Sarkar, A., Agrawal, S.B., 2010a. Identification of ozone stress in Indian rice through foliar injury and differential protein profile. Environ. Monit. Assess. 161, 205–215.

Sarkar, A., Agrawal, S.B., 2010b. Elevated ozone and two modern wheat cultivars: an assessment of dose dependent sensitivity with respect to growth, reproductive and yield parameters. Environ. Exp. Bot. 69, 328–337.

Sarkar, A., Rakwal, R., Agrawal, S.B., Shibato, J., Ogawa, Y., Yoshida, Y., Kumar Agrawal, G., Agrawal, M., 2010. Investigating the impact of elevated levels of ozone on tropical wheat using integrated phenotypical, physiological, biochemical, and proteomics approaches. J. Proteome Res. 9, 4565–4584.

Sax, K., 1921. Chromosome relationships in wheat. Science 54, 4113–4415.

Sax, K., 1927. Chromosome behaviour in *Triticum* hybrids. Verhandlungen des V Int. Kongresses für Vererbungswissenschaft (Berlin) 2, 1267–1284.

Sears, E.R., 1948. The cytology and genetics of the wheats and their relatives. Adv. Genet. 2, 239–270.

Sears, E.R., 1969. Wheat cytogenetics. Annu. Rev. Genet. 451–468.

Serbinova, E.A., Packer, L., 1994. Antioxidant properties of α-tocopherol and α tocotrienol. Methods Enzymol. 234, 354–366.

Shewry, P.R., 2009. J. Wheat. Exp. Bot. 60 (6), 1537–1553. https://doi.org/10.1093/jxb/erp058.

Shiferaw, B., Smale, M., Braun, H., Muricho, G., Smale, M., Shiferaw, B., Duveiller, E., Reynolds, M., Duveiller, E., Reynolds, M., Muricho, G., 2013. Crops that feed the world 10. Past successes and future challenges to the role played by wheat in global food security. Food Sec. 5, 291–317.

Simons, K.J., Fellers, J.P., Trick, H.N., Zhang, Z., Tai, Y.S., Gill, B.S., Faris, J.D., 2006. Molecular characterization of the major wheat domestication gene Q. Genetics 172, 547–555.

Singh, S., Agrawal, S.B., 2009. Use of ethylene diurea (EDU) in assessing the impact of ozone on growth and productivity of five cultivars of Indian wheat (*Triticum aestivum* L.). Environ. Monit. Assess. 159, 125–141.

Singh, S., Agrawal, S.B., 2010. Impact of tropospheric zone on wheat (*Triticum aestivum* L.) in the Eastern Gangetic plains of India as assessed by ethylene diurea (EDU) application during different developmental stages. Agric. Ecosyst. Environ. 138, 214–221.

Singh, P., Agrawal, M., Agrawal, S.B., 2009a. Evaluation of physiological, growth and yield responses of a tropical oil crop (*Brassica campestris* L. var. *Kranti*) under ambient ozone pollution at varying NPK levels. Environ. Pollut. 157, 871–880.

Singh, S., Agrawal, S.B., Agrawal, M., 2009b. Differential protection of ethylene diurea (EDU) against ambient ozone for five cultivars of tropical wheat. Environ. Pollut. 157, 2359–2367.

Singh, S., Agrawal, M., Agrawal, S.B., Emberson, L., Büker, P., 2010. Use of ethylene diurea for assessing the impact of ozone on mung bean plants at a rural site in a dry tropical region of India. Int. J. Environ. Waste Manag. 5 (1/2), 125–139.

Singh, A.A., Agrawal, S.B., Shahi, J.P., Agrawal, M., 2014. Assessment of growth and yield losses in two *Zea mays* L. cultivars (quality protein maize and non quality protein maize) under projected levels of ozone. Environ. Sci. Pol. 21 (4), 2628–2641. https://doi.org/10.1007/S11356-013-2188-6.

Singh, E., Rai, R., Pandey, B., Agrawal, M., 2017. Development of resistance in two wheat cultivars against constant fumigation of ozone. Proc. Natl. Acad. Sci. India B Biol. Sci. https://doi.org/10.1007/s40011-017-0849-9.

Singh, A.A., Fatima, A., Mishra, A.K., Chaudhary, N., Mukherjee, A., Agrawal, M., Agrawal, S.B., 2018. Assessment of ozone toxicity among 14 Indian wheat cultivars under field conditions: growth and productivity. Environ. Monit. Assess. 190, 190. https://doi.org/10.1007/s10661-018-6563-0.

Smirnoff, N., 2000. Ascorbate biosynthesis and function in photoprotection. Philos. Trans. R. Soc. Lond. B Biol. Sci. 355, 1455–1464.

Swaminathan, M.S., 2006. An EverGreen revolution. Crop Sci. 46, 2293–2303.

Tai, A.P., Martin, M.V., 2017. Impacts of ozone air pollution and temperature extremes on crop yields: spatial variability, adaptation and implications for future food security. Atmos. Environ. 169, 11–21.

Tang, H.Y., Liu, G., Han, Y., Zhu, J.G., Kobayashi, K., 2011. A system for free-air ozone concentration elevation with rice and wheat: control performance and ozone exposure regime. Atmos. Environ. 45, 6276–6282.

Teixeira, E., Fischer, G., van Velthuizen, H., et al., 2011. Limited potential of crop management for mitigating surface ozone impacts on global food supply. Atmos. Environ. 45, 2569–2576.

Thompson, A.M., 1992. The oxidizing capacity of the Earth's atmosphere. Probable past and future changes. Science 256, 1157–1165.

Tiwari, S., Agrawal, M., 2018. Ozone concentrations in troposphere: historical and current perspectives. In: Tropospheric Ozone and its Impacts on Crop Plants. https://doi.org/10.1007/978-3-319-71873-6_1.

Tiwari, S., Agrawal, M., Manning, W.J., 2005. Assessing the impacts of ambient ozone on growth and productivity of two cultivars of wheat in India using three rates of applications of ethylene diurea (EDU). Environ. Pollut. 138, 153–160.

Tiwari, S., Rai, R., Agrawal, M., 2008. Annual and seasonal variations in tropospheric ozone concentrations around Varanasi. Int. J. Rem. Sens. 9 (15), 4499–4514.

Tripathi, R., Agrawal, S.B., 2013. Interactive effect of supplemental ultraviolet-B and elevated ozone on seed yield and oil quality of two cultivars of linseed (*Linum usitatissimum* L.) carried out in open top chambers. J. Sci. Food Agric. 93, 1016–1025.

Van Dingenen, R., Dentener, F.J., Raes, F., Krol, M.C., Emberson, L., Cofala, J., 2009. The global impact of ozone on agricultural crop yields under current and future air quality legislation. Atmos. Environ. 43 (3), 604–618.

Varotsos, C., Cartalis, C., Vlamakis, A., Tzanis, C., Keramitsoglou, I., 2004. The long-term coupling between column ozone and tropopause properties. J. Clim. 17, 3843–3854.
Wahid, A., 2006. Productivity losses in barley attributable to ambient atmospheric pollutants in Pakistan. Atmos Environ. 40, 5342–5354.
Wahid, A., Maggs, R., Shamsi, S.R.A., Bell, J.N.B., mad Ashmore, M.R., 1995. Air pollution and its impacts on wheat yield in the Pakistan Punjab. Environ. Pollut. 88, 147–154.
Wahid, A., 2006. Influence of atmospheric pollutants on agriculture in developing countries: A case study with three new wheat varieties in Pakistan. Sci. Total Environ. 371, 304–313.
Wang, X., Zheng, Q., Yao, F., Chen, Z., Feng, Z., Manning, W.J., 2007. Assessing the impact of ambient ozone on growth and yield of rice (*Oryza sativa* L.) and wheat (*Triticum aestivum* L.) cultivar grown in Yangtze Delta, China, using three rates of application of Ethylene diurea (EDU). Environ. Pollut. 148, 390–395.
Wang, X., Zhang, Q., Zheng, F., Zheng, Q., Yao, F., Chen, Z., Zhang, W., Hou, P., Feng, Z., Song, W., Feng, Z., Lu, F., 2012. Effects of elevated O_3 concentration on winter wheat and rice yields in the Yangtze River Delta, China. Environ. Pollut. 171, 118–125.
Wang, J., Zeng, Q., Zhu, J., Chen, C., Liu, G., Tang, H., 2014a. Apoplastic antioxidant enzyme responses to chronic free-air ozone exposure in two different ozone-sensitive wheat cultivars. Plant Physiol. Biochem. 82, 183–193.
Wang, J., Zhu, J., Zeng, Q., Liu, G., 2014b. Phenolic compounds and antioxidant ability responsesto experimental free-air ozone exposure in two wheat cultivars. Bull. Environ. Contam. Toxicol. 93, 625–631. https://doi.org/10.1007/s00128-014-1375-6.
Wild, O., Fiore, A.M., Shindell, D.T., Doherty, R.M., Collins, W.J., Dentener, F.J., Schultz, M.G., Gong, S., MacKenzie, I.A., Zeng, G., Hess, P., Duncan, B.N., Bergmann, D.J., Szopa, S., Jonson, J.E., Keating, T.J., Zuber, A., 2012. Modelling future changes in surface ozone: A parameterized approach. Atmos. Chem. Phys. 12, 2037–2054.
Wohlgemuth, H., Mittelstrass, K., Kschieschan, S., Bender, J., Weigel, H.J., Overmyer, K., Kangasjarvi, J., Sandermann, H., Langebartels, C., 2002. Activation of an oxidative burst is a general feature of sensitive plants exposed to the air pollutant ozone. Plant Cell Environ. 25, 717–726.
Zhang, Y., Cooper, O.R., Gaudel, A., Thompson, A.M., Nedelec, P., et al., 2016. Tropospheric ozone change from 1980 to 2010 dominated by equatorward redistribution of emissions. Nat. Geosci. 9 (12), 875–879.
Zhao, H., Zheng, Y., Wu, X., 2018. Assessment of yield and economic losses for wheat and rice due to ground level O_3 exposure in the Yangtze River Delta, China. Atmos. Environ. 191, 241–248.
Zheng, Z., Xiaoke, W., Zhang, W., Peiqiang, H., Lu, F., Du, K., Sun, Z., 2013. Effects of elevated O_3 exposure on nutrient elements and quality of winter wheat and rice grain in Yangtze River Delta, China. Environ. Pollut. 179, 19–26.
Zhu, X., Feng, Z., Sun, T., Liu, X., Tang, H., Zhu, J., Guo, W., Kobayashi, K., 2011. Effects of elevated ozone concentration on yield of four Chinese cultivars of winter wheat under fully open-air field conditions. Glob. Change Biol. 17, 2697–2706.
Zhu, Z., Sun, X., Zhao, F., Meixner, F.X., 2015. Ozone concentrations, flux and potential effect on yield during wheat growth in the Northwest-Shandong Plain of China. J. Environ. Sci. 34, 1–9.

Past and present events of climate change: natural versus anthropogenic causes

12

Gereraj Sen Gupta, Supriya Tiwari
Department of Botany, Institute of Science, Banaras Hindu University, Varanasi, Uttar Pradesh, India

1. Introduction

The Earth is the only known planet to inhabit life and life forms due to the presence of appropriate environmental conditions, and today's environmental condition is thought to have evolved and evolving since the accretion of Earth at about 4.5 billion years ago (Drake and Righter, 2002). Interestingly, change in the Earth's climate has been observed to be very fast since the onset of the industrial revolution (Hubbe and Hubbe, 2019). The aforementioned climatic condition specifically refers to the long-term average of various meteorological conditions such as temperature, precipitation, atmospheric pressure, humidity, and wind. In a broader sense, the climate comprises not just the mean conditions but also the integrated statistics (magnitude, persistence, frequency, trends, etc.), often collaborating parameters to express events such as droughts (IPCC, 2013). Along timescale, the atmospheric conditions have been changing largely due to various natural phenomena occurring across the globe such as volcanic eruptions, plate tectonics, solar output variations, atmospheric carbon dioxide variations, variation in the Earth's orbital characteristics, and thermohaline circulation (Nwankwoala, 2015), thereby changing the climatic conditions of the Earth. This ever-changing entity over a large area in a specific period is called as climate change. As such, variations in the climatic condition over small intervals, i.e., shorter than a decade or so does not represent climate change, e.g., El Niño effects (Nwankwoala, 2015). According to the Intergovernmental Panel for Climate Change (IPCC, 2013), "the change in the state of the climate that can be identified (e.g., by using statistical tests) by changes in the mean and/or the variability of its properties, and that persists for an extended period, typically decades or longer" is defined as climate change. The climate system of the Earth is mostly governed by the radiations coming from the Sun, and approximately half of the solar energy is contributed to the visible part of the electromagnetic spectrum (IPCC, 2013). Although the Earth's climatic system is largely governed by the energy from the Sun, its variation has played a minute role in the climate change since the past few decades. According to the Royal Society (2020), direct satellite measurements have shown no net increase in output of the solar radiation since the late 1970s, while at the same time huge increments have been

Global Climate Change. https://doi.org/10.1016/B978-0-12-822928-6.00014-9

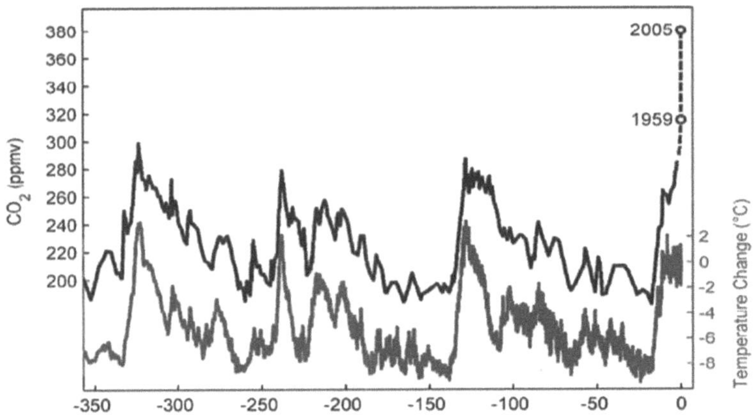

Figure 12.1 Trends in atmospheric carbon dioxide (CO_2) concentration of the past 350,000 years.
Source: Petit, J.R., et al., June 1999."Climate and atmospheric history of the past 420,000 years from the Vostok Ice Core, Antarctica." Nature 399, 429–36. Based on Vostok ice core in Antarctica. These ice core data end before the onset of anthropogenic changes. Anthropogenic emissions have now increased the CO_2 concentration to 380 ppm (as of 2005).

observed in the global surface temperatures. This abrupt increase in the global surface temperature is attributed to increase in CO_2 concentrations led by human activities. The European Project for Ice Coring in Antarctica (EPICA), a body that provides a very accurate and reliable data for CO_2 concentration, depicted an unusual and unprecedented increase in CO_2 level in the past hundreds and thousands of years (Fig. 12.1).

In the early 1960s, various physical and mathematical models have predicted a decrease in the temperature in stratosphere and a gradual increase in the tropospheric temperature triggered by human-induced CO_2 production. Furthermore, simulation studies based on computer model have shown that the tropospheric warming and subsequent cooling of the stratosphere since the past 40 years have a positive correlation with human-induced increase in atmospheric CO_2 concentration and decrease in stratospheric ozone (O_3), respectively (Royal society, 2020).

The change in climatic conditions due to unprecedented increase in the CO_2 concentration in the atmosphere and consequent elevation in atmospheric temperature have negatively affected all forms of lives on the Earth. Climate change has reshuffled various community compositions, caused shifts in species distributions (Parmesan and Yohe, 2003), and promoted the survival of the warm-adapted species, the phenomenon known as "thermophilization" (Brice et al., 2019). According to VanDerWal et al. (2013), the climate change has not only forced upward range shifts (in elevation) and poleward migration of species, rather the magnitude, direction, and timing of shifts in the species varied considerably among regions and among different taxa. Climate change in terms of temperature increase has shown adverse effects on invertebrates (Crawford, 1981), reptiles (Sinervo et al., 2010), birds (du Plessis et al., 2012), and many mammalian species (Moses et al., 2012). Decline in the population of different species and species extinction have become a common phenomenon in recent world.

It has become well known that climate change poses all round threat on all different life forms including human race, leading to either extinction or migration of the same from one place to another in search of a better habitat for survival. Humans, though an extremely powerful creation of God among all life forms, also migrated in search of better climatic condition on Earth, which can be better understood by a word called climate refugees. Comenetz and Caviedes (2002) stated that due to drastic change in the climatic condition, whole humanity, particularly those communities that are globally vulnerable and suffered from famine and drought conditions, showed an unusual displacement in their population. A paper published on the United Nations Environment Programme (UNEP) in 1985 by El-Hinnawi, where he described climate refugees as "people who have been forced to leave, temporarily or permanently, from their traditional or original habitat, because of a striking environmental disruption (natural and/or anthropogenic) that threatens their existence and/or solemnly affects the standard and grade of their life." Decades after El-Hinnawi, the climate refugee concept was also supported by Meyers in 1997, where he stated that "those who have no option other than to flee at any new place from their own territory, terrorized by climatic hazards such as drought or deforestation, with few exceptions to return back to their own homeland" are termed as climate refugees. Geographically, whole world has been categorized into various sectors according to their climate specificity, viz., rain forests, temperate region, tundra regions, etc. Similarly, climate has climatic extremes such as deserts, particularly known for low precipitation and dry weather. According to Iknayan and Beissinger (2018), the desert in the United States experienced huge decline in Mojave desert birds occupancy and lost half of the total habitat due to climate change events; particularly due to precipitation change patterns, further decline has been estimated in the avian community by the end the century as a consequence of accelerated climate change. The impacts of climate change cover all sectors of the ecosystem, including human activities, and the relationship between the climate change and economic index has long been under discussion in relation to resulting costs and benefits (Acevedo et al., 2017; Dell et al., 2014; McCarthy et al., 2001; Pearce et al., 1996; Watson et al., 1996). A very recent study by Li et al. (2015) and Jones and Olken (2010) found that the climate change has disfigured the overall pattern of international trade, whereas Barua and Valenzuela (2018) found agricultural activities, their products, and thereby the agricultural trade to be affected the most due to its greater sensitivity toward climate change.

There have been many events in the Earth's history that became the evidence of how the climate variables have changed in ancient times and the change that has occurred within few centuries. These changes have been well established by using various models to study the physical responses such as surface temperature, precipitation, atmospheric water vapor, ocean and land ice, glaciers, sea level and other severe events. The concept of "radiative forcing" has also been studied extensively to understand the climate variables, which are further found to be insufficient to understand the whole mechanism of climate change variables. Furthermore, IPCC (2013) has introduced a new concept called "effective radiative forcing" (ERF) that accounts for rapid response for climate system. ERF can be defined as "the change in net downward flux at the top of the atmosphere after allowing for atmospheric temperatures, water vapor, clouds and land albedo to adjust, but with either sea surface temperatures (SSTs) and sea ice cover unchanged or with global mean surface temperature unchanged." In relation to the simulations made on future climatic conditions, Matthews and Weaver (2010) and Wigley (2005) have found that even if the emissions caused by anthropogenic means were stopped or if the climate forcings were fixed at their present values, respectively, the adverse effects on climate will continue to occur until it comes to equilibrium state with those forcing.

2. Evidence of climate change

The impact of natural or human-induced changes in the physical, chemical, and biological Earth system can be well acknowledged and understood using various methods and techniques. The technological advancements enabled us to visualize the physical changes on the Earth that are the indicators or evidence reflecting the actual impact of climate change. The loss of polar ice sheets and mountain glaciers, ocean acidification, change in average surface temperature, and precipitation pattern and sea level rise produce full length evidence of concerned climate change.

2.1 Loss of polar ice sheets and mountain glaciers

The presence of ice covers on the mountains is of much significance. They help in reflecting back the harmful UV-B radiations coming from the sun rays, protecting various life forms on the Earth. The polar ice sheets of Greenland (North pole) and Antarctica (South pole) together comprise 96% of total world's land ice. Drastic loss in ice sheet cover has been observed in both the polar regions since the past century. It is worth noting that the ice sheets have slower melting rate as compared with mountain glaciers due to its less surface area to volume ratio. Still, a high rate of melting has been observed in Greenland ice sheets, increasing the number of lakes and the size of prevailing lakes in southern part of the ice sheets. Das et al. (2008) and Zwally et al. (2002) separately observed crevices in the ice sheets, called "moulins" that act as a passage for entry of melted water to the bottom of ice sheets, thus lubricating the lower part and helping their easy travel to the sea. Thompson in 2010 reported that an iceberg from the "Petermann" glacier of size four times greater that Manhattan broke off in Greenland within few days of his paper in the press (the cited paper, "Thompson et al., 2010"). Half of the total loss in ice sheets each year in Greenland comes from calving from their margins, which alone does not explain the climate change due to warming; rather it is coupled with the high rate of increase in global surface temperature within past few decades. Apart from the evidence showing effect of climate change at the North pole, the icebergs of Antarctica (South pole) are also experiencing huge downturn in its size and number since the past few decades. In 1978, a glacial geologist, late John Mercer, of Ohio State University predicted unfortunate breakdown of huge Antarctic ice shelves, due to global warming caused by abrupt increase in the atmospheric carbon dioxide (CO_2). The Antarctic peninsula experienced a massive increase in mean temperature by 2.5°C (4.5°F) in the past 50 years, which consequently broke down the ice shelves in a very similar way that Mercer predicted. Among all the ice sheets deteriorations, most rapid and historical downfall was observed during 2002 when an iceberg named "the Larsen B" having depth of 200 m covering an area the size of Rhode Island fell down within 31 days.

The ice covers at the North and the South pole serves as major reflecting bodies that reflects the harmful radiation striking the earth. Nonetheless, the protective ice covers of Arctic and Antarctic sea cannot be overlooked. The sea ice has been highly reflective and thus of high importance. The Arctic sea ice cover area is limited by surrounding land masses yet extends its ice cover freely during the winter season and periodically melts during the summer; in contrary, the Antarctic sea ice forms lesser ice cover as it stretches near the equatorial regions. Apart from the geographical positioning of Arctic and Antarctic Sea, the ocean currents, winds, and the temperature also play a vital role in forming the sea ice in both the regions. A very recent study has revealed that the Arctic ice cover has been decreasing more rapidly (Fig. 12.2) as compared with Antarctic ice although the Arctic Sea has better geographical positioning for higher ice cover formation. According to the NAS and Royal Society (2020), the aforementioned anomaly is due to the fact that the sea ice in the partially enclosed Arctic oceans is found to be responding directly to the warming phenomenon; however, the Antarctic Sea ice seems to be responding more to the changes in winds and to the ocean currents. Various simulation studies have predicted an increase in sea level by 64 m if all these ice sheets and icebergs melt (Church et al., 2001; Lemke et al., 2007), and further rise would be observed in world's ocean level due to thermal expansion caused by climate warming.

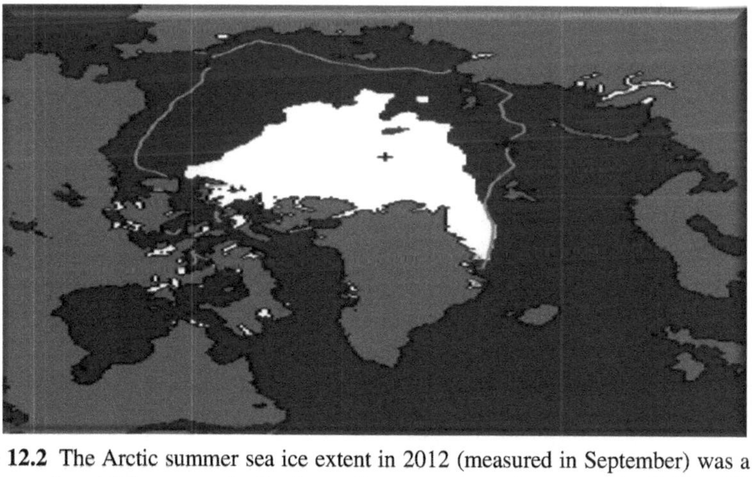

Figure 12.2 The Arctic summer sea ice extent in 2012 (measured in September) was a record low, shown (in white) compared with the median summer sea ice extent for 1979 to 2000 (in orange [light gray in printed version] outline). In 2013, Arctic summer sea ice extent rebounded somewhat but was still the sixth smallest extent on record.
Modified from the data of "National Snow and Ice Data Center".

Modern era of satellite imagery study has not only explored the upsets of polar ice sheets but also revealed the feeble condition of glaciers present on the mountain ranges. Maurer et al. (2019), in his study using the satellite images and digital elevation models, explored that the glaciers stretching 2000 km transect along the Himalayan mountain ranges have undergone depletion of -0.22 ± 0.13 m of water equivalent per year between 1975 and 2000, which has nearly doubled its value by approximately two times the loss encountered during the year 2000–16, i.e., -0.43 ± 0.14 m of water equivalent per year. This shrinkage of Himalayan glaciers has not only challenged the existence of human society rather poses difficulty for the policy makers related to the risk of outburst floods due to increase in the proglacial lakes and seasonal runoff especially in the regions where the climatic region is comparatively drier such as in the western regions.

2.2 Ocean acidification

Ocean is considered as the largest habitat for aquatic organisms that helps in maintaining food chain and thereby stability in oceanic ecosystem. It also acts as buffer system that helps in neutralizing the surplus atmospheric gases, thus balancing the gaseous composition of the atmosphere. Studies have revealed an unprecedented increase in the atmospheric carbon dioxide (CO_2) concentration by approximately 80 ppm as compared with the concentration prevailed 74,000 years back (EPICA community members, 2004; Petit et al., 1999). The unanticipated increase in atmospheric CO_2 concentration, largely induced by anthropogenic means, has led to an absorption of surplus CO_2 inside the oceanic water, increasing the CO_2 concentration at the ocean surface and consequently decreasing the overall pH of the ocean leading to oceanic water acidification (Fig. 12.3). The acidification largely depends on the mechanism of production of a weak acid such as carbonic acid. The formation of carbonic acid takes place through the following steps:

i. $CO_2 + H_2O \rightarrow HCO_3^- + H^+$
ii. $H^+ + CO_3^{2-} \rightarrow HCO_3^-$

Since the past century, the average global oceanic surface temperature was found to increase by 0.74°C, and the increase was found parallel to the increase of atmospheric CO_2 concentration, as a consequence of which the sea level rose by 17 cm (IPCC, 2007). Further reports of IPCC have revealed that the sea water carbonate concentration has depleted by ~30 μmol/kg, and 0.1 units of pH have decreased since the 20th century. It is worth noting that the carbonic acid is not only responsible for decreasing the oceanic pH but also directly involved in decomposing carbonate rich entities. Out of total anthropogenic emissions of CO_2, one-fourth parts (2.2 Pg C/year) were estimated to enter the ocean that readily reacts with the water (H_2O) to form weak carbonic acid that quickly dissociates into bicarbonate (HCO_3^-) ion and hydrogen ion (H^+), which further reacts with the carbonate ($^{2-}CO_3$) ion and forms HCO_3^- (equation i and ii), thus reducing the availability of carbonate ions to biological systems

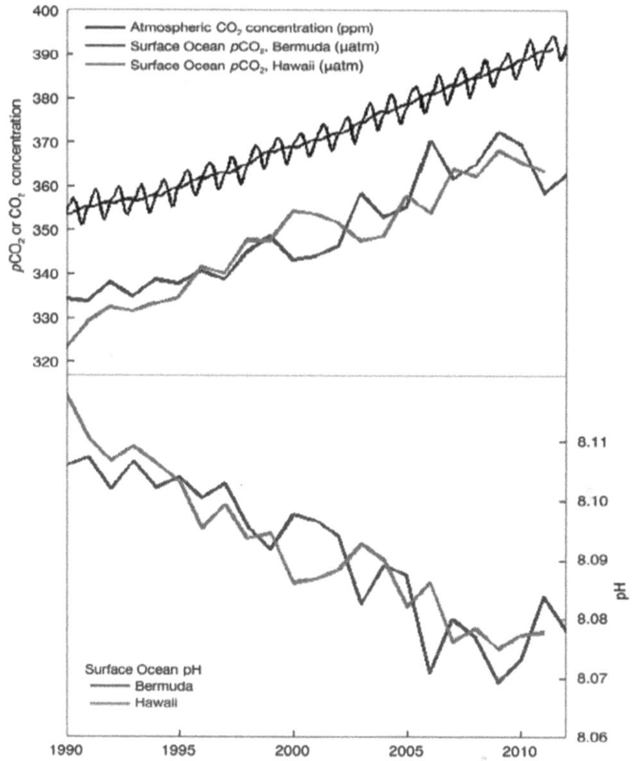

Figure 12.3 Rise in oceanic surface CO_2 concentration and consequent decrease in pH depicted in upper and lower graph, respectively, for more than past two decades.
From Bates et al. (2012) and Dore et al. (2009).

(Fig. 12.4). Marine organisms such as reef-building corals reflect utmost vulnerability as decrease in carbonate-ion concentrations reduces the rate of calcification in them, which eventually leads to erosion at ~200 mmol/kg seawater (Raven et al., 2005; Kleypas et al., 1999).

The shallow region of the ocean is largely ornamented with the presence of coral reefs that have high economic importance and are among highly biologically diverse organisms that provide great extent of economic and ecological services to human society. It also extends its benefits to fishery industries, production of new biochemical compounds, building materials, tourism, and many more (Moberg and Folke, 1999). Although corals are beneficial to various sectors, their ultimate fate has been quite miserable as a consequence of decrease in oceanic water pH. The present scenario explains that the atmospheric CO_2 has undergone such drastic change that the genotype and phenotype of coral reefs do not appear to face the sudden change and their adaptation to the changed environment becomes almost impossible. High CO_2

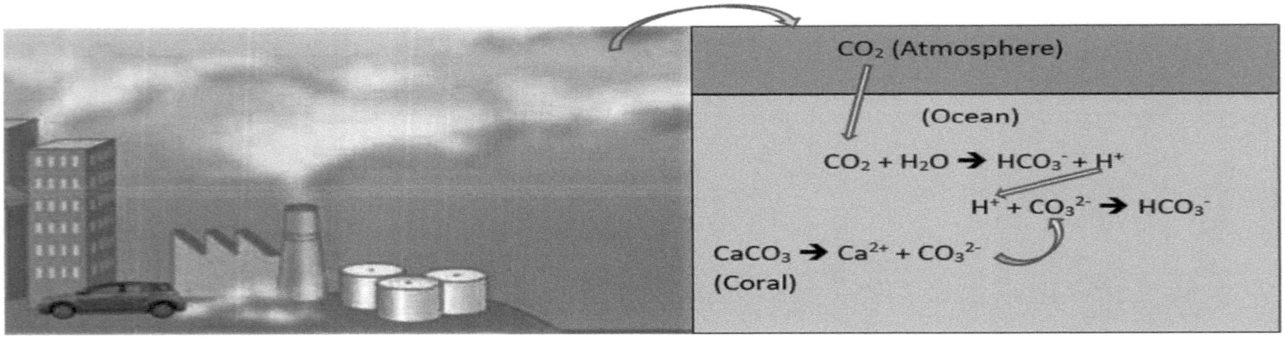

Figure 12.4 Liberation of CO_2 from various sources to the atmosphere, which dissolves in ocean water, ultimately leading to destroy the calcification capability among corals.

and low-pH-induced decrease in calcification process in coral reef have led to three different fates, as shown in the following.

Firstly	Secondly	Thirdly
The coral reefs encountered a huge decrease in their skeletal density and rate of the linear extension. Reports have shown that the Great Barrier Reef containing massive coral Porites have undergone decrease in skeletal density and in their linear extension by 0.36% and 1.02% per year (Guldberg et al., 2007).	For an instance, corals may succeed to maintain their linear extension rate but at the cost of skeletal density. Thus, their skeletal density reduces and in turn become more vulnerable and an easy food for parrotfish that prefers to remove carbonates from low density substrates. Increasingly fragile skeletons become more susceptible to storm damage (Szmant and Gassmamn, 1990) and unable to protect the coastal areas from wave	In few circumstances, they manage to sustain both, their skeletal density and growth by allocating more energy towards calcification. As a consequence, the energy to be utilized for other essential processes such as reproduction are challenged (Szmant and Gassmamn, 1990) which ultimately hampers the larval output from reefs and reduces their ability to recolonization after any disturbance.

2.2.1 Reduced calcification

Ocean acidification has brought some coral communities near verge of destruction and alternatively replaced by other communities that have never existed before. Hughes et al. (2003) explained the "ecological resilience," which describes the ability of an ecosystem to return to its original condition after a period of disturbance. Present-day changes in the scale and frequency of disturbances such as mass coral bleaching and destructive fishing along with decrease in the ability of corals to compete and grow are turning the coral reef ecosystems to algal-dominated states (Hughes et al., 2003). If it is pushed further, it may reach to a condition, often making it difficult to return from algal to a coral-dominated state. Three scenarios can be visualized because of rise in oceanic temperature and CO_2 (Figs. 12.4 and 12.5C). According to Guldberg et al. (2007), the three projected scenarios have been estimated showing the effect of rise in oceanic temperature and CO_2 levels at the Great Barrier Reef. The unfortunate transition from well-organized reef ecosystem (Fig. 12.5A) to invasion of algal communities in them (Fig. 12.5B) reveals the ill effects of unhealthy conditions of oceanic water, which finally leads to the condition reaching the third scenario having lost communities of coral reefs (Fig. 12.5C). Various laboratory experiments show that under acidic waters and high CO_2 concentration, some of the marine species have deformed shells along with decelerated rates of overall growth. According to the NAS and Royal Society (2020), the ocean acidification also hampers the normal cycling of important nutrients including many other essential compounds and elements present in the ocean, which probably shifts the advantageous environment from one species to another due to competition, which may disturb the marine food web.

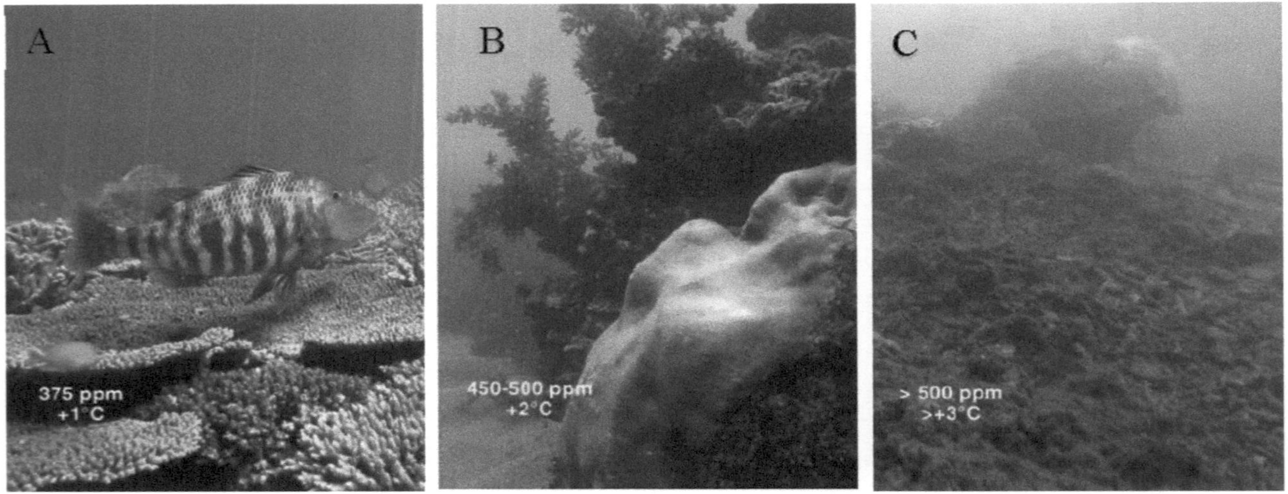

Figure 12.5 Reef communities of Great Barrier Reef at three different time scale. (A) Reef community along with other plants and fishes' dependent on corals during 375 ppm CO_2 and 1°C rise in temperature. (B) Bleached communities of corals partially replaced by algal colonies at 500 ppm CO_2 and 2°C rise in temperature. (C) Coral ecosystem with crumbling frameworks and calcareous corals at more than 500 ppm and above 3°C.
Modified from Guldberg, O.H., Mumby, P.J., Hooten, A.J., Steneck, R.S., Greenfield, P., Gomez, E., Harvell, C.D., Sale, P.F., Edwards, A.J., Caldeira, K., Knowlton, N., Eakin, C.M., Iglesias-Prieto, R., Muthiga, N., Bradbury, R.H., Dubi, A., Hatziolos, M.E., 2007. Coral reefs under rapid climate change and ocean acidification. Science (318), 14.

2.3 Change in average surface temperature and precipitation pattern

Long-term laboratory experiments have enabled the scientists to understand the physics of how the energy balance of the Earth is affected by the CO_2 in the atmosphere. Further evidence using modern technologies of satellite imagery and surface observations of absorption and emission of infrared energy by the atmosphere have strengthened the energy balance mechanism of the Earth. Since the past many decades, the unprecedented increase in CO_2 concentration has led to increased trapping of heat, which consequently rises the average surface and oceanic temperature. Interestingly, at a wavelength of 15 μm, the CO_2 has the strongest heat trapping capacity and can further extend out a few micrometers on both the sides. As CO_2 concentrations increase, the absorption at the center of the strong band is already so high that it plays a little role in causing additional warming. However, more energy is absorbed in the weaker bands and away from the center of the strong band, causing the lower atmosphere and the surface to warm further. According to Royal Society and NAS (2020), since 1900, the Earth's average surface air temperature have increased by 1.8°F or 1°C (Fig. 12.6), and almost half of the total increase was experienced during the mid-1970s. As a consequence, the increase in oceanic heat, reduction in the sea ice cover in Greenland and Antarctic regions, and poleward shift of various naturally occurring organisms such as insects, fishes, mammals, birds, and so on have been observed that altogether depicts the incontrovertible evidence of planetary-scale warming. The wide-scale warming of the Earth's atmosphere has not only affected the ice

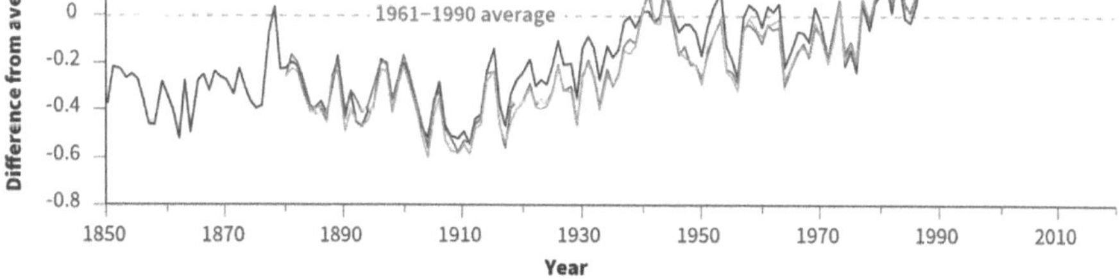

Figure 12.6 Study showing rise in global average surface temperature, as estimated by three independent measuring bodies. The data sets depict combined measurements of ocean and land from 1850 to 2019. The above changes in temperature are shown with respect to the global average surface temperature of 1961–1990.
Source: NOAA Climate. gov; data from UK Met Office Hadley Center (maroon), US National Aeronautics and Space Administration Goddard Institute for Space Studies (red), and US National Oceanic and Atmospheric Administration National Centers for Environmental Information (orange).

covers but also dehydrated the Earth's surface, leading to surface drying owing to high rate of evaporation, and as a consequence, the intensity and duration of drought have increased by many folds (Trenberth, 2011). Planetary warming coupled with high evaporation rate essentially increased the water holding capacity of the air, which can be equated as 7% rise in water holding capacity of the air, per degree increase in warming, ultimately leading to increase in water vapor in the atmosphere. Subsequently, the thunderstorms, snowstorms, or extratropical rain, coupled with the increased moisture content, have produced more intense precipitation events (Trenberth, 2011). The normal atmospheric and the surface energy budget have more influence on the hydrological cycle and the total precipitation change than to the total column water vapor (TCWV). However, according to Trenberth and Shea (2005), a strong correlation of TCWV (also called as precipitable water) has been observed separately with the sea surface temperature (over the oceans) and the precipitation mean annual cycle, as far as their variation and pattern over time (the month of January and July) is concerned (Fig. 12.7). The change in the precipitation pattern along the history of humankind till date has not experienced much variation from its mean value, and hence, a slower rate of change in the total precipitation pattern as compared with total column water vapor has been observed. Other climatic changes including the change in wind pattern also have modest influence on the total

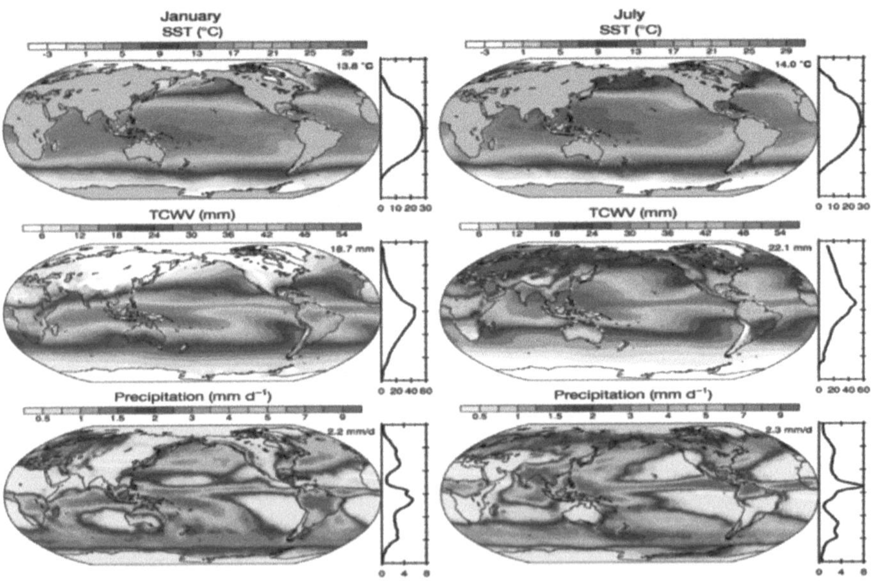

Figure 12.7 Representation of global mean sea surface temperature (SST) (at the top) followed by total column water vapor (TCWV) and total precipitation during the month of January (left side) and July (right side). Values at the top right represent global mean, and graphs at the right represent zonal means.
Source: Smith et al. (2008), Trenberth and Fasullo (2007b) and Huffman et al. (2009) for SST (between years 1979–2008), TCWV (between years 1987–2008) and Total precipitation (between years 1979–2008) respectively.

precipitation change yet in long term, it results in the dry areas becoming drier near the subtropical regions and wet areas wetter, near mid to high latitude regions, thereby justifying the phrase "poor becoming poorer and rich becoming richer." Simulation studies have projected various scenarios with evidence and ascertained that with the increase in warming, largest portion of precipitation occurs in the form of rain instead of snowfall, and thus with increased warming potential of water droplets, the snow cover melts earlier, thereby increasing the runoff and risk of flooding in many areas, particularly in the continental regions.

The climate change has deciphered variable impacts on humans and the nature, and evidently, the effects are increasingly apparent considering the unmatched frequencies of heat waves, flooding, and wildfires, which have cost billions in damages. The Royal Society and NAS (2020) have already depicted that the warming of the Earth's surface, and changes in precipitation pattern have forced various habitats to shift and altered the geographical ranges of many animal and plant species and have negatively changed their long-adjusted timing of their life cycle. In addition to the effects on climate, some of the excess CO_2 in the atmosphere is being taken up by the ocean, changing its chemical composition (causing ocean acidification). A warming atmosphere is also associated with heavier precipitation events (rain and snowstorms) through increases in the air's capacity to hold moisture. El Niño events favor drought in many tropical and subtropical land areas, whereas La Niña events promote wetter conditions in many places. These short-term and regional variations are expected to become more extreme in a warming climate.

2.4 Sea level rise

Abrupt increase in the global mean seal level over the past few decades declared it to be an important evidence of climate change (IPCC, 2013). Rise in the sea level is an interactive effect of land ice melt and the warming of ocean. Previously, for 150 years, the direct observations of rise in global mean sea level were precisely done with the help of tide gauges, which was later replaced by the satellite radar altimeters that measured with higher accuracy. The authentic measurement using tide gauge showed 1.7 ± 0.2 mm/year rate of increase in global mean sea level during the early and middle of the 20th century (Church and White, 2011) that eventually increased to 3.2 ± 0.4 mm/year since 1990 due to the inputs from land ice melts and increased thermal expansion (Church and White, 2011; IPCC, 2013). According to the US National Academy of Sciences and the Royal Society (2020), the rise of global mean sea level during the past decade (1990–2000) was 0.14 inches per year or 3.6 mm/year, and this increased rate of expansion in mean sea level was estimated using the "altimetry" from space during the year 1992. The overall observed rise in global mean sea level since 1902 was about 6 inches (16 cm approximately), and surprisingly, this rise in mean sea level was found predominantly due to the human-caused warming since 1970 (Fig. 12.8). The sea level rise has been accounted as fatal to coastal habitats as it essentially deforms the healthy food chain of marine organisms, specifically those of shallow regions. It can cause wetland flooding, destructive erosions, contamination of aquifers, and the agricultural soils with the salty oceanic

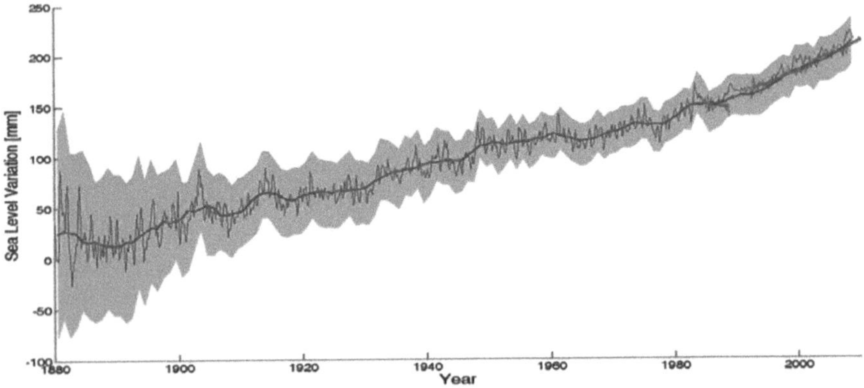

Figure 12.8 Since the late 19th century, 16 cm (6 inches) of rise in global average sea level has been observed. Sea level is rising faster in recent decades; measurements from tide gauges (blue: gray in printed version) and satellites (red: dark gray in printed version) indicate that the best estimate for the average sea level rise over the past decade is centered on 3.6 mm per year (0.14 inches per year). The shaded area represents the sea level uncertainty, which has decreased, as the number of gauge sites used in calculating the global averages and the number of data points have increased.
Source: Shum and Kuo (2011).

texture, which are responsible for loss of various communities of organisms of fishes, birds, and the diverse flora. The rise in sea level in the past few decades has aggravated and intensified, which is largely due to harsh human intervention to the astonishing mother Nature. Studies reveal that the anthropogenic inputs are largely responsible for such unpredictable change in the climate (Royal Society, 2020), and thus the humans too are not spared to endure the consequences. The uneven flooding and the storm surges associated with the rise in sea level have abruptly increased the threat level, leading to drowning, injury, and finally displacement of human population from one place to another. Increase in flooding near the coastal region and storms has also raised the risk of indoor mold growth due to excess of dampness, which further negatively impacts respiratory diseases.

3. Causes of climate change: natural versus anthropogenic

The climate of the Earth is an ever-changing entity, and the mechanics of change is simple and uncomplicated. The climatic variables such as temperature (warming or cooling), pressure, and humidity have changed since the prehistoric times; consequently, the long-term climatic condition and local weather varied. The planet cooled either due to backscattering of the energetic waves from the Sun striking the Earth's surface (mostly by land, sea ice sheets, and clouds) or due to release of energy by the Earth's atmosphere (Denchak, 2017). The short waves from the Sun turns into low frequency longer waves on reflection (Fig. 12.9), which gets trapped into the gases

Surface absorbs high energy light from the Sun

↓

Surface releases lower energy infrared light

↓

Infrared light absorbed and re-emitted by the atmospheric gases

↓

Increases gas energy and results in temperature increase

Figure 12.9 Steps of influx and outflux of solar energy.

present in the atmosphere (the greenhouse effect), keeping the planet warm and habitable for organisms.

Earlier, it was only natural factors that were thought to play a role in changing the global environmental conditions, but with the technological advancement, it was found that along with natural, it was anthropogenic factors that majorly influenced and still influencing the climatic system of the Earth. According to Royal Society and National Academy of Sciences (2020), an increased warming in the lower atmosphere and depreciation in the upper atmosphere have disclosed important insights into underlying reasons of such change in climate condition and deduced that the natural factors can never explain such expeditious changes alone. Detailed study with the help of climatic models and the comparison between them have estimates to show that human-induced increase in the concentration of "greenhouse gases" (GHGs) is the ultimate wrongdoer responsible for the recent climate change (Santer et al., 1996).

A comparison between the natural and anthropogenic sources of GHGs (Table 12.1) depicts the human intervention and its contribution to global climate change. Human intervention has led to generation of chlorofluorocarbons (CFCs) that existed in the atmosphere as extremely dangerous gas, which have never been existed before in the Earth's history.

3.1 Natural causes of climate change

The Earth has undergone various phases of warming and cooling in past times, long before the human existence. These extreme changes in the climatic condition were due to different forces including volcanic eruptions, Sun's intensity, and change in the atmospheric concentration of various atmospheric gases and effects of El Niño and La Niña. Immense evidences have shown that the small pulses of volcanism have remarkable contributions to decadal-scale climate changes during the Little Ice Age (Robock et al., 1979; Briffa et al., 1998; Mann et al., 1998; Crowley and Kim, 1999; Free and Robock, 1999). Crowley and Lowery (2000) in his study revealed that volcanism has caused large-scale cooling of the planet in the Little Ice Age and further modeling estimation showed that during the entire preanthropogenic period, only 22%–23% (approximately) of the decadal-scale variance can be ascribed to the volcanism process; however, during the 1400–1850 interval, the contribution increased to 41%–49%, which clearly depicts the importance of volcanism during the Little Ice Age. The mechanism involved in surface cooling is that since volcanic eruption releases numerous small particles (aerosols) to the stratosphere, these

Table 12.1 Various sources of natural and anthropogenic climate change.

Greenhouse gases	Natural sources	Anthropogenic sources
Carbon dioxide	Volcanoes, organic decay, and forest fire	Burning fossil fuel cars, factory emissions, and electricity generation
Methane	Wet lands, organic decay, and termites	Natural gas and oil extraction, biomass burning, rice cultivation, cattle, and refuse landfills
Nitrous oxide	Forest, grasslands, oceans, soils, cultivation, and fertilizers	Biomass burning and burning of fossils
CFCs	None	Refrigerators, aerosols spray propellants, and cleaning solvents
Water vapor	Evaporation of water from lakes, seas, rivers, and other waterbodies Transpiration in plants and volcanic eruptions	Mostly combustion and respiration
Ozone	Natural fires and soils	From NO_x (transport, power station, and industrial and combustion process) From VOCs (solvents, oils, transport, and combustion)

CFCs, clorofluorocarbons; *VOCs*, volatile organic compounds.

particles reflect the sunlight (IPCC, 2013). Recent study showed that, very minute amount of carbon dioxide is released in the volcanic eruptions (round about 1% of the outpouring rate from the combustion of fossil fuels). Another factor is the sunlight, but its variation has not played a crucial role in climate changes during the recent decades, and therefore, the effect of solar irradiance was found to be weak in deciding its role play in climate change (Stern and Kaufmann, 2014). Studies using various models have been found important to explain the emergence of various GHGs and the El Niño and La Niña effect. However, records indicate that today's climatic warming—particularly the warming since the mid-20th century—is occurring much faster than ever before and cannot be explained by natural causes alone. According to the National Aeronautics and Space Administration (NASA), "these natural causes are still in play today, but their influence is too small, or they occur too slowly to explain the rapid warming seen in recent decades."

3.2 Anthropogenic causes of climate change

New advancements in the scientific fields, almost on an annual basis, have built better understanding about the abrupt and unprecedented climate changes, viz., the increased frequency of hurricanes (record season of hurricane during 2005), higher number of floods across the globe, or the concerns related to the solidity of ice covers (Rahmstorf, 2008). Various simulation studies using different climatic models at different places have made future estimations that if it would have been the natural factors alone playing role in influencing the climate system, then modest change would be observed in case of surface warming and subsequent cooling during the 20th century and 21st century; however, inclusion of human influences on the atmospheric composition along with the natural effects has shown consistent change in temperature pattern with the actual prevailing condition (Royal society and NAS, 2020). The pre- and postindustrial change in temperature is largely due to increase in the GHGs, and CO_2 had the maximum contribution. But following minor depreciation in CO_2 concentration in the atmosphere during the period of Little Ice Age (Schmidt et al., 2012; Masson-Delmotte et al., 2013), CO_2 began to rise since the beginning of industrialization along with the increase in other GHGs. The substantial rise in radiative forcing with the effect of GHGs occurred in the past few decades, an increase which is supposed to continue to date. With the rise in burning of fossil fuels, simultaneous growth in anthropogenic aerosols began as well, with aerosol forcing estimates peaking globally around 1980, although emissions have continued increasing in East Asia and South Asia, whereas decreasing trend was observed in continents such as Europe and North America (Hegerl et al., 2019). Other simulation studies have suggested that during the 20th century and before, the aerosols have played a critical role in molding the regional climate and further asserted that small-scale aerosol distress during initial industrial years might have significantly impacted the environment in a comparatively less polluted atmosphere (Carslaw et al., 2013) with their potential consequences for best estimate of aerosol radiative forcing (Stevens, 2015; Kretzschmar et al., 2017; Booth et al., 2018). It is, however, quite burdensome to quantify the loading of natural aerosol due to the uncertainty of biomass burning and the natural sources, and the same goes for the estimation of magnitude of aerosol cloud interactions (Zelinka et al., 2014; Wilcox et al., 2015; Toll et al., 2017). According to the study by Stern and Kaufmann (2013), anthropogenic forcings such as GHGs, sulfur emissions, and black carbon was found low until 1970, while a sharp increase in their concentration was seen after the year 1970.

The GHG emissions generated by human activities are the greatest causes of rapid change in the climatic condition on the planet due to which the amount of these gases has skyrocketed in the atmosphere in the past decades. The concentrations of methane, carbon dioxide, and nitrous oxides "have increased to levels unprecedented in at least the past 800,000 years," and it has been observed that the atmospheric share of carbon dioxide (chief climate change contributor of the planet) has risen by approximately 40% since the preindustrial times (UCCE Sonoma, 2020). The burning of fossil fuels such as oil, coal, and gases for generation of electricity (31% contribution) and transportation (27% contribution) is the preliminary source of anthropogenic (human generated) emissions (Fig. 12.10). A second major source is deforestation, which releases sequestered carbon into the air. It has been estimated that clear-cutting, fires, logging, and other forms of degradation in forest contribute up to 20% of global carbon emissions. Other human activities that are the biggest contributors of atmospheric pollution

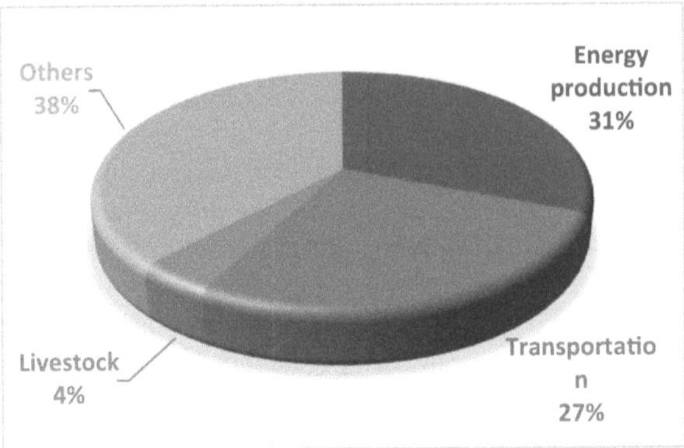

Figure 12.10 Percent contribution of different anthropogenic sources of pollutants leading to climate change.
Modified from EPA (sources of greenhouse gas emission).

Various anthropogenic sources of pollution viz. use of fertilizer, road build-up, smokes from industries, deforestation, logging, and livestock production. Modified: Various sources.

Figure 12.11 Various anthropogenic sources of pollution viz. use of fertilizer, road build-up, smokes from industries, deforestation, logging, and livestock production. Modified: Various sources.

include livestock production (sheep, cattle, goats, and buffalos are the major methane emitters), use of various fertilizers (a preliminary source of gases such as nitrous oxide), and many industrial processes that release gases that are amalgamated with fluorine and other halogens (Fig. 12.11). Activities such as agriculture and road construction can change the reflectivity of the Earth's surface, leading to local warming or cooling too. Though our planet's forests and oceans absorb GHGs from the atmosphere through photosynthesis and other processes, these natural carbon sinks cannot keep up with our rising emissions. The resulting buildup of GHGs is causing alarmingly fast warming worldwide. It is estimated that the Earth's average temperature rose by about 1°F during the 20th century. To explain the warmth of the aforementioned statement, consider this: When the last Ice Age ended and the northeastern United States was covered by more than 3000 feet of ice, average temperatures were just 5–9° cooler than they are now (NRDC, 2017).

4. Past climatic changes

A changing climate has been the norm throughout the 4.6-billion-year history of the Earth, and these changes in past climates have always been useful in predicting the future environmental conditions and their impacts. The study of whole process of prediction of past climatic conditions and simulations done for future references based on the past changes is collectively termed as paleoclimatology. To be very precise, "paleoclimatology" is the study of prehistoric climates and is related to paleobiogeography, paleoecology, and paleogeography. All of these "paleo" research foci use similar techniques to recreate past species distributions, plant physiology, and Earth's geology. The study related to past climate can be done under three different classes: the instrumental, prehistoric, and historic. The instrumental category produces a very accurate data but is limited to the timeframe during which the human society had technical instruments to record climate data; however, the data are available for past 150 years only. Furthermore, climate data can be obtained from historical data sets, which are obtained from written records, possible after the invention of precise climate-measuring instruments and methods. Proper historical data originate from official journals, letters, documents, ancient writings, and other human records (Stiepani, 2016). Lastly, the prehistoric data sets, which originate from climate proxies, reveal information on prehistoric climates (Roberts, 2014). This kind of data sets proved very useful to the scientists as it can recreate past climates spanning millions of years with the help of these proxies. The climatic data sets that are particularly derived from proxies can be very useful to the scientists in hypothesizing whether the present and the future droughts are extreme events or simply climate variation (Roberts, 2014). The change in climatic conditions, particularly the change in temperature (Fig. 12.12), has always been strongly connected to the evolution of life, to erosion and formation of rocks, and even to the generation of mountains, and all these things are equally responsible for the evolution of the prevailing climate status. Over the recent geological past, climate swings have given us repeated glaciations separated by warmer intervals.

Throughout the history of life forms that existed over the surface of the Earth evolved with their local environment, change in the environment involves the change in the composition of atmospheric gases such as CO_2 levels and consequent changes in temperature and the precipitation pattern, change in tilt and axis of the Earth, and so on. The following points describe the changes in past climatic conditions:

4.1 Evidence from ice and sediment cores reveals cycles of climate changes tens of thousands to hundreds of thousands of years long. These can be related to small but predictable changes in the Earth's orbit and in the tilt of the Earth's axis. Those predictable changes set the pace for the glacial–interglacial cycles of the Ice Age of the past 2.6 million years (Imbrie and Imbrie, 1979). In addition, the heat emitted by the Sun varies with time. Complex patterns of atmospheric and oceanic circulation cause the El Niño events and related climatic oscillations on the scale of a few years (Cronin, 2009; Houghton, 2009).

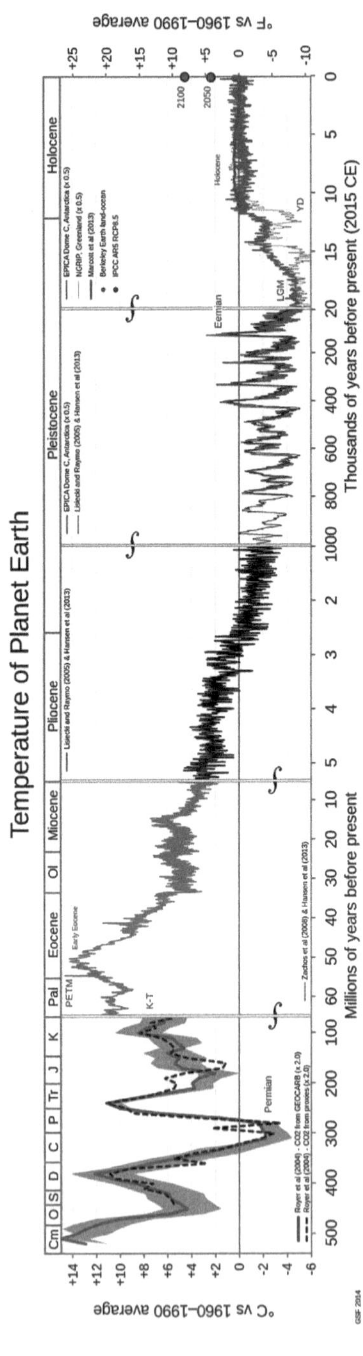

Figure 12.12 Depicts the change in temperature since the origin of the Earth.
Source: Fergus (2014).

4.2 During warmings from glacial to interglacial, temperature and CO_2 rose together for several thousand years, although the best estimate from the end of the last glacial is that the temperature probably started to rise a few centuries before the CO_2 showed any reaction. According to the paleoclimatologists, the initial warming driven by changes in the Earth's orbit and axial tilt eventually caused CO_2 to be released from the warming ocean and thus, via positive feedback, to reinforce the temperature rise (Luthi et al., 2008). Additional positive feedback reinforcing the temperature rise would have come from increased water vapor evaporated from the warmer ocean, water being another GHG, along with a decrease in sea ice, and eventually in the size of the northern hemisphere ice sheets, resulting in less reflection of solar energy back into space.

4.3 Global sea level was found to be very sensitive to changes in global temperatures. When ice sheets were at a maximum during the Pleistocene, world sea level fell to at least 120 m below where it stands today. Relatively small increases in global temperature in the past have led to sea level rises of several meters. During parts of the previous interglacial period, when polar temperatures reached 3–5°C above today's (Otto-Bliesner et al., 2006), global sea levels were higher than today's by around 4–9 m (Kopp et al., 2009). Global patterns of rainfall during glacial times were very different from today.

4.4 At the end of the Pleistocene (the earlier part of the Quaternary, from 2,600,000 to 12,000 years ago), there was a major extinction of large mammals and animals like the woolly mammoth became extinct worldwide.

4.5 During the Quaternary period of geological timescale (the past 2.6 million years), great changes in the climate were seen that caused ice sheets to advance from the poles into usually temperate places on the globe. Evidence shows that although there has been a progressive long-term trend of global cooling during the Quaternary, there have also been several different ice ages, or glaciations, separated by episodes of global warming where the climate was much warmer or more temperate than the present day. Ice has advanced and retreated repeatedly, beginning at the start of the Quaternary, and continuing to the present day. These repeated glacial episodes have resulted in significant fluctuations in sea level and caused plants and animals to migrate to more comfortable climates.

4.6 About 55 million years ago, at the end of the Paleocene, there was a sudden warming event in which temperatures rose by about 6°C globally and by 10–20°C at the pole (Kopp et al., 2009, Fig. 12.12). Carbon isotopic data show that this warming event (called by some the Paleocene–Eocene Thermal Maximum, or PETM) was accompanied by a major release of 1500–2000 billion tons or more of carbon into the ocean and atmosphere. This injection of carbon may have come mainly from the breakdown of methane hydrates beneath the deep sea floor (Kopp et al., 2009), perhaps triggered by volcanic activity superimposed on an underlying gradual global warming trend that peaked some 50 million years ago in the early Eocene. Most estimates agree that there was a significant decrease of CO_2 in the atmosphere from more than 1000 ppm at 50 million years ago (during the Eocene) to the range recorded in the ice cores of the past 800,000 years (Zachos et al., 2008). This decrease in CO_2 was probably one of the main causes of the cooling that led to the formation of the great ice sheets on Antarctica (DeConto and Pollard, 2003). Changes in ocean circulation around Antarctica may also have also played a role in the timing and extent of formation of those ice sheets (Livermore et al., 2007; Huber et al., 2004; Francis et al., 2009).

4.7 Cretaceous—One of the warmest periods in the Earth's history was the Cretaceous, from 140 to 65 million years ago. The Earth was then several degrees warmer than today and is described as having a "greenhouse" climate. As "greenhouse" temperatures were reached high levels (Fig. 12.13), the world's ice melted, which caused significant sea level rise. Two major anoxic events of the cretaceous, those of early Aptian (120 Mya), and the

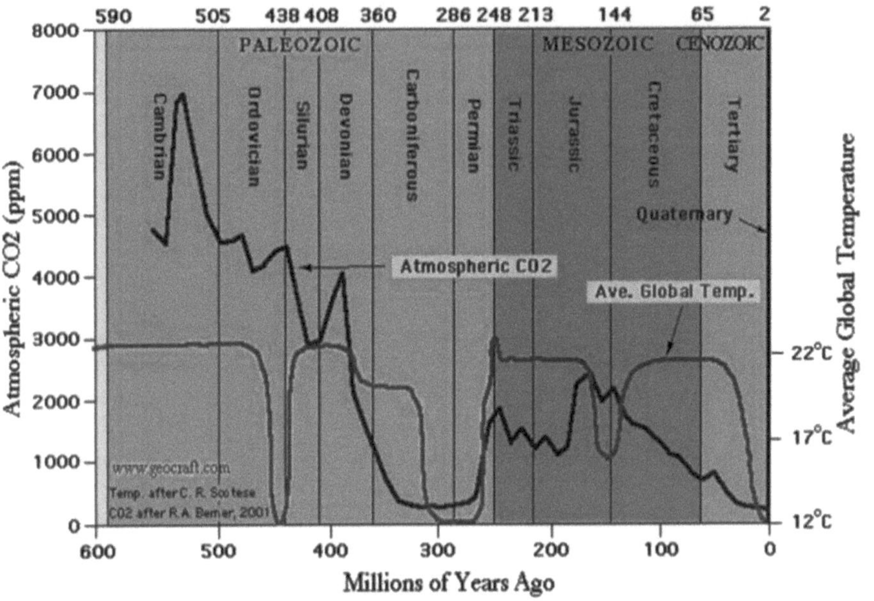

Figure 12.13 Global temperature and CO_2 levels over 600 million years.
Source: MacRae (2008)

Cenomanian–Turonian boundary (94 Mya), have been found to register some of the highest temperature reconstructions for the period of whole cretaceous, thought to be related to the input of volcanically derived CO_2 from one or more large igneous provinces.

4.8 Among the recent change, Greenland ice cores record during the last glacial stage (100,000–11,500 years ago) revealed that the temperature there alternately warmed and cooled several times by more than 10°C (Blunier et al., 2001). This was accompanied by major climate change around the northern hemisphere, felt particularly strongly in the North Atlantic region. Each warm and cold episode took just a few decades to develop and lasted for a few hundred years. The climate system in those glacial times was clearly unstable. These changes were almost certainly caused by changes in the way the oceans transported heat between the hemispheres (Jouzel et al., 2007).

5. Recent trends in climate change

Recent reports on GHG concentrations show that 2015–2019 has seen a continued increase in carbon dioxide (CO_2) levels and other key GHGs in the atmosphere to new records, with CO_2 growth rates nearly 20% higher than the previous 5 years (WMO, 2019). CO_2 remains in the atmosphere for centuries and in the ocean for even longer. According to the data produced by the World Meteorological Organization (2019), a subset of GHG observational sites for 2019 indicates that CO_2 global concentrations are on track to reach or even exceed 410 ppm by the end of this decade. In case of other gases such as methane and nitrous oxide, 60% of their presence in the atmosphere is due to the anthropogenic activities (NOAA, 2019).

5.1 Sea level rise

Over the 5-year period, May 2014–19, the rate of global mean sea level rise has amounted to 5 mm per year, compared with 4 mm per year in the 2007–16, 10-year period. This is substantially faster than the average rate since 1993 of 3.2 mm/year. The contribution of land ice melt from the world glaciers and the ice sheets has increased over time and now dominates the sea level budget, rather than thermal expansion.

5.2 Shrinking ice

Throughout 2015–18, the **Arctic's** average September minimum (summer) sea ice extent was well below the 1981–2010 average, as was the average winter sea ice extent. The four lowest records for winter occurred during this period. Multiyear ice has almost disappeared.

Antarctic February minimum (summer) and September maximum (winter) sea ice extent values have become well below the 1981–2010 average since 2016. This contrasts with the previous 2011–15 period and the long term 1979–2018 periods. Antarctic summer sea ice reached its lowest and second lowest extent on record in 2017 and 2018, respectively, with 2017 also being the second lowest winter extent. The amount of ice lost annually from the Antarctic ice sheet increased at least sixfold, from 40 Gt per year in 1979–90 to 252 Gt per year in 2009–17.

The **Greenland** ice sheet has witnessed a considerable acceleration in ice loss since the turn of the millennium. For 2015–18, the World **Glacier** Monitoring Service (WGMS) reference glaciers indicate an average specific mass change of −908 mm water equivalent per year, higher than in all other 5-year periods since 1950.

5.3 Ocean heat and acidity

More than 90% of the excess heat caused by climate change is stored in the oceans. 2018 had the largest ocean heat content values on record measured over the upper 700 m, with 2017 ranking second and 2015 third.

The ocean absorbs around 30% of the annual anthropogenic emissions of CO_2, thereby helping to alleviate additional warming. The ecological costs to the ocean, however, are high, as the absorbed CO_2 reacts with seawater and changes the acidity of the ocean. There has been an overall increase in acidity of 26% since the beginning of the industrial revolution.

5.4 Extreme events

More than 90% of the natural disasters are related to weather. The dominant disasters are storms and flooding, which have also led to highest economic losses. Heat waves and drought have led to human losses, intensification of forest fires, and loss of harvest. Heat waves were the deadliest meteorological hazard in the 2015–19 period, affecting all continents and resulting in numerous new temperature records. Almost every study of a significant heatwave since 2015 has found the hallmark of climate change, according to the report.

The largest economic losses were associated with **tropical cyclones**. The 2017 Atlantic hurricane season was one of the most devastating on record with more than US$ 125 billion in losses associated with Hurricane Harvey alone. On the Indian Ocean, in March and April 2019, unprecedented and devastating back-to-back tropical cyclones hit Mozambique.

5.5 *Wildfires*

Wildfires are strongly influenced by weather and climate phenomena. Drought substantially increases the risk of wildfire in most forest regions, with a particularly strong influence on long-lived fires. The three largest economic losses on record from wildfires have all occurred in the past 4 years.

In many cases, fires have led to massive releases of carbon dioxide to the atmosphere. Summer 2019 saw unprecedented wildfires in the Arctic region. In June alone, these fires emitted 50 megatons (Mt) of carbon dioxide into the atmosphere. This is more than that was released by Arctic fires in the same month from 2010 to 18 put together. There were also massive forest fires in Canada and Sweden in 2018. There were also widespread fires in the nonrenewable tropical rain forests in Southern Asia and Amazon, which have had impacts on the global carbon budget.

6. Conclusion

The climate change has always been a normal phenomenon, long before the existence of life on the Earth. In the pace of development, the human influence and its intrusion with various habitats have led to destruction of all parts of the nature. Recent developments have disclosed the physical mechanism involved in changing the amounts of GHGs that caused major changes in climate variables. Immense evidences are known till date that shows an increasing trend of these GHGs in the atmosphere and that most of the recent and abrupt changes in the climatic conditions are certainly due to the emissions of GHGs caused by human activities. The present scenario tells that global warming is here, and it has already adversely affected and is still affecting our climate and so prevention is no longer an option left especially for human beings. There may be three options left remaining to deal with the crisis, viz., mitigate, adapt, and suffer. Citizens and governments of all countries across the globe need to choose among several options among the following response to this information: they can either change their pattern of energy production and usage to limit emissions of GHGs and therefore the magnitude of climate changes; they can wait for changes to occur and accept the losses, damage, and suffering that arise; they can adapt to actual and expected changes as much as possible; or they can seek as yet unproven "geoengineering" solutions to counteract some of the climate changes that would otherwise occur.

References

Acevedo, S., Mrkaic, M., Novta, N., Pugacheva, E., Topalova, P., 2017. The effects of weather shocks on economic activity: what are the channels of impact? Chapter 3. In: World Economic Outlook, October 2017 - Seeking Sustainable Growth: Short-Term Recovery, Long-Term Challenges. IMF, pp. 117–183.

Barua, S., Valenzuela, E., 2018. Climate Change Impacts on Global Agricultural Trade Patterns: Evidence from the Past 50 Years. Published in the Conference Proceedings on 01 November 2018.

Bates, N.R., Best, M.H.P., Neely, K., Garley, R., Dickson, A.G., Johnson, R.J., 2012. Detecting anthropogenic carbon dioxide uptake and ocean acidification in the North Atlantic Ocean. Biogeosciences 9, 2509–2522. https://www.biogeosciences.net/9/2509/2012.

Blunier, T.T., Brook, E.J., 2001. Timing of millennial-scale climate change in Antarctica and Greenland during the last glacial period. Science 291 (5501), 109–112.

Booth, B.B.B., Harris, G.R., Jones, A., Wilcox, L., Hawcroft, M., Carslaw, K.S., 2018. Comments on 'Rethinking the lower bound on aerosol radiative forcing'. J. Clim. 319407–319412.

Brice, M-H., Cazelles, K., Legendre, P., Fortin, M-J., 2019. Disturbances amplify tree community responses to climate change in the temperate–boreal ecotone. Global Ecol. Biogeogr. 1–14. https://doi.org/10.1111/geb.12971.

Briffa, K.R., Jones, P.D., Schweingruber, F.H., Osborn, T.J., 1998. Nature (Lond.) 450.

Carslaw, K.S., et al., 2013. Large Contribution of Natural Aerosols to Uncertainty in Indirect Forcing Nature, pp. 50367–50371.

Church, J.A., Gregory, J.M., Huybrechts, P., Kuhn, M., Lambeck, K., Nhuan, M.T., et al., 2001. Changes in sea level. Climate Change 2001: The Scientific Basis: Contributions of Working Group I to the 3rd Assessment of the IPCC. Cambridge, UK, Cambridge University Press.

Church, J.A., White, N.J., 2011. Sea-level rise from the late 19th to the early 21st century. Surv. Geophys. 32, 585–602.

Comenetz, J., Caviedes, C., 2002. Climate variability, political crises, and historical population displacements in ethiopia. Environ. Hazards 4, 113–127. https://doi.org/10.1016/j.hazards.2003.08.001.

Crawford, C.S., 1981. Biology of Desert Invertebrates. Springer, New York.

Cronin, T.M., 2009. Paleoclimates: Understanding Climate Change Past and Present. Columbia University Press.

Crowley, T.J., Kim, K.-Y., 1999. Geophys. Res. Lett. 26, 1901.

Crowley, T.J., Lowery, T.S., 2000. How warm was the medieval warm period? AMBIO J. Hum. Environ. 29, 51–54.

Das, S.B., Joughin, I., Behn, M.D., Howat, I.M., King, M.A., Lizarralde, D., et al., 2008. Fracture propagation to the base of the Greenland ice sheet during supraglacial lake drainage. Science 320, 778–781.

DeConto, R.M., Pollard, D., 2003. Rapid cenozoic glaciation of Antarctica induced by declining atmospheric CO_2. Nature 421, 245–249.

Dell, M., Jones, B.F., Olken, B.A., 2014. What do we learn from the weather? The new climate–economy literature. J. Econ. Lit. 52 (3), 740–798.

Denchak, M., 2017. Global climate change: what you need to know. The lowdown on the earth's central environmental threat. NASA/Goddard Space Flight Center Scientific Visualization Studio/NASA/JPL AIRS Project.

Dore, J.E., Lukas, R., Sadler, D.W., Church, M.J., Karl, D.M., 2009. Proc. Natl. Acad. Sci. 106, 12235–12240. https://doi.org/10.1073/pnas.0906044106.

Drake, M., Righter, K., 2002. Determining the composition of the Earth. Nature 416, 39–44. https://doi.org/10.1038/416039a.

du Plessis, K.L., Martin, R.O., Hockey, P.A.R., Cunningham, S.J., Ridley, A.R., 2012. The costs of keeping cool in a warming world: implications of high temperatures for foraging, thermoregulation and body condition of an arid-zone bird. Glob. Change Biol. 18, 3063–3070.

EPICA community members, 2004. Nature 429, 623.

Fergus, G., 2014. Sudden global warming 55 m years ago was much like today. The Conversation. Academic rigour, journalistic flair.

Francis, J.E., Marenssi, S., Levy, R., Hambrye, M., Thorn, V.C., Mohr, B., Brinkhuis, H., Warnaar, J., Zachos, J., Bohaty, S., DeConto, R., 2009. From greenhouse to icehouse – the Eocene/Oligocene in Antarctica. In: Florindo, F., Siegert, M. (Eds.), Antarctic Climate Evolution, Chapter 8, Developments in Earth and Environmental Sciences, vol. 8. Elsevier, pp. 209–368.

Free, A., Robock, J., 1999. Geophys. Res. 104, 19057.

Guldberg, O.H., Mumby, P.J., Hooten, A.J., Steneck, R.S., Greenfield, P., Gomez, E., Harvell, C.D., Sale, P.F., Edwards, A.J., Caldeira, K., Knowlton, N., Eakin, C.M., Iglesias-Prieto, R., Muthiga, N., Bradbury, R.H., Dubi, A., Hatziolos, M.E., 2007. Coral reefs under rapid climate change and ocean acidification. Science 318, 14.

Hegerl, G.C., Brönnimann, S., Cowan, T., Friedman, A.R., Hawkins, E.D., Iles, C., Müller, W., Schurer, A., Undorf, S., 2019. Causes of climate change over the historical record. Environ. Res. Lett. 14, 123006. https://doi.org/10.1088/1748-9326/ab4557.

Houghton, J., 2009. Global Warming: the Complete Briefing, fourth ed. Cambridge University Press.

Hubbe, A., Hubbe, M., 2019. Current climate change and the future of life on the planet. Front. Young Minds. 7, 37. https://doi.org/10.3389/frym.2019.00037.

Huber, M., Brinkhuis, H., Stickley, C.E., Doos, K., Sluijs, A., Warnaar, J., Schellenberg, S.A., Williams, G.L., 2004. Eocene circulation of the Southern Ocean: was Antarctica kept warm by subtropical waters? Paleoceanography 9 (4), PA3026.

Huffman, G.J., Adler, R.F., Bolvin, D.T., Gu, G., 2009. Improving the global precipitation record: GPCP Version 2.1. Geophys. Res. Lett. 36, L17808.

Hughes, T.P., et al., 2003. Science 301, 929.

Smith, T.M., Reynolds, R.W., Peterson, T.C., Lawrimore, J., 2008. Improvements to NOAA's historical merged land–ocean surface temperature analysis (1880–2006). J. Clim. 21, 2283–2296.

Stern, D.I., Kaufmann, R.K., 2013. Anthropogenic and natural causes of climate change. Clim. Change 122, 257–269.

Stern, D.I., Kaufmann, R.K., 2014. Anthropogenic and natural causes of climate change. Clim. Change 122, 257–269. https://doi.org/10.1007/s10584-013-1007-x.

Iknayan, K.J., Beissinger, S.R., 2018. Collapse of a desert bird community over the past century driven by climate change. PNAS 115 (34), 8597–8602.

Imbrie, J., Imbrie, K.P., 1979. Ice Ages: Solving the Mystery. MacMillan, London.

IPCC, 2007. In: Parry, M.L., Canziani, O.F., Palutikof, J.P., van der Linden, P.J., Hanson, C.E. (Eds.), Climate Change 2007: Impacts, Adaptation and Vulnerability. Contribution of Working Group II to the Fourth Assessment Report of the Intergovernmental Panel on Climate Change. Cambridge University Press, Cambridge, United Kingdom. https://www.ipcc.ch/pdf/assessmentreport/ar4/wg2/ar4_wg2_full_report.pdf. (Accessed 18 November 2016).

IPCC, 2013. Climate change. In: Stocker, T.F., Qin, D., Platter, G.K., Tignor, M., Allen, S.K., Boschung, J., Nauels, A., Xin, Y., Bex, V., Midgeley, P.M. (Eds.), Te Physical Science Basis. Contribution of Working Group 1 to the Fifth Assessment Report of the Intergovernmental Panel on Climate Change. Cambridge University Press, Cambridge, United Kingdom and New York, NY, USA, p. 1535.

Jones, B.F., Olken, B.A., 2010. Climate shocks and exports. Am. Econ. Rev. Pap. Proc. 100, 454–459.

Jouzel, J., Masson-Delmotte, V., Cattani, O., Dreyfus, G., Falourd, S., Hoffmann, G., Nouet, J., Barnola, J.M., Chappellaz, J., Fischer, H., Gallet, J.C., Johnsen, S., Leuenberger, M., Loulergue, L., Luethi, D., Oerter, H., Parrenin, F., Raisbeck, G., Raynaud, D., Schwander, J., Spahni, R., Souchez, R., Selmo, E., Schilt, A., Steffensen, J.P., Stenni, B., Stauffer, B., Stocker, T., Tison, J.-L., Werner, M., Wolff, E.W., 2007. Orbital and millennial Antarctic climate variability over the last 800,000 years. Science 317, 793–796.

Kleypas, J.A., McManus, J.W., Meñez, L.A.B., 1999. Am. Zool. 39, 146.

Kopp, R.E., Simons, F.J., Mitrovica, J.X., Maloof, A.C., Oppenheimer, M., 2009. Probabilistic assessment of sea level during the last interglacial stage. Nature 462, 863–867.

Kretzschmar, J., Salzmann, M., Mülmenstädt, J., Boucher, O., Quaas, J., 2017. Comment on 'Rethinking the lower bound on aerosol radiative forcing'. J. Clim. 306579–306584.

Lemke, P., Ren, J., Alley, R.B., Carrasco, J., Flato, G., Fujii, Y., et al., 2007. Observations, Changes in Snow, Ice and Frozen Ground in Climate Change 2007: The Physical Science Basis. Contributions of Working Group I to the 4th Assessment of the IPCC. Cambridge University Press, Cambridge, UK.

Li, X., Clinton, N., Si, Y., et al., 2015. Projected impacts of climate change on protected birds and nature reserves in China. Sci. Bull. 60, 1644–1653.

Livermore, R.A., Hillenbrand, C.-D., Meredith, M., Eagles, G., 2007. Drake passage and cenozoic climate: an open and shut case? Geochem. Geophy. Geosyst. 8 (1), Q01005.

Lüthi, D., Le Floch, M., Stocker, T.F., Bereiter, B., Blunier, T., Barnola, J.M., Siegenthaler, U., Raynaud, D., Jouzel, J., 2008. High-resolution carbon dioxide concentration record 650,000-800,000 years before present. Nature 453, 379–382.

MacRae, P., 2008. We are long way from global- warming' oblivion', False Alarm. Retrieved from. http://www.paulmacrae.com/?p=29on. (Accessed 5 July 2015).

Mann, M.E., Bradley, R.S., Hughes, M.R., 1998. Nature 392, 779.

Masson-Delmotte, V., et al., 2013. Information from paleoclimate archives. In: Stocker, T.F., et al. (Eds.), Climate Change, 2013: The Physical Science Basis. Cambridge University Press, Cambridge, pp. 383–464.

Matthews, H.D., Weaver, A.J., 2010. Committed climate warming. Nat. Geosci. 3, 142–143.

Maurer, J.M., Schaefer, J.M., Rupper, S., Corley, A., 2019. Acceleration of ice-loss across the Himalayas over the past 40 years. Climatol. Sci. Adv. 5, eaav7266.

McCarthy, J.J., Canziani, O.F., Leary, N.A., Dokken, D.J., White, K.S., 2001. Climate Change 2001: Impacts, Adaptation and Vulnerability. Cambridge University Press.

Moberg, F., Folke, C., 1999. Ecol. Econ. 29, 215.

Moses, M.R., Frey, J.K., Roemer, G.W., 2012. Elevated surface temperature depresses survival of banner-tailed Kangaroo rats: will climate change cook a desert icon? Oecologia 168, 257–268.

NOAA, 2019. National Oceanic and Atmospheric Administration. https://oceanservice.noaa.gov/facts/eutrophication.html. (Accessed 10 May 2017).

NRDC, 2017. The lowdown on the earth's central environmental threat. NASA/Goddard Space Flight Center Scientific Visualization Studio/NASA/JPL AIRS Project.

Nwankwoala, H.N.L., 2015. Causes of climate and environmental changes: the need for environmental-friendly education policy in Nigeria. J. Educ. Pract. 6 (30). www.iiste.org. ISSN 2222-1735 (Paper) ISSN 2222-288X (Online).

Otto-Bliesner, B.L., Marshall, S.J., Overpeck, J.T., Miller, G.H., Hu, A., CAPE Last interglacial project members, 2006. Simulating arctic climate warmth and icefield retreat in the last interglaciation. Science 311, 1751–1753.

Parmesan, C., Yohe, G., 2003. A globally coherent fingerprint of climate change impacts across natural systems. Nature 421, 37–42. https://doi.org/10.1038/nature01286.

Pearce, D.W., Cline, W.R., Achanta, A.N., Fankhauser, S., Pachauri, R.K., John, R., Hohmann, N., Kane, S., 1996. Climate change and agricultural trade: who benefits, who loses. Glob. Environ. Change 4 (1), 24–36.

Petit, J.R., Jouzel, J., Raynaud, D., Barkov, N.I., Barnola, J.-M., Basile, I., Bender, M., Chappellaz, J., Davis, M., Delaygue, G., Delmotte, M., Kotlyakov, V.M., Legrand, M., Lipenkov, V.Y., Lorius, C., PÉpin, L., Ritz, C., Saltzman, E., Stievenard, M., 1999. Climate and atmospheric history of the past 420,000 years from the Vostok Ice Core, Antarctica. Nature 399, 429–436.

Rahmstorf, S., 2008. Anthropogenic Climate Change: Revisiting the Facts (Chapter 3).

Raven, J., et al., 2005. In: Acidification due to increasing carbon dioxide. Policy Document 12/05. The Royal Society, London.

Roberts, N., 2014. The Holocene: an Environmental History. Wiley Blackwell, West Sussex.

Robock, A., 1979. Science 206, 1402.

Royal Society and National Academy of Sciences, 2020. Evidences and Causes of Climate Change. royalsociety.org.

Santer, B.D., 1996. Nature 382, 39.

Schmidt, G.A., et al., 2012. Climate forcing reconstructions for use in PMIP simulations of the Last Millennium (v1.1). Geosci. Model Dev. 5185–5191.

Shum, H.T., Kuo, C.Y., 2011. Sea- Level Rise for the Coasts of California, Oregon, and Washington: Past, Present and Future. The National Academies Press, Washington, DC, pp. 190–202.

Sinervo, B., et al., 2010. Erosion of lizard diversity by climate change and altered thermal niches. Science 328, 894–899.

Stevens, B., 2015. Rethinking the lower bound on aerosol radiative forcing. J. Clim. 284794–284819.

Stiepani, J., 2016. Paleoclimatology: Examples of Ecological Impacts from Prehistoric Climate Change. The Climate Institute.

Thompson, L.G., Brecher, H.H., Mosley- Thompson, E., Hardy, D.R., Mark, B.G., 2010. Glacier loss on Kilimanjaro continues unabated. Proc. Natl. Acad. Sci. 106, 19770–19775.

Toll, V., Christensen, M., Gassó, S., Bellouin, N., 2017. Volcano and ship tracks indicate excessive aerosol-induced cloud water increases in a climate model. Geophys. Res. Lett. 4412492–4412500.

Trenberth, K.E., 2011. Changes in precipitation with climate change. Clim. Res. 47, 123–138. https://doi.org/10.3354/cr00953.

Trenberth, K.E., Fasullo, J., 2007b. Water and energy budgets of hurricanes and implications for climate change. J. Geophys. Res. 112, D23107. https://doi.org/10.1029/2006JD008304.

Trenberth, K.E., Shea, D.J., 2005. Relationships between precipitation and surface temperature. Geophys. Res. Lett. 32, L14703. https://doi.org/10.1029/2005GL022760.

UCCE Sonoma, 2020. Agriculture and natural resource. Santa Rosa, California, USA. mgsonoma@ucanr.edu.

VanDerWal, J., Murphy, H.T., Kutt, A.S., Perkins, G.C., Bateman, B.L., Perry, J.J., Reside, A.E., 2013. Focus on poleward shifts in species' distribution underestimates the fingerprint of climate change. Nat. Clim. Change 3, 239–243. https://doi.org/10.1038/nclimate1688.

Watson, R.T., Zinyowera, M.C., Moss, R.H., 1996. Climate Change 1995: Impacts, Adaptation, and Mitigation of Climate Change – Scientific-Technical Analysis. Contribution of Working Group II to the Second Assessment Report of the Intergovernmental Panel on Climate Change. Cambridge University Press, Cambridge.

Wigley, T.M.L., 2005. The climate change commitment. Science 307, 1766–1769.

Wilcox, L.J., Highwood, E.J., Booth, B.B.B., Carslaw, K.S., 2015. Quantifying sources of inter-model diversity in the Cloud Albedo effect. Geophys. Res. Lett. 421568–421575.

WMO, 2019. World Encyclopedia. Encyclopedia.com. https://www.encyclopedia.com.

Zachos, J.C., Dickens, G.R., Zeebe, R.E., 2008. An early cenozoic perspective on greenhouse warming and carbon-cycle dynamics. Nature 451, 279–283.

Zelinka, M.D., Andrews, T., Forster, P.M., Taylor, K.E., 2014. Quantifying components of aerosol-cloud-radiation interactions in climate models. J. Geophys. Res. Atmos. 119, 7599–7615.

Zwally, H.J., Abdalati, W., Herring, T., Larson, K., Saba, J., Steffen, K., 2002. Surface melt-induced acceleration of Greenland ice-sheet flow. Science 297, 218–222.

Radioecology: dissecting complexities of radionuclide transfer under climate change

13

Poonam Yadav, Bhupinder Singh
Centre for Environment Science and Climate Resilient Agriculture (CESCRA), Nuclear Research Laboratory (NRL), ICAR-Indian Agricultural Research Institute, New Delhi, India

1. Introduction

Radionuclides, also called radioisotopes, are essentially an unstable form of a chemical element that decays with the emission of nuclear radiation. Many of these radionuclides occur naturally in the environment, while many others are human-made, either produced deliberately or as by-products of nuclear reactions. These radionuclides have a characteristic rate of radiation emission, which is measured in terms of half life and is defined as the time required for half of the radioactive atoms to decay and may vary from few seconds to hundreds or millions of years. Some radionuclides are natural components of rocks and minerals of earth's crust and, through natural weathering processes and volcanic activity, are cycled through the environment. In general, 12 radionuclides are commonly used in medical, commercial, or military purposes (Table 13.1). Radionuclides or radioisotopes to a common man, in general, mean a nuclear bomb, destruction, war, and deformed people. However, these radionuclides are immensely important and have several peaceful applications for the society at large. Paramount peaceful application of radionuclides is in the energy sector where nuclear power or nuclear energy, produced via nuclear fission, suffices 16% of world electricity demand. Nuclear fission is a process in which slow-moving neutrons are absorbed by the nucleus of the uranium-235 atom, which in turn splits into fast-moving lighter element (fission product) and free neutrons. This process creates large amount of energy in the form of γ-rays and in the form of kinetic energy. Safety of these nuclear plants is ensured as per the globally abided international regulations; however, at times in the past, natural calamities have caused nuclear accidents such as Chernobyl or Fukushima disaster to accidently release the radionuclides such as ^{137}Cs and ^{125}I in the environment. Once in the environment, the dispersion and precipitation of the radionuclide, which can occur over several thousand miles, depend on the weather attributes such as wind direction, wind speed, rainfall, and so on. Furthermore, the naturally occurring and precipitated radionuclide can potentially enter the animal and human food chain chiefly through plant-derived feed and economic produce.

Global Climate Change. https://doi.org/10.1016/B978-0-12-822928-6.00013-7

Table 13.1 List of commonly used radionuclides for medical, commercial, and military purposes.

Radionuclide	Half-life	Principle radiation emitted	Main occurrence
Ac^{227}	21.8 years	B	Natural
Am^{241}	4.33×10^2 years	α, γ	Nuclear reactors
C^{14}	5.7×10^3 years	β^-	Natural and nuclear reactors
Cs^{134}	2.06 years	β^-, γ	Nuclear reactors
Cs^{137}	30 years	β^-, γ	Nuclear reactors and nuclear weapons
I^{131}	8 days	β^-, γ	Nuclear reactors and nuclear weapons
K^{40}	1.3×10^9 years	β^+, γ	Natural
Np^{237}	2.14×10^6 years	α, γ	Nuclear reactors
Pu^{239}	1.44×10^4 years	α, γ, SF	Nuclear reactors and nuclear weapons
Ra^{226}	1.62×10^3 years	α, γ	Natural
Rn^{222}	3.82 days	A	Natural
Sr^{89}	52 days	β^-	Nuclear reactors and nuclear weapons
Sr^{90}	28 years	β^-	Nuclear reactors and nuclear weapons
Th^{232}	1.41×10^{10} years	α, γ	Natural
U^{235}	7.04×10^8 years	α, γ, SF	Nuclear reactors and nuclear weapons
U^{238}	4.47×10^9 years	α, γ	Natural

However, the extent to which a radionuclide can contaminate the food chain, in addition to its half-life, shall also depend on physicochemical characteristics of the growth medium, growing environment, agronomic practices, crops, and their management. Among different broad categories of agricultural products, i.e., foods, fibers, fuels, and raw materials, specific foods such as cereals (grains), vegetables, fruits, oils, meats, and spices are important for radionuclide contamination of the food chain. Fibers such as cotton, wool, hemp, silk and flax, and raw materials include lumber and bamboo may act as biological fixers for radionuclide. Other plant-derived materials such as resins, dyes, drugs, perfumes, biofuels, and ornamental products, such as cut flowers and nursery plants, chiefly lead to radionuclide contamination of the environment. Agricultural parameters, viz., soil type and related physico-chemical attribute, use of biological manure, genetically modified organism, green manure, conservation agricultural practices, fertilizer and irrigation level method schedule, abiotic stresses, crops and crop variety interference in terms of uptake and induced rhizosphere modification, crop rotation, and other attributes further increase the complexity. Recent focus on effect of global warming on agriculture adds another dimension, which has to be researched to understand and get indepth insight into the radionuclide transfer dynamics under climate change scenario. Limited isolated information is available on effect of some agricomponents on transfer dynamics of radionuclides marred with contradictions and uncertainty. These contradictions essentially originate from difference in the soil physico-chemical characteristics, growth environment and crop behavior, tissue, and stage of sampling which are either not determined or spelled out clearly. This chapter attempts to critically review the available information on radionuclide dynamics in soil—plant system and discusses the discrepancies in light of the complex network of interaction between various agricultural attributes and those operating at the plant level.

2. Radionuclides in environment

The use of radionuclides for sufficing energy requirement is a powerful technology favored by both developing and developed countries around the globe. However, safety of these nuclear plants is sometimes compromised by human/technical/natural calamities, leading to nuclear accident. Chernobyl or Fukushima disasters are one of the most reminded nuclear accidents, which lead to accidental release of the radionuclides such as ^{137}Cs and ^{125}I in the environment. Once in the environment, the dispersion and precipitation of the radionuclide can occur over several thousand miles and depends on the weather attributes such as wind direction, wind speed, rainfall, and so on. Furthermore, these naturally occurring and/or precipitated radionuclides can potentially enter the animal and human food chain chiefly through plant-derived feed and economic produce. Among different environmental radionuclides, i.e., airborne (99)Mo, (99m)Tc, (129m)Te, (129)Te, (132)Te, (131)I, (132)I, (133)I, (133)Xe,(133m)Xe, (135)Xe, (134)Cs, (136)Cs, (137)Cs, and (140)La, radiocesium (^{137}Cs) is a major long-lived contaminant, and following Fukushima accident, the in situ half-lives of the long-lived radionuclides such as ^{134}Cs and ^{137}Cs were used as reflection of environmental dispersal (Nagaoka et al., 2012). Owing to nuclear weapon tests and incidents at nuclear power plants, several soils have been contaminated by radiocesium (Cs). Areas subjected to extreme or diffuse emissions by Cs have become a major environmental concern. Pollution from Cs poses a significant risk to human health, as Cs can reach the human body through the food chain. Owing to the relative long half-life of its two isotopes (30.2 years for ^{137}Cs and 2.1 years for ^{134}Cs) and potassium-like chemical activity (K), Cs is one of the most dangerous radiopollutants present in the world. Similarity between K and Cs in terrestrial ecosystems suggests a substantial risk of contamination of the vegetation above ground, for these two elements appear closely linked as root uptake mechanisms.

3. Accidental radiation contamination of the environment

The explosion destroyed the Chernobyl 4 reactor and killed 30 operators and firemen within 3 months and several more deaths later. Initially, 237 people on site were diagnosed with acute radiation syndrome (ARS) and associated with cleanup and subsequently confirmed in 134 cases. Of those, 28 died as a result of ARS within weeks of the incident. No one off-site suffered from acute radiation effects although a small yet unknown fraction of the thyroid cancers diagnosed after the accident in patients who were children at the time was likely to be triggered by radioactive iodine exposure. The Chernobyl incident was a rare occurrence and the only accident in the history of commercial nuclear power in which radiation-related deaths occurred.

The nuclear disaster in Fukushima Daiichi (almost, Fukushima Dai-ichi [pronunciation] genshiryoku hatsudensho jiko) was a nuclear accident in 2011 at the Fukushima Daiichi Nuclear Power Plant in Fukushima, Fukushima. It was the most serious nuclear accident since the Chernobyl disaster of 1986 and the only other disaster to receive the International Nuclear Event Scale Level 7 event classification (WHO, 2006). ^{133}Cs mean soil concentration is between 0.3 and 25 μg/g dry soil (Coughtrey et al., 1983). However, due to the absence of any biological role of this element, and due to the fact that it is not toxic at these concentrations for plants and soil organisms, ^{134}Cs ($t1/2 = 2.1$ years) and ^{137}Cs ($t1/2 = 30.2$ years) were produced by the first nuclear fission reactions and aboveground nuclear weapon testing (ca. 3%–7% fission yield). But this curiosity soon developed into a significant and global concern when the Chernobyl nuclear facility's fourth reactor burned down in 1986 (Bell et al., 1988). Indeed, this catastrophe led to the release into the atmosphere of enormous quantities of several radionuclides, and especially of radiocesium (Cs).

A general consensus emerged from research on the environmental fate of Cs that Cs exhibits a very similar biogeochemical activity to that of potassium (K). This means that in terrestrial ecosystems, Cs shows low mobility in the soil profile due to its adsorption on clay minerals and a relatively high mobility in biological systems (Smith and Elder, 1999; Rosén et al., 1999; Delvaux et al., 2001). Low mobility in soil is problematic because Cs appears to remain in upper layer of soil, explored widely by plant roots and soil microorganism. While the primary route of Cs entering the food chain in terrestrial ecosystems is the soil-to-plant. Thus, plant absorption of Cs and their accumulation are the primary causes of human exposure to these radionuclides (Carter et al., 1993; Ashraf et al., 2014).

4. Effect of radionuclide contamination on human population

Radionuclide-mediated effects on human population may depend on the nature of radionuclide decay, which could be alpha, beta, and/or gamma decay (Table 13.1). Extent of damage depends on the energy dissipation and penetration characteristics of the decay components, viz., alpha particles that can travel only a few inches in the air and lose their energy almost as soon as they collide with anything and thus are hazardous only when they are inhaled or swallowed or beta particles that can travel in the air for a distance of a few feet and can damage skin but are most hazardous when swallowed or inhaled or γ-rays, which are waves of pure energy, similar to X-rays, that travel at the speed of light through air or open spaces and can present an extreme external hazard or neutrons, which are small particles, with no electrical charge, that can travel long distances in air and are released during nuclear fission and, like γ-rays, can present an extreme external hazard. These effects can be deterministic or stochastic. Health effects in humans have been most studied for radiocesium, which is a major source of environmental and food chain radiation pollution. β- and γ-radiations are emitted by ^{137}Cs and ^{134}Cs. Skin and body tissues are penetrated by β-radiation because it travels short distances, and γ-radiation penetrates entire body because it travels long. The radiation exposure from these radionuclides may be categorized as either external (if radiation source is outside the body) or internal (if radiation source is inside the body). β-radiation released outside of the body is usually of little health concern if the radioactive material does not touch the skin. Tissue damage and disruption of cellular function are induced by β- and γ-radiations. Half-lives of ^{134}Cs and ^{137}Cs are 2 and 30 years, respectively. Owing to the constant emission of radiation, people may be exposed to naturally emitted radiation from ^{137}Cs or ^{134}Cs (Anonymous, 2006; U.S. Nuclear Regulatory Commission; Regulatory Guide, 2006).

4.1 Food and agriculture response to nuclear emergency

No, only events that cause harmful material to be released into the environment or sources of water outside the plant may have a direct impact on food and resources. Contamination of radionuclides can happen in agricultural products and commodities. Radiation can also have adverse biological effects on agricultural plants and animals (Fig. 13.1).

Iodine and cesium are the primary radionuclides of concern in cases where significant quantities of radionuclides have been released from nuclear power plants. Radioactive I-131 has an 8-day half-life and will naturally degrade within weeks. Thus, while it is alarming, it may only be present in the atmosphere for several weeks after release. Radioactive cesium, on the other hand, has a half-life longer than radioactive iodine (Cs-134 has a half-life of approximately 2 years and ^{137}Cs has a half-life of approximately 30 years), and as a result, radioactive cesium can remain in the atmosphere for longer periods of time. Traces of cesium-137, for example, have been detected in the atmosphere for many years after the accident at the Chernobyl power station, and this is also the case after the accident at the Fukushima Daiichi Nuclear Power Plant. Codex guidance standards for radionuclides in globally traded food during a nuclear radiological emergency are agreed globally. The Codex Alimentarius Commission established these Guideline Standards for 20 radionuclides, and they are included in the General Standard for Pollutants and Toxins in Food and Feed (CODEX, 1995).

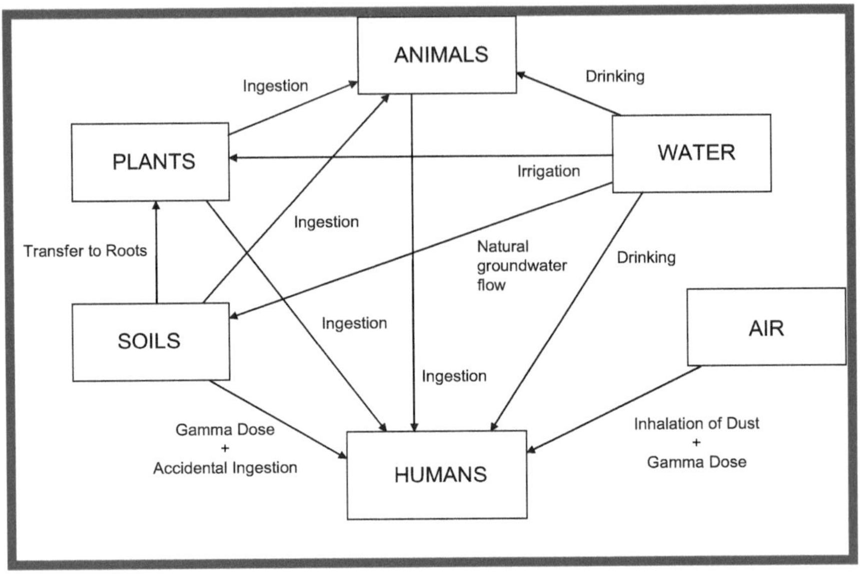

Figure 13.1 Routes of radiation exposure from radionuclide contaminants.

5. Radionuclide transfer in plant–soil environment: an agricultural perspective

Agriculture is the mainstay and catalyst of human existence on earth. Globally, 1.6 billion ha area is cultivated by crops. However, sustaining the productivity levels to feed the burgeoning population is a herculean task in light of variation in soils, agriclimate, crops, and cropping systems. The following sections briefly outline the agricultural challenges that directly or indirectly influence soil–plant–environment transfer dynamics of radionuclide.

The factors that cause variation in radionuclide uptake by crops and types of soil and the time of radionuclide that how much time it has been in the soil are also very important, and other factors that affect radionuclide uptake are organic matter, pH, and crop variety.

6. Effect of soil pH on radiocesium transfer in the environment

A detailed analysis of Cs interactions in natural systems (Hokkanen, 1991). The main chemical characteristics of Cs are as follows: (1) inorganic cation is normally present; (2) its action is the same as $K^+$and NH'; (3) their conduct is modified in a competitive manner by the existence of K^+; (4) it is fixed in clay products, or absorbed almost irreversibly, probably by trapping platelets of clay particles; and (5) mostly, ^{137}Cs is the result of nuclear tests in the atmosphere; its presence in the soil is thus due to the relatively recent deposition in the atmosphere. Following deposition, quantities of ^{137}Cs in surface soils decreased due to leaching, degradation, and radioactive decay. Overall, the pH-to-Cs mobility relationship is perhaps a fundamental phenomenon of cation exchange (Biotechnologie, Agronomie, Société et Environnement, 2019). At low pH, Cs are substituted to a greater extent by hydrogen at soil exchange sites than at high pH (Nutrient and Water Management Practices, 2005; Barberi et al., 2001). Soil acidification would potentially activate Cs. Based on the current information, this is very difficult to determine.Using a simple cation exchange model, a tenfold shift in hydrogen ion concentration (1 pH unit) could displace 90% of the exchangeable Cs, which shows analogy with other compounds solubilized by acid rain. Assuming that if pH controls the mobility of an element in the laboratory, variations in plant uptake will result in acid rain conditions in the field, is simplistic. In their latest analysis of acid rain's effect on soils, studies have shown an effect of acid rain on the uptake of metal by field plants. Ulrich et al. determined Al toxicity through soil acidification of beech roots. In general, the mobile element increasingly represents the number of radioactive decay per second or Bq (Becquerel). The exact fate of the activated object becomes a function of the hydraulic regime on earth. But an evaluation of the fate of radionuclides emitted by acid rain needs to pay close attention to the system's water flow (Sheppard and Sheppard, 1988).

7. Effect of organic matter on radionuclide uptake

Chernobyl accident has shown that a strong soil-to-plant transfer of ^{137}Cs occurs over the years in organic soils. In several of these soils, the particular sites in clays regulate the adsorption of radiocesium, organic compounds having an indirect effect. Adsorption occurs mainly on nonspecific sites in organic soils with more than 95% of organic matter content and marginal clay content. Two key factors that account for the high transfer after a contamination event: the low solid-liquid distribution coefficient due to low clay content and high concentration of NH_4^+ in the soil solution, and the low availability of K^+ that enhances root uptake. The calculation of the reversibly adsorbed fraction by means of desorption protocols agrees with the previous conclusions, since it cannot be associated with the content of organic matter and demonstrates the lack of specificity of the organic phase adsorption. In addition, the time-dependent pattern of the exchangeable fraction is related to the dynamics of transfer of soil plants (Rigol et al., 2002).

8. Soil characteristic and radionuclide distribution

Sustainable agricultural production requires a healthy physical and biochemical soil characteristics. Different soil types posses varied soil physical and chemical characteristics (Fig. 13.2) all of which can alter the soil distribution and variability of nutrients and chemical elements including radionuclides such as Cs. The following section

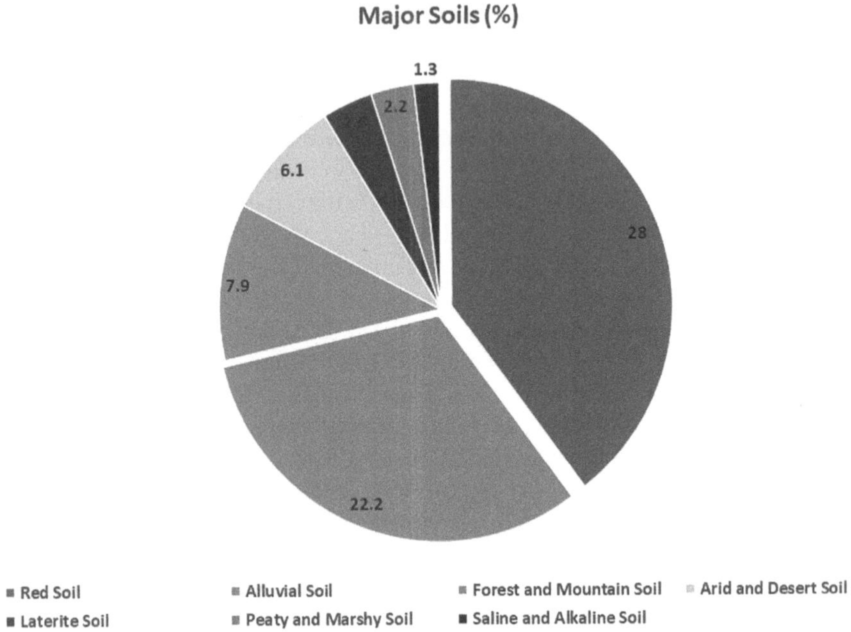

Figure 13.2 Indian soil types and their characteristics.

outlines and critically discusses various soil attributes that can limit or induce soil availability, adsorption and desorption kinetics, and bioavailability of radionuclides.

Soil is our primary natural, economic resource. In India, the composition and structure of the soils differs. Prominent types of soils found in India are as follows:

Alluvial soils ($\sim$43%): They are rich in humus and very fertile. They are found in Great Northern plain, lower valleys of Narmada and Tapti, and Northern Gujarat.

Black soils ($\sim$15%): These soils occur in parts of Maharashtra, Chhattisgarh, Madhya Pradesh, Gujarat, Andhra Pradesh, and Tamil Nadu.

Red soils ($\sim$18%): They cover almost the whole of Tamil Nadu, Andhra Pradesh, Chhattisgarh, Karnataka, Maharashtra, and parts of Orissa.

Acid soils ($\sim$11%): Soils are commonly found in Kerala, Tamil Nadu, Maharashtra, and Chhattisgarh, as well as the Orissa and Assam mountain regions.

Mountain soils ($\sim$8%): These soils are common in the Himalayas and vary in different regions depending on the altitude.

Desert soils ($\sim$4%): These soils are not well formed and are found in the desert regions of Rajasthan. These soils are normally sandy and are deficient in organic matter.

9. Soil characteristics

Transfer of radioactive elements from soil to plant is a complex process that is controlled by soil's geochemical features, including the pH, clay, silt, and organic matter content that reduce the bioavailability of soil radionuclides to plants, which in turn reduces the risk of radionuclide contamination from biota (Hegazy et al., 2013).

The absorption of uranium, thorium, radium, lead, and polonium from the soil by plant roots is affected by a variety of factors, including polluted soil properties, local climate conditions, and plant biosynthetic behavior. The main factor for the availability of these elements in soils to plants is likely the solubility and thermodynamic behavior of uncomplexed ions (Jensen and Thomas, 2010, Luoma and Jenny, 1977). For root absorption to occur, for a finite time, ionic forms that are soluble in the soil solution must exist adjacent to the root membrane (Cataldo and Wildung, 1978). The soluble species' chemical composition would have a significant impact on their survival in the soil solution, their stability in soil, and the rate and degree of plant absorption. Based on studies with different trace metals, it can be inferred that the solubilization of radionuclides is primarily determined by the shape and concentration of the involved radionuclides, as well as by the physicochemical characteristics of the soil.

10. Plant attributes

10.1 Crop and cropping system

All inclusive agricultural systems are often based on optimizing the productivity of monocultures. In these scheme, crop diversity is limited to one or very few species that are generally genetically homogeneous plants, the planting pattern is standardized and symmetrical, and external inputs are often supplied in large quantities. These processes are widely criticized today for their harmful effects on the environment, such as soil erosion and depletion, chemical pollution, biodiversity loss, and the use of fossil fuels (Perera, 2018; Tilman et al., 2002). In comparison, multispecies crop systems may often be regarded as the realistic implementation of ecological concepts based on biodiversity, plant interactions, and other mechanisms of natural regulation. They are believed to have potential benefits in terms of productivity, output stability, disruptive resilience, and ecological sustainability, although they are sometimes considered more difficult to manage (Vandermeer, 1989). A majority of the world's farmers, particularly those located in tropical regions, still depend for their food and income on multispecies agricultural systems, i.e., the cultivation of a variety of crops on a single piece of land (Malezieux et al 2009). These systems, mostly without synthetic inputs and focused on integrated management of local natural resources and, in many cases, on sound biodiversity management, theoretically provide numerous ecological benefits (Malézieux et al., 2007).

10.2 Crop rotation

Another important agricultural practice is crop rotation wherein crops are so rotated on a farm to obtain maximum annual production without any negative impact on soil quality. Crop rotation is reported to improve soil structure and soil fertility and helps in control of pests and protection against losses due to biotic and/or abiotic stresses. Rotating crops is one of the key principles of conservation agriculture. It helps in improving the soil structure, increasing the soil fertility, controlling weeds, and reducing the risk of pests and diseases. Liu et al. (2005) determined the activity concentration of Cs and Sr in soil and cereals at 17 monitoring sites in Japan and found that the average concentration of ^{137}Cs has steadily decreased and reached 1/10th and 1/100th of the peak, respectively, in soil and brown rice. These findings can be used as precedent to estimate radionuclide deposition at the Fukushima Daiichi Nuclear Power Plant.

11. Management of Farm Resources and radionuclide Transfer

11.1 Water Management

Mass flow and diffusion is the process by which solutes are transported to plant roots (Barber, 1962). Along with the mass flow of solutes, the absorption of root water reacts to transpiration. Depletion of the solvent at the root–soil interface creates a concentration gradient that initiates additional diffusional transport of the solvent to the roots, if the root absorption rate of a solution exceeds the mass flow rates. As a result, a depletion zone forms around the absorbent base, which in the long term decreases the solute uptake rates (Tinker and Nye, 2000). Nutrients and trace substances taken by plant will be affected by the depletion of nutrient in rhizosphere. This effect can be significant in the nutrient-poor soils (Delvaux et al., 2000; Rengel and Marschner, 2005). Solvent transport to roots affected by root transport, both diffusion and mass flow, decreases by reduction, but concentrations of exchangeable cations in soil solution increase (Nye and Tinker, 1977). Impact of changes in moisture on plant uptake processes can be even more complex, under natural conditions, as the root system physiology and morphology are often affected by differences in water content (Basu et al., 2016; Smucker and Aiken, 1992). Concentrations of ^{90}Sr and ^{137}Cs in pasture vegetation growing at various types of soil were negatively associated with soil moisture (Greger, 2004). They intersperse this result as the rise in solvent concentrations with decreasing soil humidity is of primary importance for the root uptake rates of both radionuclides (Delvaux et al., 2000).

11.2 Fertilizer management

To study the transfer factor, competition between ions is not an important aspect. The most significant conceptual disadvantage of the transfer factor method is that it does not recognize ion rivalry. Soil-to-plant transfer factors are most frequently measured for trace substances whose activity in the soil–plant system largely depends on the concentrations of macronutrients. Plant absorption of radionuclides and heavy metals is affected by a variety of naturally occurring substances in soils (Tangahu et al., 2011; Wallace, 1989; Desmet et al., 1991; Lorenz et al., 1994). These competitive effects form the basis for variation in soil -plant transfer of radioactive cesium and strontium. The concentration of a trace substance accumulating in plants may not primarily depend on its absolute concentration in the soil—plant system but on the concentration ratio to other micro- and macronutrients (Wallace, 1989; Desmet et al., 1991; Konoplev et al., 1993).

11.3 Weed and insect pest management

Weed in agriculture system is one of the major problems around the world, reducing crop yield and income of farmers. In industrialized countries, despite the availability of high-tech solutions (e.g., selective herbicides and genetically modified herbicide-resistant crops), the proportion of crop loss due to weeds does not appear to decrease significantly over time (Cousens and Mortimer, 1995; Green and Owen, 2011). In developing countries, farmers also need to rely on alternative weed management methods because herbicides are rarely reachable at a reasonable cost. The limited global success in weed control is likely the result of oversimplification in solving the problem. Great emphasis has been put on developing weed control methods (especially synthetic herbicides) as the "solution" to any weed problem, with the importance of incorporating different tactics (such as preventive, cultural, mechanical, and chemical methods) into growing crops. The weed control approach on a machine basis has always been overlooked.

Integrated weed management is focused on knowledge of the biological and ecological characters of weeds to understand how cultivation practices can modulate their presence. The farmer must first develop a detailed weed management strategy within his cash crop sequence based on this information and then choose the best method for direct weed control during crop growth cycles. In addition, it must be ensured that weed control is always strictly embedded in the plant management itself. As such, due account must be taken of the relations between cannabis control and other cultural activities. For example, using harvest cover crops is an interesting way to combine weed control with nutrient management in systems with little external feedback, with additional benefits for other essential agroecosystem properties, e.g., soil fertility, soil moisture retention, biodiversity, and so on (Barberi et al., 2001; Bàrberi and Cascio, 2001).

11.3.1 Use of pesticides

Pest management needs to be managed in such a way to optimize pest control while minimizing pesticide loss to the environment. Integrated pest management (IPM) offers a broad range of services to help farmers control pest issues (Fig. 13.3). This approach is intended to allow growers to accurately determine whether there is an economically justifiable pest management plan, good growing practices, proactive pest management programs, and a regular pest management program. Integrated pest control uses a systematic method for identifying the missing link in the cycle

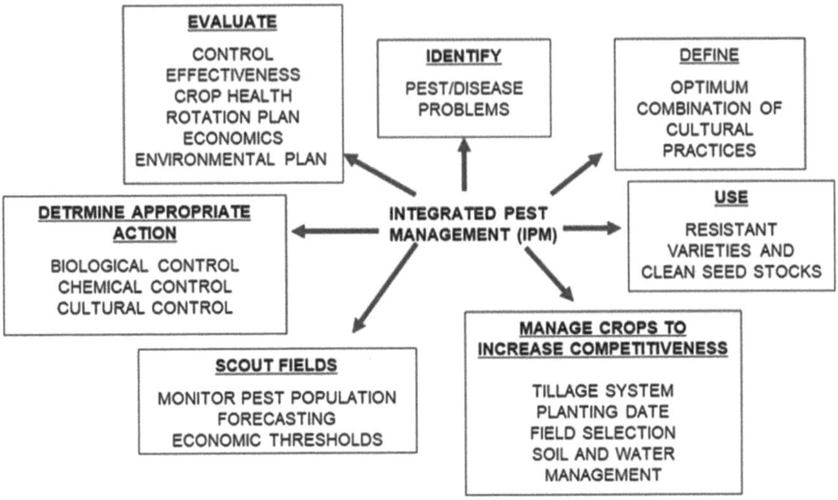

Figure 13.3 IPM practices that may impact the atmospheric radionuclide transfer. *IPM*, integrated pest management.

of pest life and may mediate or interfere with the natural movement or transfer of radionuclides in the atmosphere. Sound pest control systems do not aim at eliminating pests but rather at controlling them to minimize crop economic losses. It involves the practices:

- Pick plants and varieties that are resistant to pesticides
- Timing of planting and harvesting dates, crop rotation to reduce damage to pest
- Monitoring of species of pests and natural enemies
- Usage of valuable insects and other environmental controls

The theory behind the IPM approach is to establish unfavorable pest accumulation conditions by enhancing crop vigor and protecting natural enemies that help manage populations of pests (Hokkanen, 1991).

11.4 Trap cropping

For centuries, insect trap control theory has been developed and is still used explicitly or implicitly in many traditional farming systems. During this century, disease control ideas have emerged as noted in the entomological literature but have rarely resulted in practical applications. For example, in 1860, Curtis (2015) suggested controlling the parsnip worms *Depressaria depressella* and *Depressaria daucella* on carrots when placing parsnip plants 2–3 m apart between carrots. Moths tend to suck and lay eggs on carrots, the trapping crop in which larvae cease growth and die. Similarly, by 1900, only a few years after pest identification in the United States (Sawidis,

1988), the management of the cotton boll weevil by trap cropping was already suggested. In reality, however, common, successful uses of trap cultivation are a phenomenon of the past two decades. To date, there have been very few practical applications of trap cropping in modem farming. Approximately, 11 plague species (seven organism classes) are or have been successfully controlled in four rising media where trap crops played a major role (Table 13.1): cotton, soybeans, potatoes, and cauliflower. Globally, the cotton and soybean crop systems are of the greatest importance. In addition to agricultural use, the trap cultivation principle was used in control of forest pests. Trap trees, or just tree trunks, were long developed in Europe to catch certain types of plagues. For example, control of the spruce bark beetle, *Ips typographus*, was practiced this way more than 200 years ago (Hokkanen, 1991).

12. Environmental stresses and radionuclide transfer

Plant growth and development depends on provisioning of optimum experiments, which should be free of biotic and abiotic stresses.

12.1 Biotic stresses

The natural environment of plants is composed of a complex set of abiotic and biotic stresses. The response of plants to these stresses is complex. Systems biology takes the molecular parts of an organism (transcripts, proteins, and metabolites) and attempts to incorporate them into functional networks or models developed to describe and predict the dynamic activities of the organism in various environments (Ashraf et al., 2014). Three varieties of (*Triticum aestivum* L.) cultivars (Mironovskaya 808, Polesskay 70, and Kiyanka) in greenhouse experiment were affected with powdery mildew and brown rust. Collected in the exclusion zone of Chernobyl, it was 1.5–2.0 times higher than the plants developed from seed control. In the Chernobyl zone, tolerance of wheat plant to biotic stress has been diminished. Production of the disease was enhanced, on plots with increased radiation history at artificial infection with brown rust. Under low dose of chronic irradiation, a decreased activity of plant protease inhibitor is clearly one of the mechanism causing the decline in phytoimmunity. Compared to unirradiated control, the observed decline in protase inhibitor activity with radiation varied between 35–60% in grains of wheat and rye. The microevolutionary processes occurring in both plants and their pathogens, thus, must be studied under technogenic stress conditions (Dmitriev et al., 2011).

12.2 Abiotic stresses

Abiotic and biotic stresses exert environmental pressures on crop plants. Temperatures, water, salinity, and alkalinity are various abiotic stresses on crop plant, which have worldwide negative effect on crop productivity. These abiotic constraints are intermingled in the form of osmotic stress, which causes the cell to disrupt ionic distribution and homeostasis. Growth rates and productivity are affected by expression of a group of genes. To understand the stress response mechanism in plants, therefore, it is extremely important to learn about some of the genes that are subject to abiotic stress. The transmission of radionuclides into the soil–plant system may differ from that of abiotic stress.

12.2.1 Low temperature stress

Productivity of agricultural crops is affected by cold stress, one of the major abiotic stresses. Immobile plants change their metabolism to withstand such stress. In cold acclimatization process, temperate plants become sweat and freeze tolerant when exposed to sublethal cold stress. However, many agriculturally important crops are incompetent with cold acclimatization. Cellular activity in plants is affected by cold stress. Signal transduction component pathways are transduced by cold stress. Cascades of Ca^{2+}, ROS, protein kinase, protein phosphatase, and lipid signaling are essential constituents. Cold stress is also mediated by ABA. Different species of plants, which depend on the gene expression, concerned to alter their physiology, metabolism, and growth withstand cold stress in a different way. Various changes such as kinase expression associated with signal transduction, osmolyte accumulation, and membrane lipid composition are associated with cold stress.

12.2.2 Salinity

Salinity in soil is a global issue and a significant challenge to global agriculture, as it reduces crop yields in affected areas. Crop growth and yields are restricted by salinity stress in various ways. Two major consequences placed by plants are toxicity of ion and osmotic stress.

In plant cells under normal condition, the osmotic pressure is higher than that in soil solution. The plant cells use this increased osmotic pressure to absorb water and essential minerals from the soil solution into the root cells. In soil solution, the osmotic pressure goes beyond under salt tension. Osmotic pressure in plant cells limits the ability of plants to absorb water and minerals in the presence of more salts and enter the cells and have direct toxic effects on cell membranes and metabolic activity. Some side effects such as material assimilation, reduced cell expansion, decreased cytosolic metabolism, and ROS production and membrane function are primary effects of salt stress. The key mechanisms against ionic stress during salt stress include reducing the absorption of toxic ions such as Na and Cl into the cytosol and sequestration of these toxic ions into either the vacuole or the apoplast.

12.2.3 Water stress

Temperature, atmospheric CO_2, and pattern of rainfall are affected by climate change. Prolonged low rainfall is result of drought. Intermittent drought conditions affect the growth and development of plants but are generally not lethal, but extreme drought conditions cause the supply of soil water in plants to decrease gradually and cause premature death of plants. Varieties, i.e., tolerant to drought, succor high yields and growth during drought condition, the product of the plant's efforts to withstand stress. It is said that the plant is acclimatized to drought, if tolerance is restricted to that particular generation. The plant genotype that persists for decades is believed to be adapted to the drought conditions. Arrested growth is first reaction under drought stress. Decreased metabolic demand of the plant is result of reduced shoot growth under drought and mobilizes metabolites to synthesize defensive compounds required for osmotic adjustment.

12.2.4 High temperature stress

Increased temperature is threat for world with serious consequences for plant growth and development, particularly crops. With higher-temperature fluctuations, crop plant vulnerability will increase in the near future. Therefore, much attention needs to be paid to the heat stress response mechanism in plants, particularly in crops, and the cultivation of heat-tolerant breeds. Yield and gas exchange attributes decrease with heat stress. Increased temperatures impede pollen grain swelling during flowering, leading to weak release of pollen grains and anther indehiscence. For plants to withstand heat stress, multiple molecular and physiological pathways have been created.

12.2.5 Heavy metal stress

Growing agriculture's dependence on chemical fertilizers and wastewater irrigation and rapid industrialization have introduced toxic metals to agricultural soils, which have had adverse effects on the soil environment. For instance, cadmium (Cd) is the principal metal. Contamination of heavy metals stays on the soil for thousands of years, which is considered a major environmental problem of the agricultural (Kumar, 2013). Radionuclide transfer dynamics is not only influenced at the macro, i.e., crop, but also at micro, i.e., individual plant, level. How this might occur is briefly discussed to encompass aspects of radionuclide uptake, role of plant root, rhizosphere modification including chelates, gas exchange characteristic viz., transpiration stream, stomatal conductance and carbon metabolism wherein a high photosynthetic carbon fixation means more vegetative growth to induce a higher demand and uptake of soil minerals including radionuclides.

13. Crop management and radionuclide transfer in the environment

13.1 Radionuclide mobilization

A variety of issues arise when organic complexants, nutrients, or selectives are added to improve mobilization of metals and radionuclides, by applying silt and clays to exploit electrokinetic processes. For example, the interaction of the injected material with the local bacterial population may affect its homogeneous penetration by building up biomass, which obstructs the injection wells because the bacteria inevitably proliferate faster than the material injected moves. Slower-reacting additive mechanisms and programmed microbe introduction are solutions to overcome these kinetic constraints. In addition, for the effective modeling and implementation of engineering design, a clear understanding of the microbial pathways at a given location, including the concentrations and transformation mechanisms of the injected material, is important. Information on the speciation of complexes with low radionuclide concentrations, including their solubility and stability constants, is limited. Further inconsistencies in literature report exist toward knowledge of the limits to toxicity, acceptability, and bioavailability of radionuclides under variable environmental conditions (Anna and Murthy, 1999).

13.2 Radionuclide translocation

Radionuclide uptake by plants in the atmosphere can occur in two ways: uptake by root soil and plant foliar absorption by leaves and shoot. Data can be collected from analyses of fission product radionuclides accumulated as global fallout and from plant physiology tracer studies on radionuclide plant accumulation. Anatomy and physiological factors of leaf determined the transport of radionuclides via the cuticle and epidermis of plant leaves. The foliar absorption of fallout radionuclides, ^{99}Sr, ^{131}I, and ^{137}Cs, is example from scientific literature. For example, ^{131}I's environmental half-life is considerably shorter than its physical half-life due to physical and biological factors, which can produce as short a half-life as 0.23/day. Radionuclides ^{99}Sr and ^{137}Cs are readily absorbed by the leaf, but production of fruit and translocation of seeds by ^{137}Cs is more than ^{99}Sr. Chemical and physical soil factors that can selectively maintain a radionuclide, such as ^{137}Cs, are primarily influenced by soil–root absorption. Soil and plant absorption of radionuclide ^{99}Sr and ^{137}Cs would have impact by the presence of organic matter, inorganic colloids (clay), and competing components. γ-radioactivity measured from a wide range of plant samples showed that radioactivity in plants depended on the area of the exposed surface, the plants' developmental season, and the external morphology of their parts above ground. Furthermore, autoradiographs of many plants revealed the presence of radioactivity in some spots (hot spots) that were randomly distributed on the plant surface and varied in size and activity (Sawidis, 1988).

13.3 Radionuclide partitioning

Partitioning is important as when the radionuclide is retained in the root or the leaf or the stem and/or is not transported to the grain or the edible economic produce in quantities over and above the permissible limit; then the said farm produce can be used as feed or food.

14. Transfer of radiocesium in plant and soil environment: a case study

14.1 Agriculture attributes affecting radiocesium transfer

To a greater or lesser degree, any soil used for agriculture anywhere in the world produces radionuclides. Typical soils (IA89a) contain about 300 kBq/m^3 of 40 K at 20 cm depth. Radionuclide concentration of between 50 and 150 Bq/kg in food is taken up by crops. Route of ingestion of radionuclides in food is one of the routes leading to internal retention and triggering human exposure from natural sources. Level of radionuclide increase in food due to imprudent contamination of agricultural land, as in a serious accident, is a major concern. Most important radionuclide pollutants in agriculture are those with relatively high crop intake, high transfer rates to animal products, and long radiological half-lives. However, the ecological processes leading to crop contamination and the radionuclide's radioecological actions are complex and are influenced not only by the radionuclide's physical and chemical properties but also by factors such as soil type, crop system (like tillage), climate, season, and, where relevant, biological half-life within animals. In agriculture, the main radionuclides of concern after a nuclear reactor accident are ^{131}I, ^{137}Cs, ^{134}Cs, and ^{90}Sr. The key cause of pollution of agricultural produce in temperate regions is direct deposition on plants. Cesium and ^{90}Sr isotopes are relatively static in soil; the plant deposition is more significant than absorption of roots. Clay mineral composition, organic matter content, tillage practice, and environment all affect proclivity to move into groundwater. The factor that regulates the soil solution concentration also affects the availability of plant. Pickup of cesium and strontium by plant is same as potassium and calcium, respectively; the degree of their absorption depends on the element availability. Increase levels of potassium fertilization can thus decrease the absorption of cesium, and liming can reduce the absorption of strontium (Agricultural and Environmental Impacts, 2002).

14.2 Physicochemical effect of soil on radiocesium

Matsunaga et al. (2013) describe that soil mineralogical research, including the detection of clay minerals, suggested that the exchangeable fraction of ^{137}Cs could be reduced by smectite and mica. In the upper soil layer, there was, however, no direct correlation between the mineral composition and the retention of ^{137}Cs. ^{137}Cs extractability and properties of soil such as pH, organic matter content, finer particle size, and cation exchange capability show positive correlation. These findings indicate that the mineralogical impact of firm fixation of 137Cs on soil constituents can be masked by the nonspecific adsorption provided by the soil's physicochemical properties. Transfer factors from soil to plant are widely used to estimate the movement of radionuclides into the food chain. Their concept assumes that the concentration of a radionuclide in a plant is linear only in relation to concentration in the soil's root zone (Ehlken and Kirchner, 2002). The range of transfer factors (TFs) for radiocesium and radiostrontium is broader when comparing individual data for different types of soil.

The largest average difference in TF between soils found in one inventory of transfer data is a factor of 50 for radiostrontium and of 40 for radiocesium. The variability of TF on agricultural (i.e., fertilized) mineral soils, however, is a maximum factor of 10, while the hardness of these soils varies considerably. The less possible impact fertilization may have decreased the transfer of radiocesium and radiostrontium from soil to plant, when the amount of nutrient mineral in the soil is higher (Lembrechts, 1993).

14.3 Absorption and deabsorption of radiocesium

From polluted land, root absorbs radiocesium and transfers into the food chain. On both the initial nonrhizosphere soil and rhizosphere soil, Cs adsorption in dilute soil suspension has been studied. Cs desorption in solutions containing different stable C concentrations was calculated by resuspending subsamples of contaminated soil. The proportion of Cs collection, i.e., not readily desorbable, was calculated by a comparison of adsorption and desorption isotherms. Different soil shows variation in uptake of radiocesium, and diffusion did not seem to be restricted. Study shows that transfer of radionuclide from root to shoot did not differ for the soils. Kd's value was elevated to a magnitude order. A large proportion of adsorbed Cs was found to be fixed; on desorption, the Kd was up to seven times higher than adsorption, suggesting that up to 80% of adsorbed Cs were not readily exchangeable. Action of root did nothing to repair the fraction (Guivarch et al., 1999).

14.4 Biochemical regulation of radiocesium transfer in soil

Study on the uptake of cesium in plants dates back to the early 1940s (Collander, 1941; Zhu and Smolders, 2000); research is based on variations in solution absorption of ions and evaluation of the suitability of using radioactive isotopes to monitor nutrient intake in plants such as K^+ (Epstein and Hagen, 1952; Menzel, 1954; Kiser et al., 2008). In the 1950 and 1960s during nuclear weapon tests, the radiocesium deposition provided chance for research in radiocesium mechanics via foliar and root absorption pathways (Russell, 1966). Broader area of Europe and former Soviet Union was accumulated with radiocesium after Chernobyl Nuclear Power Plant accident in April 1986. Three areas are mainly focused during the research: (1) radiocesium mechanics for plant uptake; (2) radiocesium regulation for plant uptake, i.e., characteristics of plant, soil potassium, and mycorrhizas; and (3) formulating agricultural countermeasures for and remediating radiocesium-contaminated soils.

14.5 Soil to plant transfer factor of cesium

The mechanics regulating and determining the transfer of ^{137}Cs from the contaminated soil into an edible farm produce i.e., the absorption of radionuclide by the plant root, and its transfer to the above ground vegetative and reproductive part of the plant (Guivarch et al., 1999; Marčiulionienė et al., 2017; Rai and Kawabata, 2020) ultimately govern the long-term internal radiation exposure in humans. According to Zhu and Smolders (2000), the genetic difference in absorption of ^{137}Cs is obvious from the wide variety of transfer factors present in different plants. In one species developing on the same soil, the developmental stage and mycorrhizal symbiosis are important for the efficiency of Cs uptake (Zhu and Shaw, 2000; Berreck and Haselwandter, 2001). Typically, <3% of the total Cs is consumed in the soil under natural conditions (Lasat et al., 1998; Sahr et al., 2005). As the time increases, the availability of cesium decreases in soil (Jagetiya et al., 2014). The main route of transfer of radiocesium from soil to human diet via plant absorption analyzes the factors regulating radiocesium uptake. The greatest effect on the uptake by Cs, as observed under the nutrient solution culture, is exerted by the potassium (K) supply. K^+ transporter and the K^+ channel pathway are two transport pathways on plant root cell membranes that regulate the uptake of radiocesium. On isolated K-channels or K carriers, the cationic relationship between K and Cs is consistent with studies involving intact plants. Less discrimination against Cs^+ is demonstrated by K^+ transporter operating at low external potassium concentration (often < 0.3 mM), whereas high discrimination against Cs^+ is demonstrated by K^+ channel at high external potassium concentration. Inside plants, radiocesium is most likely taken up within the plant by the K transport systems, Nonuniform ratio of Cs:K is present within the plant. Differences between different plant species grown under similar conditions in internal Cs can vary by a factor of 20. While phytoremediation can be an alternative to decontaminating radiocesium-contaminated soils, its main downside is that it takes too long (10 years) and produces significant quantities of waste.

The transfer factor for soil-to-plant (TF, usually expressed as Bq/kg dry weight plant tissue/Bq/kg dry weight soil) describes absorption of radiocesium from soil. An insight into various soil physico-chemical attributes that are likely to effect the transition of radionuclides in a soil-plant continum, reveal a major regulatory influence of the soil texture and the potassium status, both in the soil and the plant on the soil-plant radionuclide transfer (Guivarch et al., 1999). For mineral soils with loam or clay texture, the TF values are within the $0.001-1$ range, whereas the TF value for fertile or sandy soils may be more than 1 or even as high as 28.5 for agricultural crops (Sanchez et al., 1999; Nisbet and Woodman, 2000). The relationship between TFs for radiocesium and soil exchangeable K may be defined in some cases by a negative power function (Nisbet et al., 1999; Zhu and Smolders, 2000). A large database was analyzed by Nisbet and Woodman (2000) who discovered that TFs of radiocesium are independent of concentration of soil radiocesium. This validates the use of the TF method to forecast radiocesium plant uptake in agricultural systems (Zhu and Smolders, 2000).

15. Conclusion

Climate change is eminent and is regarded as a major environmental concern that may limit the agricultural productivity and thus the survival of human and animal population. It will not only cause a rise in temperature and carbon dioxide but also alter various soil and plant traits mentioned earlier to alter the radionuclide transfer. A change in root exudation properties of the plant owing to an altered source efficiency for carboxylation in particular or the carbon metabolism in general or the source to sink transfer of fixed carbon may alter the soil pH and the biological activities to influence the availability and uptake of mineral elements that include radionuclides of natural or manmade origin. Climate change is also likely to cause an increase in the frequency of the extreme weather events such as flooding, drought, or high or low temperature stress and may also alter the interplay of different biotic components in agriculture to change the dynamics of biotic stress under the climate change scenario. No study has yet been reported on combined effects of high CO_2 and temperature on radionuclide transfer dynamics in soil–plant system. It will be interesting to not only investigate the aforementioned relationship but also understand the agricultural factors that get indirectly affected under climate change scenario, such as soil temperature and gas exchange characteristics to impact radionuclide transfer.

References

Agricultural and environmental impacts, 2002. Chapter VI, Chernobyl: Assessment of Radiological and Health Impact.

Anna, P., Murthy, A.V., 1999. Natural and accelerated bioremediation program, Workshop report US. Department of Energy, Virginia, pp. 1–38.

Anonymous, October 12, 2006. Multi-: Chapters 10-17 and Appendix F. (July 2004) NUREG-1576, EPA 402-B-04-001B, NTIS PB2004-105421. Available at: http://www.nrc.gov/reading-rm/doc-collections/nuregs/staff/sr1576/sr1576v2.pdf.

Ashraf, M.A., Khan, A.M., Ahmad, M., Akib, S., Balkhair, K.S., Bakar, N.K.A., 2014. Retracted article: release, deposition and elimination of radiocesium (137 Cs) in the terrestrial environment. Environ. Geochem. Health 36 (6), 1165–1190.

Barberi, P., Lo Cascio, B., 2001. Long-term tillage and crop rotation effects on weed seedbank size and composition. Weed Res. 41 (4), 325–340.

Barber, S.A., 1962. A diffusion and mass flow concept of soil nutrient availability. Soil Sci. 93 (1), 39–49.

Bàrberi, P., Bonari, E., Mazzoncini, M., 2001. Weed density and composition in winter wheat as influenced by tillage systems. In: Proc. Ist World Congress on Conservation Agriculture, Madrid (Spain), 1–5 October, pp. 451–455.

Basu, S., Ramegowda, V., Kumar, A., Pereira, A., 2016. Plant adaptation to drought stress. F1000Res. 5.

Bell, J.N.B., Minski, M.J., Grogan, H.A., 1988. Plant uptake of radionuclides. Soil Use Manag. 4 (3), 76–84.

Biotechnologie, Agronomie, Société et Environnement, 2019. 1370-6233 1780-4507 Facteur d'impact: 0,457, citation.

Berreck, M., Haselwandter, K., 2001. Effect of the arbuscular mycorrhizal symbiosis upon uptake of cesium and other cations by plants. Mycorrhiza 10, 275–280.

Carter, M.W., Burns, P., Munslow-Davies, L., 1993. Radiotoxicity hazard classification- the basis and development of a new list. In: Technical memorandum, 38. Australian Government Publishing Service, Canberra, p. 23.

Cataldo, D.A., Wildung, R.E., 1978. Soil and plant factors influencing the accumulation of heavy metals by plants. Environ. Health Perspect. 27, 149–159.

Codex, 1995. General Standards for contaminants and toxins in food and feed, 193-1995. GSCTFF Codex Stan, pp. 33–37.
Collander, R., 1941. Selective absorption of cations by higher plants. Plant Physiol 16, 691–720.
Coughtrey, P.J., Jackson, D., Thorne, M.C., 1983. Radionuclide distribution and transport in terrestrial and aquatic ecosystems. Crit. Rev. Data Vol. 3 (AA Balkema).
Curtis, J., 2015. 1 860. Farm Insects, Blackie & Son, Glasgow, Scotland.
Cousens, R., Mortimer, M., 1995. Dynamics of Weed Populations. Cambridge University Press, Cambridge, U.K, p. 332.
Delvaux, B., Kruyts, N., Cremers, A., 2000. Rhizospheric mobilization of radiocesium in soils. Environ. Sci. Technol. 34 (8), 1489–1493.
Delvaux, B., Kruyts, N., Maes, E., Smolders, E., 2001. Fate of radiocesium in soil and rhizosphere. In: Trace Elements in the Rhizosphere, pp. 61–91.
Desmet, G.M., Van Loon, L.R., Howard, B.J., 1991. Chemical speciation and bioavailability of elements in the environment and their relevance to radioecology. Sci. Total Environ. 100, 105–124.
Dmitriev, A.P., Grodzinskii, D.M., Gushcha, N.I., Kryzhanovskaya, M.S., 2011. Effect of chronic irradiation on plant resistance to biotic stress in 30-km Chernobyl nuclear power plant exclusion zone. Russ. J. Plant Physiol. 58 (6), 1062.
Ehlken, S., Kirchner, G., 2002. Environmental processes affecting plant root uptake of radioactive trace elements and variability of transfer factor data: a review. J. Environ. Radioact. 58 (2–3), 97–112.
Epstein, E., Hagen, C.E., 1952. A kinetic study of the absorption of alkali cations by barley roots. Plant Pathol. 27 (3), 457–474.
Green, J.M., Owen, M.D., 2011. Herbicide-resistant crops: utilities and limitations for herbicide-resistant weed management. J. Agric. Food Chem. 59 (11), 5819–5829.
Greger, M., 2004. Uptake of Nuclides by Plants. Department of Botany, Stockholm University. SKB TR-04-14, Svensk Kärnbränslehantering AB, p. 70.
Guivarch, A., Hinsinger, P., Staunton, S., 1999. Root uptake and distribution of radiocaesium from contaminated soils and the enhancement of Cs adsorption in the rhizosphere. Plant Soil 211 (1), 131–138.
Hegazy, A.K., Afifi, S.Y., Alatar, A.A., Alwathnani, H.A., Emam, M.H., 2013. Soil characteristics influence the radionuclide uptake of different plant species. Chem. Ecol. 29 (3), 255–269.
Hokkanen, H.M., 1991. Trap cropping in pest management. Annu. Rev. Entomol. 36 (1), 119–138.
Jagetiya, B., Sharma, A., Soni, A., Khatik, U.K., 2014. Phytoremediation of radionuclides: a report on the state of the art. In: Radionuclide Contamination and Remediation through Plants. Springer, Cham, pp. 1–31.
Jensen, D., Thomas, L., 2010. Soil pH and the Availability of Plant Nutrients. IPNI Plant Nutrition Today, Fall, No. 2. www.ipni.net/pnt. http://www.ipni.net/publication/pnt-na.nsf/0/013F96E7280A696985257CD6006FB98F/$FILE/PNT-2010-Fall-02.pdf.
Kiser, M.R., Reid, C.D., Crowell, A.S., Phillips, R.P., Howell, C.R., 2008. Exploring the transport of plant metabolites using positron emitting radiotracers. HFSP J. 2 (4), 189–204.
Konoplev, A.V., Viktorova, N.V., Virchenko, E.P., Popov, V.E., Bulgakov, A.A., Desmet, G.M., 1993. Influence of agricultural countermeasures on the ratio of different chemical forms of radionuclides in soil and soil solution. Sci. Total Environ. 137 (1–3), 147–162.
Kumar, M., 2013. Crop plants and abiotic stresses. J. Biomol. Res. Therapeut. 3 (1).

Lasat, M.M., Baker, A.J.M., Kochian, L.V., 1998. Altered Zn compartmentation in the root symplasm and stimulated Zn absorption into the leaf as mechanisms involved in Zn hyperaccumulation in Thlaspi caerulescens. Plant Physiol. 118, 875–883.

Lembrechts, J., 1993. A review of literature on the effectiveness of chemical amendments in reducing the soil-to-plant transfer of radiostrontium and radiocaesium. Sci. Total Environ. 137 (1–3), 81–98.

Liu, S., Ogawa, N., Senda, T., Hasebe, A., Miyashita, K., 2005. National Institute for Agro-Environmental Sciences.

Luoma, S.N., Jenne, E.A., 1977. Biological Implications of Metals in the Environment. In: Drucker, H., Wildung, R. (Eds.), Avail as CONF-750929. U.S. NTIS, Springfield, VA, p. 213.

Lorenz, S.E., Hamon, R.E., McGrath, S.P., Holm, P.E., Christensen, T.H., 1994. Applications of fertilizer cations affect cadmium and zinc concentrations in soil solutions and uptake by plants. Eur. J. Soil Sci. 45, 159–165.

Malézieux, E., et al., 2009. Mixing Plant Species in Cropping Systems: Concepts, Tools and Models: A Review. In: Lichtfouse, E., Navarrete, M., Debaeke, P., Véronique, S., Alberola, C. (Eds.), Sustainable Agriculture. Springer, Dordrecht.

Malézieux, E., Crozat, Y., Dupraz, C., Laurans, M., Makowski, D., Ozier-Lafontaine, H., Rapidel, B., de Tourdonnet, S., Valantin-Morison, M., November 30, 2007. Mixing Plant Species in Cropping Systems: Concepts, Tools and Models (A Review).

Marčiulionienė, D., Lukšienė, B., Montvydienė, D., Sakalauskas, V., Sevriukova, O., Druteikienė, R., Jefanova, O., Žukauskaitė, Z., 2017. Radiocesium phytotoxicity to single cell and higher plants. In: Impact of Cesium on Plants and the Environment. Springer, Cham, pp. 209–230.

Matsunaga, T., Koarashi, J., Atarashi-Andoh, M., Nagao, S., Sato, T., Nagai, H., 2013. Comparison of the vertical distributions of Fukushima nuclear accident radiocesium in soil before and after the first rainy season, with physicochemical and mineralogical interpretations. Sci. Total Environ. 447, 301–314.

Menzel, R.G., 1954. Competitive uptake by plants of potassium, rubidium, caesium, calcium, strontium, and barium from soils. Soil Sci. 77 (6), 419–425.

Nagaoka, K., Hanioka, N., Ikushiro, S., Yamano, S., Narimatsu, S., 2012. The effects of N-glycosylation on the glucuronidation of zidovudine and morphine by UGT2B7 expressed in HEK293 cells. Drug Metab. Pharmacokinet. 27, 388–397.

Nisbet, A.F., Woodman, R., 2000. Soil-to-plant transfer factors for radiocesium and radiostrontium in agricultural systems. Health Phys. 78, 279–288.

Nisbet, A.F., Woodman, R.F.M., Haylock, R.G.E., 1999. Recommended soil-to-plant transfer factors for radiocaesium and radiostrontium for use in arable systems. NRPB-R304. NRPB, Chilton, U.K.

Nutrient and Water Management Practices for Increasing Crop Production in Rainfed Arid/ Semi-arid Areas, 2005. 0205-Soil Fertility and Irrigation IAEA-TECDOC-1468, ISBN 92-0-109405-1.

Nye, P.H., Tinker, P.B., 1977. Solute movement in the soil-root system. Sudies in Ecology, 4. Blackwell Scientific Publications, Oxford, p. 342.

Perera, F., 2018. Pollution from fossil-fuel combustion is the leading environmental threat to global pediatric health and equity: solutions exist. Int. J. Environ. Res. Publ. Health 15 (1), 16.

Rai, H., Kawabata, M., 2020. The dynamics of radio-cesium in soils and mechanism of cesium uptake into higher plants: newly elucidated mechanism of cesium uptake into rice plants. Front. Plant Sci. 11, 528.

Rengel, Z., Marschner, P., 2005. Nutrient availability and management in the rhizosphere: exploiting genotypic differences. New Phytol. 168 (2), 305–312.

Rigol, A., Vidal, M., Rauret, G., 2002. An overview of the effect of organic matter on soil–radiocaesium interaction: implications in root uptake. J. Environ. Radioact. 58 (2–3), 191–216.

Rosén, K., Öborn, I., Lönsjö, H., 1999. Migration of radiocaesium in Swedish soil profiles after the Chernobyl accident, 1987–1995. J. Environ. Radioact. 46 (1), 45–66.

Russell, R.S., 1966. Radioactivity and human diet, 1st edn. Pergamon Press, Oxford.

Sahr, T., Voigt, G., Paretzke, H.G., Schramel, P., Ernst, D., 2005. Caesium-affected gene expression in *Arabidopsis thaliana*. New Phytol. 165 (3), 747–754.

Sanchez, A.L., Wright, S.M., Smolders, E., Naylor, C., Stevens, P.A., Kennedy, V.H., Dodd, B.A., Singleton, D.L., Barnett, C.L., 1999. High plant uptake of radiocesium from organic soils due to Cs mobility and low soil K content. Environ. Sci. Technol. 33 (16), 2752–2757.

Sawidis, T., 1988. Uptake of radionuclides by plants after the Chernobyl accident. Environ. Pollut. 50 (4), 317–324.

Sheppard, S.C., Sheppard, M.I., 1988. Modeling estimates of the effect of acid rain on background radiation dose. Environ. Health Perspect. 78, 197–206.

Smith, J.T., Elder, D.G., 1999. A comparison of models for characterizing the distribution of radionuclides with depth in soils. Eur. J. Soil Sci. 50 (2), 295–307.

Smucker, A., Aiken, R., 1992. Dynamic root responses to water deficits. Soil Sci. 154, 281–289.

Tangahu, B.V., Sheikh Abdullah, S.R., Basri, H., Idris, M., Anuar, N., Mukhlisin, M., 2011. A review on heavy metals (As, Pb, and Hg) uptake by plants through phytoremediation. Int. J. Chem. Eng. 2011.

Tilman, D., Cassman, K.G., Matson, P.A., Naylor, R., Polasky, S., 2002. Agricultural sustainability and intensive production practices. Nature 418, 671–677.

Tinker, P.B., Nye, P.H., 2000. Solute Movement in the Rhizosphere. Oxford University Press.

U.S. Nuclear Regulatory Commission; Regulatory Guide 8.34, September 25, 2006. Monitoring Criteria and Methods to Calculate Occupational Radiation Doses. 1992. Available at: http://www.nrc.gov/reading-rm/doc-collections/reg-guides/occupational-health/active/8-34/index.html.

Vandermeer, J., 1989. The Ecology of Intercropping. Cambridge University Press, Cambridge, p. 237.

Wallace, A., 1989. Effect of liming on trace-element interactions in plants. Soil Sci. 147 (6), 416–421.

WHO, 2006. World Health Report 2006. Working together for health, p. 237

Zhu, Y.G., Shaw, G., 2000. Soil contamination with radionuclides and potential remediation. Chemosphere 41, 121–128.

Zhu, Y.G., Smolders, E., 2000. Plant uptake of radiocaesium: a review of mechanisms, regulation and application. J. Exp. Bot. 51 (351), 1635–1645.

Further reading

Bennett, B., Repacholi, M., 2006. Health Effects of the Chernobyl Accident and Special Health Care Programmes. Report of the UN Chernobyl forum expert group“health”. World Health Organization.

Cawse, P.A., 1983. The accumulation of caesium-137 and plutonium-239+ 240 in soils of Great Britain, and transfer to vegetation. In: Ecological Aspects of Radionuclide Release.
Cowart, J.B., Burnett, W.C., 1994. The Distribution of Uranium and Thorium.
Cramer, G.R., Urano, K., Delrot, S., Pezzotti, M., Shinozaki, K., 2011. Effects of abiotic stress on plants: a systems biology perspective. BMC Plant Biol. 11 (1), 1−14.
n.d. Decay-series radionuclides in the environment—a review. J. Environ. Qual. 23 (4), 651-662.
ExitCare®, 2004. Patient Information ©2014 ExitCare, LLC, © 2015 Baylor Scott & White Agency Radiological Laboratory Analytical Protocols Manual Volume II Health.
Knight, M.J., 1983. Uptake by Plants of Radionuclides from FUSRAP Waste Materials (No. ANL/EIS–19). Argonne National Lab.
Lemke, K.A., 2013. klemke@uwsp.edu Last revised October 24.
Nafziger, E., 2009. Cropping systems. In: Illinois Agronomy Handbook, vol. 24, pp. 49−63.
Palmisano, A., Vairavamurthy, M.A., October 1999. Natural and Accelerated Bioremediation Program, Virginia.
Participants, A., 2007. Bio-Soils Interdisciplinary Science and Engineering Initiative.
Shaw, G., Bell, J.N.B., 1994. Plants and Radionuclides. In: Plants and the Chemical Elements: Biochemistry, Uptake, Tolerance and Toxicity, pp. 179−220.
Tributh, H.V., Boguslawski, E.V., Lieres, A.V., Steffens, D., Mengel, K., 1987. Effect of potassium removal by crops on transformation of illitic clay minerals. Soil Sci. 143 (6), 404−409.

Climate change and flood risk, global climate change

14

Sanchayita Rajkhowa[1], *Jyotirmoy Sarma*[2]
[1]Department of Chemistry, Jorhat Institute of Science & Technology (JIST), Jorhat, Assam, India; [2]Department of Chemistry, Kaziranga University, Jorhat, Assam, India

1. Introduction

Global climate change is unambiguously witnessed over the past several decades, and its inevitable consequences include rise in sea level, global warming, seasonal irregularities, drought, and flood, thus enforcing the societies to adapt with this alteration of the environment (Intergovernmental Panel on Climate Change, 2007a). Impacts of climate change, although, are geographical specific; the attention of global, national, and local concerning bodies must be assured. An international organization, the Intergovernmental Panel on Climate Change (IPCC), was established in the year 1988 by the United Nations Environment Programme (UNEP) and the World Meteorological Organization (WMO) aiming to provide policy makers with comprehensive assessment on climate change: impacts, vulnerability, adaption, its implications, and potential future environmental risks. Since then, a multiple numbers of reports on the emerging issues on climate change have been published. In 2007, the IPCC published three volumes under a common umbrella entitled *Climate* Change 2007 as (1) Climate Change 2007—The Physical Science Basis, (2) Climate Change 2007—Impacts, Adaptation and Vulnerability, and (3) Climate Change 2007—Mitigation of Climate Change (Intergovernmental Panel on Climate Change, 2007a). In a "Special Report on Managing the Risks of Extreme Events and Disasters to Advance Climate Change Adaptation (SREX)" of the IPCC, climate change and its impact were critically assessed, which was a collaborative effort made by authors from multinational and multidiscipline backgrounds with the assistance of academic and government pools. Prioritization of domains has been established based on the information, hypotheses, and opinions assessed in the report (Intergovernmental Panel on Climate Change, 2012). The concept of SREX is to assess the exposure and vulnerability to climate change and weather that ultimately determines the impacts and disaster risk. It also examines the role of development in connection with disaster risk, management and adaptation to climate change, and implementation of climate extremes for socioeconomic as well as sustainable development. However, strategic crafting of such comprehensive adaptation responses integrated with environmental and socioeconomic upliftment may be a challenging task for the policy makers.

Extreme weather and unprecedented climate events are the results of change in intensity, frequency, duration, spatial extent, and timing of natural events. Few climate

Global Climate Change. https://doi.org/10.1016/B978-0-12-822928-6.00012-5

extremes, for instance, droughts and floods, may be the cumulative outcome of several extreme weathers and/or climate events. Natural variability greatly influences many of the extreme weather and climate events in addition to the anthropogenic interferences. The Intergovernmental Panel on Climate Change (2018) report reiterated the urgency to reduce average global temperatures below 1.5°C within 12 years target and the impacts related to global greenhouse gas (GHG) emission pathways in Paris Agreement (Intergovernmental Panel on Climate Change, 2018). Meanwhile, the Fourth National Climate Assessment estimated the rise in average global temperature, which could reach upto 5°C by the end of the 21st century if emission was not being controlled (National Climate Assessment (NCA), 2018). According to the World Economic Forum (WEF), "extreme weather events" and "failure of climate-change mitigation and adaptation" are the inextricably melded high-risk factors by likelihood in relation to environmental catastrophe (World Economic Forum, 2019). Urbanization of population at a rapid rate accelerates the biodiversity loss resulting in unprecedented consequences—global warming, rising sea levels, extinction of variety of species (flora and fauna) across the globe, declination of natural resources, ultimately exacerbating ecological imbalance. Perpetual degradation of ecosystem reflects on the socio-economic fragility. Costanza et al. (2014) approximated the global value of ecosystem services to average at US$125 trillion/year that is equivalent to more than two-thirds of global GDP accounting for safe drinking water, protection against flood, and conservation of landscapes. Equilibration among the critical components of natural assets society, individuals, economy, and ecosystem embodies sustainability with human well-beings for a desirable future. A number of extreme events have taken place during the year 2017—extreme flooding in South Asia, severe droughts in Eastern and Horn of Africa and South Asia; a series of hurricanes, viz., Maria, Harvey, and Irma in the Caribbean and the United States; storm in Netherlands; major flood in Paris along the River Seine; heavy rainfall and flood hit in several parts of California; and heat waves in Europe. Slow progress in reducing emissions of GHGs is unequivocally responsible for an increased number of global extreme weather events. However, all regions do not witness same degree of devastation with equal frequency and intensity.

Extreme precipitation events due to climate change will cause local pluvial flood that turns into fluvial flood without implementing proper flood risk management projects. For instance, spatiotemporal precipitation pattern, rate of evaporation, snow fall, melting of glaciers, and so on have to be taken into consideration while projecting future flood risk in large rivers (Yamazaki et al., 2018). In IPCC SREX report, flood may be defined as "*the overflowing of the normal confines of a stream or other body of water or the accumulation of water over areas that are not normally submerged. Floods include river (fluvial) floods, flash floods, urban floods, pluvial floods, sewer floods, coastal floods, and glacial lake outburst floods*". Assessment of climate change at local level must be designed to mitigate any drastic climate change adapting the infrastructure in tune with the environment. Global projections of temperature rise and change in precipitation pattern, though uncertain, increase the risk of extreme flood events. The National Oceanic and Atmospheric Administration (NOAA) Atlas 14 provides hydrologic and hydraulic model to develop design, plan, and estimate precipitation frequency in the United States using intensity–duration–frequency relationships although no explicit interpretation to delineate the link between temperature rise, rainfall intensification, and flooding (National Oceanic and Atmospheric Administration (NOAA) Atlas 14, 2018). The society can play a vital role in mitigating inundation by posing resilience to potential climate change. Society's ability toward climate change resilience greatly influences the vulnerability of national critical infrastructure[1] (CI, i.e., energy, transport, agriculture, energy, public health, telecommunication, water supply, economic sector, etc.); thus, inspite of having longevity, climate change may reduce CI performances. It is, therefore, worthwhile to understand climate change and develop effective strategies on society's climate adaptation in context of flood management infrastructure to mitigate surface runoff with simultaneous construction of storage during extreme flood outbursts. Besides climate change, several other possible drivers such as land use and land cover shifts, construction of dams or reservoirs, and urbanization attribute to the overall flood hazards. Rapid urbanization in developing countries with marginalized population contributes a significant share in economic activities with fewer resources, and thus, this portion of society is more fragile and exposed to the vulnerability of disaster events. An international charity organization, Climate Outreach, focuses on bringing researchers, environmentalists, policy makers, and civil society under a common roof to discuss and propose solution to address climate change (Messling et al., 2015).

Although the complexed correlation between greenhouse forces and enhanced inundation has not been well described in literature with either several models or empirical analysis, human activities causing an increment in flood occurrences cannot be ignored. This chapter discusses the impact of climate change on flood, risks associated, and mitigation policies in great detail.

2. Climate change scenarios

The global average surface temperature has been rising, especially since about 1950. With global warming, both the water holding and evaporation capacity of the atmosphere have increased favoring climate variability to higher notch with amplifying the degree of precipitation and more drought conditions, thus directly influencing the hydrological cycle. Global intensity of precipitation is expected to rise every year; however, there are several instances where downscaling of precipitation is witnessed. Irregular precipitation pattern due to climate change often leads to certain alteration in the components of global freshwater sources albeit extreme temperature changes in accordance with atmospheric warming. There is a widespread reduction in the duration of frost days in the central latitudinal regions, and multiple occasions of warm extremes with lesser number of cold extremes are observed in rest of the world. From the latter half of 20th century, duration of heat weaves has extended where the western and central Europe hit with record-breaking heat waves in 2003,

[1] Critical infrastructure (or critical national infrastructure [CNI] in the United Kingdom) is a term used by governments to describe assets that are essential for the functioning of a society and economy—the infrastructure.

making the warmest summer since 1807. Dry surface during spring in Europe is another consequential event of extreme temperature that year (Kundzewicz et al., 2014). Likewise, other geographical regions have also been victimized under frequent heat waves (Solomon et al., 2007). Anthropogenic forces accelerate the rate of sea level rise during the second half of 20th century, whereas natural forces are likely to increase the same in the early 20th century and bring down later, hence making a steady rise in the 20th century. Rio Earth Summit 1992 objectified "*Stabilization of greenhouse gas concentrations in the atmosphere at a level that would prevent dangerous anthropogenic interference with the climate system. Such a level should be achieved within a time-frame sufficient to allow ecosystems to adapt naturally to climate change, to ensure that food production is not threatened and to enable economic development to proceed in a sustainable manner*" in the United Nations Framework Convention on Climate Change (Article 2 UNFCCC, 1992). Resource managers and policy makers predict the potential influence of climate change based on the information about average climate variability data. However, anticipation of future climate patterns is a real challenge since the anthropogenic forces highly depend on several assumptions and uncertainties, e.g., population, exploitation of resources, industrial–economic development, policies, and concern toward environment. Due to these factors, development of climate scenarios is extremely substantial for investigating the potential outcomes of anthropogenic climate change to make climate forecast as accurate as possible. Therefore, climate science is believed to contribute immensely in the context of understanding present vulnerabilities and project scenarios of long-term climate changes derived from global climate models (GCMs) and regional climate models (RCMs).

3. Flood risk and causes of flooding in terms of climate change

As per IPCC SREX report, floods can be defined as the overflowing of the normal water body and streams as well as the accretion of water over areas that are not normally inundated. The various classes of floods such as river (fluvial) floods, flash floods, urban floods, pluvial floods, sewer floods, coastal floods, and glacial lake outburst floods are produced by different mechanisms.

As per Kundzewicz et al. (2012), increase in flood risk is dependent on several factors such as improvement in reporting of disasters, greater exposure of population and assets, and the magnitude of flood hazard. Flood risk is driven by anthropogenic changes in climatic systems (Kundzewicz and Schellnhuber, 2004). The anthropogenic drivers lead to flood probability, exposure, and vulnerability. The flood risk is dependent on these as follows:

Flood risk = (flood probability) × (exposure) × (vulnerability) (Fig. 14.1).

Flood risk is enhanced by various characteristics of the climatic system, most importantly precipitation and temperature patterns (Bates et al., 2008). The intensity, duration, amount, timing, and phase (particularly rain or snow) of precipitation are of great concern. Phenomena such as soil freezing, snow, and ice melt and ice jam

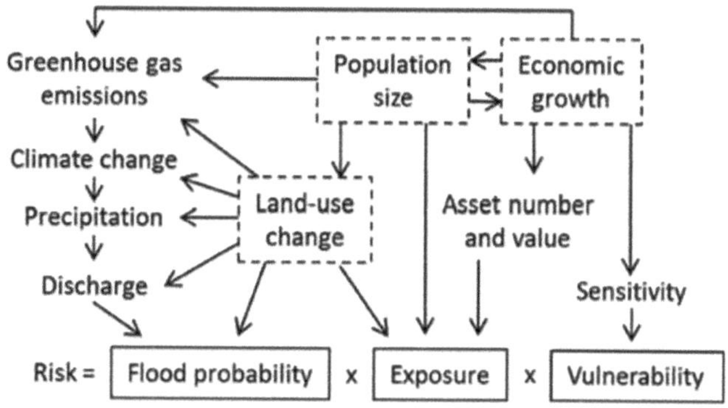

Figure 14.1 Drivers of changes in flood risk (Kundzewiczm et al., 2014, Open access).

formations are dependent on temperature patterns. Moreover, occurrence of destructive flooding is also a result of global warming and increased rate of urbanization. Large river basins undergo fluvial floods as a result of intense and long-lasting precipitation. However, floods in high-latitude regions are related to snowmelt floods. On the other hand, in case of small basins, floods are caused by short-duration, highly intense rainfall.

According to IPCC SREX glossary, floods are known as "climatological normal" occurrences. Generally, flood occurs in every spring during snowmelt (Kundzewicz, 2012) or during an annual monsoon (Field et al., 2012). There are evidences for increase in flood risk over time because of changes in heavy precipitation, river flow, and floods. The "IPCC" and the "SREX" assessment reported about significant increases in the number of heavy precipitation events in more regions than that of significant decreases as per statistics. However, these particular trends vary from region to region. Global and regional scale changes are rather different (Zhang et al., 2007). Anthropogenic climate change contributes to the hydrological cycle causing floods that include both mean precipitation (Zhang et al., 2007) and heavy precipitation (Barnett et al., 2008). However, as per the SREX report (S12), there is no evidence of any statistical link observed between anthropogenic climate change and trends in the magnitude/frequency of floods. Annual runoff is caused by seasonal snow storage and melting leading to flood risk. Changes in temperature and timing of spring peak flows in snowmelt and glacier-fed rivers disturb the hydrological regime. Identification of climate-driven trends in floods is sometimes recognized to be a tough task as changes in land use and river engineering also hamper this (Rosenzweig et al., 2007). Moreover, climate models do not sufficiently represent long-term persistence of flooding as observed in real data, and the outcome is naturally ignored in climate studies (Koutsoyiannis and Montanari, 2007). Acceptance of the long-term persistence hypothesis prevents the detection of climate-driven trends. The gathering or collection of instrumental, historical, and paleoinformations about flood events is an important task in understanding flood hazard and climate variability of flood occurrence. Paleohydrology or paleoflood hydrology deals mainly with documentary evidence about hydrological events (Brázdil and Kundzewicz, 2006).

With respect to the observed trends in costs of flooding as per the SREX report, it can be inferred that economic losses from weather- and climate-related disasters have increased gradually. The reason behind this can be attributed to the expanding exposure of assets at risk (Miller et al., 2008). However, it has also been mentioned that the long-term trends in losses are not the outcome of natural or anthropogenic climate change (Pielke and Downton, 2000). On the contrary, a partial relationship has also been observed to exist between increases in flood losses and changes in intense rainfall events as a result of climate change (Fengqing et al., 2005).

Flood risk is dependent on some principal climate factors such as water containing capacity of the atmosphere related with the phenomena "evapotranspiration" and the characteristics of intense precipitation (Koutsoyiannis, 2012). Intense precipitation along with its amount and distribution in space and time has played a significant role in flood risk. Most importantly, the climate-induced factors such as snowmelt followed by evapotranspiration, phase of precipitation, and sequence of temperature are noteworthy in this context. Nevertheless, the preconditioning of surface hydrological variables such as soil moisture content, groundwater, and surface-water levels also plays a very important role. Climate change can alter many of the aforementioned flood-affecting factors. As an example in case of precipitation, volume, and timing, amount of precipitations falling as snow and rain are significant parameters in contributing to flood risk. Additionally some other related factors such as snow cover, forerunner soil moisture content, surface water levels, sea level, glacial lake conditions, and vegetation are all climate driven and hence dependent on climate conditions. Alteration in land use change because of urbanization as well as alteration in river regulation is also having an insightful relationship with climate change. Model-based analysis is found helpful for demonstration of land use effects on flood risk (Gerten et al., 2008).

3.1 Perspective on future flood risk

A prediction on the future changes in flood risk can be achieved by a combination of climate change (especially precipitation), catchment conditions, and acquaintance to loss (Montanari et al., 2013). Proper study about the possible future changes in precipitation, river flow, and changes in flood frequency is essential prior to flood exposure and cost forecasting.

In the 21st century, many areas across the globe, specifically, in summer, high latitudes and tropical regions and, in winter, the northern and midlatitudes, experience an increase in precipitation frequency leading to flood risk (SREX report). The SREX report has also mentioned about the existence of a connection between heavy rainfall and the global warming. Certainly the global warming plays a significant role in climate change and hence floods risk. The long-term heating of the earth's climate system is continuing since preindustrial period because of several anthropogenic activities, leading to emission of heat-trapping GHGs in the earth's atmosphere. With industrial development and subsequent urbanization, the concentration of such GHGs increases many fold in the earth's climate system, which in turn is responsible for certain flood risk and associated loss. In some regions, heavy precipitation is found to increase in spite of the decrease in projected total precipitation in tune with global warming. The SREX assessment reported about higher emissions of GHGs in the 21st century than that of the 20th century climate. Nevertheless, increase in the frequency and magnitude of heavy precipitation events has also been observed with a 20-year return period because of the rise in GHG emissions (Christensen et al., 2007). The SREX assessment (S12) involves, with large uncertainties in the projection of precipitation extremes, climate models and natural variability (Nicholls and Seneviratne, 2013). Extensive work has been carried out in the estimation of projected return periods of annual maximum daily precipitation rates under the different GHG emission conditions (Kharin et al., 2007). Maintenance of daily precipitation index is an effective measure in the context of flood mitigation planning. This study carried out across different regions of the globe reveals a tendency in the decrease in return periods. Hence the flood generating precipitation analysis reflects the strong regional dependency of the projected changes. High-latitude and tropical regions experience larger changes compared with that of drier regions. Nevertheless, flood occurrence at the regional scale is dependent on the projected future changes in heavy precipitation events.

Projected changes in river flows and floods are very limited. In the 21st century, with the help of regional climate model (RCM) or global climate model (GCM), daily river discharges have been estimated for flood prediction (Hirabayashi et al., 2008). Outputs of such climate model along with global river routing model help in the evaluation of spatially distributed to flood risk across the globe (Hirabayashi et al., 2013). Determination of potential changes in future flood exposure and costs is also an essential task in the context of flood risk. Millions of people worldwide live in flood-prone areas and on average a good percentage of people across the globe exposed to floods each year (Peduzzi et al., 2011). Accuracy in the estimation of flood impact is dependent upon the accuracy level of flood frequency and magnitude determination. Prediction of the trends in the number of exposed people is achievable from study of projected population increase in flood-prone areas (UNISDR, 2011). Exposure to floods basically consists of human exposure and economic exposure. Human exposure can be articulated as the population living in flood-prone areas. More specifically, it is the percentage of total population exposed per year (Kleinen and Petschel-Held, 2007). Correspondingly, economic exposure is related with the value of the assets located in flood-prone areas. Primarily, it talks about the average GDP (gross domestic product) exposed per year. One flood event may result in a huge loss in GDP. On the other hand, a range of approaches have been adopted for estimation of the future costs of flooding (Bouwer, 2013). Changes in flood hazard have been accounted based on the projected climate change and use of loss model estimates. Anthropogenic climate change resulting from GHG emissions significantly contributes to the economic flood losses. Rapid urbanization with subsequent human modifications of watersheds changes their runoff characteristics that in turn lead to flood risk and losses.

4. Anticipated situations in the future

Future flood risks are presupposed to be a cumulative impact of intensity of precipitation due to climate change, catchment conditions, exposure, and extent of loss. A new scientific decade 2013–22 of the International Association of Hydrological Sciences

(IAHS), entitled "PantaRhei-Everything Flows,[2]" is dedicated to the elaborated interpretation of such changes (Montanari et al., 2013). Koutsoyiannis clearly discussed the idea of multiscale changes that support the Panta Rhei concept, which is a manifestation of the intensified frequency of long-term changes, and these changes are highly unpredictable and uncertain in the long run (Koutsoyiannis, 2013). His approach is the key to perceive this multiscale changes to imply a model for future uncertainty and risk with a future perspective in which change and uncertainty are inevitable considerations.

Floods are predominantly affected by intensity, duration, frequency, quantity, and phase of precipitation and also by the temperature patterns. Several other factors may include the drainage basin systems such as water level in rivers and oceans: ice coverings, soil characteristics (soil moisture, permeability, water holding capacity etc), urbanization, storm surge or extreme tide, and presence of reservoirs, dams, and dikes. The SREX report predicted the likelihood of increase in proportion or frequency of total rainfall events in the 21st century globally especially in tropical regions, high latitudinal areas, and in winter in the northern central latitudes (O'Brien et al., 2012). This report is also opinionated about the probability of rising tropical cyclonic incidents and thereafter heavy rainfall if global temperature continues to rise. There is a possibility of several regions likely to receive more heavy precipitation than that has been predicted. Higher the emission of CO_2, the stronger will be the probability of extreme events with reducing intermittent period by the end of 20th century and augmentation in precipitation magnitude with an intermittent period of 20 years. The projected rise in precipitation would be thus a major contributory factor to increase regional flooding in several catchments, and therefore, the predictions of regional climate variations play a key role in mitigating inundation and its consequences. With increasing precipitation, projections expected variations in river floods associated with large-scale rainwater discharges into river basins. Magnitude and frequency of flooding is likely to increase in the 21st century presumed by RCMs/GCMs coupled with hydrological models. Both population and flooding are escalating each year; over 11% of global population (about 800 million) worldwide is living in flood-prone regions, while 1%, i.e., about 70 million people are exposed to these floods every year (UNISDR, 2011). These numbers, however, would expect to grow up to 940 million and 86 million/year, respectively, even at constant hazardous situation due to demographic changes solely. With insufficient information and data on yesteryears' flood events and the associated uncertainties with predicted flood trends, it is extremely difficult to estimate their future impacts. Nevertheless, it may be possible to evaluate the flood-affected population using the projected population growth in flood-prone regions. Several studies have also addressed an increased economic loss that would cause from floods and anthropogenic climate change (due to emission of GHGs). All these studies carried out by various groups and organizations attempt to recognize the future flood risk trends, their assessment, and impacts.

[2] "PantaRhei—Everything Flows," is dedicated to research activities on change in hydrology and society. The purpose of PantaRhei is to reach an improved interpretation of the processes governing the water cycle by focusing on their changing dynamics in connection with rapidly changing human systems.

5. Approaches for flood mitigation

According to the Federal Emergency Management Administration (FEMA), mitigation is defined as "sustained action taken to reduce or eliminate long-term risks to people and property from hazards" (FEMA, 2013). There are two basic approaches: structural and nonstructural for flood mitigation based on the kind of methods employed. Structural approaches are established on permanent solutions or facilities to mitigate flood, whereas nonstructural ones may not entirely permanently focus on urban flood problems. Combination of both these approaches turns out to be an economically viable and environmentally benign way to combat floods.

5.1 Structural approaches

Structural methods are further divided into two categories as extensive and intensive measures. The extensive measures include soil conversion, restructuring of surface land, flow delay, and rise of infiltration, while the latter measures primarily focus on construction of levees, reservoirs, and dams; embankments to protect flood-prone regions; and improvement of channel flow capacities for adequate discharge of flooding. Runoff delays, flood attenuation, groundwater control, rise in infiltration, and downstream discharge controls are few notable advantages of structural approaches. A list of disadvantages of that approach includes lowering of floodplain fertility, morphological changes, high risk of ecological impact, and land subsidence. One of the ancient structural approaches to reduce the flood impact is polder system that has been practiced in Holland about 10 centuries ago to protect the land from sea floods by building a special kind of dam. Later, this approach was made extended to protect land from lakes, rivers, and shallow reservoirs. Apart from controlling flood water, polders can also be used in agriculture, fisheries, industrial environments, and tourist spot and recreation centers. On the other hand, the dependency on pump and maintenance expenses limits its popularity.

5.2 Nonstructural approaches

The nonstructural approaches can be classified into three broad categories: (1) regulation of floodplain proofing, i.e., coding and zoning[3]; (2) awareness and education programs, warning, forecasting, flood proofing, evacuation, and relocation of affected mass to reduce flood impacts; and (3) development of fiscal strategies such as insurance policies, relief grants, funding, and so on. Policies and regulations for zoning, viz., restrictions of development beyond a threshold, clustering, transfer of development rights, and density bonuses can be a favorable step toward mitigating flood impacts by shifting the development plans to less prone zones. Proactive strategies for land use planning and zoning bypass the projected developments to less vulnerable

[3] Zoning is a method of urban planning in which a municipality or other tier of government divides land into areas called zones, within which certain land uses are permitted or prohibited.

Table 14.1 Flood risk management with structural and nonstructural approaches.

Risk = hazard × exposure × vulnerability		
Modify hazard	**Modify exposure**	**Modify vulnerability**
■ Flood control via dams, levees, dikes	■ Land use or zoning ■ Acquisition of properties	■ Flood monitoring, warning, and forecast
■ Detention of basins	■ Planning and development controls	■ Emergency plans
■ River channel improvements	■ Flood-resistant buildings	■ Community education, awareness, and preparedness
■ Flood diversion channels	■ Building codes	■ Postflood recovery, reconstruction, and flood insurance
■ Watershed management		
← *Structural*		*Non-Structural* →

regions that reduce the flood damages as well as balance the ecosystem and maintain water quality. Education and awareness in the form of printed materials, workshops, websites, and so on is another wing of nonstructural approaches. Flood forecast and warnings are announced by local authorities to predict the flood events and consequences and assess and gather data. Several instances of allocation of local government funds to local jurisdictions during the time of crisis and catastrophe are common in fiscal strategic frameworks. The advantage of this approach is that flood impacts and detriments are effectively dealt without compromising with environmental integrity and also improved interorganization relations. However, floodplain invasion, rise in the value of properties, and wider insurance coverage are few negative features of this approach (Anita, 2013).

Merging of both structural and nonstructural approaches can be effective measures in flood mitigation, e.g., floodplain zoning supplemented with flood forecast will be an improved and cost-effective scheme. Table 14.1 explains the flood risk in terms of hazard, exposure, and vulnerability with structural and nonstructural approaches for flood damage and mitigation (Anita, 2013).

6. Limitations related to the uncertainties of future climate prediction

Anthropogenic activities, especially emissions and land use, promote changes of climate forces in the coming years. Various methods have been developed for future emissions and land use projections though certain limitations resulting in uncertainties associated with the future projection of climate change. Variations in natural processes such as volcanism and changeability in climate system are more difficult to predict. Nevertheless, projection of this natural forcing facilitates the policy maker to manage future climate change. All the projected models having diverse climate subsets such as solar variability, GHG emission, aerosol emission, land use, and land cover change in their framework must be skillfully predicted for future global/regional climate change, and addressing of these scenarios is a challenging assignment with the key focus on environmental sustainability and societal vulnerability (National Research Council, 2005).

6.1 *Solar variability*

According to empirical evidences, it has been suggested that about 11-year activity cycle of the sun is presumed to modulate any future scenario in solar irradiance of the order of 1 W/m^2 or less in total. Unfortunately, the present database neither suffices the purpose of detecting the long-term solar irradiation patterns nor predicts the future solar activities. Little information is available on the variation of solar activity in the near future and the duration of this variation, if ever exists. From the sunspot numbers as well as cosmogenic isotopes, the solar irradiation magnitude is not expected to increase significantly from the current levels, which are already at the peak. This database is confirmed by spectral synthesis of cosmogenic isotope report for projection of future solar activities, and these values in 2100 will become equivalent to those in 1990. Meanwhile, combined three solar cycle projections of 11, 88, and 208 years reflect that the solar activity will form a climate forcing of about +0.45 W/m^2 top-of-the-atmosphere (TOA) within 2000–30, followed by a decrease up to −1 W/m^2 by the year 2090. A similar kind of study based on numerical model contradicts this projection, suggesting that there will be 0.05% decrease in irradiance with −0.1 W/m^2 climate forcing by 2020. There is a complete failure of understanding the two opposing factors of solar activities, i.e., blocking the sunspot and facular brightening, which limits and makes the future projections uncertain (National Research Council, 2005).

6.2 *Emissions of greenhouse gases and aerosols*

Estimation of GHGs, various aerosols, and their precursors emitted in the atmosphere is the basic aspect of future climate change analysis. The emission data are the prerequisites to design various climate models for projecting the fate of the atmosphere, biosphere, cryosphere, and hydrosphere. Anthropogenic activities contribute the major share of emissions, and the more the information on such activities, the better will be the policy analysis to predict the future emission trends. Besides, environmental status, economic−industrial development, and demographic patterns are other pillars for underlying the socioeconomic development and environmental vulnerability assessment strategies. With the extensive industrial growth over the years, unpredictability associated with the estimation of absolute anthropogenic emissions compels the analysts to rely on "scenarios" that offer possible future emission routes. Concentration of GHGs and cost-effectiveness are taken into account under these emission scenarios. Policy makers focus on the stabilization scenarios associated with long-term goals of atmospheric benefits. Emission scenarios development is parallel to the growing concern over energy production, transportation and consumption, exploitation of natural resources, and alteration of landscapes in the future century. Synchronization of all facets is a complex and multifold task, which is incorporated with inherent uncertainties. Unforeseen events such as technological revolution and/or geopolitical paradigm shift are additional burden to future emission scenarios. Another major factor that decides the outcome of these scenarios is the global population growth, which is accelerating at a rate by more than 80 million/year and is only continuing to grow for the next century (Population Matters, 2020). With rising population, CO_2 emission rate is also growing accordingly. In principle, combining projection of population and socioeconomic growth, technological boom will depict the actual picture for future emission rates. Such projections of world economics and energy distributions introduced "decarbonizing" as an integral agenda of their policies, which means reducing net carbon or CO_2 emission (quantitative) in addition to the stabilization of short-lived GHGs (methane) emission per unit of energy produced and/or economic output without compromising the overall development (Fay et al., 2015). Emission of CO_2 more than the nature can hold or absorb in forests, other vegetation, and oceans; its concentration will continue to rise making the climate warm. Zeroing the net emission of CO_2 by 2100 would stabilize the climate change by approximately 2°C rise, which is the maximum permissible limit agreed upon by international organizations. Future population projections may be available; however, information on other long-term socioeconomic and technological parameters for emission predictions is not well documented.

Short-lived compounds are another category of materials that contribute to the future emissions with significant uncertainties. These compounds include carbon monoxide (CO), nitrogen oxides (NO_x), methane (CH_4), and volatile organic compounds (VOCs) that are known as tropospheric ozone precursors: sulfur dioxide (SO_2; precursor for sulfate aerosols), dust, carbonaceous aerosols, and sea salt. Concentrations of aerosols as well as ozone are continuously varying; thus, the assessment of their emission and spatial distribution should be made accurately to simulate their distribution over the atmosphere and predict the impacts, but the current source inventories are complex and thus require rigorous future excavation. To have a better picture for future spatial variability of emission, population projections must exist at national and subnational levels to project emission scenarios. Emission of biogenic VOCs from anthropogenic sources and vegetation is sensitive to climate change whose rate depends on radiation, temperature, and so on. The emission gets doubled with every 7K rise in temperature, but the solar radiation has a weaker impact on emission

(Guenther et al., 2000). Biogenic VOCs emission response to variation in vegetation type is highly uncertain; hence, vegetation dynamic models are necessary to quantify the emission with several limitations associated for future projection (National Research Council, 2005).

NO_x emission by thermal and chemical processes during lightning is one of the most significant precursors of tropospheric ozone and highly uncertain sources of NO_x as well. This type of emission is considerably increased in urban regions due to high emission of aerosols. The future projection of dust emissions can be estimated by several land use models where dust emission is found to be enhanced by 11% over the past century and expected to reduce by 20% by the end of the 21st century. This decreased projection is attributed to the changes that would expect to occur in soil moisture and winds over the surface in arid regions although human interventions in landscape degradation have been ignored. Nonetheless, extensive research and information is required for precise dust emission projections for the later years. On the other hand, sulfate and other aerosols generate a radiative imbalance and increase surface temperature with reducing concentrations. Due to the lack of understanding the correlation between aerosol concentrations, hydrological cycle, and vegetation, climate change prediction becomes incomplete and complex.

6.3 Land use and land cover change

With continual conversion of land for urbanization, agricultural and industrial developments and various other applications due to population growth alter the dynamics of land use and involve climate and socioeconomic forcings/feedbacks (Lambin et al., 2003). For example, construction of huge infrastructures and highway projects alters the vegetation and thereby the climate. The satellite monitoring facility of land use change provides detailed data on climate forcing, i.e., surface radiation, aerosol, and trace gas fluxes. Although very limited reports have been recorded effect of land use change on climate forcing as compared with GHG forcing, there is significant effect on regional climate due to land use change during the 20th century (Kabat et al., 2004). Sometimes, the magnitude of regional effects is equivalent to those due to doubling the concentration of CO_2 in atmosphere. Another noticeable impact on global climate is land use change via teleconnections. Prolonged exploitation of landscape will have adverse effects on surface energy, water budget, and ecosystem imbalances. Previous simulation models on climate forcing of future landscape estimated the response of the environment in terms of GHG sources and sinks rather than moisture and heat fluxes by the landscape itself (Snyder et al., 2004).

6.4 Abrupt climate change

Abrupt change is when natural or anthropogenic means force the climate system to act beyond a certain threshold; a transition to a different state has been triggered at a much faster rate than the cause ascertained by the system itself (National Research Council, 2002). These abrupt changes are so rapid and unexpected that ecosystem cannot have quickening response to adaptation. If these changes persist for longer duration, then their impacts will reflect on significant climate variability affecting the large regions. Climate change and variability of any extent can be termed abrupt if the duration, intensity, and frequency are extreme. For instance, single case of floods, storm, or volcanic eruptions may not be abrupt unless their impact does not lead to new climatic state; however, their frequent and persistent event of storm or flood might cause an abrupt climate change. Abrupt changes alter several environmental parameters including temperature, precipitation, drought, snow fall, and so on. Apparent behavioral change of El Niño toward wetter and warmer conditions in the year 1976 is a perfect example of region-specific abrupt change. Ecosystem's vulnerability toward abrupt changes is attributed to the fact that these changes are relatively constant, long-lived as well as the low adaptability of ecosystem. Their susceptibility is further challenged by human activities that alter the ecosystem and its composition of flora and fauna and interrupt species migration. According to paleoclimate records, frequent abrupt changes have taken place in the past at high magnitude those created difficulty for environmental and human adaptation. Although efforts to develop several model-based scenarios based on the understanding of magnitude, frequency, and probability of such changes are continuously made, there is little research on the mechanism, socioeconomic and environmental influences of abrupt changes. Therefore, it is remarkably important to enhance our knowledge and assessment on climate processes, abrupt changes, and their impacts on overall ecosystem.

7. Flood risk tools

The FEMA outlines tools to the societies for better understanding the risks associated with natural hazards and to protect the communities from disasters and their after effects. These tools can equip the communities in designing disaster mitigation plans, enhancing the communication between the vulnerable section of residents with government and organizations, and creating awareness and discuss mitigation strategies with community leaders to reduce the risks. Flood risk tools are primarily nonregulatory products that include database, data sets, maps, and reports. These products are meant to be prepared for an entire range of communities within the frame of Discovery Watershed and submitted at the end of Discovery effort.

Administrators, organizations, and local governments working toward this cause are provided these tools to reinforce their design, plan, and management for minimizing the damages due to flood or any other natural catastrophe. These tools will be beneficial for the following:

(i) Preparation and upgradation of hazard mitigation plans
(ii) Floodplain Management and Community Rating System
(iii) Comprehensive and/or investment planning
(iv) Public outreach
(v) Flood risk minimization along with financial support
(vi) Recovery and relocation planning

7.1 Flood risk report and flood risk map

The stakeholders entirely depend on the flood risk report and get a better understanding of flood hazards and risks to which a community and demography is exposed. This report can be compared to the Flood Insurance Study (FIS) report in the aspect of assessment methods and results of flood risk. The information, data, and maps provided in this report surely enrich the local communities with better preparedness to combat the hazardous situation before it arises. The report also delivers risk assessment information at community as well as watershed levels aiming at small/large-scale economic status during flooding. Other emphases by the report are on the risk mitigation activities, plans, and risk assessment information supply to every strata of a community. Besides, the flood risk maps included in the report correlate flood risk data with risk reduction activities within a flood-prone region and projection of impacts within and beyond that site.

7.2 Flood risk database

This is the key source where all collected information are stored, developed, and extracted during flood risk assessment processes. This flood risk database is analogous to Flood Insurance Rate Map database in various aspects where database is formed by creating, collecting, and analyzing at project level. FEMA is authorized to maintain and publish such data at national, state, and regional standards. A modern viewing tool is being developed that will help to visualize and understand the many-fold data sets without virtually accessing the Geographic Information System (GIS) application.

7.3 Flood risk assessments

This particular tool highlights those risk mitigation activities where the probability of investment return is the highest. For example, in HAZUS[4] Annualized Flood Loss (AAL) study, estimation of loss due to flood was made by using default HAZUS building stock information. Analysis of HAZUS and other risk assessment is achievable only if the data on local building environment are available. Enhanced analyses are supported by community funding if FEMA provides several additional GIS data. Finally, the results obtained from basic along with enhanced HAZUS analyses must be integrated with disaster mitigation plans.

8. Conclusion

Flood events, being natural hazard, are inevitable part of our nature existing from time immemorial and only continue to exist. Flood events due to direct or indirect influence

[4] *HAZUS* is a geographic information system-based natural hazard analysis tool developed and freely distributed by the Federal Emergency Management Agency (FEMA).

of anthropogenic activities such as dam failure, landslide, urbanization, global warming, and so on are exceptional incidents witnessed by the earth. Based on the rate of hydrological and meteorological processes, flash floods or slow-developing riverine floods can occur that leads to increase in flood risks. Rapid climate change augments the frequencies and intensities of flooding. The IPCC and NCA projected an average rise of global temperature from 1.4°C at present to 5.8°C by the end of the 21st century. Serious impacts of this climate change have been reflected on the irregular seasons, precipitation pattern, melting of ice caps, rise of sea level, drought, and humidity. This results in increasing flooding probability and drying up of rivers in arid regions. The flooding events get worsening due to rise in sea level that makes the river water to unable to drain into sea. Deterioration of human health and vegetation, economic loss, property damage, and so on are few notable consequences of flooding. Mitigation of flood with nonstructural methods is proved to be more efficient in terms of future sustainability from the perspective of flood protection, prevention, and mitigation. Nonetheless, structural methods are the primary tools that reduce the vulnerability of human health and provide safety to all living and nonliving properties.

The statistical calculations from various climate models concluded about the significant increases in the number of heavy precipitation events in more regions than that of the significant decreases. Moreover, this is associated with strong regional and subregional variations. Causes of flood are significantly different at both global and regional scales. Frequency of heavy precipitation or the proportion of total rainfall is region and season specific or more explicitly climate specific. Changes in floods are directly proportional to the changes in atmospheric temperature and precipitation. However, projection of flood frequency and magnitude is associated with significant uncertainty. Rain-generated local floods (e.g., flash flooding and urban flooding) are related with heavy rainfall or in turn to the climate change. Changes in floods are dependent upon several factors such as total precipitation, frequency and intensity of precipitation, snow cover, snowmelt, and soil moisture content. Compensation of the increase in flood exposure stabilizes the global mortality rate from floods. Flood losses and causalities are the outcome of increasing exposure of population and assets along with the anthropogenic climate change. Rapid urbanization, deforestation, building of flood walls and dams, and artificial land drainage are responsible for flood risk and losses. The study of climate change for flooding is very essential but limited with a lack of scientific understanding. Improvements in "empirical and model-based science" might be a valuable and encouraging step.

Conservation of nature and management of landscapes: Both these aspects must be taken care of during postflood scenarios. Potential adaptation measures, viz., land use, land coverage, and zoning can be well incorporated in various flood mitigation models to execute the possible approaches of risk assessment to policy maker for comfortable future. In addition, every individual should furnish one's responsibility toward the flood precautionary measures. Public awareness and participation must go hand-in-hand with scientific research, exchange of knowledge, and education on flooding. Finally, appropriate tools have to be designed, developed, and implemented with reliable information, predictions, and forecast at local and national levels by administrations, governments, or any other competent voluntary organizations to prevent flooding, minimizing their impacts with maximum benefit to humankind and environment.

References

Anita, J., November 2013. Structural and non structural Approaches as flood protection Strategy in Muara Angke Settlement, North Jakarta. In: The Second International Conference on Sustainable Infrastructure and Built Environment. SIBE), pp. 1–9.

Article 2, UNFCCC, 1992. United Nations Framework Convention on Climate Change. https://unfccc.int/resource/docs/convkp/conveng.pdf. (Accessed 6 September 2020).

Barnett, T.P., et al., 2008. Human-induced changes in the hydrologyof the western United States. Science 319 (5866), 1080–1083.

Bates, B.C., et al., 2008. Climate Change and water.Technical Paperof the Intergovernmental Panel on Climate Change. IPCC Secretariat, Geneva.

Bouwer, L.M., 2013. Projections of future extreme weather lossesunder changes in climate and exposure. Risk Anal. 33 (5), 915–930.

Brázdil, R., Kundzewicz, Z.W., 2006. Historical hydrology—editorial.Hydro. Sci. J. 51 (5), 733–738.

Christensen, J.H., et al., 2007. Regional climate projections. In: Solomon, S., et al. (Eds.), *Climate Change 2007: The Physicalscience basis.*Contribution of Working Group I to the FourthAssessment Report of the Intergovernmental Panel on ClimateChange. Cambridge University Press, Cambridge, pp. 847–940.

Costanza, R., et al., 2014. Changes in the global value of ecosystem services. Global Environ. Chang. 26, 152–158.

Fay, M., et al., 2015. Decarbonizing Development: Three Steps to a Zero-Carbon Future. International Bank for Reconstruction and Development/The World Bank World Bank Group, Washington, DC.

FEMA, 2013. Local Mitigation Planning Handbook. FEMA Federal Emergency Management Administration, Washington,DC.

Fengqing, J., et al., 2005. Magnification of flood disasters and its relationto regional precipitation and local human activities since the 1980sin Xinxiang, Northwestern China. Nat. Hazards 36, 307–330.

Field, C.B., et al. (Eds.), 2012. Managing the Risks of Extreme Events and Disasters to Advance Climate Change adaptation.Special Report of Working Groups I and II of the Intergovernmental Panel on Climate Change (IPCC). Cambridge University Press, Cambridge.

Gerten, D., et al., 2008. Causes of change in 20th century global riverdischarge. Geophys. Res. Lett. 35, L20405.

Guenther, A., Geron, C., Pierce, T., Lamb, B., Harley, P., Fall, R., 2000. Natural emissions of non-methane volatile organic compounds, carbon monoxide, and oxides of nitrogen from North America. Atmos. Environ. 34, 2205–2230.

Hirabayashi, Y., et al., 2008. Global projections of changing risks offloods and droughts in a changing climate. Hydrol. Sci. J. 53 (4), 754–772.

Hirabayashi, Y., et al., 2013. Global flood risk under climate change. Nat. Clim. Change 3, 816–821. https://doi.org/10.1038/nclimate1911.

Intergovernmental Panel on Climate Change, 2007c. Climate Change 2007- — Mitigation of Climate Change:Contribution of Working Group III to the Fourth Assessment Report of the IPCC. UK: Cambridge University Press, Cambridge, UK.

Intergovernmental Panel on Climate Change, 2007. Climate Change 2007-The Physical Science Basis: Contribution of Working Group I to the Fourth Assessment Report of the Intergovernmental Panel on Climate Change. UK: Cambridge University Press, Cambridge, UK.

Intergovernmental Panel on Climate Change, 2007. Climate Change 2007 – Impacts, Adaptation and Vulnerability:Contribution of Working Group II to the Fourth Assessment Report of the IPCC. UK: Cambridge University Press, Cambridge, UK.

Intergovernmental Panel on Climate Change, 2012. Managing the Risks of Extreme Events and Disasters to Advance Climate Change Adaptation:Special Report of the Intergovernmental Panel on Climate Change. UK: Cambridge University Press, Cambridge, UK.

Intergovernmental Panel on Climate Change, 2018. Global Warming of 1.5°C. IPCC, Switzerland.

Kabat, P., et al. (Eds.), 2004. Vegetation, Water, Humans and the Climate—A New Perspective on an Interactive System. Springer Verlag, Berlin.

Kharin, V., et al., 2007. Changes in temperature and precipitationextremes in the IPCC ensemble of global coupled model simulations. J. Clim. 20 (8), 1419–1444.

Kleinen, T., Petschel-Held, G., 2007. Integrated assessment ofchanges in flooding probabilities due to climate change. Clim. Change 81 (3), 283–312.

Koutsoyiannis, D., Montanari, A., 2007. Statistical analysis ofhydroclimatic time series: uncertainty and insights. Water Resour. Res. 43 (5), W05429.

Koutsoyiannis, D., 2012. Reply to the Comment by T. López-Arias on"Clausius-Clapeyron equation and saturation vapour pressure: simpletheory reconciled with practice". Eur. J. Phys. 33, L13–L14. https://doi.org/10.1088/0143-0807/33/3/L13.

Koutsoyiannis, D., 2013. Hydrology and change. Hydrol. Sci. J. 58 (6), 1177–1197.

Kundzewicz, Z.W., Schellnhuber, H.-J., 2004. Floods in the IPCC TAR perspective. Nat. Hazards 31, 111–128.

Kundzewicz, Z.W. (Ed.), 2012. Changes in Flood Risk in Europe. IAHS Press, Wallingford.

Kundzewicz, Z.W., et al., 2012. Flood risk in Europe—setting the stage. In: Kundzewicz, Z.W. (Ed.), Changes in Flood Risk inEurope. IAHS Press, Wallingford, pp. 11–26.

Kundzewiczm, Z.W., et al., 2014. Flood risk and climate change: global and regional perspectives. Hydrol. Sci. J. 59 (1), 1–28.

Lambin, E.F., Geist, H.J., Lepers, E., 2003. Dynamics of land-use and land-cover change in tropical regions. Annu. Rev. Environ. Resour. 28, 205–241.

Messling, L., et al., 2015. Communicating Flood Risks in a Changing Climate. Climate Outreach, Oxford. http://www.climateoutreach.org/resources. (Accessed 10 September 2020).

Miller, S., Muir-Wood, R., Boissonnade, A., 2008. Anexploration of trends in normalised weather–related catastrophelosses. In: Diaz, H.F., Murnane, R.J. (Eds.), Climate Extremes and Society. CambridgeUniversity Press, Cambridge, pp. 225–247.

Montanari, A., et al., 2013. "Panta Rhei—everything flows": change in hydrology and society—the IAHS scientific decade 2013–2022. Hydrol. Sci. J. 58 (6), 1256–1275. https://doi.org/10.1080/02626667.2013.809088.

National Climate Assessment (NCA), 2018. Fourth National Climate Assessment:Volume II: Impacts, Risks, and Adaptation in the United States. U.S. Global Change Research Program, Washington, USA.

National Oceanic and Atmospheric Administration (NOAA) Atlas 14, 2018. Precipitation-Frequency Atlas of the United States. National Weather Service NOAA, Maryland.

National Research Council, 2005. Radiative Forcing of Climate Change: Expanding the Concept and Addressing Uncertainties. Uncertainties Associated with Future Climate Forcings. The National Academies Press The National Academies of Sciences, Engineering, Medicine, Washington, DC.

Nicholls, N., Seneviratne, S.I., 2013. Comparing IPCCAssessments: how do the AR4 and SREX assessments of changes in extremes differ? Clim. Change. https://doi.org/10.1007/S10854-013-018-0.

O'Brien, K.M., et al., 2012. Toward a sustainable and resilient future. In: Carlo Giupponi, N.M.((Ed.), Managing the Risks of Extreme Events and Disasters to Advance Climate Change Adaptation. A Special Report of Working Groups I and II of the Intergovernmental Panel on Climate Change (IPCC). Cambridge University Press, Cambridge, UK, pp. 437–486.

Peduzzi, P., et al., 2011. Preview Global Risk Data Platform,[online], UNEP/GRID Geneva, UNISDR, World Bank. http://preview.grid.unep.ch/index.php?preview=tools&cat=1&lang=eng. (Accessed 18 October 2012).

Pielke Jr., R.A., Downton, M.W., 2000. Precipitation and damagingfloods: trends in the United States, 1932–1997. J. Clim. 13 (20), 3625–3637.

Population Matters, 2020. Population Matters. https://populationmatters.org/the-facts/the-numbers?gclid=CjwKCAjw19z6BRAYEiwAmo64LaRa3Wk3yYxIXWGAX3Rd1HqZDLBfHsQV_mLhq1RVFLJeZGQVzL51ahoCUrgQAvD_BwE. (Accessed 9 September 2020).

Rosenzweig, C., et al., 2007. Assessment of observed changes andresponses in natural and managed systems. In: Parry, M.L., et al. (Eds.), *Climate Change 2007: Impacts, Adaptation Andvulnerability*. Contribution of Working Group II to the FourthAssessment Report of the Intergovernmental Panel on ClimateChange. Cambridge University Press, Cambridge.

Snyder, P.K., Foley, J.A., Hitchman, M.H., Delire, C., 2004. Analyzing the effects ofcomplete tropical forest removal on the regional climate using a detailed three-dimensional energy budget: an application to Africa. J. Geophys. Res. Atmos. 109 (D21), D21102. https://doi.org/10.1029/2003JD004462.

Solomon, S., Qin, D., Manning, M., 2007. Climate Change 2007: The Physical Science Basis. Working Group I Contribution to the Intergovernmental Panel on Climate Change Fourth Assessment Report. IPCC.

UNISDR (United Nations International Strategy for DisasterReduction), 2011. Revealing Risk, Redefining Development. UNISDR, Geneva.

World Economic Forum, 2019. The Global Risks Report. World Economic Forum, Geneva.

Yamazaki, D., Watanabe, S., Hirabayashi, Y., 2018. Global flood risk modeling and projections of climate change impacts. In: Guy, J.-P., Schumann (Eds.), Global Flood Hazard: Applications in Modeling, Mapping, and Forecasting, Geophysical Monograph 233, first ed. John Wiley & Sons, Inc., USA, pp. 185–203.

Zhang, X., et al., 2007. Detection of human influence on twentieth-century precipitation trends. Nature 448 (7152), 461–465.

Role of phytopesticides in sustainable agriculture

15

Dipti Bharti[1], *Rahul Singh*[2], *Dhruv Arora*[3], *Charu Arora*[4]
[1]Department of Applied Sciences, Greater Noida Institute of Technology, Greater Noida, Uttar Pradesh, India; [2]G.L. Bajaj Institute of Technology and Management, Greater Noida, Uttar Pradesh, India; [3]Department of Food Technology, Gautam Budha University, Greater Noida, Uttar Pradesh, India; [4]Department of Chemistry, Guru Ghasidas Vishwavidyalaya, Bilaspur, Chhattisgarh, India

1. Introduction

According to the US Environmental Protection Agency (EPA, 2003), "Biopesticides are the certain types of pesticides derived from natural materials such as animals, plants, bacteria, and certain minerals." Biopesticides are derived from the plants' secondary metabolites, microorganisms, and insects. Substances from plants and animals have been used to manage diseases in crops, animals, and humans as well (Kumari et al., 2014). In the field of agriculture, there are various problems of pests such as insects, fungi, bacteria, and weeds from the ancient period resulting in decrease in yield and productivity of crops. Use of plant-based compounds was overtaken by the use of synthetic chemicals due to their efficacy, reliability, and quick knockdown effect with low side effect. However, synthetic pesticides have become health hazards for humans and environment due to their toxicity and various types of pollution (Halimatunsadiah et al., 2016). Biopesticides are potential alternatives to synthetic pesticides. Sources of biopesticides are readily available with the recycled material, are easily biodegradable, do not cause any kind of pollution, exhibit various modes of action, are less expensive, and have low toxicity to humans and environment. *Emblica officinalis*, *Allium sativum*, *Artemisia absinthium*, and *Citrullus colocynthis* are known sources of botanical pesticides. The microbial species of *Trichoderma*, *Bacillus*, *Pseudomonas*, and *Beauveria* have been commercialized as antimicrobic pesticides (Ndakidemi et al., 2016). Biopesticides are, however, facing with great challenges of formation, formulation, registration, commercialization in market, and most important acceptance and adoption.

Use of synthetic pesticides in agriculture may result in several adverse effects on environment including various kinds of pollution. Biopesticides have immense scope in the sustainable agriculture. In India, biopesticides cover only 2.91% of the overall pesticide market, and it is expected to exhibit an annual growth rate of about 2.4% in coming years (Gasic and Tanovic, 2013).

The use of synthetic fungicides in agriculture is a relatively very recent phenomenon, and most of the developments have taken place during the past 80 years. This has been the best way of fungal disease control in the world during the past. It is playing an

Global Climate Change. https://doi.org/10.1016/B978-0-12-822928-6.00011-3

important role in crop protection today too (Isman, 2000). However, the chemical residues remain on the plant or within its tissues and soil during fungicidal treatment. Fungicide residues in plants and their fruits pose a great health risk to the consumer, soil, and groundwater and lead to the search for safe alternatives to synthetic fungicides. Plant extract is a group of various biological substances extracted from different parts of plants, which contain a great many of compounds with antimicrobial properties. These compounds can be obtained from roots, barks, seeds, buds, leaves, and fruits (Dar et al., 2014).

2. Classification of biopesticides

Biopesticides can be classified into the following categories:

2.1 Plant-based biofungicides

Aromatic plants are very rich in those chemical substances that have been used in foods as flavoring agent. Most of these compounds are present as terpenes and have fungicide properties. Plant extracts can be directly or indirectly used as substances, which are responsible for the antimicrobial properties (Odhiambo et al., 2009). The extract of *Aloe vera* is found to have antifungal activity against four common postharvest pathogens, namely, *Penicillium digitatum*, *Penicillium expansum*, *Botrytis cinerea*, and *Alternaria alternata*. Few studies have also been focused on the mechanism by which plant extracts and their essential oils (EOs) inhibit the growth of microorganisms (Vidyasagar and Tabassum, 2013). Alcoholic plant extracts of various medicinal plant extracts such as *Curcuma longa*, *Citrus limon*, *A. sativum*, *Capsicum frutescens*, and *Zingiber officinale* were reported to be significantly effective against antimycotic activity of the fungus, namely, *Alternaria solani*, *Pythium ultimum*, *Rhizoctonia solani*, and *Fusarium oxysporum* (Maksymiv, 2015). The terpenes present in the EOs act as a primary antimicrobial agent. Few most active terpenes, viz., eugenol, thymol, and carvacrol, are phenolic in nature. Therefore, it can be assumed that their modes of action might be related to those of other phenolic compounds. EOs may also inhibit enzyme systems in yeasts, including those involved in energy production in cells and synthesis of their structural components (Maksymiv, 2015).

Eugenol (2-methoxy-4-(2-propenyl)-phenol) is a substance that is obtained from *Syzygium aromaticum*, which contain 95% of eugenol as the main volatile oil. Cinnamic aldehyde (3-phenyl-2-propenal) is another antimicrobial substance obtained from *Cinnamomum zeylanicum*. This plant also contains eugenol only 8% of volatile oil and 75% of the first one. In addition, *Syzygium aromaticum* extract has exhibited antifungal properties too. It has inhibited growth of *Aspergillus* and *Penicillium* species (Semeniuc et al., 2017; Andrews, 1992).

The genus *Allium* contains various compounds, which inhibit the growth of fungal pathogens. Some phenolic compounds are constitutive inhibitors such as catechol that is present in the outer bulb layers of pigmented *Allium cepa* cultivars. It confers resistance to *Colletotrichum*. *A. cepa* produces a class of cyclopentane phytoalexin upon pathogen infection, which accumulates in bulb scales at infection sites during incompatible interactions with *B. cinerea*. Tsibulins inhibited spore germination and germ tube elongation of *B. cinerea*. Plant extracts of *A. cepa* and *A. sativum* contain volatile antimicrobial substances, which are reported to inhibit the growth of *Aspergillus flavus*, *Aspergillus parasiticus*, *Candida albicans*, *Cryptococcus*, *Penicillium*, and *Trichosporon* species. The antimicrobial sulfur compounds of *A. cepa* contain phenolic compounds protocatechuic acid and catechol, which contribute to its antimicrobial activities (Lacey et al., 1980; Anukwuorji et al., 2012).

Thymol, the key component of the thyme oil and its derivatives, were found to have antimycotic activity against *Rhizoctonia solani* (Gao et al., 2016). Since plant-based chemicals are considered as safe chemicals, there is great scope to use these and their synthetic derivatives against *R. solani* and other fungal pathogens. Among the six thymol derivatives evaluated, thymol acetate was considered as the most suitable commercially viable plant-based fungicide due to its superior efficacy as well as lowest residue (Hu and Coats, 2008) (Fig. 15.1) (Table 15.1).

Tephrosia vogelii is known to biosynthesize rotenoid, a class of isoflavonoids, which is already well known for the insecticidal and antipathogenic properties, and therefore *T. vogelii* is used as a pest control agent. Chemical analysis of the species bore this out and identified various kinds of isoflavonoids (Stevenson et al., 2012), of which deguelin and tephrosin were the most abundant and most important

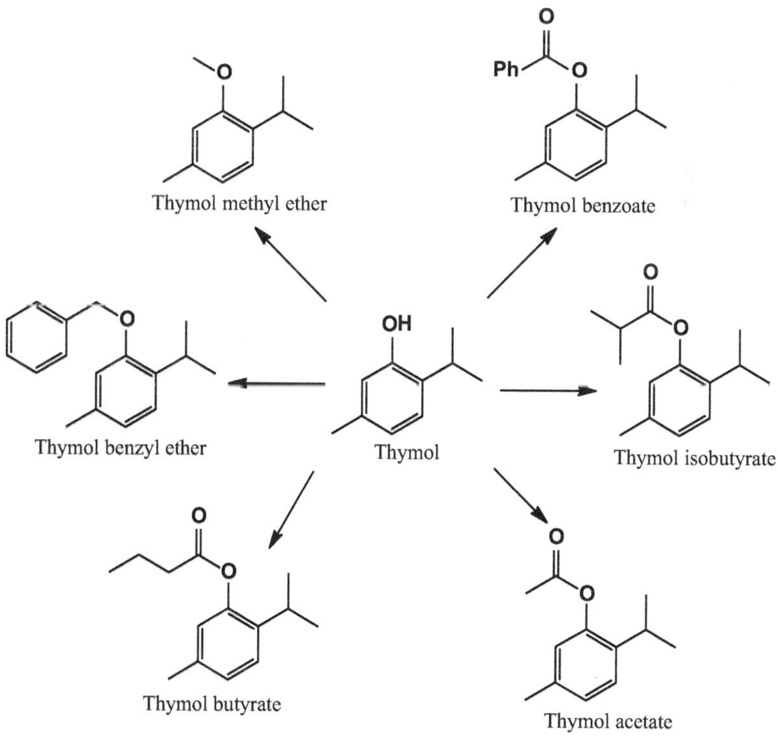

Figure 15.1 Thymol derivatives having antifungal properties.

Table 15.1 Selected biopesticides with fungicidal properties.

S. No	Name of plant	Family	Biologically active constituents	Target pests	Types of activity	References
1.	*Ageratum conyzoides*	Asteraceae	Phenolic chromene, hydroxyl ethyl chromene	*Oncopeltus fasciatus* (Dallas)	Anti-JH activity-suppressed egg laying	Kong et al. (2001)
2.	*Allium sativum*	Amaryllidaceae	Diallyl sulfide, diallyl trisulfide	*Culex pipiens* Linnaeus *Spodoptera litura* (Fabricius)	Repellent, antifeedant, toxicant	Ali et al. (2017)
3.	*Azadirachta indica*	Meliaceae	Lemnoideaeazadirachtin	*Helicoverpa armigera* (Hubner); *Plutella xylostella* (Linnaeus	Antifeedant, deterrent, physiological toxicant Oviposition, behavior, and disturbance	Ridha et al. (2018)
4.	*Citrus limon (L) Burm. (Lemon)*	Rutaceae	Limonin, nomilin, obacunone	*Helicoverpa zea* (Boddie)	Antifeedant, toxicant	Pan et al. (2016)
5.	*Ginkgo biloba (L)*	Ginkgoaceae	Salicylic acid, derivatives, bilobalide	*Popillia japonica* Newman *Pieris rapae* (Linnaeus)	Feeding deterrent	Ali Esmail (2018)
6.	*Hedera helix (English ivy)*	Araliaceae	Saponin	*Reticulitermes flavipes* (Kollar)	Antifeedant, toxicant	Wieslaw et al. (1990)
7.	*Medicago sativa* L. *(Alfalfa)*	Fabaceae	Butyric acid, succinic acid, xanthophyll, coumarins	*Bruchophagus rodi* (Gussakovsky), *Hypera postica* (Gyllenhaal)	Repellant, toxicant	Ho Dung and Ong Thi (2020)
8.	*Mentha spicata* L. *(Spearmint)*	Labiatae	Carvone, pulegone, menthol	*Microcerotermes crassus*	Antifeedant, toxicant	Souvic and UnTaek (2000)

9.	*Nicotiana tabacum* L.	Solanaceae	Nicotine	Soft bodied insects	Contact toxicant, fumigant	LohSiew et al. (2011)
10.	*Piper nigrum* L. (black pepper)	Piperaceae	Piperine, other amides	*Helicoverpa zea*; *Anthonomus grandis* Boheman	Toxicant, oviposition deterrent	Sreelatha et al. (2009)
11.	*Plumbago capensis* Thunb.	*Plumbaginaceae*	Plumbagin	*Helicoverpa* spp. *Pectinophora gossypiella* (Saunders)	Antifeedant, chitin synthesis inhibitor	TounouAgbéko et al. (2019)
12.	*Pyrethrum/ Pyrethrins*	Asteraceae	Chrysanthemic acid products, derivative of pyrethrin II	Aphids, roaches, fleas, flies, and ticks	Antifeedant, toxicant	Neli et al., 2014
13.	*Ricinus communis* L. (*Castor bean*)	Euphorbiaceae	Ricinine	*Cydia pomonella* (Linnaeus)	Toxicant, oviposition deterrent	Stevenson et al. (2012)
14.	*Ryania*	Salicaceae	Ryanodine and 9,21 didehydroryanodine	Caterpillars (European corn borer, corn earworm, and others) and thrips	Repellent, antifeedant	Murray, 1994
15.	*Tephrosia vogelii* Hook F.	Fabaceae	Rotebone, tephrosin	*Spodoptera exempta* (walker)	Feeding deterrent	David et al. (1988)
16.	*Vitex negundo* L.	Verbenaceae		Stored grain pest	Insecticidal, repellant, fumigant	Zhi et al. (2009)
17.	*Zanthoxylum monophyllum* Lam.	Rutaceae	Zanthophylline	*Hypera postica*	Feeding deterrent	Zhi et al. (2009)

compounds due to their higher levels of bioactivity against insects (Belmain et al., 2012). Identification of several others less abundant rotenoids that partly contributed to activity including rotenone, sarcolobine, and α-toxicarol has been reported (Sola et al., 2014; Anukwuorji et al., 2016) (Fig. 15.2).

2.2 Microbial pesticides

These are naturally occurring microorganisms that include bacteria, fungi, protozoa, or viruses. All are capable to control many different kinds of pest, although each active ingredient is specific for its target pests and pathogens, for example, *Bacillus thuringiensis*, *Baculoviruses*, *Metarhizium anisopliae*, *Verticillium lecanii*, and so on. *Baculoviruses* are target agents for the control of Lepidopteran, Hymenopteran, and Coleopteran pests (Campbell and Madden, 1990). The pathogenic fungi are an important group of pest management organisms that grow in both terrestrial and aquatic habitats and, when specifically associated with insects, are known as entomopathogenic fungi (Table 15.2).

Biopesticides are covered into 12 classes within six phyla and belong to four major groups such as Laboulbeniales, Pyrenomycetes, Hyphomycetes, and Zygomycetes. Common microbial biopesticides species include *Beauveria bassiana*, *Metarhizium anisopliae*, *Nomuraea rileyi*, *Paecilomyces farinosus*, and *V. lecanii*. Microbial pesticides can be used as an alternative source of chemical pesticides, as indiscriminate pesticide use is detrimental to the environment as well as human health and also increases insect resistance to pesticides and pathogens (Leger and Wang, 2010).

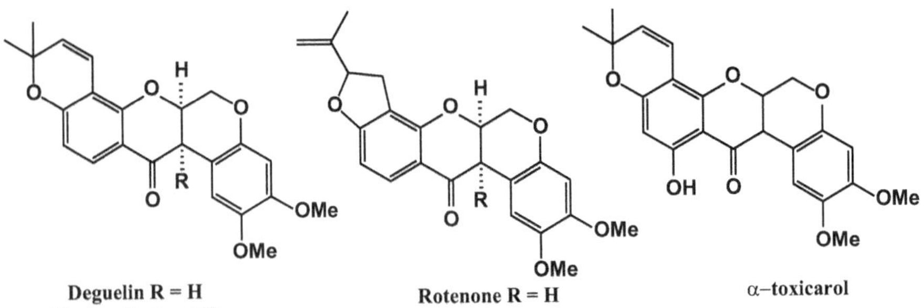

Deguelin R = H
Tephorsin R = OH

Rotenone R = H
12α-hydroxyroten one R = OH

α–toxicarol

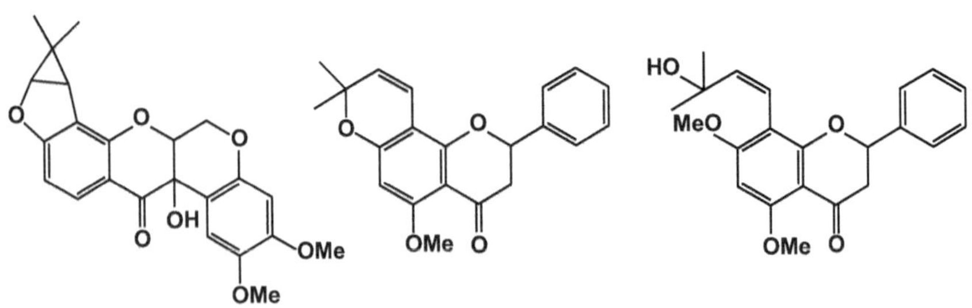

Sarcolobine

Obovatin 3-methylether

Z-tephrostachin

Figure 15.2 Rotenoids and flavanones from *Tephrosia vogelii* hemo types.

Table 15.2 List of some commonly used microbial pesticide (Pekrul and Grula, 1979).

Crop pest	Microbial pesticide
Colorado potato beetle	*Bacillus thuringiensis subsp. tenebrionis*
Lepidoptera	*Bacillus thuringiensis* subspecies *kurstaki*, *B. thuringiensis* subspecies *aizawai*
Diptera	*Bacillus moritai*
Coleoptera: Scarabaeidae, *Popillia japonica*	*B. thuringiensis* subspecies *japonensis* *Paenibacillus popilliae*
Entomopathogenic fungi	
Coleoptera (Scarabaeidae)	*Beauveria brongniartii*
Hemiptera, Thysanoptera	*Conidiobolus thromboides* Acari
Hemiptera	*Aschersonia aleyrodis*, *Lecanicillium longisporum*
Coleoptera, Diptera, Hemiptera, Isoptera	*Metarhizium anisopliae* sensu lato
Lepidoptera	*Nomuraea rileyi*
Phytophagous mites	*Verticillium lecanii*

The microbial pesticides are being used as alternative pest management sources in several constraints such as development of stable formulations, lack of registration procedures, and so on, which are associated with their introduction and promotion in most of the developing countries. The demand for microbial pesticides is growing steadily in all around the world. By combining performance and safety both, biopesticides perform efficaciously with minimum application restrictions and eco-friendly environmental safety benefits. In the future, role of microbial pesticides will be more beneficial in agriculture and horticulture fields (Kunimi, 2007).

2.3 Plant-based weedicides

It was reported that several ornamental plants have the potential to suppress growth of weeds. *Nerium oleander* plant extract has herbicidal property. Extract of *N. oleander* flower inhibits the growth of Italian ryegrass. Aqueous extract of *N. oleander* leaves is reported to exhibit inhibition on the germination and seedling growth of purple nut sedge (*Cyperus rotundus* L.) (Grange and Ahmed, 1988). It was also reported that *Thevetia peruviana* has the inhibitory effect on *Parthenium hysterophorus* while marigold has allelopathic effects on hoary cress. Aqueous extracts of *Chrysanthemum morifolium* showed growth inhibitory effects against weeds (Seigler, 1998).

Extracts of *Sorghum* and maize are used as organic herbicides. Aqueous extract of sorghum at higher concentration reduced the spore germination of *Trianthema portulacastrum*. The water extract of leaves of *Sorghum* and sunflower could be used to inhibit the growth of weeds in wheat cultivation during growing season. The report also found that allelopathic potential of maize could be used for weed control as well. Aqueous extracts of *Salvia moorcroftiana* and *Verbascum thapsus* have also found significant weed-suppressing effects (Kumar, 2012).

Aqueous extracts of *Houttuynia cordata* (whole plant) have the ability to inhibit the growth of rice weeds, namely, *Echinochloa* and *Monochoria*. Leaves extract has also shown the weed-suppressing effects. It was reported that methanol extract of teak leaves could be considered as a potential preherbicide for controlling the wild rice weeds. *Azadirachta indica* bark and leaves extracts inhibit the growth germination of *Echinochloa* crus-galli, *Monochoria vaginalis*, and *Aeschynomene indica* L. in a bioassay and in soil. It was also reported that concentration of phenolic compounds was higher in bark of the tree. Several other weedy plants extract also has herbicidal activities (Rathore and Nollet, 2012). *Chenopodium album* shoot residue has the ability as a growth inhibitor of shoots and tuber formation of *Cyperus esculentus*. The aqueous extracts of *Cirsium arvense* and *Ageratum conyzoides* have the weedicidal activity against both the weeds which are *Phalaris minor* and *Poa annua*. During the past two decades, more than 300 species of insects have been reported to be affected by the *Azadirachta indica* component (Suman and Dikshit, 2010).

Use of herbicides may increase crop yields by allowing weed control agents and harvest management. Glyphosate is the most widely used herbicide active ingredient, with $11 billion spent annually on glyphosate-containing products applied to >350 million hectares worldwide, using about 8.6 billion kg of glyphosate. The herbicidal effectiveness of glyphosate can depend upon the time of day of spraying. It has been shown that the plant circadian clock regulates the effectiveness of glyphosate. It has been identified a daily and circadian rhythm in the inhibition of plant development by glyphosate, due to interaction between glyphosate activity, the circadian oscillator, and potentially auxin signaling. It is observed that the circadian clock controls the timing and extent of glyphosate-induced plant cell death. Furthermore, the clock controls a rhythm in the minimum effective dose of glyphosate. The concept of agricultural chronotherapy, similar in principle to chronotherapy in medical practice, has also been proposed (Mangels et al., 1991).

Plants and microbes contain a largely unexploited storehouse of phytotoxins that might be used directly or as structural indicators for new synthetic herbicides. Several factors have increased the interest of the herbicide industry in this source of natural products as herbicides. Biopesticides have been proved to be very effective insecticides, whereas relatively minor successes have been obtained with these compounds as herbicides. The number of options that must be considered in the discovery and development of a natural product as a herbicide is larger than for a synthetic herbicide. Furthermore, the molecular complexity, limited environmental stability, and low herbicidal activity of many phytotoxic natural products are discouraging. However, advances in chemistry and biotechnology are increasing the speed and ease with which we can discover and develop natural products as herbicides (Battaglin et al., 1998).

Nonselective or broad-spectrum herbicides are used to kill all vegetation in an area. Broad-spectrum herbicides do not discriminate between garden plants, lawn, and weeds. Some of the more popular broad-spectrum herbicides, such as Roundup and Brush Be Gone, contain a substance called glyphosate, which has been banned from use in some areas. A nonselective herbicide should not be used very often without extreme caution.

Few acetic acid–based broad-spectrum herbicides are also known. These cause burn in the leafy portion of the plant but do not always kill the roots and the plants eventually. The best use for these organic products is on annual weeds and when treating weeds growing near cement, on a hot, sunny day. The heat from the cement intensifies the burning caused by the acid.

The imidazolinone is a well-known herbicide used since the 1980s and 1990s. These imidazolinone herbicides were widely used because of their low application rate, reduced environmental pollution risk, and the most important selectivity in a wide range of cropping systems. On the abiotic degradation of the imidazolinone, herbicides imazapyr, imazethapyr, and imazaquin ring will form under controlled conditions (Fig. 15.3).

All imidazolinone herbicides are mixture of two enantiomers, which is derived from the chiral center of the imidazolinone ring. The inhibitory effect exhibited that the R-enantiomer is nearly eight times greater than that of the S-enantiomer on target weeds (Barkania et al., 2005).

2.4 Plant-based antiviral compounds

Various medicinal plants have been reported to possess antiviral activity. Antiviral activity of *Boerhavia diffusa* has shown a very good effect in inhibiting the hypersensitive and systemic hosts. The inhibitory effect of these extracts was found to be very sensitive to actinomycin D. The aqueous extract of root of this plant inhibited yellow mosaic disease of *Vigna radiata* caused by mung bean yellow mosaic virus. To resist and treat viral infection in many susceptible crops, the extracts obtained from the roots of *B. diffusa* was used. This antiviral agent was active against *Nicotianata bacum* and *Crotalaria juncea* (Razvi et al., 1995; Vanden Berghe and Vlietinck, 1986).

3. Challenges in the use of conventional pesticides

The harmful effects associated with the use of synthetic pesticides are toxicity and poisoning of the crops. Synthetic pesticides also lead to environmental pollution due to the nonbiodegradable nature of their constituent compounds, and they persist for longer time in environment. Parliman has observed that metam-sodium and other

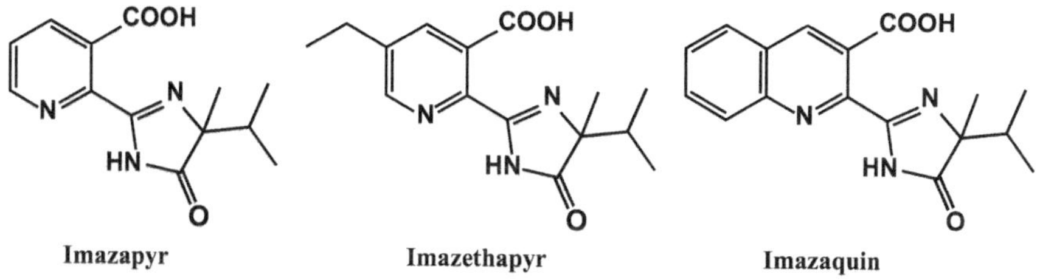

Figure 15.3 Chemical structures of some herbicides.

fumigants take over 6 months to last up after their application. In a report by pesticides action network, metam-sodium pollutes the air and soil, thereby affecting the population of natural enemies in the soil as well as air (Dhaliwal et al., 1996). Methyl bromide has been banned from agricultural use due to its negative impact on the soil and air. It also persists for longer time in environment. It is also involved in the depletion of ozone layer, which contributes significantly to climate change and global warming. The constituent compounds of chemical pesticides contaminate soil and make it unsuitable for next crop production. They also pollute surface and groundwater and make potable water unfit for consumption. Use of dichlorodiphenyltrichloroethane (DDT) contributes for biological magnification, which leads to poisoning of birds, water species, and human beings. Few pesticides known to be carcinogenic are banned for agricultural use (Harada et al., 2016). Continuous use of synthetic pesticides leads to development of resistant plant pathogen strains, and to get rid of this problem, farmers apply more and more chemicals in an effort to eradicate such pests. In the process of managing target pests, synthetic pesticides also kill nontarget beneficial organisms too such as pollinators, predators, and antagonists; thereby, they also disrupt biodiversity. Consumption of such crops, which is treated by synthetic chemicals, poses chronic health problems to humans due to the accumulation of toxins in the body. Pesticides containing malathion and trichlorfon have been reported to cause reproductive complications in humans as well as in other animal species (Cragg et al., 1997). Some pesticides have also been reported to reduce resistance of animals to disease-causing pathogens. Continuous exposure to pesticides such as chlorpyrifos causes DNA breakage, genetic mutations, reproductive problems, and chronic diseases such as asthma, hypertension, and cancer (Dey et al., 2015). Thus, pesticides adversely affect human population and environment directly as well as indirectly.

4. Origin of biopesticides

On the basis of method of extraction, botanical pesticides can either be plant extracts of bioactive plants or EOs possessing pesticidal properties. They are obtained from various parts of plants such as leaves, barks, flowers, roots, rhizomes, bulbs, seeds, cloves, or fruits which are either fresh or ripe (ChughBharti, 2012; ChughMehta and Dua, 2012; Arora et al., 2003; Kaushik et al., 2004). Dried plant parts are more effective as it reduces water concentration resulting in higher yield of bioactive constituents (Sinclair and Dhingra, 1995).

Several extraction, isolation, and purification techniques are used for standardization of biopesticides. The technique such as gas chromatography–mass spectrometry (GC-MS) analysis was carried out on *Citrus sinensis*. D-Limonene and myrcene were found as the major constituents of the isolated oil. The quality of extracts and oils is highly dependent on the solvent used and method of extraction. The solvents should be of low toxicity, able to dissolve as many compounds as possible and evaporate easily at low temperatures (Mazid et al., 2011).

5. Production, formulation, and commercialization of biopesticides

Botanical pesticides are prepared from various parts of plants such as leaves, stem, roots, flowers, and so on, which are cleaned of dirt or foreign materials, and then material is shade dried. Material is extracted either using solvents or distillation to obtain extracts or EOs, respectively. The extracts are then screened for biological activities in vitro against different pests using different methods such as disc diffusion, agar well diffusion, agar dilution, and poisoned food technique (Gasic and Tanovic, 2013). The most active plant botanicals are then evaluated for efficacy in managing pests and diseases under field conditions. The bioactive constituents of the active plant extract are then identified by using different chemical techniques for optimum formulation. The laboratory and various field trials are carried out to ensure that the most effective combination of the bioactive compounds, carrier materials, and optimization of other components used in pesticidal formulation. After getting the reports from the laboratory, field trials are carried out to request for registration of the product from the pest control products body (Giongo et al., 2016).

It is clear from the aforementioned report that the biopesticides are the key for sustainable agriculture and ecofriendly development (Balog et al., 2017). There is urgent need to put more attention and more scientific efforts to identify the thrust area of research for the development of eco-friendly production technology with the use of biopesticides as a pest control agent.

However, several factors may affect the concentration of bioactive ingredients of plant species, which is discussed in the following.

6. Effect of climatic conditions on activity and active ingredients of plant based pesticides

Phytoconstituents of various medicinal plants are bound to fluctuate with season and variation in different climatic and geographic conditions. Plants face different local climatic conditions as well as different edaphic factors all over the world. To predict how different environmental factors affect chemical constituents of several species, the abundance of populations and chemotypes as well as the content of specific compounds can be a valuable tool to understand plant variation in chemical features. It can also facilitate prospecting plants with high amounts of specific compounds for nutrition, pharmaceutical, industrial, and agricultural uses. In most cases, plant EOs are characterized by a strong aroma, which is mainly produced by secondary metabolites present in the plant species (Thompson, 2005).

Medicinal plants with their wide variety of chemical constituents offer a promising source of various agents with medicinal properties. Besides forming the backbone of traditional medicine, these plants continue to support human and other animal's health.

In recent years, attention is directed toward herbal medicines, natural food, and food ingredients from plants that are believed to be safer and healthier than their synthetic counterparts (Sivarajan and Indira, 1994). Agroclimatic conditions and environment can influence the chemical composition and medicinal properties of plant species. Earlier findings have suggested that temperature, soil type, and other environmental factors are responsible for the variations in the chemical constituents and various medicinal properties of the plants from different geographical locations (Prasad and Tyagi, 2005).

Few studies are reported to examine the yield components, fatty acid, EO compositions, phenolic contents, fruit EO composition, and the antioxidant activities of cumin (*Cuminum cyminum* L.) seeds under drought condition (Kunle, 2012). Fatty acid composition analysis of this plant indicated that petroselinic acid was the major fatty acid (55.9%) followed by palmitic (23.82%) and linoleic (12.40%) acids. Water deficit enhanced the palmitic acid percentage and affected the double bound index of the fatty acid pool and thus the oil quality too. The EO yield was 1.64% based on the dry weight and increased by 1.40 folds under moderate water deficit (MWD). Nevertheless, it decreased by 37.19% under severe water deficit (SWD) in comparison with the nontreated seeds. Drought results on the modification of the EOs chemotype from γ-terpinene/phenyl-1,2 ethanediol in the control seeds to γ-terpinene/cuminaldehyde in stressed ones. Besides it, total phenolic contents were higher in the treated seeds (MWD and SWD). Results suggest that water deficit treatment may regulate the production of bioactive compounds in cumin seeds, influencing their nutritional and industrial values too (Rubaya Sultana et al., 2018).

Moringa oleifera is a major tropical plant with high nutritive and medicinal properties. Antioxidant activities of extracts of *M. oleifera* flowers collected from four different geographical locations of Tamilnadu have been carried out. Free radical scavenging activity was evaluated using diphenyl-picryl-hydrazyl (DPPH) assay (Inbathamizh and Padmini, 2013). Other methods, viz., hydroxyl radical scavenging assay, superoxide anion radical scavenging assay, metal chelating assay, phosphomolybdenum assay, ferric thiocyanate (FTC) method, and thiobarbituric acid (TBA) method were also adopted. The differences were observed in the antioxidant machinery of the extracts. Though the presence of phytochemicals among the samples remained the same, their quantity varied as noticed from their phenol and flavonoid contents, which correlated to their antioxidant activities. The methanol extract of *M. oleifera* flowers from the geographical location with loamy red soil, moderate monsoon, and less pollution exhibited the highest antioxidant potential. Phenol and flavonoid content was also higher as compared with the other samples studied in present investigations (Anna et al., 2011).

Crude aqueous extracts of *Aloe vera* collected from different climatic conditions in India were screened for the phytochemical analysis and antioxidant activity. Phytochemical analysis performed with the help of Fourier transform infrared (FTIR) spectroscopy study showed the presence of various phytoconstituents present in samples collected from different *A. vera* samples. All antioxidant assays revealed that highland and semiarid zone samples possessed higher antioxidant activity, whereas tropical zone samples possessed minimum antioxidant activity. We can conclude that different agroclimatic conditions have effects on phytochemical diversity and antioxidant potential of *A. vera* plant (Kumar et al., 2017).

The studies reported that EO of medicinal plants component shows a rhythmic increase in oil production throughout the growing season and then a steady decline toward the winter and hence recommended late summer as the best collection time. In contrast, other researchers concluded the winter as the best season for harvesting of EOs containing plant parts. Not only EO is influenced by the season, but also other important plant constituents such as polyphenol, flavonoids, glycosides, alkaloids, terpenoids, and so on are also influenced significantly by seasons (Khaleefa and Irshad, 2015). Although various studies have been published indicating the effect of seasons on secondary metabolites of plants, there is no compiled information representing the optimized season for the collection of pharmaceutically important constituents from different plant sources (Brian, 2003).

Centella asiatica belonging to the family Apiaceae is a small herb, which is used traditionally as brain tonic for neuroprotection, Alzheimer's disease, Parkinson's disease, and decreasing oxidative stress. To establish impact of geographical variation on this plant, the samples were collected from different geographical regions by performing pharmacognostical and phytochemical screening under pharmacognostical standardization to determine the macroscopical and microscopical characters such as transverse section of leaf, stomatal number, stomatal index, vein-islet number, powder microscopy, and proximate analysis such as ash values, extractive values, moisture content, and lastly fluorescence analysis. The massive changes in pharmacognostical characters were found in plant samples collected from Sikkim and West Bengal region. So it can be concluded that species character changes with geographical variation (Thomas et al., 2008).

Plant kingdom is a potential source of medicinal and pharmaceutically active substances. Medicinal plants depict a variation in bioactive constituents during different climatic conditions. Synergistic effects due to UV radiation, humidity, and hot temperatures can explain seasonal trends of plant chemical constituents' accumulation according to the current aspect. Since active principles of some medicinal plants vary quantitatively in different climatic condition of the year so that therapeutic efficacy of medicinal plants is also likely to vary during different climatic seasons of the year and the majority of medicinal plant materials are usually best collected during season when the herbs are at peak growing stage and concentration (Yang et al., 2018).

Various geographic location and climatic condition connected to environmental and edaphic factors such as temperature, rainfall, soil type, and composition influence the presence and the total chemical constituents of specific plant compounds. Geographic, edaphic, and climatic information can be utilized to predict the presence of specific compounds from medicinal or aromatic plants, which is also evaluated in some investigations (Mammen et al., 2010). Studies on effect of environmental factors on EO content of *Zataria multiflora* and the composition and distribution of chemotypes have been reported. The species growing in areas rich in calcium, iron, potassium, and aluminum, with mean rainfall of 40.46–302.72 mm/year and mean annual temperature of 14.90–28.80°C, are found to possess higher content of EO. EO content ranged from 2.75% to 5.89%. Carvacrol (10.56%–73.31%), thymol (3.51%–48.12%), linalool (0.90%–55.38%), and *p*-cymene (1.66%–13.96%) were the major constituents, which classified 14 populations into three chemotypes.

Corresponding to the phytochemical cluster analysis, the hierarchical cluster analysis (HCA) based on NIR data also recognized the carvacrol, thymol, and linalool chemotypes. Hence, NIR has the potential to be applied as a useful tool to determine rapidly the chemotypes of *Z. multiflora* and similar herbs. EO and EO constituent content were correlated with different geographic location, climate, and edaphic factors (Karimi et al., 2020). The structural equation models (SEMs) approach revealed direct effects of soil factors (texture, phosphor, pH) and mostly indirect effects of latitude and altitude directly affecting, e.g., soil factors. The approach of identifying environmental predictors for EO content, chemotype, or presence of high amounts of specific compounds can help to select regions for sampling plant material with the desired chemical profile for direct use or for breeding.

Pseudarthria viscida, a well-known medicinal plant, is widely used for the treatment of various diseases and ailments. Root of this plant is an important ingredient in more than 70 various ayurvedic formulations. The variation in anatomical, physicochemical, and phytochemical characters in *P. viscida* is observed due to change in different geographical and climatic conditions. Variation in phytochemicals taking place due to various seasons and different regions was studied with the help of high-performance thin-layer chromatography (HPTLC) densitometric quantification of lupeol in methanol extract of roots. The variation was observed in the quantity of cell inclusions such as starch and tannin, number of fibers, and wall thickness of lignified cells. Physicochemical parameters and HPTLC densitometric quantification have also shown variation in the quantity of chemical constituents according to season and place of collection (Jayanthy et al., 2013).

Similar studies are required to select best climatic and geographical zone to grow the raw material for preparation of biopesticidal formulations.

7. Future prospects

Nowadays, sustainability in agricultural production system is must to meet the growing demand of food with limited availability of resources in which plant protection is one of the major key factors. To get rid of this problem, biopesticides seem to be the best alternatives to synthetic chemicals due to various advantages and because of eco-friendly nature. Though there is increasing use of biopesticides day-by-day but for sustaining the availability of food to increasing population, whole system of agriculture across the world should adopt biopesticides as control agent of pest. In all over the world, both private and government organizations should make a joint effort in production, manufacturing, and sale of eco-friendly biopesticides. This effort may lead to the step toward the organic farming, which is free from all types of contaminations. Regulation regarding registration and formation of biopesticides needs to be made easy so that the commercialization and availability of biopesticides are enhanced in the market.

Various biomolecular and biotechnological tools and techniques are being used for the production of biopesticides in a safe and sustainable manner. In the biotechnological technique, DNA recombinant technology is also being used for enhancing efficacy of biopotential pesticides. Fusion of protein is being designed to develop next-generation biopesticides in this technology that allows selected toxins (not toxic to higher animals) to be combined with a carrier protein, which makes them toxic to insect pests when consumed orally, while they were effective only when injected into a prey organism by a predator not on other organisms.

Globally, there are 175 registered biopesticide active ingredients and 700 products available in the market. The global market for biopesticides was valued at around US $1.3 billion in 2011. However, in India, biopesticides are not very much represented; it is only 4.2% of the overall pesticide market and is expected to exhibit an impressive annual growth rate of about 10% in the coming years. In India, only few categories of biopesticides production is taking place, i.e., neem-based pesticides, *Bacillus thuringiensis*, nuclear polyhedrosis virus (NPV), and *Trichoderma* are some of the major biopesticides produced, while a total of 287 pesticides have been registered so far (Olson, 2015; Pavela and Singh, 2014).

It is clear from the various reports that the biopesticides are the key for sustainable agriculture and there is urgent need to put more and more scientific efforts to identify the new technique to thrust area of research for the development of eco-friendly production technology with the use of biopesticides as a pest control agent.

References

Ali Esmail, A.S., 2018. Pharmacological and therapeutic activities of Hedera helix- a review. IOSR J. Pharm. 8 (5), 41–53.

Ali, A.M., Mohamed, D.S., Shaurub, E.S.H., Elsayed, A.M., 2017. Antifeedant activity and some biochemical effects of garlic and lemon essential oils on *Spodoptera littoralis* (Boisduval) (Lepidoptera: noctuidae). J. Entomol. Zool. Stud. 5 (3), 1476–1482.

Andrews, J.H., 1992. Biological control in the phyllosphere. Annu. Rev. Phytopathol. 30, 603–635.

Anna, S., Cezary, P., Max, H., 2011. Influence of environmental abiotic factors on the content of saponins in plants. Phytochem. Rev. 10 (4), 471–491.

Anukwuorji, C.A., Obianuju, C.M., Ezebo, R.O., Anuagasi, C.L., 2016. Antimicrobial effects of four plant extracts against post harvest spoilage fungi of yam (*Dioscorea rotundata* poir). Int. J. Phys. Soc. Sci. 12, 1–10.

Anukwuorji, C.A., Putheti, R.R., Okigbo, R.N., 2012. Isolation of fungi causing rot of cocoyam (*Colocasiaesculenta* (L.) Schott) and control with plant extracts: *Allium sativum*, L., *Garcinia kola*, heckle., *Azadirachta indica*, L. and *Carica papaya*, L.). Global Adv. Res. J. Agric. Sci. 1, 33–47.

Arora, C., Kaushik, R.D., Kumar, A., Garg, G.K., 2003. Fungicidal potential of Kumaon and Tarai region plants against mushroom fungal pathogens. Allelopathy J. 11 (1), 63–70.

Balog, A., Hartel, T., Loxdale, H.D., Wilson, K., 2017. Differences in the progress of the biopesticide revolution between the EU and other major crop-growing regions. Pest Manag. Sci. 73, 2203–2208.

Barkania, H., Catastinia, C., Emmelina, C., Sarakhac, M., ElAzzouzi, M., Chovelon, J.M., 2005. Study of the photo transformation of imazaquin in aqueous solution: a kinetic approach. J. Photochem. Photobiol. Chem. 170, 27–35.

Battaglin, W.A., Furlong, E.T., Burkhardt, M.R., Peter, C.J., 1998. Occurrence of sulfonylurea, sulfonamide, imidazolinone, and other herbicides in rivers, reservoirs and groundwater in the Midwestern United States. Sci. Total Environ. 248, 123–133.

Belmain, S.R., Amoah, B.A., Nyirenda, S.P., Kamanula, J.F., Stevenson, P.C., 2012. Highly variable insect control efficacy of *Tephro siavogelii* chemotypes. J. Agric. Food Chem. 60, 10055–10063.

Brian, M., 2003. The Wealth of India, Raw Materials, vol. 7. Council for Scientific and Industrial Research, pp. 69–70.

Campbell, C.L., Madden, L.V., 1990. Introduction to Plant Disease Epidemiology, first ed. Wiley-Interscience, New York, NY, USA, p. 523.

Chugh, A.C., Bharti, D., 2012. Phytochemical analysis and screening of antibacterial activity of some selected Indian medicinal plants. Int. J. Phytomed. 4 (2), 261–265.

Chugh, A.C., Mehta, S., Dua, H., 2012. Phytochemical screening and evaluation of biological activities of some medicinal plants of Phagwara, Punjab. Asian J. Chem. 24 (12), 5903–5905.

Cragg, G.M., Newman, D.J., Sander, K.M., 1997. Natural products in drug discovery and development. J. Nat. Prod. 60, 52–60.

Dar, A.S., Nisar, A.D., Mudasir, A.B., Mudasir, H.B., 2014. Prospects, utilization and challenges of botanical pesticides in sustainable agriculture. International challenges of botanical pesticides in sustainable agriculture. Int. J. Mol. Biol. Biochem. 1, 1–14.

David, B.V., Sukumaran, D., Kandasamy, C., 1988. The Indian privet *Vitex negundo* Linn – a plant possessing promising pesticidal activity. Pesticides 22, 27–29.

Dey, K.R., Choudhury, P., Dutta, B.K., 2015. Impact of pesticide use on the health of farmers: a study in Barak valley, Assam (India). J. Environ. Chem. Ecotoxicol. 10, 269–277.

Dhaliwal, G.S., Arora, R., Dilawari, V.K., 1996. Botanical pesticides in insect pest management: emerging trends and strategies. In: Narwal, S.S., Tauro, P. (Eds.), Allelopathy in Pest Management for Sustainable Agriculture. Scientific Publishers, Jodhpur, India, pp. 93–119.

Gao, T., Zhou, H., Zhou, W., Hu, L., Chen, J., Shi, Z., 2016. The fungicidal activity of thymol against *Fusarium graminearum* via inducing lipid peroxidation and disrupting ergosterol biosynthesis. Molecules 21, 770.

Gasic, S., Tanovic, B., 2013. Biopesticide formulations, possibility of application and future trends. J. Pest. Phytomed. 2, 97–102.

Giongo, A.M.M., Vendramim, J.D., Forim, M.R., 2016. Evaluation of neem-based nanoformulations as alternative to control fall armyworm. Cienc. Agrotecnol 40, 26–36.

Grange, M., Ahmed, S., 1988. Handbook of Plants with Pest Control Properties. John Wiley & Sons, New York.

Halimatunsadiah, A.B., Norida, M., Omar, D., Kamarulzaman, N.H., 2016. Application of pesticide in pest management: the case of lowland vegetable growers. Int. Food Res. J. 1, 85–94.

Harada, T., Takeda, M., Kojima, S., Tomiyama, N., 2016. Toxicity and carcinogenicity of dichlorodiphenyltrichloroethane (DDT). Toxicol. Res. 32, 21.

Ho Dung, M., Ong Thi, T., 2020. Larvicidal and repellent activity of *Mentha arvensis* L. essential oil against *Aedesa egypti*. Insects 11 (3), 198–201.

Hu, D., Coats, J., 2008. Evaluation of the environmental fate of thymol and phenethyl propionate in the laboratory. Pest Manag. Sci. 775–779.

Inbathamizh, L., Padmini, E., 2013. Effect of geographical properties on the phytochemical composition and antioxidant potential of *Moringa oleifera* flowers. Bio. Med. Rx 1 (3), 239–247.

Isman, M.B., 2000. Plant essential oils for pest and disease management. Crop Protect. 19, 603−608.

Jayanthy, A., Prakash Kumar, U., Remashree, A.B., 2013. Effect of spacial and temporal variations in cellular characters and chemical contents in *Pseudarthria viscida* (l.)W. &Arn. − a medicinal plant. Int. J. Pharmacogn. Phytochem. Res. 5 (3), 177−182.

Karimi, A., Krahmer, A., Herwig, N., Hartwig, S., Javad Hadian, J., Meiners, T., 2020. Variation of secondary metabolite profile of *Zataria multiflora* boiss. Populations linked to geographic, climatic, and edaphic factors. Front. Plant Sci. 03, 1−15.

Kaushik, R.D., Sharma, G., Arora, C., 2004. Biocidal potential of leaf extracts of *Vitex negundo* Linn. Allelopathy J. 14 (2), 227−234.

Khaleefa, A., Irshad, A., 2015. Altitudinal variation in some phytochemical constituents and stomatal traits of *Primula denticulate*. Int. J. Adv. Sci. Res. 02, 93−101.

Kong, C.H., Huang, S.S., Hu, F., 2001. Allelopathy of *Ageratum conyzoides*. V. biological activities of the volatile oil from ageratum on fungi, insects and plants and its chemical constituents. Acta Ecol. Sin. 21, 584−587.

Kumar, S., 2012. Biopesticides: a need for food and environmental safety. J. Biofert. Biopestic. 3, 4−3.

Kumar, S., Yadav, M., Yadav, A., Yadav, J.P., 2017. Impact of spatial and climatic conditions on phytochemical diversity and in vitro antioxidant activity of Indian *Aloe vera* (L.) Burm. f. South Afr. J. Bot. 111, 50−59.

Kumari, A., Kumar, K.N.R., Rao, C.H.N., 2014. Adverse effects of chemical fertilizers and pesticides on human health and environment. Proceedings National seminar on impact of toxic metals, minerals and solvents leading to environmental pollution. J. Chem. Pharmaceut. Sci. 3, 150−151.

Kunimi, Y., 2007. Current status and prospects on microbial control in Japan. J. Invertebr. Pathol. 95, 181−186.

Kunle, K., 2012. Standardization of herbal medicines, a review. Int. J. Biodivers. Conserv. 3, 101−112.

Lacey, J., Hill, S.T., Edwards, M.A., 1980. Microorganisms in stored grains, their enumeration and significance. Trop. Stored Prod. Inf. 39, 19−33.

Leger, R.J., Wang, C., 2010. Genetic engineering of fungal biocontrol agents to achieve greater efficacy against insect pests. Appl. Microbiol. Biotechnol. 85, 901−907.

LohSiew, F., Rita, M., Dzolkifli, O., Mawardi, R., 2011. Insecticidal properties of *Piper nigrum* fruit extracts and essential oils against *Spodoptera litura*. Int. J. Agric. Biol. 13 (4), 517−522.

Maksymiv, I., 2015. Pesticides: benefits and hazards. J. Vasyl Stefanyk Precarpathian Natl. Univ. 1, 70−76.

Mammen, D., Daniel, M., Sane, R.T., 2010. Seasonal and geographical variations in chemical constituents of *Leptadenia reticulate*. Int. J. Pharmaceut. Sci. Rev. Res. 4 (2), 228−235.

Mangels, G., 1991. Behavior of the imidazolinone herbicides in soil: a review of the literature. In: Shaner, D.L., O'Connor, S.L. (Eds.), The Imidazolinone Herbicides. CRC Press, Boca Raton, pp. 191−210.

Mazid, S., Kalita, J.C., Rajkhowa, R.C., 2011. A review on the use of biopesticides in insect pest management. Int. J. Sci. Adv. Technol. 1, 169−178.

Murray, B., Isman, 1994. Botanical Insecticides and Antifeedants: New Sources and Perspectives. Pesticide Research Journal 6 (1), 11−19.

Ndakidemi, B., Mtei, K., Ndakidemi, P.A., 2016. Impacts of synthetic and botanical pesticides on beneficial insects. Agric. Sci. 7, 364−372.

Neli, P., Harro, J., Bouwmeester Maarten, A., Jongsma, 2014. Comparative antifeedant activities of polygodial and pyrethrins against whiteflies (Bemisia tabaci) and aphids (Myzus persicae). Pest Manag Sci 70 (4), 682−688.

Odhiambo, A.J., Siboe, G.M., Lukhoba, C.W., Dossaji, F.S., 2009. Antifungal activity of crude extracts of selected medicinal plants used in combination in lake Victoria basin, Kenya. Plant Prod. Res. J. 13, 35−43.

Olson, S., 2015. An analysis of the biopesticide market now and where is going. Outlooks Pest. Manag. 26, 203−206.

Pan, L., Ren, L., Chen, F., Feng, Y., Luo, Y., 2016. Antifeedant activity of *Ginkgo biloba* secondary metabolites against *Hyphantria cunea* larvae: mechanisms and applications. PLoS One 11 (5), 39−46.

Pavela, R., 2014. Limitation of plant biopesticides. In: Singh, D. (Ed.), Advances in Plant Biopesticides. Springer Publishing, New Delhi, India, pp. 347−359.

Pekrul, S., Grula, E.A., 1979. Mode of infection of the corn earworm (*Heliothiszea*) by *Beauveria bassiana* as revealed by scanning electron microscopy. J. Invertebr. Pathol. 34, 238−247.

Prasad, S., Tyagi, A.K., 2005. Traditional medicine: the goldmine for modern drugs. Adv. Tech. Biol. Med. 108, 90−105.

Rathore, S.H., Nollet, M.L., 2012. Pesticides Evaluation of Environmental Pollution. CRC Press, pp. 599−600.

Razvi, E.S., Welsh, R.M., 1995. Apoptosis in viral infections. In: Maramorosch, K., Murphy, F.A., Shatkin, A.J. (Eds.), Advances in Virus Research, vol. 45. Academic Press, San Diego, London, pp. 1−60.

Ridha, K., Nurdin, S., Rosnani, N., 2018. Antifeedant activity from neem leaf extract (*Azadirachtaindica* A Juss). J. Nat. 18 (1), 47−53.

Rubaya Sultana, R., Neelofar Majidb, N., Saduf Nissar, B., Aabid, M.R., 2018. Seasonal variation of phytochemicals. IJPBSTM 8 (3), 987−990.

Seigler, D.S., 1998. Plant Secondary Metabolism. Kluwer Academic Publishers, Norwell, Massachusetts, USA, p. 114.

Semeniuc, C.A., Pop, C.R., Rotar, A.M., 2017. Antibacterial activity and interactions of plant essential oil combinations gainst gram-positive and gram-negative bacteria. J. Food Drug Anal. 25, 403−408.

Sinclair, J.B., Dhingra, O.D., 1995. Basic Plant Pathology Methods. CRC Press, Boca Raton, FL, p. 448.

Sivarajan, V.V., Indira, B., 1994. Ayurvedic Drugs and Their Plant Sources. Oxford and IBH Publishing Co. Pvt. Ltd, Delhi.

Sola, P., Mvumi, B.M., Ogendo, J.O., Mponda, O., Kamanula, J.F., Nyirenda, S.P., Belmain, S.R., Stevenson, P.C., 2014. Botanical pesticide production, trade and regulatory mechanisms in sub-SaharanAfrica: making a case for plant-based pesticidal products. Food Security 6, 369.

Souvic, S., UnTaek, L., 2000. Extract of *Nicotiana tabacum* as a potential control agent of *Grapholita molesta* (Lepidoptera: tortricidae). PLoS One 23, 1−20.

Sreelatha, T., Hyma, H., Suresh, B., Janaswamy, M.R., 2009. Synthesis and insect antifeedant activity of Plumbagin derivatives with the amino acid moiety. J. Agric. Food Chem. 57 (14), 6090−6094.

Stevenson, P.C., Kite, G.C., Lewis, G.P., Nyirenda, S.P., Forest, F., Belmain, S.R., Sileshi, G., Veitch, N.C., 2012. Distinct chemotypes of *Tephrosia vogelii* and implications for their use in pest control and soil enrichment. Phytochemistry 78, 135−146.

Suman, G., Dikshit, A.K., 2010. Biopesticides: an ecofriendly approach for pest control. J. Biopestic. 3 (1 S), 186–188.

Thomas, E., Vandebroek, I., Goetghbeur, P., Sabino, S., Susana, A., Patrick, V.D., 2008. The relationship between plant use and plant diversity in the Bolivian Andes, with special reference to medicinal plant use. Hum. Ecol. 36 (6), 861–879.

Thompson, J.D., 2005. Plant Evolution in the Mediterranean. Oxford University Press Inc, New York, United States.

TounouAgbéko, K., Mawussi, G., Amadou, S., Agboka, K., Gumedzoe, Y., Mawuena, D., Sanda, K., 2019. Bio-insecticidal effects of plant extracts and oil emulsions of Ricinuscommunis L. (Malpighiales: euphorbiaceae) on the diamondback, *Plutellaxylo stella* L. (Lepidoptera: plutellidae) under laboratory and semi-field conditions. J. Appl. Biosci. 43, 2899 – 291.

U.S. Environmental Protection Agency., 2008. Office of Pesticide Programs. Biopesticides Fact Sheet, EPA 731-F-08-009.

Vanden Berghe, D.A., Vlietinck, A.J., 1986. Van Hoof, L. Plant products as potential antiviral agents. Bull Instit. Pasteur. 84, 101–147.

Vidyasagar, G.M., Tabassum, N., 2013. Antifungal investigations on plant essential oils: a review. Int. J. Pharm. Pharmaceut. Sci. 2, 19–28.

Wieslaw, O., Keith, R., Price, I.J., Colquhoun, M.J., Michal, P., Roger, F.G., 1990. Isolation and identification of alfalfa (*Medicago sativa* L.) root saponins: their activity in relation to a fungal bioassay. J. Agric. Food Chem. 38 (9), 1810–1817.

Yang, L., Chen, P., Zhou, M., Wang, T., Fang, S., Shang, X., Fu, X., 2018. Geographic variation in the chemical composition and antioxidant properties of phenolic compounds from *Cyclocarya paliurus* (Batal) Iljinskaja leaves. Molecules 23 (10), 2440.

Zhi, L., Sha, S.C., Guo, H.J., 2009. Feeding deterrents from *Zanthoxylum schini folium* against two stored-product insects. J. Agric. Food Chem. 57, 21–24.

A review on the fatal impact of pesticide toxicity on environment and human health

16

Iqbal Ansari[1], *Maha M. El-Kady*[2], *Charu Arora*[3], *Muniyan Sundararajan*[1], *Deblina Maiti*[1], *Aarif Khan*[4]

[1]Academy of Scientific and Innovative Research, CSIR-Central Institute of Mining and Fuel Research, Dhanbad, Jharkhand, India; [2]Self-Pollinated Vegetable Crops, Department of Horticulture Research Institute, Agriculture Research Centre, Giza, Giza Governorate, Egypt; [3]Department of Chemistry, Guru Ghasidas Vishwavidyalaya, Bilaspur, Chhattisgarh, India; [4]Department of Environmental Science, Central University of Rajasthan, Ajmer, Rajasthan, India

1. Introduction

Many challenges confront humans to protect their crops since agriculture establishment during New Stone Age, which occurred between 7000 and 10,000 such as insects, plant diseases, weeds, and destructive animals. A wide scope of organic and nonorganic substances got from animals, plants, minerals, and microorganisms has been reported to be effective for pest control. Plant extracts (the natural pesticides) have a longer history than the other types of pesticides; the earliest agricultural natural pesticide was the nicotine which was used in the 17th century to control plum beetle. Due to the maturing chemical industry by the mid-20th century, affordable and operative synthetic pesticides were manufactured (Ujváry, 2001). In 1935, the usage of synthetic pesticides started worldwide and diffused after World War II. Pesticides were set to augment crop yields by the year of 1950, far beyond levels before World War II (Nazir et al., 2007).

According to the World Health Organization (WHO) and Food and Agricultural Organization (FAO), pesticide definition is an item or mixture of items intended to kill, repel, prevent, or lessen any pest including insects, fungi, microorganisms such as bacteria and viruses, nematodes, mice, snails and slugs, unwanted plants (weeds), birds and animals during production stages and storage, transportation, distribution, and elaboration of food and agricultural products or animal foods. Most pesticides are broad-spectrum and can kill all insects in particular area and may kill other animals and birds. Every pesticide-formulated product contains active and inert components. Active ingredients, viz., organophosphate (OP), carbamate, organochlorine, pyrethroid, and so on, are the chemical part in the pesticide that affects the target pest. While inert ingredients are inactive components of a pesticide formulation, viz., emulsifiers, solvents, carriers, aerosol propellants, fragrances, and dyes, they are

Global Climate Change. https://doi.org/10.1016/B978-0-12-822928-6.00017-4

used to dilute or extend the pesticide, make it more effective, or make it safer or easier to handle. Inert ingredients have no direct effect on target pests. Pesticides have been considered a grace to developing countries in their efforts to enucleate endemic disease which carried and spread by insects, to protect its forests and produce sufficient food for population. Still, dispute exists over worldwide due to dependence of such factors, given their redundant misuse, their spread and volatility, long distance transport, and eventual environmental contamination in colder regions (Ecobichon, 2001). The toxic effect of pesticide exposure mainly depends on amount of pesticides as well as persistence time. Most of pesticides are extremely toxic to humans' consequential undesirable effects with only a few drops in the mouth or on the skin. But less toxic pesticides can also cause harmful effects by long term of exposure (Singh et al., 2018).

2. Pesticides classifications

Pesticides are applied in various settings, viz., agricultural, veterinary, domestic, and institutional. Containers of pesticides are made up of plastic or glass bottles, metal flasks, and bags made of plastic or paper. With reference to concentrations, it starts from 2% to 80% of active ingredients (WHO, 2008).

Pesticides classification according to commercial form: Pesticides are categorized to liquid, dust, granules, pellets, chalk, powder, liquid, baits, gel, and paste.

Liquid: A liquid concentrate that is added to water to create pesticide solution, e.g., Roundup and Vantage. Dust: pesticide prepared as dry fine powder, e.g., boric acid and sulfur.

Granules: Pesticides prepared as large dry particles, e.g., Permetrol and carbofuran.

Wettable powders: Finely crushed particles with other substances that enable the powder to be mixed with water which results in a suspension, e.g., blast-off.

Emulsion: A pesticide consists of one liquid dispersed (as droplets) in another. Each retains properties, and agitation is usually necessary to prevent separation, e.g., Buctril.

Dry bait: Pesticide mixed with edible products to form dry pellets, to attract pests, e.g., Klerat. Tablet: dry material, formulated in "pill" form, e.g., aluminum phosphide.

Smokes: The pesticide is mixed with an oxidant and combustible material, which generates hot gas, e.g., mosquito coil.

Fumigants: Pesticides that exist as a vapor at chamber temperature and might be utilized as fungicides, insecticides, or rodenticides, particularly in closed storage places as they kill every living organism. Due to their physical properties, they are highly toxic, quickly spread ecologically, and absorbed by animals or humans, e.g., cyanide, aluminum phosphide, and methyl bromide (WHO, 2008). The commercial form of pesticides classification has been shown in Fig. 16.1.

3. Classification of pesticides by targeted pest species

Insecticides—Insect repellents: Insecticides are used for killing insects, e.g., neem and carbamates, insect repellents such as citronella (essential oil [EO]).

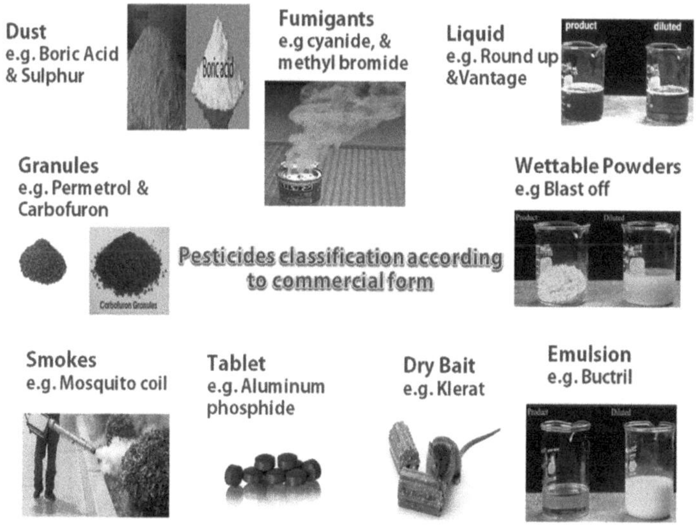

Figure 16.1 Pesticides classification according to commercial form.

Herbicides: These are weed killers, e.g., glyphosate.

Rodenticides: These are applied to kill rodents (rats and mice), e.g., warfarin.

Bactericides: Bactericides are used to kill bacteria, e.g., carbolic acid.

Fungicides: Fungicides are applied to kill fungi, e.g., Captan.

Larvicides: Larvicides kill larvae, e.g., Lavimax.

Molluscicides: Molluscicides are used to kill snails and slugs, e.g., Snailtox 70 WP.

Nematicides: These are useful to kill nematodes (very small worms feed on plant roots), e.g., Neemate-10G.

Ovicides: Ovicides kill eggs of insects and mites, e.g., Hexygon DF.

Pheromones: Pheromones are used to disrupt the mating behavior of insects to control them.

Algaecides: These are used to kill algae, e.g., Argos.

Miticides: Miticides kill moths, e.g., Terro moth trap.

Acaricides: Acaricides kill ticks, e.g., A-one.

4. Classification of pesticides by chemical structure (according to WHO (2008)

On the basis of chemical structure, the pesticides are categorized as follows:

Insecticides: Pyrethroids, organophosphorus, carbamates, organochlorine, and manganese compounds.

Insect repellents: Diethyltoluamide.

Herbicides: Bipyridyls, chlorophenoxy, glyphosate, acetanilides, triazines.

Rodenticides: Warfarins, indanediones.

Fungicides: Thiocarbamates, dithiocarbamates, cupric salts, tiabendazoles, triazoles, dicarboximides, dinitrophenols, organotin compounds, etc.

Fumigants: Aluminum and zinc phosphide, methyl bromide, ethylene dibromide.

5. Pesticides classification on the basis of persistency

Pesticides can be classified into highly persistent, intermediate, low persistent, and nonpersistent.

Highly persistent pesticides: Organochlorines, viz., dieldrin, DDT, and aldrin, persist in the environment for longer period of up to 15 years.

Intermediate: OPs, viz., malathion, carbaryl, and parathion, have persistency for several months.

Low persistent: Carbamates, viz., Tenik, Zectran, and Zineb, fall in this category, which possess low persistency of around 2 weeks.

Synthetic pyrethroids, which are nonpersistent, such as cypermethrin and permethrin, are suitable for a wide range of crops and insects.

According to source of origin, there are synthetic (chemical) and natural pesticides (biopesticide).

Synthetic (chemical) pesticide: Based on chemical structure, pesticides are further classified into four main groups, namely:

Organophosphate: OPs are esters of phosphoric acid. The OP pesticide asserts its effects through irreversible inactivation of the enzyme acetylcholine, which is essential for nervous system functions in living organisms (insects, animals, and humans). OP degrades rapidly by hydrolysis on exposure to light, air, and soil (Jayaraj et al., 2016).

Carbamate: Organic components made from carbamic acid (NH_2COOH) are kept in this category. The functional group present in insecticides is carbamate ester. Its mechanism of action is by reversible inactivation of the enzyme acetylcholine (Jayaraj et al., 2016).

Organochlorine (OC): OC is a group of chlorinated compounds having high persistency. OC insecticides were earlier successfully used in control of malaria and typhus, yet they are banned in almost all the advanced countries (Aktar et al., 2009). The statistics on the use of different pesticides show that 40% of all utilized pesticides belong to the organochlorine (Gupta, 2004; FAO, 2005). Due to their lower prices and the need against various pests, organochlorine insecticides such as aldrin, hexachlorocyclohexane (HCH), DDT, and dieldrin are among the most widely used pesticides in developing countries of Asian continent (FAO, 2005; Gupta, 2004; Lallas, 2001).

Pyrethroid: Synthetic products are developed from pyrethrin, which is natural substance obtained from chrysanthemums flowers. The detail of various classifications of pesticides is shown in Fig. 16.2.

Natural (biopesticides): Natural, biopesticides, or green pesticides consist of compounds such as pheromones, hormones, herbal extracts, and toxins from natural origin besides covering numerous sides of pest management such as microbial agents, entomopathogenic nematodes, and metabolites obtained from microorganisms, and genetically transformed crops having resistance against pests and pathogens. In recent years, the chemical products that are biodegradable can be used in pest management and are considered as green pesticides (Dhaliwal and Koul, 2007).

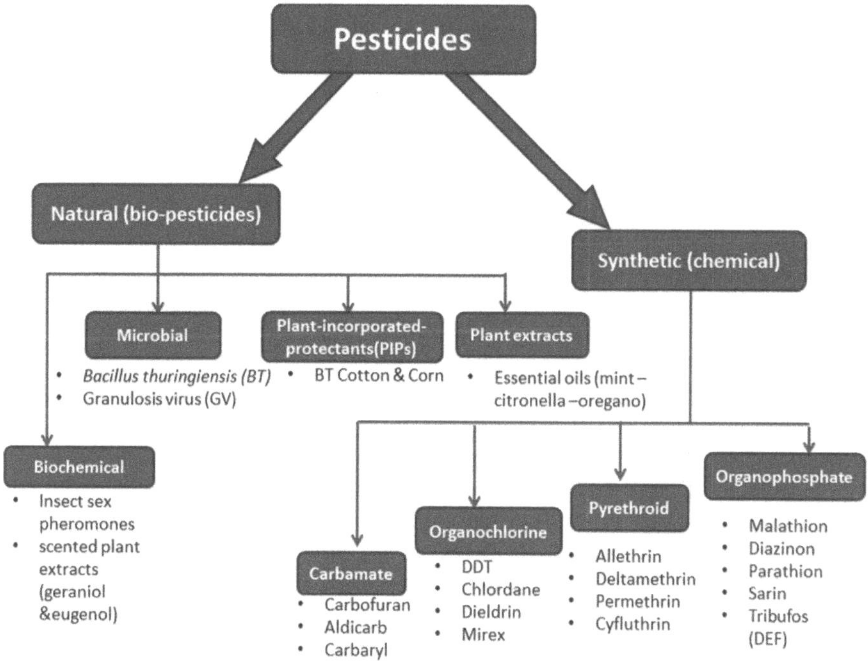

Figure 16.2 Classification of pesticides.

Biopesticides are categorized within three major groups:

Plant-incorporated protectants (PIPs): These are pesticidal compounds that plants produce from genetic materials (genes from resistant plant species), which have been added to the plant. In recent years, scientists promote transgenic crops, which contain BT genes that express a microbial proteins (crystal proteins called: δ-endotoxins or Cry proteins) from the bacterium *Bacillus thuringiensis*; when the insects consume plant parts, the insecticidal proteins can immobilize and kill the insect, for example, BT cotton and corn.

Biochemical pesticides: Natural compounds that control the pathogens by nontoxic methodology, insect sex pheromones, and scented plant extracts that attract insect pests to traps, for example, geraniol and eugenol are used as lures in traps for the Japanese beetle *Popillia japonica* (Vargas et al., 2000).

Microbial pesticides: consist of a microorganism (bacteria, fungus, virus, or protozoan) as active ingredient. Numerous different kinds of pests could be managed by microbial pesticides, although every single active ingredient is particularly specific for its target pest, for example, the most successful and widely used bacterial pathogens that presently occupy about 2% of the total insecticidal market (Bravo et al., 2011). Each one of *B. thuringiensis* strains leads to production of different types of toxins that kills one or more species of insects (*Bt* ssp. *kurstaki* and *aizawai* for lepidopteran larvae and *Bt* ssp. *tenebrionis* for coleopteran larvae). *Bt* ssp. *israelensis* is specific to mosquitoes (Kachhawa, 2017). Other examples of microbial pesticides are granulosis virus (GV) and nuclear polyhedrosis virus (NPV), which are considered the ultimate promising for insect management especially for Lepidoptera and Diptera because of their specificity. NPVs have been effectively applied for devastating pest management such as *Spodoptera* spp. and *Heliothis* spp. on vegetables, fruits, and cotton crop in many countries (Rosell et al., 2008).

Plant extracts: Plant EOs and extracts have shown a wide spectrum activity against insects–pests and plant pathogenic fungi, and their mode of action includes insecticidal, repellent, oviposition obstructive, antifeedant, growth regulatory, and antifungal activities (Vrushali et al., 2001, Kaushik et al., 2004; Arora et al., 2003; Bharti et al., 2013; Arora and Bharti, 2012).

Plant compounds that can be used as pesticides are produced in 17,500 species of higher plants, the majority of which are aromatic, belonging mostly to a few families, including the Myrtaceae(myrtle), Rutaceae(citrus), Lauraceae (cinnamon), Apiaceae (carrot), Lamiaceae (mint), and Asteraceae (Regnault- Roger et al., 2012; Arora Chugh et al., 2012). Among the top well-known fundamental EOs with bioactivity against insects and different kinds of pests are rosemary oil (1,8-cineole), thyme oil (thymol, carvacrol), clove oil (eugenol), mint oil (menthol, pulegone), oil of oregano (carvacrol), lemongrass oil (citronellal, citral), and cinnamon oil (cinnamaldehyde) (Isman and Machial, 2006). For example, for almost 50 years, DEET (*N,N*-diethyl-m-toluamide) was the most intensively utilized and trustworthy insect repellent for personal protection against arthropods bites particularly mosquitoes (Peterson and Coats, 2001). Preparations that contain oil of citronella (lemongrass oil) consider as one of the few applicable alternatives to DEET for using on human skin. As a natural alternative to DEET, a number of such products are marketed in the United States containing citronella that is often mixed with mint oil (Fradin and Day, 2002).

Huang et al. (2010) showed that the EOs extracted from fruits of star anise (*Illicium verum* Hook.) and *trans*-anethole from anise could be developed as natural fungicides (i.e., fumigants) for plant disease management in fruit and vegetable preservation. However, the mechanism of action of these EOs against fungal pathogens is unknown but may be attributed to their general ability to disrupt the integrity of cell walls and membranes (Isman, 2006; Isman and Machial, 2006; Lopez and Pascual-Villalobos, 2010).

6. Nanopesticides

The drawbacks associated with the classical pesticide formulation could be mitigated by the help of nanotechnology. Pesticides have some serious major drawbacks. Nontargeted useful species such as insect pollinators, birds, or natural foes of pests can be threatened by the application of nonselective pesticides. Also, pesticides affect the ecological balance if they destroy useful species that control pests. Moreover, some kinds of pesticides do not break down in the environment and may penetrate the food chain. It leads to the retained concentration of pesticides by the living organism, which increases at every chain food level (bioaccumulation), finally becoming a severe menace for animals and humans health. Furthermore, there is some reason to make pesticides ineffective because they may do not reach the target organism, taken away by wind or being washed off and spreading out in the environment with that ways and contaminating groundwater and soil in nontargeted areas. Environmental factors, such as light, humidity, temperature, and the presence of decomposing microorganisms can cause pesticides degradation before reaching their targets. In fact, it estimated that up to 90% of used pesticides are lost during and postapplication (Ghormadeet al., 2011).

Due to the small size and massive surface area-to-volume ratio, nanopesticides are expected to exhibit unique useful features as compared with their bulk counterparts. Nanoformulation can increase solubility and dispersion of lipophilic pesticides in water. Using appropriate surface active agents, stable aqueous nanodispersions of lipophilic pesticides can be prepared, thereby increasing pesticide bioavailability on one hand and avoiding the use of toxic organic solvents on the other hand. Such stable aqueous nanodispersions can be utilized for simultaneous application of two or various dissonant pesticides that could interact with each other in the same solution. However, in a stable nanodispersion, chemical reaction is prevented at least temporarily between the incompatible pesticides since pesticides are coated away from each other (Carvalho, 2006; Storm et al., 2001). Furthermore, nanoencapsulation provides control to the release rate of pesticide, thus preserving the effective concentration of pesticides during a longer period of time. Furthermore, nanoencapsulation also protects from premature degradation of pesticides. Nanodelivery systems can be designed to defeat the numerous natural ecological and biological barriers, which keep classical pesticide formulations from reaching their purposed targets. Site-specific delivery of pesticides would reduce potential pollution from undelivered components and greatly improve their efficiency (Chen et al., 2015). Consequently, such smart nanosized delivery systems allow a more discreet pesticides use since the amounts required for efficient pest control of pesticides have been significantly reduced, waste management and the production costs of pesticide are decreased, and simultaneously the chance of resistance emerging decreases too. Nanocomposites made of groups of photocatalytically active TiO_2 deposited on the surface of pesticide-loaded microcapsules, which decrease ecological risks by promoting pesticide residuals photocatalytic degradation of that were not taken by their targeted organisms (Kahet al., 2013; Guan et al., 2011). Moreover, nanotechnology can be utilized to manage pesticide losses taking place due to runoff and leaching. Special additives used to prepare loss control pesticides (LCPs) boost gluey properties causing nanocarriers to get attached to crop leaves and that way reducing pesticide losses as well as environmental pollution (Jia et al., 2014; Xiang et al., 2014).

7. Types of pesticide nanocarriers

Nanopesticide formulations can be categorized following various grounds. For example, they can be categorized on the basis of their intended objectives such as (1) formulations that raise the solubility of sparingly soluble pesticides in water; (2) formulations that ease up the releasing rate of the encapsulated pesticide; and (3) formulations that provide protection against early degradation and achieve targeted delivery (Kah et al., 2013). Furthermore, nanoformulations will be divided into different categories based on the chemical nature of the nanocarrier, such as organic polymer–based formulations; lipid-based formulations; nanosized metals and metal oxides; clay-based nanomaterials; layered double hydroxides; and silica nanoparticles. Additionally, depending on the nanoparticles manufacturing methods falling into any of the aforementioned classes, we can have different morphologies and compositions with various properties such as solid–lipid nanoparticles, nanospheres, nanocapsules, micelles, nanofibers, nanogels, and nanoliposomes (Iulia et al., 2017). Active ingredients converted to nanomaterial could be synthetic chemicals such as OP, carbamate, organochlorine, or pyrethroid or natural chemicals such as BT microbial proteins, eucalyptus extract, nirgundi extract, neem, or citronella extracts.

8. Pesticides production, consumptions, and market

Pesticides are chemicals or mixed materials proposed for destruction, prevention, mitigation, or repulsion of any pest. It is applied for preventing crops damaged by pests such as fungi, bacteria, weeds, insects, and so on and thus increases the agricultural production.

The Indian pesticides market was worth INR 197 billion in 2018. The market is further projected to reach a value of INR 316 billion by 2024, growing at a CAGR of 8.1% during 2019–24. The significance of pesticides has been rising over the past few decades catalyzed by the requirement to enhance the overall agricultural production and the need to safeguard adequate food availability for the continuously growing population in the country (IPM report, 2019). Approximately, 2 million tons of pesticides are used annually, which consists of 47.5% of herbicides, 29.5% of insecticides, 17.5% of fungicides, and 5.5% of other pesticides (Kumar et al., 2014). Moreover, it has been assessed that in 2020, the worldwide use of pesticides will increase up to 3.5 million tons (Zhang, 2018). The pesticides market, portioned into herbicides, insecticides, and fungicides, is estimated at $84.5 billion in 2019. In most of the developing countries, the application of pesticides in agricultural activities is increasing, particularly in Southeast Asia (Kunstadter, 2007; Schreinemachers and Tipraqsa, 2012). According to the WHO, about 20% of pesticides are utilized in developing countries with increasing rate of usage. A yearly rise in pesticides import is reported as 10% for Vietnam, 55% for Laos, and 61% for Cambodia (Schreinemachers et al., 2015). The details of annual consumptions of pesticides by different nations have been shown in Table 16.1. Maximum consumption of herbicides has been reported among three major categories of pesticides such as insecticides, fungicides/bactericides, and herbicides during 2007–08 (Wen et al., 2011). The use of fungicides/bactericides increased rapidly and ranked the second. Europe became the largest pesticide worldwide consumer, followed by Asia. Countries such as the United States, China, Japan, France, and Brazil ranked largest in producers of pesticides, consumers, and traders in the world.

Most of the pesticides are used for vegetables and fruit crops. Use of herbicides is very high for protection of maize.

9. Benefits of pesticides

Although most of the literature deal with the cons and negative view of pesticides, but no one can deny the significant role of pesticides through historical events such as wars, natural disasters, and overrun pandemics. DDT was an ideal example of

Table 16.1 Annual pesticide consumption in different Asian countries (FAO, 2017).

S. No.	Pesticides used in tons	Country
1	China	1,807,000
2	India	56,120
3	Malaysia	49,199
4	Pakistan	27,885
5	Thailand	21,800
6	South Korea	19,788
7	Vietnam	19,154
8	Bangladesh	15,833
9	Myanmar	5583
10	Nepal	454
11	Bhutan	12

pesticides role in history. In the mid-20th century and after World War II, numerous nations started to apply DDT by squirting in streets, houses, and citizens to reduce spreading of contagious diseases transmitted by insects in the purpose of saving more lives. Furthermore, DDT had used in the same period of time to increase agricultural production since many countries suffered from shortages of food, especially cereal crops. Pesticides assume three principal roles: (1) the urgent and major role is preserving agricultural and food production to support the growing world population demands, whereas worldwide population development is about 1.3% yearly, and it is currently 7.8 billion people in 2020 and expected to reach 9.8 billion in 2050 (UN 2020). Petrusheva (1975) credited crop yields increment of 1.5–2 times to the utilization of pesticides. Fruit damages by the apple worm went reduced to 1%–2%, and salable percentage increased to 80%–90% of production when pesticides were utilized. In the same direction, Keiserukhsky, (1975) and Chenkin, (1975) asserted that the financial expenses on pesticides were reimbursed four to six times in yields increasing. Without herbicides use, US ruins due to weeds have increased from $4 billion to $20 billion (Bridges, 1992). Moreover, in Britain, Webster et al. (1999) said that "considerable economic losses" could take place without pesticide use and quantitative yield increment of 50% has resulted from pesticide use in British wheat production. Not only do pesticides prevent losses of existing crops, but also it maximizes the use of agricultural lands by broadening the range of crop cultivation that farmers can grow at particular times of the year. An ideal example for what Zimbabwe farmers do is that they can only grow tomatoes in the rainy season in Zimbabwe by utilizing fungicides to prevent late blight, in case of not utilizing it, that commonly causes total crop failure (Cooper and Dobson, 2007). Moreover, there is an indirect benefit from using pesticides by providing more food in communities which allow better nutrition level, and people become more healthy and productive. In addition to its role in preserving food production and health maintenance, there is social role from using pesticides including obtaining abundant crop yields, which means higher income for farmers by controlling pests that provide good education level and better healthcare for their families (Aktar et al., 2009). Controlling contagious diseases transmitted by insect vectors and microorganisms is the second role of pesticide benefits. In conditions of warm and high humidity climate, insects can spread human diseases such as malaria, Zika virus, river blindness, and a high range of serious fevers and deformation illnesses (Cooper and Dobson, 2007). For instance, treated bed nets with insecticide fundamentally decreased newborn children mortality in western Kenya with no increased mortality in older children through belated acquisition of immunity to malaria. Lindblade et al., (2004) reported that insecticides are frequently the only functional way to control vectors; however, it has not been sufficiently exploited (Townson et al., 2005). Houseflies and cockroaches are considered as vectors for numerous microorganisms that cause diarrheal illnesses, which was estimated by the UNICEF as the number one killer of children in the age group of less than 5 years. The influence of biting flies is not limited to the developing countries, but also developed countries such as Canada and the United States utilize pesticides to control mosquitoes, blackflies, ticks, and other insects with the goal that individuals can live more relief and make the most of their pastime undisturbed. In the case that such bugs were not controlled, there would be a drastic effect on life quality, the travel industry levels, and income (Cooper and Dobson, 2007).

Third benefit of pesticide is minimizing the negative effects on human activities and environmental safety. Various pests have a negative impact on human activities and environment, for instance, the utilization of herbicides conserves expenses and time goes on using mechanical weed control, besides that it has long-term environmental benefits of using less nonrenewable energy sources, losing soil humidity from tillage, hindering soil erosion, and soil disintegration (Bates and Denton, 2007; Anon, 2005). There is a hidden role for pesticides with reducing road accidents or driving disturbances through using herbicides by the transport sector to guarantee that roads and railways are kept liberated of vegetation that may cause disturbance or risk. For instance, if vegetation on roadsides is permitted to cultivate excessively tall, it lessens the drivers' seeing at intersections, precipitation of branches, or vegetation onto the roads that may be an obstacle or make it slippery (Cooper and Dobson, 2007).

10. Impact of pesticides on environment

Anthropogenic activities have led to the high level of pollution and increase tremendously by the passage of time. Maintaining the food supply at speedy pace enhances the use of pesticides in agricultural activities at mounting rate (Morillo and Villaverde, 2017). During the past decade, due to agricultural and industrial development, large quantity of pesticides and other related pollutants has been released in atmosphere (Morillo and Villaverde, 2017; Varjani et al., 2018). Pesticides are used to control and eliminate disease vectors, to get better agricultural production and protect stored agricultural products. They consist of insecticides, herbicides, nematicides, fungicides, insect and plant growth regulators, fumigants, attractants, and repellants (Osibanjo et al., 1980). Pesticides are being largely used in the modern agricultural practices, in different quantities, across the world. Though it is helpful in crop yield improvement, it provokes serious issues to environments, health, and safety-linked concerns for terrestrial as well as living biomes including humans, animals, and plants (Sana et al., 2018). Application of pesticides into environment results in a broad spectrum of effects on target as well as nontarget organisms. Most of pesticides have a wide range of biological activities extending beyond the specific functions intended by the manufacturer. This is due to its impacts and the inadvertent or nontarget activities of certain compounds that could also change the behavior of pathogen influencing the induction of severity of disease (Rodriguez-Kabana and Curl, 1980). Hence, due to the toxic nature of such substances, several pesticides such as DDT (organochlorine), and recently the glyphosate (organophosphorus), have been banned in many countries (Lan et al., 2019). Moreover, in some regions of Africa, still DDT is used to control the larvae of malaria disease, whereas glyphosate is used in Latin America and the United States for agriculture purposes (). However, the excessive uses of agrochemicals in agriculture have led to contagion to diverse environmental matrices, such as air, soil, and aquatic ecosystem.

11. Adverse effects on fish health

Freshwater ecosystems are facing pollution threats caused by pesticides across the world known to be a major threat (Madeleine et al., 2015). However, these fresh ecosystems provide an important range of ecosystem services, such as clean potable water, irrigation facility, water for industrial requirements, water storage, water recreation, and water bodies and other valuable foods. An invertebrate helps in food chains from one tropic level (primary producers) to higher tropic levels for energy transfer and nutrients (Madeleine et al., 2015). Fishes get exposed to pesticides and their residues through different routes such as runoffs or spray drifting from gardens, agricultural fields, and orchids. Among aquatic organisms, fish is considered one of most affected organisms by pollutant present in aquatic medium. Pesticides present in aquatic bodies deteriorate its quality by altering physicochemical parameters and threat for aquatic organisms (Sarwar et al., 2007; Sabae et al., 2014). A significant figure of insecticides and their compositions is documented by the pesticides action network (UK PAN, 2009). Globally, various pesticides are being produced in a range of ratios, e.g., herbicides (15%), fungicides (1.46%), and insecticides (80%) (Marigoudar, 2012). Cypermethrin is an extensively used pesticide in agricultural activities as well as in households, and their concentration is reported in aquatic ecosystem. Its existence causes a severe threat to humans as well as other nontarget terrestrial and aquatic livings in the environment (Sana et al., 2018). The effects of cypermethrin toxicity in fish have been illustrated in Fig. 16.3.

Pesticides alter in various physiological and biological changes in freshwater organisms by influencing the activities of several of several enzymes. Such alterations present in natural aquatic environment in the form of chemical compositions often affect

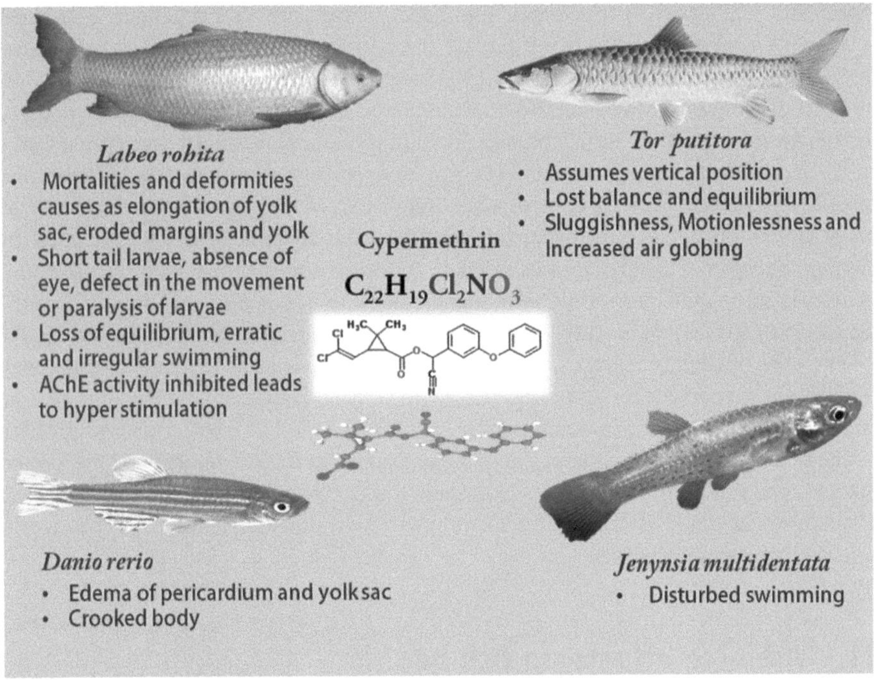

Figure 16.3 The adverse effects of cypermethrin toxicity in fish health.

physiological systems along with behavior of the inhabitants, mostly of the fishes (Radhaiah et al., 1987). These biochemical alterations produced by pesticidal stress lead to inhibition of important enzymes, metabolic disturbances, retardation of growth and decline in the richness, and prolonged existence of the organisms (Murty, 1986). The organs of fish such as gills, liver, kidney, and brain are the most susceptible when exposed to toxicity containing medium (Jana and Bandyopadhyaya, 1987). A study by Calado et al. (2019) reveals the presence of glyphosate in the samples of water and different parts of fish *Astyanax altiparanae*. The reduction in the acetylcholinesterase (AChE) activity was pragmatic in exhibiting that the urban and agriculture contaminants together can cause anticholinesterasic effects. Such sublethal effects can be dangerous for human health and environment. Ewing, (1999) reported that sublethal concentrations of agrichemicals can affect various aspects of salmon fish biology, including change in behavioral pattern. Examples of sublethal effects of the OP insecticide phenthoate are synergized by carbaryl in fish, which results in AChE inhibition (Rao et al., 1989) and change in morphology and behavior (Rao et al., 1987).

12. Undesirable effects of pesticides on human health

Pesticides are such chemical substances used to kill animals, insects, plant, and fungal pests in agricultural activities and domestic and institutional settings (Health and the environment, 1998). Pesticides possess unique intrinsically toxic chemicals, which are used in environment to kill off pests. They consist of various different categories of chemicals whose toxicity is increasingly reduced as less toxic compounds benefit from stricter regulations. It is estimated that about 5.2 billion pounds pesticides were applied globally in 2006 (US-EPA, 2011). However, merely 1% of this quantity reaches the target pests at fatal doses (Gavrilescu, 2005). According to a WHO/UNCEP report, developing countries use 20% of the total pesticides, whereas the developed countries use more than 80%; however, death rate is 13 times higher in developing nations due to improper or indiscriminate application of pesticides (WHO/UNCEP, 1990). Several researchers have reported that pesticides are certainly fatal to man not only at high doses, responsible for acute poisonings, but even in low doses, as are mixtures of pesticides (Tsatsakis et al., 2009; Zeliger, 2011). But long-term exposures may lead to collections of various health effects including cancer and neurodegenerative diseases (Bassil et al., 2007; Kanavouras et al., 2011; Parrón et al., 2011), reproductive and developmental toxicity (Hanke and Jurewicz, 2004), and even respiratory effects (Hernández et al., 2011). Epidemiologic study reveals a positive association with pesticides exposure and solid tumors. Some pesticides such as maneb, paraquat, and rotenone result in reactive oxygen species (ROS)–mediated oxidative stress and thus cause neurodegenerative disorders in the body (Sabarwal et al., 2018). Parkinson's disease (PD) is basically connected with excessive ROS production from MET (mitochondrial electron transport) complexes and causes severe damages to mitochondrial DNA and other related biomacromolecules at the cellular level (Nandipati and Litvan, 2016). Meanwhile, there is a close link between pesticides exposure and heath disorders, viz., malignancies such as leukemia, PD, Hodgkin's disease, non-Hodgkin's lymphoma, Burkitt lymphoma, respiratory, reproductive, and endocrine diseases (Wang et al., 2014; Luo et al., 2016; Brouwer et al., 2017; Sabarwal et al., 2018). The steadiest associations were observed in brain and prostate cancer. Some other associations were also found at workplace of parents exposed to pesticides mainly cancer of kidney in their children (Bassil et al., 2007). Moreover, the extreme use of these pesticides/chemical fertilizers in modern agricultural activities caused contamination to different matrices, including air, soil, and aquatic systems. The adverse effect of contaminated environmental matrices on health of human with nontargeted animals in various ways has been shown in Fig. 16.1.

Furthermore, Venkataraman et al. (1994) have observed the fatal effects of pesticides on some enzymes, especially in case of cholinesterase. They observed a close correlation between RBC cholinesterase with methyl parathion, which indicates that pesticides inhibited the AChE activity. A biochemical and hematological study reveals the presence of organochlorine (OC) residue in blood samples in both genders as a part of pesticides. The OC compounds such as α-endosulfan (1.565 mg/kg) and β-endosulfan (1.233 mg/kg) were found in higher quantity. Results indicate that the bioconcentration and biomagnifications of such chemicals are frequent phenomena observed in the city of Karachi and may responsible for various diseases and health hazards (Khwaja et al., 2013). Similarly in a study by Undeger and Basaran (2005), the genotoxic effects have been observed by using pesticides, viz., dimethoate and methyl parathion from the OP class, propoxur and pirimicarb from carbamates, and

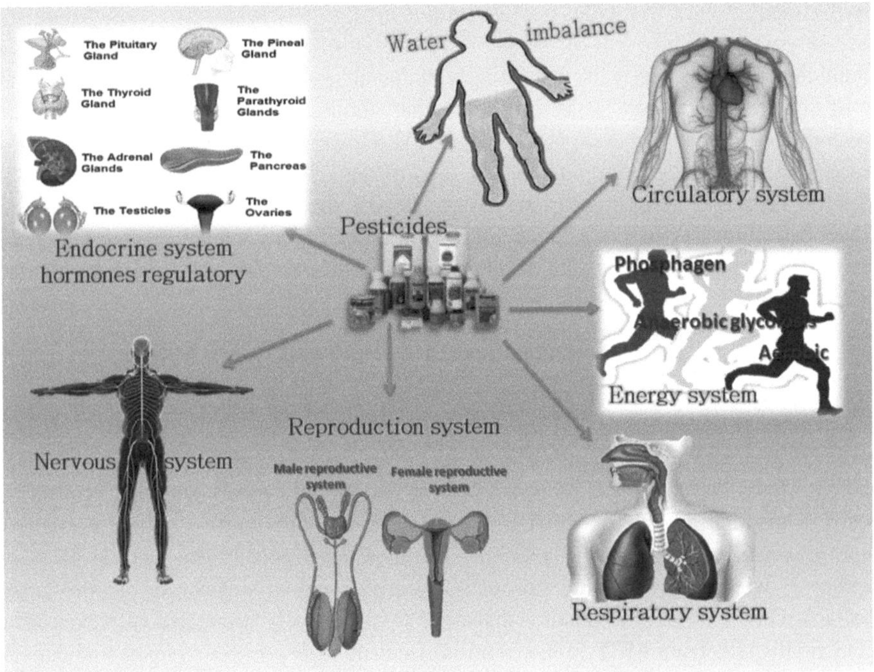

Figure 16.4 Pesticides' adverse effects on human health related as different living systems. Modified and adapted from Bilal, M., Iqbal Hafiz, M.N., & Barceló D. (2019). Persistence of pesticides-based contaminants in the environment and their effective degradation using laccase-assisted biocatalytic systems. Sci. Total Environ. 695, 133896.

cypermethrin and permethrin from pyrethroids. Results indicate that only dimethoate and methyl parathion at 100 and 200 μg/mL; propoxur at 50, 100, and 200 μg/mL; and pirimicarb, cypermethrin, and permethrin at 200 μg/mL have significantly increased DNA damage in human lymphocytes. This genotoxic effect was examined by using single-cell gel electrophoresis (comet) assay in freshly isolated human peripheral lymphocytes. Major toxic effects of such pesticides on human-related heath have been discussed in the following.

12.1 Neurological effects

Acute and chronic occupational exposure to a variety of pesticides can result in mild to severe decline in neurologic function that may be unalterable (Ecobichon et al., 1990). Such chronic neurologic effects have been associated with exposures to OP, organochlorine, and carbamate insecticides, as well as a variety of fungicides (tnercurials, diphenyl. hexachlorobenzene, hexachlorophene) and fumigants (methyl bromide, carbon disulfide, sulfuryl fluoride) (Ecobichon et al., 1990; Peters et al., I982; Anger et al., 1986; Savage et al., 1988; White et al., 1990; Rosenstock et al., 1990). OP pesticides broadly used in agricultural activities until recently and slowly have replaced the organochlorine pesticides for several years. OP pesticides poisoned to insects and mammals by phosphorylation of AChE are found in nerve endings. The effect on organ becomes overstimulated by surplus acetylcholine (ACh), the impulse transmitting substance in the nerve ending, causes a loss of AChE activity. Acute exposure to OP may lead to vomiting, diarrhea, abdominal cramps, dizziness, eye pain, blurred vision, confusion, numbness, twitching, paralysis, and death. Several health effects have been correlated with chronic low dose exposure to OP pesticides. Various epidemiological studies reveal such associations linked between OP exposure and neurobehavioral effects and delayed neurologic diseases such as PD (Colosio et al., 2003; Hancock et al., 2008; Manthripragada et al., 2010; Rohlman et al., 2011). Children are more susceptible to pesticide neurotoxicity as their brains develop quickly (Eskenazi et al., 1999; Rice and Arone, 2000). Moreover, children get espoused to greater extent as compared with adults as they consume and inhale larger quantities of food and air, respectively, per body weight unit. Some animal studies reveal that exposure of OP pesticides in prenatal is being also associated with neurodevelopment (Eskenazi et al., 1999). However, Cherniak (1988) and Kaloyanova (1991) opined that some OPs are responsible for a delayed-onset peripheral neuropathy in several weeks after exposure, which primarily affects the extremities. Symptoms may endure or gradually decrease. This peripheral neuropathy results from severe inhibition of neuropathy target esterase, causing axonal degeneration.

Michalis et al. (2012) reviewed the harmful impacts on prenatal exposure of OP pesticides on male reproductive system and neurodevelopment. This neurologic effect on adults, DNA damage, and defects in birth have been observed in humans and were also associated with exposure to OP pesticides. Literature opined that pesticide exposure may raise risk of PD (Le Couteur et al., 1999). Increased risk of Parkinson has been reported in population of rural area drinking well water, and depend on farming as livelihood (Priyadarshi et al., 2001). Various studies have concerned the herbicide paraquat (Hertzman et al., 1990; Liou et al., 1997), which releases selective degeneration of neurons involved in PD (McCormack et al., 2002). Several case studies report that PD occurs in individual due to exposed of OP (Bhatt et al., 1999; Davis et al., 1978); to herbicides including glyphosate (Barbosa et al., 2001), paraquat (Sanchez-Ramon et al., 1987), and diquat (Sechi et al., 1992); and to fungicides including maneb (Meco et al., 1994) and other dithiocarbamates (Hoogenraad, 1988). Experimental data reveal additive/synergistic effects of various pesticides, which are active on diverse targets within dopaminergic system. Pesticides such as paraquat, maneb and other dithiocarbamates, pyrethroids, rotenone, and dieldrin cause neurotoxic effects and are also responsible for development of PD-like syndrome in animals (Moretto and Colosio, 2011). These pesticides causes Parkinson's like diseases, which has been illustrated in Fig. 16.5.

Moreover, chlorpyrifos (O,O-diethyl O-3,5,6-trichloro- 2-pyridyl phosphorothioate) is an example of OP pesticide broadly used in both agricultural and residential pest control. It exhibits moderate acute toxicity, voluntarily inhibiting plasma cholinesterase at low doses and inhibiting red blood cell cholinesterase at high doses (Richardson, 1995). Chlorpyrifos causes acute poisoning with well-known symptoms such as diarrhea, diaphoresis, and increased amount of urination, miosis, salivation,

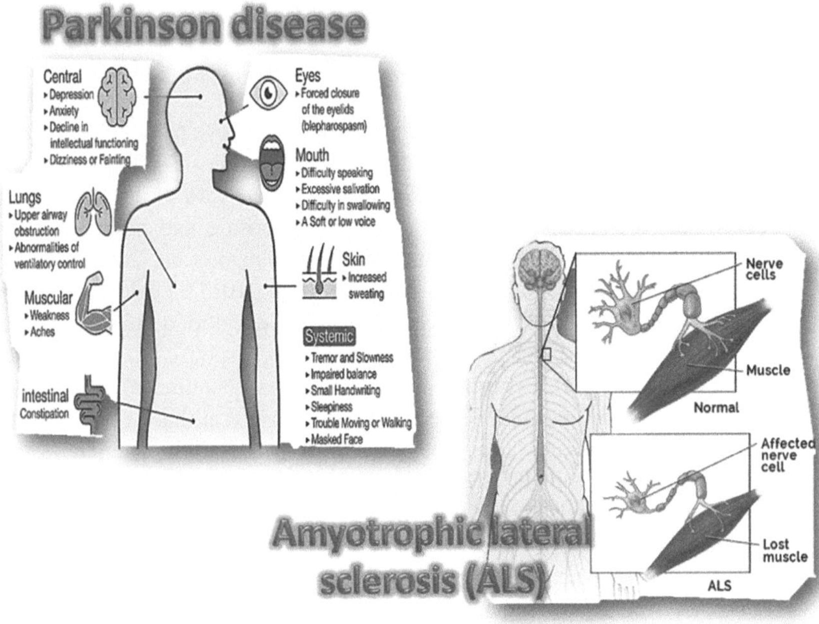

Figure 16.5 Organophosphate and other related pesticides causing Parkinson's like diseases in animals.

and lacrimation (O'Malley, 1997). Chlorpyrifos is fairly toxic OP pesticide, which is detected in the urine of 82% in US adults in the form of chlorpyrifos metabolite (3, 5, 6-trichloro-2-pyridinol; TCP). In US population, the average urinary TCP level of 65 numbers of exposed applicators was 629.5 pg/L, observed when it is compared with 4.5 pg/L. These exposed subjects were significantly showing more symptoms, including emotional states, loss of muscle strength, memory problems, and fatigue (Kyle et al., 2000). Some related studies with pesticide neurotoxicity, which has typically evaluated either for the long-term sequel of pesticide poisoning or the effects of occupational exposure, have been shown in Table 16.2.

12.2 Reproduction abnormalities

The adverse effects of chemical pesticides have decreased the fertility rates in both males and females and antiandrogenic (demasculinizing) effects, increased rates of miscarriage, altered sex ratios, and altered maturity. Some pesticides such as OPs and carbamates are toxic to the nervous system; similarly, some pyrethroids are also being toxic for the reproductive systems and disruptive to endocrine function. Studies show that endosulfan exposures cause delay in sexual maturity and obstruction through sex hormone synthesis in male children (Clark et al., 2002; Narayana et al., 2004). A study by Meeker et al. (2006) has observed contrary relations between pesticides and male testosterone levels. The study found that elevated levels of the urinary

Table 16.2 Studies of chronic pesticide exposure and neurotoxicity exposure measurement.

Exposed population	Chemical	Exposure measure	References
Hospital patients	OP	Poisoning	Miranda et al. (2002)
Commercial applicators	OP	Apply pesticide	Misra et al. (1985)
Farm workers	OP	Poisoning	Rosenstock et al. (1991)
Farmers	Multiple	Apply pesticide	Smit et al. (2003)
Flower bulb farmers	FNG	Apply pesticide	Ruijten et al. (1994)
Farmers, spouses	Multiple	Poisoning	Stallones and Beseler (2002)
Commercial applicators	Multiple	Apply pesticide	Sack et al. (1993)
Pesticide registry	OP	Poisoning	Steenland et al. (1994)
Registry	OP	Poisoning	Savage et al. (1988)
Greenhouse workers	OP	Work with plants	Bazylewicz-Walczak et al. (1999)
Factory workers	OP, CAR	AChE inhibition	Bellin and Chow (1974)
Fumigators	Fumigants	High pesticide use	Calvert et al. (1998)
Farmworkers	Multiple	Self-report; AChE inhibition	Ciesielski et al. (1994)
Farmworkers	Multiple	Years of work	Kamel et al. (2003)
Fruit farm applicators	OP	Cumulative exposure	London et al. (1998)
Occupational exposure	OP	High cumulative exposure	Korsak and Sato (1977)
Pesticide applicators	OP	Current pesticide use	Levin et al. (1976)

AChE, acetylcholinesterase; *CAR*, carbamates; *FNG*, fungicides; *OP*, organophosphate; *PYR*, pyrethroids.

metabolites of chlorpyrifos (TCPY) and carbaryl along with naphthalene (1N) show a relationship directly with low levels of testosterone in male. Epidemiological studies and laboratory experiments reveal a strong relationship or linkages between sterility rates amount of pesticide exposure.

In reported literature, it was found that pesticide affects the reproductive health of human. Several authors have reported a progressive turn down in semen quality of persons exposed to pesticides, especially in sperm concentration or count. An animal study reveals the presence of PYR (pyrethroid), and its contact can have an effect on sperm concentration, their motility, and morphology (Elbetieha et al., 2001; Kumar et al., 2004; Mani et al., 2002). Michalis et al. (2012) have reported that the exposure of PYR mainly affects male reproductive systems in sperm quality, sperm DNA damage, and reproductive hormone disorders. In their review, findings give an evidence to support on the hypothesis that PYR exposure is negatively linked with effects on the reproductive system of male.

12.3 Cancer

The associations linked between pesticides and cancer are a matter of concern for a long time. Whereas agricultural activities have usually been tied to pesticide-related illness, 19 out of 30 commonly used lawn pesticides and 28 out of 40 used school pesticides are interlinked to cancer. Several published experimental studies based on animal toxicity data and human cells/tissue and their laboratory results reveal that pesticides are known or suspected to be carcinogenic. Some review articles reported a close association in between cancer and pesticides used in agricultural activities and also the urban environment such as public places, homes, and schools (Kagan et al., 2010). A positive association was observed in a study related to non-Hodgkin's lymphoma and leukemia on pesticide exposure. Several have shown dose–response associations, and a few are capable to recognize presence of pesticides precisely. In some studies on children and pregnant women, positive association was found in cancer and exposure to pesticides. This is due to exposure of their parents to pesticides at workplace. Another studies showed a positive association between pesticides exposure and solid tumors (Bassil et al., 2007).

The pesticide along with xenobiotic estrogens enhances the threat of breast cancer. Estradiol metabolism proceeds by hydroxylation at one of two mutually exclusive sites at C-2 and C-16a. The catechol pathway results in the production of the weakly estrogenic 2-hydroxyestrone (2-OH E1), which inhibits breast cell proliferation. On the other hand, the alternative pathway produces the genotoxic 16a-hydroxyestrone (16a x-OHE1), which leads to breast cell growth, increases unscheduled DNA synthesis, oncogene, and virus expression, and increases anchorage-independent growth (Leon et al., 1995).

13. Pesticides poisoning incidents in developing countries

Globally, pesticides poisoning has become a major public health issues as well as in some developing nations due to accidental or unintentional ingestion and unsafe use of pesticides products during occupational activities (Caldas, 2011). In Brazil, National Poisoning Information System (SINITOX) reported that pesticide is the third highest reason of exogenous poisoning, responsible for about 7%–11% of the cases yearly from 2000 to 2012. The Brazilian Mortality Information System (SIM) reported

10,666 number of fatalities occurred after pesticides poisoning, mainly after self-poisoning, and the main pesticides were OPs and carbamates (Caldas, 2011). An average of 7000 cases of pesticide poisoning yearly reported in Brazil. In Central America, a widespread use of OPs, carbamates, endosulfan, and paraquat as perilous pesticides continues. In 2000, 12 toxic pesticides were banned in seven Central American countries. Due to their adverse effects in 2004, nearly 4 years later, restrictions have been implemented in El Salvador and in Nicaragua (Catharina et al., 2005). Similarly in Egypt, on use of DDT and several other pesticides, organochlorine in agricultural activities has been restricted. Due to long persistency of pesticide molecules, compounds are detected in various environmental samples such as water, fish, milk, fruits, foodstuffs, vegetables, and sediments (Mansour, 2004). The percentage of incidents by pesticide applications in various occupational actives has been shown in Table 16.3.

Table 16.3 Percentage of farmers poisoned during pesticide application (selected studies) (PAN Germany report, 2012).

Country/ regions	Segments of populations	Year	Percentage of poisoned farmers
Pakistan	105 tobacco farmers	2009	The majority reported headaches, dizziness, nausea, shortness of breath, muscle weakness, skin rashes, in connection with pesticide use; reduction of cholinesterase activity was measured for 55%
Vietnam	190 rice farmers	2004	35% pesticide poisonings (21% chronic poisonings) through blood tests
Myanmar	123 women exposed to pesticides (application and handling)	1999	40% chronic poisoning
China (Sichuan)	Families in two rural districts	2001	20% reported pesticide poisonings
Vietnam	50 farmers	2000/ 2001	61% of the pesticide applications are associated with mild symptoms of illness; 31% are associated with at least one definite poisoning symptom (self-reported)

Continued

Table 16.3 Percentage of farmers poisoned during pesticide application (selected studies) (PAN Germany report, 2012).—cont'd

Country/regions	Segments of populations	Year	Percentage of poisoned farmers
Africa, Asia, and Latin America	2220 people exposed to pesticides	2007–09	21%–59% of those interviewed who have been exposed to pesticides subsequently suffer from symptoms such as headaches, dizziness, impaired vision, increased sweating.
India	Female cotton farmers in three villages	2003/2004	83.6% of pesticide exposures (application, mixing, or work in the fields) were linked to poisoning symptoms; of these, 10% were linked to three or more symptoms that typically occur with organophosphates; 6% with extremely severe poisonings; no one consulted a physician

14. Climate change and pesticide use

The general concept is that the effect of pesticides will increase due to global warming (Noyes et al., 2009; Noyes and Lema, 2015). Three factors are considered responsible for its (1) higher toxicity of the same concentration of most of the pesticides at higher temperature (Noyes et al., 2009; Holmstrup et al., 2010; Noyes and Lema, 2015), (2) rise in pest population at high temperature that leads to increased pesticide application at higher doses of pesticides in agriculture and changes in crops potentially because of the use of additional agricultural areas (Bloomfield et al., 2006; Kattwinkel et al., 2011), and (3) increased pesticide runoff in ponds in agricultural areas due to increased precipitation (Kattwinkel et al., 2011). With a longer growing season and a warmer climate, weeds and insect pests proliferate, most likely leading to more pesticide use—which is itself responsible for harmful emissions that further exacerbate climate change.

However, some studies have shown that climate change will reduce environmental concentrations of pesticides due to a combination of increased volatilization and accelerated degradation of pesticide molecule (Delcour et al., 2015).

15. Pesticides control and regulations

Pesticides are controlled and regulated by the Central Insecticides Board and Registration Committee in India, whereas the Food Safety and Standard Authority of India approved the MRLs (maximum residue limits) of registered pesticides (Bedi et al., 2015; Nambirajan et al., 2018).

In China, the Institute for Control of Agrochemicals regulates pesticides (Han et al., 2018; Handford et al., 2015). China follows Codex MRLs as controlled by the National Pesticide Residues Committee under the Ministry of Agriculture (MOA) (Yang et al., 2018). In China, the MOA and the Ministry of Health develop pesticides residue limits, which are associated with testing methods and procedures (Buijs et al., 2018; Jia and Jukes, 2013). However, some nations in Southeast Asia do not possess their pesticides legislation and regulations (Shahid et al., 2016).

16. Preventive strategies

To decrease the pesticide usage, the following steps should be implemented during agricultural practices (Pierpaolo, 2017):

1. Agronomic practices

Suitable agronomic practices are essential to achieve healthy crops and to prevent diseases pest and weed pressure, providing proper plant nutrition, soil fertility management, appropriate irrigation management, crop rotation, proper timing of sowing or planting, to reduce the pest attack. Intercropping along with use of variety mixtures can limit and prevent the spread of pests and diseases. This method also provides food and shelters for the natural enemies of pests.

2. Practices with resistant crops

Growing crops may be suitable for use of local conditions, and therefore, selecting suitable crop varieties has become fundamental for the preventive pest management system. Using disease-resistant varieties along with rotations of nonsusceptible crop provides sustainably bound pest upsurge within a field.

3. Natural pesticides and biocontrol

Natural pesticides and biocontrol—Biocontrol provides pathogens (bacteria, fungi, and viruses), insect predators or parasitoids, insect traps, and pheromones that control the population of pest low.

4. Integrated pest management (IPM)

IPM is an ecosystem approach that does not seek to eradicate pests but rather to manage them. The most fundamental line of defense against pests and diseases in agriculture is a healthy agroecosystem, in which the biological processes that underpin production are protected, encouraged, and enhanced. A strong focus is on pest prevention by applying good agronomic practices and using resistant varieties, pest identification, and monitoring biological pest control.

5. Agroecology

In agroecology, the main aim is to maintain natural balance in the ecosystem. The pest control seeks to strengthen interactions of pests with natural enemies.

6. Organic agriculture.

Organic agriculture sustains the health of soils, as well as ecosystems and people. Use of synthetic pesticides and chemical fertilizers is totally prohibited. Crop protections in organic agriculture are done by the use of organic green manures, disease-resistant varieties of seeds, and biocontrol to prevent the attack from pests, disease, and so on.

7. Use of less toxic pesticides

There are various systems to classify pesticides as per their toxicity for humans as well as environment. Highly hazardous pesticides should be replaced by ecofriendly and less harmful pesticides. Pesticides should be applied within safe and recommended dose to minimize its impact on people as well as environment. Use of plant-based pesticides should be encouraged to control agricultural pests and pathogens. There are several reports on pesticidal activities of plant extracts or compounds (Khan et al., 2017; Arora Chugh and Bharti, 2014).

Furthermore, Ron (2020) has suggested some other ways to minimize the extreme usage of pesticides, and hence, there should be minimum health risk. **(1) Biological controls**—Introducing or conserving of natural enemies (biological control agents) can prevent the rise of certain pests. Some beneficial mites feeding on mite pests in orchards can be used as biological control agent. Another example is the milky spore disease-causing organism, which controls harmful soil grubs, and *Encarsia formosa,* a kind of wasp which parasitizes the greenhouse whitefly. Choose and use pesticides wisely to conserve indigenous or released natural enemies of insect and mites. **(2) Pheromones** (natural insect scents) can be used in pest management. Artificially pheromone sprayed can be used to confuse males and prevent mating. This technique is very useful for controlling from oriental fruit moth, codling moth, the grape berry moth pink bollworm, tomato pinworm. **(3) Pest-resistant cultivars** are relatively less vulnerable than other varieties to certain insects and diseases. Planting disease-resistant crops provide the simplest method of reducing disease management. Resistant varieties often do need to apply as many pesticides as with susceptible varieties. Apple growers can save up to eight fungicide applications a year by growing liberty and freedom cultivars, which resist diseases. Farmers growing alfalfa and wheat keep several pests at bay by planting resistant varieties.

17. Conclusion

Contaminations of pesticide toxicity in natural resources may impose significant risks on environment and untargeted organisms of different verities from helpful soil micro-organisms to insects, aquatic animals, birds, humans, and plants. The long-term exposure to pesticides causes chronic acute toxicity and affects human, animal, and plant health. The pesticides are hysterically used across the world in developing countries such as India as pesticides are considered to be a rapid, simple, and cheaper solution for controlling ruining means of crop production. As a result, the chronic toxicity caused to humans, animals, and plants from contaminated natural resources such as air, soil and water has hazardously entered the food chain. Therefore, the approach for controlling weeds and insects in crop production should be green technology with scientific judgment and not on commercial considerations. This chapter concludes with limited inferential knowledge that the incessant use of pesticides goes on enforcing the people to undergo a certain ambiguity with lifelong issues. It emphasizes on the need to develop a health awareness package with alerting knowledge on the fatal outcomes of extreme exposure to pesticides. IPM along with herbal pesticides may be a suitable alternative as herbal pesticides are biodegradable, environmentally friendly, and less/nontoxic to nontargeted animals. It is difficult to develop resistance against herbal pesticidal formulations, as these possess several active ingredients although in low concentration, it requires several simultaneous mutations to take place in pests to overcome all ingredients.

References

Aktar, M.W., Sengupta, D., Chowdhury, A., 2009. Impact of pesticides use in agriculture: their benefits and hazards. Interdiscipl. Toxicol. 2 (1), 1–12.

Anger, W.K., Moody, Burg, J., et al., 1986. Neurobehavioral evaluation of soil and structural fumigators using methyl bromide and sulfury1 tluoride. Nemotoxicology 7, 137–156.

Anon, 2005. Conservation Technologies and the Plant Science Industry: Managing Natural Resources Sustainably. Crop Life International, Belgium, p. 42pp.

Arora, C., Kaushik, R.D., Kumar, A., Garg, G.K., 2003. Fungicidal potential of Kumaon and Tarai region plants against mushroom fungal pathogens. Allelopathy J. 11 (1), 63–70.

Arora Chugh, C., Bharti, D., 2014. Chemical characterization of antifungal constituents of *Emblica officinalis*. Allelopathy J. 34 (1), 155–178.

Arora Chugh, C., Bharti, D., 2012. Phytochemical analysis and screening of antibacterial activity of some selected Indian medicinal plants. Int. J. Phytomed. 4 (2), 261–265.

Arora Chugh, C., Mehta, S., Dua, H., 2012. Phytochemical screening and evaluation of biological activities of some medicinal plants of Phagwara, Punjab. Asian J. Chem. 24 (12), 5903–5905.

Barbosa, E.R., Leiros da Costa, M.D., Bacheschi, L.A., Scaff, M., Leite, C.C., 2001. Parkinsonism after glycine-derivate exposure. Mov. Disord. 16, 565–568.

Bassil, K.L., Vakil, C., Sanborn, M., Cole, D.C., Kaur, J.S., Kerr, K.J., 2007. Cancer health effects of pesticides: systematic review. Can. Fam. Physic. 53, 1704–1711.

Bates, G., Denton, H.P., 2007. No-till Establishment of Forage Crops Agricultural Extension Service. The University of Tennessee.

Bazylewicz-Walczak, B., Majczakowa, W., Szymczak, M., 1999. Behavioral effects of occupational exposure to organophosphorous pesticides in female greenhouse planting workers. Neurotoxicology 20, 819–826.

Bedi, J.S., Gill, J.P.S., Aulakh, R.S., Kaur, P., 2015. Pesticide residues in bovine milk in Punjab, India: spatial variation and risk assessment to human health. Arch. Environ. Contam. Toxicol. 69 (2), 230–240.

Bellin, J.S., Chow, I., 1974. Biochemical effects of chronic moderate exposure to pesticides. Res. Commun. Chem. Pathol. Pharmacol. 9, 325–337.

Bharti, D., Gupta, S., Arora, C., 2013. Antimicrobial activity of medicinal plants against some pathogenic microbial strains. Int. J. Phytomed. 5 (2), 154–158.

Bhatt, M.H., Elias, M.A., Mankodi, A.K., 1999. Acute and reversible parkinsonism due to organophosphate pesticide intoxication: five cases. Neurology 52, 1467–1471.

Bloomfield, J., Williams, R., Gooddy, D., Cape, J., Guha, P., 2006. Impacts of climate change on the fate and behaviour of pesticides in surface and groundwater—a UK perspective. Sci. Total Environ. 369, 163–177.

Bravo, A., Likitvivatanavong, S., Gill, S.S., Soberón, M., 2011. *Bacillus thuringiensis*: a story of a successful bio-insecticide. Insect Biochem. Molecul. Biol. 41, 423–431.

Bridges, D. (Ed.), 1992. Crop Losses Due to Weeds in the United States. Weed Sci. Soc. Am. Champaign, IL.

Brouwer, M., Huss, A., van der Mark, M., Nijssen, P.C., Mulleners, W.M., Sas, A.M., van Laar, T., de Snoo, G.R., Kromhout, H., Vermeulen, R.C., 2017. Environmental exposure to pesticides and the risk of Parkinson's disease in The Netherlands. Environ. Int. 107, 100–110.

Buijs, J., van der Meulen, B., Jiao, L., 2018. In: China's Food Safety Law. Legal Systematic Analysis of the 2015 Food Safety Law of the People's Republic of China (No. 2018). European institute for food law, Wageningen.

Calado, S.L.M., Andrade, M.C., Garrido, F., Pelanda, A., Lirolla, J., Salgado, L.D., Cestari, M.M., de Assis, H.C.S., 2019. Biochemical and genotoxicity assessment of a polluted urban river using the native fish *Astyanax altiparanae* Garutti & Britski (*Teleostei, Characidae*). Ecotoxicol. Environ. Contam 14 (1), 73–77.

Caldas, E.D., 2011. Pesticides poisoning in Brazil. Encyclopedia Environ. Health 419–427.

Calvert, G.M., Mueller, C.A., Fajen, J.M., Chrislip, D.W., Russo, J., Briggle, T., et al., 1998. Health effects associated with sulfuryl fluoride and methyl bromide exposure among structural fumigation workers. Am. J. Publ. Health 88, 1774–1780.

Carvalho, F.P., 2006. Agriculture, pesticides, food security and food safety. Environ. Sci. Pol. 9, 685–692.

Catharina, W., Marianela, C., Viria, B., 2005. Acute pesticide poisoning and pesticide registration in Central America. Toxicol. Appl. Pharmacol. 207 (2), 697–705.

Chen, M., Jensen, S.P., Hill, M.R., Moore, G., He, Z., Sumerlin, B.S., 2015. Synthesis of amphiphilic polysuccinimide star copolymers for responsive delivery in plants. Chem. Commun. 1–5.

Chenkin, A.F., 1975. Economic effects of plant protection in the Russian Federated Republic. In: Papers at Sessions, VIII International Congress of Plat Protection, vol. 2, p. 27. Moscow, USSR.

Cherniak, M., 1988. Toxicological screening for organophosphorus-induced delayed neurotoxicity. Neurotoxicology 9 (2), 249–272.

Ciesielski, S., Loomis, D.P., Mims, S.R., Auer, A., 1994. Pesticide exposures, cholinesterase depression, and symptoms among North Carolina migrant farm workers. Am. J. Publ. Health 84, 446–451.

Clark, G., Garry, V., Burroughs, B., Holland, S., Chu, M., Brown, D., 2002. Endocrine disruptors. Relationships between exposure to dioxin-like chemicals, testosterone levels, and sex of the children of pesticide applicators. Organohalogen Compd. 56, 73–76.

Colosio, C., Tiramani, M., Maroni, M., 2003. Neurobehavioral effects of pesticides: state of the art. Neurotoxicology 24, 577–591.

Cooper, J., Dobson, H., 2007. The benefits of pesticides to mankind and the environment. Crop Prot. 26, 1337–1348.

Davis, K., Yesavage, J., Berger, P., 1978. Single case study. Possible organophosphate induced parkinsonism. J. Nerv. Ment. Dis. 166, 222–225.

Dhaliwal, G.S., Koul, O., 2007. Biopesticides and Pest Management: *Conventional and Biotechnological Approaches*. Kalyani Publishers, New Delhi.
Delcour, I., Spanoghe, P., Uyttendaele, M., 2015. Literature review: impact of climate change on pesticide use. Food Res. Int. 68, 7–15.
Ecobichon, D.J., 2001. Pesticide use in developing countries. Toxicology 160, 27–33.
Ecobichon, D.J., Davies, J.E., Doull, J., et al., 1990. Neurotoxic effects of pesticides. In: Baker, S.R., Wilkinson, C.F. (Eds.), The Effects of Pesticides on Human Health. Advances in Modern Environmental Toxicology, vol. 18. Princeton Scientific, Princeton, NJ, pp. 131–199.
Elbetieha, A., Da'as, S.I., Khamas, W., Darmani, H., 2001. Evaluation of the toxic potentials of cypermethrin pesticide on some reproductive and fertility parameters in the male rats. Arch. Environ. Contam. Toxicol. 41, 522–528.
Eskenazi, B., Bradman, A., Castorina, R., 1999. Exposures of children to organophosphate pesticides and their potential adverse health effects. Environ. Health Perspect. 107 (Suppl. 3), 409–419.
Ewing, R.D., 1999. Diminishing Returns: Salmon Decline and Pesticides. Funded by the Oregon Pesticide Education Network, Biotech Research and. Consulting, Inc., Corvallis, OR, p. 55.
FAO, 2017. Pesticide Residues in Food. Joint FAO/WHO Meeting on Pesticide Residues. Food and Agriculture Organization of the United Nations, WHO, Rome.
FAO, 2005. Proceedings of the Asia Regional Workshop. Bangkok: Regional Office for Asia and the Pacific.
Fradin, M.S., Day, J.F., 2002. Comparative efficacy of insect repellents against mosquito bites. NEJ Med. 347, 13–18.
Gavrilescu, M., 2005. Fate of pesticides in the environment and its bioremediation. Eng. Life Sci. 5, 497–526.
Ghormade, V., Deshpande, M.V., Paknikar, K.M., 2011. Perspectives for nanobiotechnology enabled protection and nutrition of plants. Biotechnol. Adv. 29, 792–803.
Guan, H.N., Chi, D.F., Yu, J., Zhang, S.Y., 2011. Novel photodegradable insecticide W/TiO2/A vermectin nanocomposites obtained by polyelectrolytes assembly. Colloids Surf. B Biointerfaces 83, 148–154.
Gupta, P.K., 2004. Pesticide exposure-Indian scene. Toxicology 198, 83–90.
Han, Y., Song, L., Liu, S., Zou, N., Li, Y., Qin, Y., Pan, C., 2018. Simultaneous determination of 124 pesticide residues in Chinese liquor and liquor-making raw materials (sorghum and rice hull) by rapid multi-plug filtration cleanup and gas chromatography-Tandem mass spectrometry. Food Chem. 241, 258–267.
Hancock, D.B., Martin, E.R., Mayhew, G.M., Stajich, J.M., Jewett, R., Stacy, M.A., Scott, B.L., Vance, J.M., Scott, W.K., 2008. Pesticide exposure and risk of Parkinson's disease: a family-based case-control study. BMC Neurol. 8, 6.
Handford, C.E., Elliott, C.T., Campbell, K., 2015. A review of the global pesticide legislation and the scale of challenge in reaching the global harmonization of food safety standards. Integ. Environ. Assess. Manag. 11 (4), 525–536.
Hanke, W., Jurewicz, J., 2004. The risk of adverse reproductive and developmental disorders due to occupational pesticide exposure: an overview of current epidemiological evidence. Int. J. Occup. Med. Environ. Health 17, 223–243.
Hernández, A.F., Parrón, T., Alarcón, R., 2011. Pesticides and asthma. Curr. Opin. Allergy Clin. Immunol. 11, 90–96.
Hertzman, C., Wiens, M., Bowering, D., Snow, B., Calne, D., 1990. Parkinson's disease: a case-control study of occupational and environmental risk factors. Am. J. Ind. Med. 17, 349–355.

Holmstrup, M., Bindesbøl, A.-M., Oostingh, G.J., et al., 2010. Interactions between effects of environmental chemicals and natural stressors: a review. Sci. Total Environ. 408, 3746–3762.
Hoogenraad, T., 1988. Dithiocarbamates and Parkinson's disease [letter]. Lancet 1, 767.
Huang, Y., Zhao, J., Zhou, L., Wang, J., Gong, Y., Chen, X., Guo, Z., Wang, Q., Jiang, W., 2010. Antifungal activity of the essential oil of *Illiciumverum* fruit and its main component trans-anethole. Molecules 15 (11), 7558–7569.
Isman, M.B., 2006. Botanical insecticides, deterrents, and repellents in modern agriculture and an increasingly regulated world. Annu. Rev. Entomol. 51 (1), 45–66.
Isman, M.B., Machial, C.M., 2006. Pesticides based on plant essential oils: from traditional practice to commercialization. Natural. Occurr. Bioact. Compound. 29–44, 2.
Iulia, G., Balaure, P.C., Gudovan, D., 2017. Nanopesticides: a new paradigm in crop protection. New Pest. Soil Sens. 129–192, 4.
Jana, S., Bandyopadhyaya, S., 1987. Effect of heavy metals on some biochemical parameters in the freshwater fish *Channa punctatus*. Environ. Ecol. 5, 488–493.
Jayaraj, R., Megha, P., Sreedev, P., 2016. Organochlorine pesticides, their toxic effects on living organisms and their fate in the environment. Interdiscip.Toxicol. 9 (3–4), 90–100.
Jia, C., Jukes, D., 2013. The national food safety control system of China-a systematic review. Food Contr. 32 (1), 236–245.
Jia, X., Sheng, W.-B., Li, W., Tong, Y.-B., Liu, Z.Y., Zhou, F., 2014. Adhesive polydopamine coated a vermectin microcapsules for prolonging foliar pesticide retention. ACS Appl. Mater. Interfaces 6, 19552–19558.
Kachhawa, D., 2017. Microorganisms as a biopesticides. J. Entomol. Zool. Studies 5 (3), 468–473.
Kagan, O., Jay, F., John, K., 2010. Wide range of diseases linked to pesticides. *Pesticides and You.* A Quart. Publicat. Beyond Pest. 30 (2), 1–21.
Kah, M., Beulke, S., Tiede, K., Hofmann, T., 2013. Nanopesticides: state of knowledge, environmental fate, and exposure modeling. Crit. Rev. Environ. Sci. Technol. 43, 1823–1867.
Kaloyanova, F., El Batawi, M., 1991. Human Toxicology of Pesticides. CRC Press, Boca Raton, Fla.
Kamel, F., Rowland, A., Park, L., Anger, W., Baird, D., Gladen, B., et al., 2003. Neurobehavioral performance and work experience in Florida farm workers. Environ. Health Perspect. 111, 1765–1772.
Kanavouras, K., Tzatzarakis, M.N., Mastorodemos, V., Plaitakis, A., Tsatsakis, A.M., 2011. A case report of motor neuron disease in a patient showing significant level of DDTs, HCHs and organophosphate metabolites in hair as well as levels of hexane and toluene in blood. Toxicol. Appl. Pharmacol. 256 (3), 399–404.
Kattwinkel, M., Kühne, J.-V., Foit, K., Liess, M., 2011. Climate change, agricultural insecticide exposure, and risk for freshwater communities. Ecol. Appl. 21, 2068–2081.
Kaushik, R.D., Sharma, G., Arora, C., 2004. Biocidal potential of leaf extracts of *Vitex negundo* Linn. Allelopathy J. 14 (2), 227–234.
Keiserukhsky, M.G., Kashirsky, O.P., 1975. Economics of plant protection in the USSR. In: VIII International Congress of Plant Protection, Papers at Sessions, p. 12. Moscow, USSR, 2.
Khan, S., Taning, C.N.T., Bonneure, E., et al., 2017. Insecticidal activity of plant-derived extracts against different economically important pest insects. Phytoparasitica 45, 113–124.
Khwaja, S., Mushtaq, R., Yousuf, M., Attaullah, M., Tabbassum, F., Faiz, R., 2013. Monitoring of biochemical effects of organochlorine pesticides on human health. Health 5, 1342–1350.
Korsak, R.J., Sato, M.M., 1977. Effects of chronic organophosphate pesticide exposure on the central nervous system. Clin. Toxicol. 11, 83–95.

Kumar, S., Gautam, A.K., Agrawal, K.R., Shah, B.A., Saiyad, H.N., 2004. Demonstration of sperm head shape abnormality and clastogenic potential of cypermethrin. J. Environ. Biol. 25 (2), 187–190.

Kunstadter, P., 2007. Pesticides in Southeast Asia: Environmental, Biomedical, and Economic Uses and Effects. Silkworm Books, Ms Trasvin.

Kyle, S., Robert, B.D., Ronald, J.H., David, W., Chrislip, C.H., Thomas, M.R., Everett, L., Patty, L., Edward Jr., F.K., Charles, K., 2000. Neurologic Function among termiticide applicators exposed to chlorpyrifos. Environ. Health Perspect. 108 (4).

Lallas, P., 2001. Reproductive effects in birds exposed to pesticides and industrial chemicals. The Stockholm convention on persistent organic pollutants. Am. J. Int. Law 95, 692–708.

Lan, J., Jia, J., Liu, A., Yu, Z., Zhao, Z., 2019. Pollution levels of banned and non-banned pesticides in surface sediments from the East China Sea. Mar. Pollut. Bull 139, 332–338.

Le Couteur, D.G., McLean, A.J., Taylor, M.C., Woodham, B.L., Board, P.G., 1999. Pesticides and Parkinson's disease. Biomed. Pharmacother. 53, 122–130.

Leon, H., Bradlow, D., Lee, D., Gong, L., Daniel, S., Raj, T., 1995. Effects of Pesticides on the Ratio of 160 α/2-Hydroxyestrone: a biologic marker of breast cancer risk. Environ. Health Perspect. 103 (Suppl. 7), 147–150.

Levin, H.S., Rodnitzky, R.L., Mick, D.L., 1976. Anxiety associated with exposure to organophosphate compounds. Arch. Gen. Psychiatr. 33, 225–228.

Lindblade, K.A., Eisele, T.P., Gimnig, J.E., Alaii, J.A., Odhiambo, F., terKuile, F.O., Hawley, W.A., Wannemuehler, K.A., Phillips-Howard, P.A., Rosen, D.H., Nahlen, B.L., Terlouw, D.J., Adazu, K., Vulule, J.M., Slutsker, L., 2004. Sustainability of reductions in malaria transmission and infant mortality in western Kenya with use of insecticide-treated bed nets: 4 to 6 years of follow-up. J. Am. Med. Assoc. 291 (21), 2571–2580.

Liou, H.H., Tsai, M.C., Chen, C.J., Jeng, J.S., Chang, Y.C., Chen, S.Y., et al., 1997. Environmental risk factors and Parkinson's disease: a case-control study in Taiwan. Neurology 48, 1583–1588.

London, L., Nell, V., Thompson, M., Myers, J., 1998. Effects of long-term organophosphate exposures on neurological symptoms, vibration sense and tremor among South African farm workers. Scand. J. Work. Environ. Health 24, 18–29.

Lopez, M.D., Pascual-Villalobos, M.J., 2010. Mode of inhibition of acetylcholinesterase by monoterpenoids and implications for pest control. Ind. Crop. Prod. 31 (2), 284–288.

Luo, D., Zhou, T., Tao, Y., Feng, Y., Shen, X., Mei, S., 2016. Exposure to organochlorine pesticides and non-Hodgkin lymphoma: a meta-analysis of observational studies. Sci. Rep. 6, 25768.

Madeleine, C., David, K., Edward, A.D.M., Christy, A.M., Dominique, A.N., Jeroen, P., Van der, S., 2015. Risks of large-scale use of systemic insecticides to ecosystem functioning and services. Environ. Sci. Pollut. Res. 22, 119–134.

Mani, U., Islam, F., Prasad, A.K., Kumar, P., Kumar, V.S., Maji, B.K., Dutta, K.K., 2002. Steroidogenic alterations in testes and sera of rats exposed to formulated Fenvalerate by inhalation. Hum. Exp. Toxicol. 21, 593–597.

Mansour, S.A., 2004. Pesticide exposure-Egyptian scene. Toxicology 198, 91–115.

Manthripragada, A.D., Costello, S., Cockburn, M.G., Bronstein, J.M., Ritz, B., 2010. Paraoxonase 1, agricultural organophosphate exposure, and Parkinson disease. Epidemiology 21, 87–94.

Marigoudar, S.R., 2012. Cypermethrin Induced Pathophysiological and Some Biochemical Changes in the Freshwater Teleost, *LabeoRohita* (Hamilton). Department of Zoology, Karnatak University, India. PhD thesis.

McCormack, A.L., Thiruchelvam, M., Manning-Bog, A.B., Thiffault, C., Langston, J.W., Cory-Slechta, D.A., et al., 2002. Environmental risk factors and Parkinson's disease: selective degeneration of nigral dopaminergic neurons caused by the herbicide paraquat. Neurobiol. Dis. 10, 119–127.

Meco, G., Bonifati, V., Vanacore, N., Fabrizio, E., 1994. Parkinsonism after chronic exposure to the fungicide maneb (manganese ethylene-bis-dithiocarbamate). Scand. J. Work. Environ. Health 20, 301–305.

Meeker, J.D., et al., 2006. Exposure to nonpersistent insecticides and male reproductive hormones. Epidemiology 17 (1), 61–68.

Michalis, K., Andreas, T., Aristidis, T., Christos, H., Christo, d., 2012. Systematic review of biomonitoring studies to determine the association between exposure to organophosphorus and pyrethroid insecticides and human health outcomes. Toxicol. Lett. 210, 155–168.

Miranda, J., Lundberg, I., McConnell, R., Delgado, E., Cuadra, R., Torres, E., et al., 2002. Onset of grip- and pinch-strength impairment after acute poisonings with organophosphate insecticides. Int. J. Occup. Environ. Health 8, 19–26.

Misra, U., Nag, D., Bushan, V., Ray, P.K., 1985. Clinical and biochemical changes in chronically exposed organophosphate workers. Toxicol. Lett. 24, 187–193.

Moretto, A., Colosio, C., 2011. Biochemical and toxicological evidence of neurological effects of pesticides: the example of Parkinson's disease. Neurotoxicology 32, 383–391.

Morillo, E., Villaverde, J., 2017. Advanced technologies for the remediation of pesticide contaminated soils. Sci. Total Environ. 586, 576–597.

Murty, A.S., 1986. Toxicity of Pesticides to Fish. (I & II). C.R.C Press Inc., p. 355pp, 483.

Nambirajan, K., Muralidharan, S., Manonmani, S., Kirubhanandhini, V., Ganesan, K., 2018. Incidences of mortality of Indian peafowl *Pavo cristatus* due to pesticide poisoning in India and accumulation pattern of chlorinated pesticides in tissues of the same species collected from Ahmedabad and Coimbatore. Environ. Sci. Pollut. Control Ser. 1–9.

Nandipati, S., Litvan, I., 2016. Environmental exposures and Parkinson's disease. Int. J. Environ. Res. Publ. Health 13, 881.

Narayana, K., Narayan, P., D'Souza, U.J., 2004. Is our drinking water a slow poison? Indian J. Med. Sci. 58, 528–530.

Nazir, J.S.R., Gowen, M., Inam-ul-Haq, K.A., Shahina, F., 2007. Systemic and persistent effect of neem (*Azadirachtaindica*) formulations against root-knot nematodes, *Meloidogynejavanica* and their storage life. Crop Protect. 26 (7), 911–916.

Noyes, P.D., Lema, S.C., 2015. Forecasting the impacts of chemical pollution and climate change interactions on the health of wildlife. Curr. Zool. 61, 669–689.

Noyes, P.D., McElwee, M.K., Miller, H.D., Clark, B.W., Van Tiem, L.A., Walcott, K.C., Erwin, K.N., Levin, E.D., 2009. The toxicology of climate change: environmental contaminants in a warming world. Environ. Int. 35, 971–986.

O'Malley, M., 1997. Clinical evaluation of pesticide exposure and poisoning. Lancet 349, 1161–1166.

Osibanjo, O., Jensen, S., 1980. Ecological and environmental health perspectives of pesticide pollution. In: Omueti, J.A., Imevbore, A.M.A. (Eds.), Proceedings of the First National Conference on Water Pollution and Pesticide Residues in Foods. Federal Ministry of Water Resources, Lagos, pp. 206–220. Akinyele 10.

Parrón, T., Requena, M., Hernández, A.F., Alarcón, R., 2011. Association between environmental exposure to pesticides and neurodegenerative diseases. Toxicol. Appl. Pharmacol. 256, 379–385.

Pestizid Aktions-Netzwerk e.V. (PAN Germany), 2012. Pesticides and Health Hazards Facts and Figures. https://www.pan-germany.org/download/Vergift_EN-201112-web.pdf.

Peters, H.A., Levine, R.I., Matthews, C.C., et al., 1982. Carbon disulfide induced neuropsychiatric changes in grain storage workers. Am. J. Ind. Med. 3, 373–391.

Peterson, C., Coats, J.R., 2001. Insect repellents-past,present and future. Pestic. Outlook 12, 154–158.
Petrusheva, N.I., 1975. Protection of horticultural crops in the USSR and future perspectives. Paper. at Sess., VIII Int. Congress of Plant Protect., Moscow, USSR 2, 124.
Pierpaolo, bacciu, 2017. Seven Strategies to Reduce Pesticide Usage. Agronomy Food Safety. http://www.casaqa.com/2017/11/01/seven-strategies-to-reduce-pesticide-usage.
Priyadarshi, A., Khuder, S.A., Schaub, E.A., Priyadarshi, S.S., 2001. Environmental risk factors and Parkinson's disease: a meta analysis. Environ. Res. 86, 122–127.
Radhaiah, V., Girija, M., Rao, K.J., 1987. Changes in selected biochemical parameters in the kidney and blood of the fish, *Tilapia mossambica* (Peters), exposed to heptachlor. Bull. Environ. Contam. Toxicol. 39, 1006–1011.
Rao, K.R.S.S., Rao, J.C., 1987. Independent and combined action of carbaryl and phenthoate on snake head, *Channa punctatus* (Bloch). Curr. Sci. 56, 331–332.
Rao, K.R.S.S., Rao, K.V.R., 1989. Combined action of carbaryl and phenthoate on the sensitivity of the acetylcholinesterase system of the fish. Ecotoxicol. Environ. Saf. 17, 12–15.
Regnault-Roger, C., Vincent, C., Arnasson, J.T., 2012. Essential oils in insect control: low-risk products in a high-stakes world. Annu. Rev. Entomol. 57, 405–425.
Rice, D., Barone, J., 2000. Critical periods of vulnerability for the developing nervous system: evidence from humans and animal models. Environ. Health Perspect. 108, 511–533.
Richardson, R., 1995. Assessment of the neurotoxic potential of chlorpyrifos relative to other organophosphorus compounds: a critical review of the literature. J. Toxicol. Environ. Health 44, 135–165.
Rodriguez-Kabana, R., Curl, E.A., 1980. Non-target effects of pesticides on soil borne pathogens and disease. Annu. Rev. Phytopathol. 18, 311–332.
Rohlman, D.S., Anger, W.K., Lein, P.J., 2011. Correlating neurobehavioral performance with biomarkers of organophosphorous pesticide exposure. Neurotoxicology 32, 268–276.
Ron, G., 2020. Pesticides Environmental Stewardship. Taking Action (Control Measures. Centre for Integrated Pest Management. https://pesticidestewardship.org/ipm/taking-action/assessed on 20/08/2020.
Rosell, G., Quero, C., Coll, J., Guerrero, A., 2008. Biorational insecticides in pest management. J. Pestic. Sci. 33, 103–121.
Rosenstock, I., Daniel, W., Barnhart, S., et al., 1990. Chronic neuropsychological sequelae of occupational exposure to organophosphate insecticides. Am. J. Ind. Med. 18, 321–325.
Rosenstock, L., Keifer, M., Daniell, W.E., McConnell, R., Claypoole, K., 1991. Chronic central nervous system effects of acute organophosphate pesticide intoxication. Lancet 338, 223–227.
Ruijten, M., Salle, H.J.A., Verberk, M.M., Smink, M., 1994. Effect of chronic mixed pesticide exposure on peripheral and autonomic nerve function. Arch. Environ. Health 49, 188–195.
Sabae, S.Z., El-Sheekh, M.M., Khalil, M.A., Elshouny, W.A.E., Badr, H.M., 2014. Seasonal and regional variation of physicochemical and bacteriological parameters of surface water in El-Bahr El-Pherony, Menoufia, Egypt. World J. Fish Mar. Sci. 6 (4), 328–335.
Sabarwal, A., Kumar, K., Singh, R.P., 2018. Hazardous effects of chemical pesticides on human health–Cancer and other associated disorders. Environ. Toxicol. Pharmacol. 63, 103–114.
Sack, D., Linz, D., Shukla, R., Rice, C., Bhattacharya, A., Suskind, R., 1993. Health status of pesticide applicators: postural stability assessments. J. Occup. Med. 35, 1196–1202.
Sana, U., Amina, Z., Mahmoud, A., Mayada, R.F., Maryam, D., Kumara, G.K., Ruchi, T., Kuldeep, D., Hafiz, M.N.I., 2018. Cypermethrin induced toxicities in fish and adverse health outcomes: its prevention and control measure adaptation. J. Environ. Manag. 206, 863–871.

Sanchez-Ramon, J., Hefti, F., Weiner, W.I., 1987. Paraquat and Parkinson disease [letter]. Neurology 37, 728.

Sarwar, S., Ahmad, F., Khan, J., 2007. Assessment of the quality of Jehlum river water for irrigation and drinking at district Muzaffarabad Azad Kashmir. Sarhad J. Agric. 23 (4), 1041.

Savage, E., Keefe, T.J., Mounce, L.M., Heaton, R.K., Lewis, J.A., Burcar, P.J., 1988. Chronic neurological sequelae of acute organophosphate pesticide poisoning. Arch. Environ. Health 43, 38–45.

Schreinemachers, P., Tipraqsa, P., 2012. Agricultural pesticides and land use intensification in high, middle and low income countries. Food Pol. 37 (6), 616–626.

Schreinemachers, P., Afari-Sefa, V., Heng, C.H., Dung, P.T.M., Praneetvatakul, S., Srinivasan, R., 2015. Safe and sustainable crop protection in Southeast Asia: status, challenges and policy options. Environ. Sci. Pol. 54, 357–366.

Sechi, G., Agnetti, V., Piredda, M., Canu, M., Deserra, F., Omar, H.A., et al., 1992. Acute and persistent parkinsonism after use of diquat. Neurology 42, 261–263.

Shahid, M., Ahmad, A., Khalid, S., Siddique, H.F., Saeed, M.F., Ashraf, M.R., et al., 2016. Pesticides pollution in agricultural soils of Pakistan. In: Soil Science: Agricultural and Environmental Prospective, pp. 199–229.

Singh, N.S., Sharma, R., Parween, T., Patanjali, P.K., 2018. Pesticide contamination and human health risk factor. In: Modern Age Environmental Problems and Their Remediation, pp. 49–68. Cham.

Smit, L.A.M., van Wendel de Joode, B.N., Heederik, D., Peiris-John, R.J., van der Hoek, W., 2003. Neurological symptoms among Sri Lankan farmers occupationally exposed to acetylcholinesterase- inhibiting insecticides. Am. J. Ind. Med. 44, 254–264.

Stallones, L., Beseler, C., 2002. Pesticide illness, farm practices, and neurological symptoms among farm residents in Colorado. Environ. Res. 90, 89–97.

Steenland, K., Jenkins, B., Ames, R.G., O'Malley, M., Chrislip, D., Russo, J., 1994. Chronic neurological sequelae to organophosphate pesticide poisoning. Am. J. Publ. Health 84 (5), 731–736.

Storm, R.M., Price, D.C., Lubetkin, S.D., 2001. Aqueous Dispersion of Agricultural Chemicals. US Patent, 20,010,051,175.

Townson, H., Nathan, M.B., Zaim, M., 2005. Exploiting the potential of vector control for disease prevention. Bull. World Health Organ. 83 (12), 942–947.

Tsatsakis, A.M., Zafiropoulos, A., Tzatzarakis, M.N., Tzanakakis, G.N., Kafatos, A., 2009. Relation of PON1 and CYP1A1 genetic polymorphisms to clinical findings in a cross-sectional study of a Greek rural population professionally exposed to pesticides. Toxicol. Lett. 186, 66–72.

U.S. Environmental Protection Agency, 2011. Pesticide Market Estimates: 2006–2007. Office of Pesticide Programs, Washington, DC.

Ujváry, I., 2001. Transforming natural products into natural pesticides-experience and expectations. Phytoparasitica 30 (5), 439–442.

UK PAN (Pesticides Action Network), 2009. List of Lists: A Catalogue of Lists of Pesticides Identifying Those Associated with Particularly Harmful Health or Environmental Impacts. P.A. Network.

Undeger, U., Basaran, N., 2005. Effects of pesticides on human peripheral lymphocytes in vitro: induction of DNA damage. Arch. Toxicol. 79, 169–176.

Vargas, R.I., Stark, J.D., Kido, M.H., Ketter, H.M., Whitehand, L.C., 2000. Methyl eugenol and cuelure traps for suppression of male oriental fruit flies and melon flies (*Diptera: tephritidae*) in Hawaii: effects of lure mixtures and weathering. J. Econ. Entomol. 93, 81–87.

Varjani, S.J., Joshi, R.R., Kumar, P.S., Srivastava, V.K., Kumar, V., Banerjee, C., Kumar, R.P., 2018. Polycyclic aromatic hydrocarbons from petroleum oil industry activities: effect on human health and their biodegradation. In: Varjani, S.J., Gnansounou, E., Gurunathan, B., Pant, D., Zakaria, Z.A. (Eds.), Waste Bioremediation. Springer Nature, Singapore, pp. 185–199.

Venkataraman, B.V., Rani, M.A., Andrade, C., Joseph, T., 1994. Correlation of time course of blood cholinesterase activity and toxic manifestations of acute methyl parathion in antidote treated rats. Indian J. Physiol. Pharmacol. 38, 214–216.

Vrushali, T., Tare, V., Shushil, K., 2001. Bioactivity of some medicinal plants against chosen insect pests/vectors. Proceedings of the National Seminar on the Frontiers of Res and Dev in Med. Plants, CIMAP, Lucknow 22.

Wang, A., Cockburn, M., Ly, T.T., Bronstein, J.M., Ritz, B., 2014. The association between ambient exposure to organophosphates and Parkinson's disease risk. Occup. Environ. Med. 71 (4), 275–281.

Webster, J.P.G., Bowles, R.G., Williams, N.T., 1999. Estimating the economic benefits of alternative pesticide usage scenarios: wheat production in the United Kingdom. Crop Protect. 18, 83.

Wen, J.Z., Fu, B.J., Jian, F.O., 2011. Global pesticide consumption and pollution: with China as a focus. Proc. Int. Acad. Ecol. Environ. Sci. 1 (2), 125–144.

White, R.F., Feldman, R.G., Travers, P.H., 1990. Neurobehavioral effects of toxicity due to metals. solvents, and insecticides. Clin. Neuropharmarol. 12, 13392–13394.

WHO, 2008. Training Package for the Health Sector, Children's Health and the Environment (July version).

WHO/UNCEP, 1990. Public Health Impact of Pesticides Used in Agriculture. Working Group, Geneva.

World wide pesticide use. In: Kumar, A., De, A., Bose, R., Mozumdar, S. (Eds.), 2014. Targeted Delivery of Pesticides Using Biode-Gradable Polymeric Nanoparticles. Springer, Berlin, pp. 5–6.

Xiang, Y., Wang, M., Sun, X., Cai, D., Wu, Z., 2014. Controlling pesticide loss through Nanonetworks. ACS Sustain. Chem. Eng. 2, 918–924.

Yang, X., Luo, J., Duan, Y., Li, S., Liu, C., 2018. Simultaneous analysis of multiple pesticide residues in minor fruits by ultra high-performance liquid chromatography/hybrid quadrupole time-of-fight mass spectrometry. Food Chem. 241, 188–198.

Yohannes, Y.B., Ikenaka, Y., Nakayama, S.M.M., Mizukawa, H., Ishizuka, M., 2017. DDTs and other organochlorine pesticides in tissues of four bird species from the Rift Valley region, Ethiopia. Sci. Total Environ. 574, 1389–1395.

Zeliger, H.I., 2011. Human toxicology of chemical mixtures. In: Toxic Consequences. William Andrew, *Elsevier*, Waltham, MA, USA.

Zhang, W., 2018. Global pesticide use: profile, trend, cost benefit and more. Proc. Int. Acad. Ecol. Environ. Sci. 8 (1), 1–27.

Impact of climate change on wetlands, concerning Son Beel, the largest wetland of North East, India

17

Moharana Choudhury[1], Anu Sharma[2], Pardeep Singh[3], Deepak Kumar[4]
[1]Voice of Environment (VoE), Guwahati, Assam, India; [2]Department of Environmental Science, Government Degree College Bhaderwah, University of Jammu, Jammu, India; [3]Department of Environmental Studies, PGDAV College University of Delhi, New Delhi, India; [4]United Nation Developmental Program (UNDP), New Delhi, India

1. Introduction

Water is a vital source of life for the planet Earth, not only to perform the necessary metabolic cycle but also to run and regulate all the functions of day-to-day life. Stable economic growth and development for every nation or country are powered by water, as its availability impacts agriculture, manufacturing, environment, sustainable development, etc. The availability of water directly linked to the amount of precipitation, which is well-taken care by the hydrological cycle. The changing climate followed by enhanced levels of temperature leads to disturbance in this cycle. The observation of the world's wetland day on February 2 every year highlights the significance of wetlands as ecosystems. This observance began in 1971 when the Ramsar convention was convened at Ramsar in Iran. Wetlands play a substantial role in pollution control and detoxification. Because of their exceptional performance as the filters, the absorption of pesticides and chemicals, and removing harmful waste from the water, they are given the Earth's kidneys' status. Nature plans its natural systems to enhance or even replace historically gray infrastructure functionalities (WBCSD, 2017). Wetlands cover around 6% of shares in the global surface they play a unique role in the biogeochemical and water cycle. They are home to a large part of global biodiversity—tremendous pressure on the ecosystem in the form of land reclamation, extreme exploitation of the resources. Alteration in hydrology, pollution threats are accruing from many sources on the wetlands on almost every continent.

Further stress to wetlands is expected by climate models, primarily due to changes in hydrology, rising temperatures, and increasing sea level (Junk et al., 2013). Twentieth-century climate data show that the United States is in a wetter, warmer climate pattern. Even other climate forecasts indicate that over the next 100 years, this phenomenon will continue and potentially worsen. Wetlands are likely to be affected by the rising sea level trend because of elevated carbon dioxide levels.

Global Climate Change. https://doi.org/10.1016/B978-0-12-822928-6.00006-X

Potential effects range from shifts to ecological roles in the group system (Burkett and Kusler, 2007). The increase in sea levels has the enormous potential of harming the coastal wetlands, and the global climate change will also put immense pressure on domestic wetlands. Such climate changes can cause a great deal of distress, resulting in permanent changes to the overall wetland ecosystem. The impacts of palaeoecological reports on wetlands of the longer-term climate change are evident from various parts of Australia. These are to highlight both the magnitude of changes in wetlands and the difficulty of distinguishing between the different effects of climate change (Finlayson, 2013). The open plains of southwestern Sweden lost 67% of their wetlands and reservoirs due to flooding during the last 50 years (Dillehay et al., 2007). The significant loss of services to humans from wetland habitats has now led to the rehabilitation and development of wetlands as priorities, especially to increase biodiversity and minimize eutrophication at national and European Union level, for example by the Habitats and Water Framework Directives. In the Baltic Sea, about 50% of the total nitrogen inputs and 30% of the total phosphorus are from agriculture. Therefore, farmers are key players in achieving the goal of wetland formation and eutrophication reduction.

On the other hand, Lundqvist (2001) and Blomqvist (2004) found that many farmers did not see why they should contribute to and pay for public advantages and benefits. The nature of dispersed nutrient leakage means that the polluter is hard to track. Site and design are critical concepts for developing efficient wetlands for the retention of nutrients. Wetlands generated on one farm could reduce outflows of nutrients from other farms or polluters—a problem that could lead to conflict between local farmers. The creation of wetlands may not be an appropriate or otherwise feasible solution from a farmer's point of view. Therefore, the mitigation mechanism itself includes multistakeholder trade-offs and decisions. For example, it may mean land sacrifices in agricultural production to try to find a suitable location for the development of wetlands contributing to public ecological services for nutrient conservation and biodiversity protection. An essential activity for the management of catchment is identifying trade-offs and goals at local and catchment levels so that priority can be provided to which steps to be taken. As Hansson et al. (2012) suggested, farmers needed to monitor the environmental measures applied to agriculture. A large number of wetland ecosystem services are intangible and have "public good" characteristics. Their actual value in the process of development and environmental decision-making is therefore generally not taken into consideration that leads to low investment and poor management (Barbier et al., 1997). Because wetland values are poorly appreciated and underpinned, few mechanisms exist on the market to either capture their benefit or price them according to their real scarcity value. This situation led to ill-informed management and planning decisions that help maintain rapid erosion, conversion, and wetland degradation. Wetlands are also environmentally sensitive and adaptive. Wetlands tend to be lost in the world despite acknowledging the need for their protection.

Wetlands act as a natural carbon sink system, but such ecosystems are largely responsible for generating methane in India as they support conducive condition for paddy production. In one of the studies by Mingkui et al. (1998), emissions from high-latitude wetlands and rice paddies contribute large methane emissions. Still, this impact reversed due to the effect of soil moisture depletion by more significant temperature increases. For the sequestration and the long-term storage of carbon dioxide from the atmosphere, the wetland ecosystems are an optimum natural environment. Still, they are natural sources of greenhouse gas emissions, mainly methane. Within tropical and subtropical wetlands, much of this carbon accumulation occurs. All wetlands are net radiative sinks as they balance carbon and methane emissions. Thus, they can be generated and restored to provide carbon sequestration and other ecosystem services without establishing net radiative environment sources due to methane emissions (Mitsch et al., 2013).

The climate disrupted by extreme droughts and floods, impacting all kinds of habitats. In the 1930s, drought caused significant changes in grassland ecosystems, killing millions of riparian forests and causing severe financial losses to agriculture (Albertson and Weaver, 1942). Floods fill lakes and wetlands in the 1990s, including destroying fields, towns, including roads (Winter and Rosenberry, 1998; Johnson et al., 2004; Shapley et al., 2005). Such meteorological extremes are essential, particularly in the long-term productivity and biodiversity of semipermanent prairie wetlands, which are more profound than the related seasonal and temporary wetland groups (Van der Valk and Davis, 1978). The Wetland simulation model (WETSIM) used by Poiani and Johnson (1991) showed that a likely future climate would yield lower breeding conditions for waterfowl. Similar habitat loss for breeding waterfowl in Central America was predicted by Larson (1995). An analysis of documented environmental effects of tropical and hurricane storms has suggested that storm timing, frequency, and intensity can alter the hydrology of coastal wetlands, geomorphology, biotic systems, energy efficiency, and nutrient cycling (William et al., 1997). Wetlands are vulnerable to climate change depending on their location within hydrological ecosystems. The flood characteristics of groundwater and surface water and the relationship of atmospheric waters, surface waters, and groundwater for any given locality or area determine hydrological landscapes. There are ten general hydrological landscapes: mountainous, mountains, highlands, large drainage basins, riverine, plains, coastal, hummocking glacier, and dune. Assessment of these ecosystems indicates that all wetlands are sensitive to climate change between two severe conditions. Those primarily dependent on precipitation for their supply of water being extremely vulnerable, and those most dependent on drainage systems are the least susceptible, owing to the high buffering ability of large groundwater systems (Winter, 2007).

Globally, there are 2300 Ramsar sites spread over more than 2.1 million square kilometers of area. Highest number of sites is located in the United Kingdom (175 sites). In India, there are 37 Ramsar sites till date. The description of all the 37 sites is given in the table below.

2. 41 Ramsar wetland sites in India

In South Asia apart from India, the countries having Ramsar sites are Pakistan (19 wetlands, 1,343,807 sq. km), Nepal (10 sites, 60,561 sq. km), Bhutan (3 sites, 1225 sq. km), and Bangladesh (2 sites, 611,200 sq. km). Ramsar Handbook released in 2016 as the *Introduction to the Wetlands Convention* mentions Conference of Parties Recommendation 5.7 and Strategic Plan 2016–24, both urge Contracting Parties to set up National Ramsar Committee that can provide a more comprehensive national context for the alteration of the Convention, including the establishment of National Ramsar Committees.

The summary given on Ramsar's official website as below:

Kolleru Lake: It was designated as a Ramsar site in 2002. It is located in the Krishna and Godavari districts of Andhra Pradesh. It occupies an enormous regime of 90100 hectares. Its part is designated as wildlife sanctuary. Andhra Pradesh Forest Department looks into the matter of administration and management of this 'Wetlands of international Importance'. It is categorized under lakes, marsh and aquaculture ponds. It occupies 224 species of birds, 75 species of fishes and 34 species of plants. This wetland is a conservation habitat for Spot-billed Pelican notably, endangered Greater Adjutant and vulnerable Grey Pelican and Lesser Adjutant.

Hokera Wetland: It was designated as a Ramsar site in 2005. It is located in Budgum and Sri Nagar districts of Jammu and Kashmir. It occupies an area of 1375 hectares. It is a conservation reserve, managed under Department of Wildlife Protection, Jammu and Kashmir. It is categorized under Marshes. It occupies 68 species of birds and 38 species of plants. It is a conservation habitat for Eurasian Otter, Schizothorax sp. And Little Grebe notably, endangered Palla's Fish-Eagle and vulnerable Common Carp and Common Pochard.

Wular Lake: India's largest fresh water lake with extensive emergent and floating vegetation marshes, especially water chestnut. It provides a significant source of income for the government and the people of the area. This body of water supports large-scale agriculture and is a good source of agricultural and domestic water. The area is essential for birds to winter, spring, and breed. Human practices include rice production and tree production. It is situated in Kashmir.

Ashtamudi Wetlands: Vast estuarine network of exceptional relevance for its hydrological functions, ecosystems, and fish conservation, the second largest in the state of Kerala. The habitat has 57 species of plants and animals, including six migratory birds, and is responsible for several mangrove species and more than 40 affiliated plant species. Nearly 100 fish species sustain a growing agriculture industry, with the inlet of their standards of living which directly supported by thousands of fishers. The site is threatened by increasing population and environmental pressure, including the pollution of hundreds of fishing vessels and surrounding factories caused by oil spills and extraction of natural resources for development.

Bhoj Wetland: Two neighboring manmade lakes—the "Upper Lake" and "Lower Lake." The lower one was built almost 200 years ago mainly from the Upper Canal spillage and was supported by the town of Bhopal, built in the 11th century by the construction of an earthen dam across the Kolans River. Wetlands are extremely rich in flora and fauna, especially phytoplanktons, zooplanktons, fishes, habitant and migratory birds, insects, reptiles, and amphibians, both of which are organic and harvested. A variety of bird species were reported, which had little to no visit to the area, following the implementation of the quality management action plan, which began in 1995 with the financial assistance of the Japanese Government.

Deepor Beel: A bird sanctuary and a significant lake in a former Brahmaputra river drain, of great biological significance as well as being the major storm-water reservoir in Guwahati city. The water body is a migrating bird's flyway processing site, and one can see some of Assam's largest concentrations of aquatic birds, particularly in winter. Help is given to some globally endangered species. Almost 50 species of the fish present here provide livelihood security. *Lily euryale* seeds provide essential sources of money to local markets for a variety of neighboring villages. Overfishing and hunting pressure on water-birds, pesticide, and fertilizer runoff, as well as water hyacinth infestation of *Eichhornia crassipes*, are possible threats. The potential consequences of the decision to create a wastewater canal directly from the town to the water body are considered disastrous.

East Calcutta Wetland: World-renowned as an intrause wetland model. The lake offers a public service for the treatment of urban wastewater and the use of treated water for fish farming activities by efficient nutrient recovery. Liquid flows into approximately 4000 ha of water bodies, and ponds serve as solar reactors. Thus, its system is characterized as "one of the rare sources of environmental protection and expansion planning in which local farmers have introduced a complex ecological approach to managing the recovery of resources" (RIS). The wetland supplies some 150 tons of natural fruits and veggies per day, and also some 10,500 tons of table fish annually, which gives direct and indirect livelihood opportunities to some 50,000 people. In some instances, in trade unions and in others, cooperative organizations whose contract rights are under violation of law also fish ponds by workers' cooperatives. The potential hazard is identified by recent misuse by the wastewater runoff system, which adds metals to the sewage effluent and affects the fresh vegetable edible quality.

Harike Wetland: Wildlife sanctuary is a small body of water with 13 islands at the confluence of two rivers. Seventy percent of the lake is covered by thick floating vegetation. This area is an ideal landmark for the growing, nesting, and staging of birds, holding more than 200,000 Anatidae during migration (ducks, swans, etc.). The full reservoir is allocated annually to commercial fisheries groups.

Loktak Lake: A vast yet decreasing freshwater lake produced by many channels and associated marshlands. The signature features are dense, floating mats of soil covered with weeds (phumdis). The lake is widely used as a water source for irrigation purpose by local people. It also has a vital role to play in flood control. In 1993, this wetland was included in the Montreux Record register due to ecological problems, such as forest depletion of the river catchment, overcrowding, and environmental degradation from a dam for hydropower generation, and water management exacerbated the local extinction of many native fish species.

Nalsarovar: It is a fresh-water wetland (residual water). It is a vast natural waterbody in the biodiversity region of the Thar Desert and a complex network with a range of salt and precipitation content. The site is home to 210 bird species, with an average of 174,128 people recorded in the winter and 50,000 in the summer. It is a vital layout in the Central Asian estuary, with critically endangered species like *Vanellus gregarius* and the marbled teal, vulnerable to social threats. In the summer when other marine habitats were dry, sarus crane (*Grus antigone*) was immune. The lake also acts as a support system in the dry period for the endangered Indian wild ass population (*Equus hemionus khur*). The water body is the place for potable water and irrigation water for the local communities as well as a significant revenue stream from catla (*Catla catla*) and rohu (*Labeo rohita*) fishing. Around 75,000 tourists visit the wetland every year.

Sambhar Lake: A massive salt lake sensitive to seasonal variations of surface water. It is accompanied by a seasonal sandy plain and a dry splasher of thorny bushes. The place is critical and known for flamingos. The operations of living things include salt and farm livestock.

Sasthamkotta Lake: In the district of Kollam (Kerala), there are approximately 27 species of aquatic organisms. This is the large freshwater lake in the southern part of the country. Sastha's ancient temple is a vital worship center here.

Tsomoriri Lake: The wetland reserve is freshwater to brackish in nature at 4.595 m altitude, with lush greenery along the shorelines and borax-laden lakes. The area is said to display one of the China's most threatened cranes, the gray-necked crane, and India's only feeding ground. The Tibetan great goat, argali, or wild donkey of Tibet, *Equus Kiang*, and *Ammon hodgsonian ovis* are native to the mountains of the Himalayas, the most of the western portion of which seems to be Changthang. The Korzok barley fields are recognized as the world's highest grown ground. The evaporation induces varying salinity levels without outflows under arid steppe conditions. The region is passing by ancient trading routes, and now large trekking tracks. The Korzok Monastery receives a lot of visitors, and local Buddhists consider the estuary sacred, so they do not use the water. Tsomoririo is a holy gift for the Living Earth. The newly opened region has a rapidly growing appeal toward western tourism (currently 2500 in the summer), which offers an "untouched destination" with uncluttered high desert landscape and vibrant cultural events.

Vembanad-Kol Wetland: Situated just southwest coast of India. It is the giant saltwater, tropical wetland ecosystem fed by 10 rivers reflecting massive aquatic ecosystems on the west coast of India, best known the third largest with its clam's migratory bird community in India during colder months. The Kol zone contains more than 90 resident bird species and 50 aquatic oat species. Flood protection is considered a massive benefit in the densely populated regions of Kerala's three districts. Groundwater reloading contributes to the region's water supply, and the network is vital for trade and local transport.

Rudrasagar Lake: A reservoir lies in the northeast hills, flowing through the Gomti River through three seasonal channels. The tank is filled with precious commercial freshwater fish such as *Botia* sp., *Mystus* sp., *Pabda*, *Bata labeo*, and freshwater *scampi*. The lake is mostly inundated 4–5 times a year by an annual mean because of increased rainfall (2500 mm) and downstream topography that helps drain groundwater. Rarely fractionally aquatic weeds consist of freshwater plants. The land is state-owned, and the existing fishing community rents annual water zones, and paddy is cultivated for local seasonal species. Thanks to deforestation, agricultural land growth, industrial agriculture, and the transformation of soils into people's pressures, the main threats are growing silt loads. At the Vijaya Dashamí, one of the most popular Hindu celebrations, for various sports competitions, tourists and enthusiasts cross at least 50,000 per year.

Sundarban Wetland: The Sundarban's ecosystem, which comprises of hundreds of islands, rivers, and streams on the Ganga and Brahmaputra Delta, the Bavarian Bay of Bengal in India and Bangladesh, is part of the world's most massive mangrove environment. The Sundarban of India, occupying more than 50% of the global mangrove forest and 90% of the Indian population mangroves, comprises the south-western part of the delta. Mangrove forests shield the hinterland from hurricanes, cyclones, mare waves, and saltwater flows and tidal waters. They are molluscs fertilizing nurseries, which help all the East Coast fishing industry. Inland within the city, the Sundarban tiger center has been deemed a "sensitive tiger sanctuary" by national law and an essential worldwide tiger conservation landscape. The only mangrove forest that endorses the tiger population and real possibilities underwater hunting abilities is that of the Sundarban Tiger Reserve. A variety of rare, global, and extinct Irrawaddy dolphin and sensitive cat (*Prionailurus viverrinus*) are animals such as the northern river terrapin (*Batagur Baska*), the endangered dolphin of Irrawaddy are still present on site. Also, there are 2 of the 4 horseshoe crab populations in the world and 8 of India's 12 kingfish populations. Area protection and preservation are a priority, considering the intrinsic importance of the ecosystem and its biodiversity.

Mangroves Bhitarkanika: Natural reserve has been a well-known wildlife refuge, one of the wealthiest conservation mangrove forests along the Indian coast, for 25 years. Gahirmatha on this location, with a half-million nests annually, is said to be the world's leading recognized olive ridley beach with saltwater crocodile, nearly 700 *Crocodylus porosus*. The locations have a high density of saltwater. It is the largest nursery for brackish water and sea fish fauna on the east coast and is a rearing and wintering vital site for many native as well as migratory water-birds. The dense shoreline forests provide essential protection from devastating cyclones, and tidal surges for millions of humans—of 58 confirmed mangrove species from India, 55 of which are more diverse than the Sundarbans in Bhitarkanika. The law has predominantly been the viable harvest of nutrition, medicinal products, tannins, wood and equipment fuel and building, and especially sweet and fish, but population pressures and interference jeopardize this equilibrium.

Chandertal Wetland: A high-altitude wetland in the upper valley of Chandra, which flows to the western Himalayan river Chandra (4337 m). Over the last few years, these species have evolved unusual physiological characteristics such as cold arid adaptation, massive radiation, and depletion of oxygen. When nomadic shepherds kill the ecosystem, about 65% of the more significant captures are degraded forests and 35% of grasslands. Many detrimental factors to this delicate and scarce vegetation are spring trekking, littering waste, and inadequate sanitation facilities. After the site was declared National Historic Wetland in 1994, authorities have sponsored the natural tourism infrastructure.

Chilika Lake: Brackish reservoir, isolated by a long sandy ridge from the Bay of Bengal and subject to an exchange of seawater, has resulted in significantly different seasons in salinity in the various parts of the lake. Sweetwater areas provide habitat for aquatic algae. The location is an exciting area for 33 species of water-birds for breeding, wintering, and staging. A total of 118 fish species, including valuable species for trade, are also funded. Many people rely on the lake's richness.

Kanjli: It is a permanent lake, which was converted into a water-storage area for irrigation by building a small dam in 1870, because of its efforts to foster a significant biogeographical variety for marine, mesophytic, and terrestrial wildlife. The site meets Criteria 3. Also, it serves as the principal regulator for groundwater drainage and seasonal recharge. By this way, and by the local population directly withdrawing water for irrigation, the place is a significant part of agriculture, with less water pressure than anywhere else in Punjab in fertile lands that cover it. Invasive water hyacinth is widespread and must slowly be phased out, and rising pollution and riverbed destruction are seen as potentially threatening. The flux is considered, from a religious point of view, the most important in the state, since this stream is linked to Shri Guru Nanak Dev, the first Sikh mentor. The channel itself and the woods adjacent are the private property of the provinces and neighboring regions. This site assumes importance as an ecotourist destination and a picnic spot.

Keoladeo National Park: Montreux Record added on July 4, 1990, heritage site; Bird Sanctuary; 10 permanent, seasonal reservoirs in a densely populated area of different sizes. Greenery is open grassland and mosaic grassland that offers support for breeding, winter, and migratory birds. Five ungulate, four cat, and two primate species are found. The Canal provides both household and farm water. The livestock and the sea are grazing cattle on the farm. The invasive Paspalum grass growth altered the ecological characterization of the site's broad areas, reduces its suitability for some species of water birds, notably Siberian cranes.

Bird Sanctuary Point Calimere: Sanctuary for wildlife coastal region comprising deep shoreline and long sandy beaches, subtidal habitats, and nearshore forests, primarily mangrove and seasonal, sometimes salt lagoons and developed sites. Some 257 bird species, including 119 water-birds were found, with some 30,000 larger and lower flamingos, and the endangered species. The site is used as a nursery or breeding ground for several commercially important fish species and crabs. About 35,000 fishers and farmers help their families at the sanctuary boundaries. Illegal logging of firewood and forest products such as firewood fruits (from treetroop branches) and proliferation of *Prosopis chilensis* are all considered potentially alarming, as well as increasingly brackish groundwater from the extensions of historical salts and diminishing inflows. For leisure and pilgrimage, the tourists come to the place.

Pong Dam Lake: A wildlife refuge and a reservoir formed in 1975 on the Beas River, Indo-Gangetic plain on the north side on the low foothills of the Himalayas. The avian habitats created by the formation of the Pong Dam assume great importance. The Trans-Himalayan Flyway area has reported as possessing more than 220 bird species and 54 waterfowl species. The hydrological values include monsoon and downstream water management, soil preservation, trapping of sediment, and the prevention of land degradation; electricity is generated in and near areas; irrigation waters are flowing through fertile regions of the Punjab and the Rajasthan desert. The place is known for fishing with 27 species of fish. Yield is growing each year significantly. Approximately 1800 fishermen are now in the direct work in fishing farms, and it gives benefit to 1000 families directly and indirectly. There is training on Nature Conservation Center at Ramsar. New technology for management has moved from law enforcement to more participatory approaches and community knowledge, which is suitable for urban tourism development.

Renuka Wetland: Refuge for animals and a forest reserve. A natural wetland with surface water inlets and away from the coast subterranean karst formations supplied by a shallow river sloshing out of the lower Himalayas into the River Giri. The water body represents an approximate habitat of 443 wildlife species and 19 animal and insect wildlife types, which are lacustrine habitats, including *Puntius*, *Labeo*, and *Rasbora*. Significant greenery ranges from dry lime to hydrophytes, including *Shorea robusta*, *Dalbergia sissoo*, and *Terminalia tomentosa*. There are 103 bird species, 66 of those are native, all breasted with crimson. The area is also abundant among the ungulates of sambar, barking deer, and ghorals. The reservoir named after the Hindu guru's mother Parshuram and toured by thousands of tourists. It has an essential historical relevance. To date, environmental concern and the reduction of inflows of wet slopes and 50 ha included in conservation projects. The site is governed by Himachal Pradesh, Forest Department of Shimla, for its major farming reservoir.

Ropar: Human-made estuary wetland created by the construction of the irrigation and drinking supply dam in 1952. It is an essential reproductive site. Across the food safety chain, there are up to 40 fish species, and about 150 birds provided locally and migrants. Local harvesting is economically significant and harvested with wheat, corn, cane, and sorghum in the nearby region. Potential threats and invasive wetlands are deforested local hills, which lead to silage and rising industrial development triggering environmentally damaging capital flows. The place is visited in large numbers by nature lovers, bird-watchers, swimmers, and boaters.

Surinsar-Mansar Lakes: Sanctuary of the wildlife and a sacred Hindu site. CITES and IUCN red-listed animals are there at the shore. This hybrid lake has several essential nutrients for which attractive habitats, breeding grounds, and nurseries are the migratory waterfowl, including *Fulica atra*, *Podiceps nigricollis*, *Gallinula chloropus*—specific *Anas* species. The site has social and cultural significance due to its legendary origin from the Mahabharata period, with many temples nearby. While the lakes are home to a wide range of species, fishing for religious values is discouraged. Growing visitors, the practices of crop runoff, bathing, and cremation are the main challenges. Conservation is planned to raise consciousness.

Upper Ganga: A deep extend of the beautiful Ganges River. Dolphin, gharial, alligator, six tortoise and otter's varieties, 82 fish species, and 100 bird species are present along the river. The Red List lists significant plants of the Ganges River some of which are of high value. This stretch of the river is of sacred significance to thousands of pilgrims and used for funeral service and holy baths for spiritual insight. Extreme concerns include water pollution, crop surfaces, and diverse and costly fisheries. Planting, organic farming, and lobbying against commercial fishing are conservation initiatives undertaken to avoid river erosion.

Beas Conservation Reserve: The Beas ecology forest is a 185-km big leap of Beas River situated primarily in the northwest region of Punjab. The river flows through some canals from the foothills of the Himalayas into the Harike Headworks. This section includes over 500 bird species and over 90 fish species. The only reported Indian population of endangered Indus dolphins (*Platanista gangetica minor*) also lives in the Reserve. Certain threatened species include the listed mash (*Tor putitora*) and hog deer (*Axis porcinus*) and the endangered otters (*Lutrogale perspicillata*). In 2017, a campaign initiated to recover *Gavialis gangeticus*, an endangered species with 47 people released into the river 30 years after they had disappeared. The main threats include commercial and residential sweeping and agricultural impacts across most of the river's reach. Scientific protection of wetlands is taken out by the Department of Environment and Fauna Preservation, Punjab.

Samaspur Bird Sanctuary: The Samaspur conservation area is a seasonal swamp that is standard of the Indo-Gangetic Plains near Raebareli district of Uttar Pradesh. The six related lakes are of significance to the early rains of the moon. More than 75,000 birds regularly found in annual counts, with more than 250 native and migrant species recorded. The Sanctuary is the destination to endangered species such as the threatened birds and the fish eagle of Pallas (*Haliaeetus leucoryphus*) and more than 1% of the vulnerable common pochard (*Aythya ferina*) population of South Asia. About 46 aquatic fish species use the wetland, and some of them arrive with the monsoon floods from nearby rivers. The site provides the security and conservation for the food and agricultural feed products. Nevertheless, its ecological character is challenged by invasive species, more than 40% of which are native. The shrine co-operated by the Office of Conservator of Forest (Wildlife) and the authorities from the ministry of environment and forests for making the conservation efforts a success.

Parvati Agra Bird Sanctuary: Parvati Agra Bird Sanctuary is a protected fresh-water area comprising of two water bodies. The wetlands of Uttar Pradesh are characteristic of and offer outstanding water-bird habitats, providing over 100,000 birds recorded in annual counts for both roosting and breeding sites. The Sanctuary is a shelter for some endemic species like Indian vulture birds. These are reported as endangered white vultures and other endangered vultures. This location is also crucial for the conservation of water supplies, the restoration, and drainage of groundwater. Critical threats are known for invasive species, such as the common water hyacinth (*E. crassipes*). The Sanctuary Officers are responsible for the overall responsibility.

Sarsai Nawar Jheel: Sarsai Nawar Jheel is a seasonal swamp located within Etawah district of Uttar Pradesh. This conventional Indo-Gangetic floodplain body of water is faded by the rains of the monsoon in the southwest from the precipitation runoff. The lake is an example of human and wildlife cohabitation. Agricultural practices around a large part of the site play a significant role in the protection of habitats of water birds. The threatened species like sarus crane, a considerable flock of 400 individuals are particularly beneficiary here. The name of the area originated from that kind of class of nonmigratory crane. Such organisms at risk are the nearly extinct white-haired vulture (*Gyps bengalensis*) and the fragile wool-haired stork. The water body is also directly related to religious and spiritual importance every year. Thousands of pilgrims visited the nearby temple of Hajari Mahadev. Droughts and drainage may threaten the ecological character of the site. Birdlife International recognizes this as a major bird area.

Nangal Wildlife Sanctuary: The extremely eco-sensitive Nangal Wildlife area located in the Shiwalik highlands of Punjab, promoting productive animals and plants, including invasive species such as endangered animals. This site was part of the 1961 Bhakra-Nangal project as a human reservoir. Here is a historical site as the first Indo and Sino Govt. formalized the "Five basic mutual cooperation Principles." The shrine is managed by the Forestry and Wildlife Conservation Ministry of Punjab (Division Rupnagar for wildlife).

Keshopur-Miani Reserve Community: In the middle of Punjab, the Reserve is situated. It is covered by annual rainwater by ecological swamps, aquaculture ponds, and agricultural waterways. It is strongly affected by human beings and includes several well-maintained fish ponds and other crops of lotus and chestnut. The management of this species contributes to a variety of flora with 344 plants in the region. The area is an instance of the wise use of a community-run wetland, which provides food to the family and protects local ecosystems. *A. ferina* and the rare turtle are endangered species. A Management Committee is set by the Department of Forestry and Wildlife Conservation, Punjab.

Sandi Bird Sanctuary: It is a lake situated in Hardoi in Uttar Pradesh. It is a freshwater lake. The water body of the Indian-Gangetic Plain is typical and receives much of its water from the rainy season. The site is abundant in aquatic plants and offers a safe habitat for sea oats with over 40,000 people in 2018. It contains more than 1% of the common teal (*Anas crecca*), red-crested pottery (*Netta rufina*), and ferruginous duck (*Aythya nyroca*) populations of South Asia. At the same time, the endangered sarus cranes (*G. antigone*) are present in the shrine. A large number of individuals of sarus cranes in this area help Bird Life International's classification in categorizing it as a significant bird area. The lake is a famous tourist and leisure destination and supports local farmers as a food source. Drought is a threat; from 2014 to 2015, the sanctuary dried up, and the water birds died. The Forest Management Office manages the site and is also in partnership with community, wildlife, and forest officials.

Nawabganj Bird Sanctuary: A shallow marshland in Uttar Pradesh, 45 km from Lucknow. Monsoon rains feed this complex wetland while additional water comes from the Sarda Canal. The sanctuary promotes leisure, tourism, and local biodiversity. It is a bird sanctuary regularly recorded with 25,000 water birds and registered 220 species resident and migratory. The vulnerable small adjutant (*Leptoptilos javanicus*) and the wool-necked stork (*Ciconia episcopus*) include endangered Egyptian vulture (*Neophron percnopterus*) and palla fish eagle (*H. leucoryphus*). Conservation and forestation have led to an increase in global species diversity. Highly invasive common water hyacinth (*E. crassipes*) and wood removal from forests pose a threat. The sanctuary usually assisted by State Forest Officers and the Forest Conservator's (Wildlife) Department.

Saman Bird Sanctuary: Saman conservation area in the Uttar Pradesh district of Mainpuri is a summer oxbow reservoir on the Ganges region. It is hugely dependent on the arrival in July and August of the southwestern monsoon, which accounts for the vast majority of annual rainfall. The sanctuary frequently facilitates more than 50,000 water birds (187 species of birds have recorded) and is especially significant for many migrants as a wintering site, including the *Anser anse*, with more than 1% of the South Asian population present during the winter. Species such as sarus cranes (*G. antigone*) and more spotted eagles (*Aquila clanga*) are also at risk. The environmental services offered include providing freshwater agriculture, as well as leisure and location-related tourism focused on a wide variety of birds. Intrusion and salinization of settlements pose risks. The Forest Conservator's Office (Wildlife) oversees the conservation of the soil.

Nandur Madhameshwar: The site of the Deccan Plateau is a mosaic of lakes, swamps, and wetland vegetation. Building it at the confluence of Godavari and Kadwa, we have succeeded in creating productive watersheds. The site designed to resolve the scarcity of water at first in the local area now serves as a flood control and a hot spot for biodiversity. Its diverse habitats with 536 species reported, contrast with its semi-arid surroundings caused by the rainy shade of the Western Ghats. There are several of India's most common species on the Internet, including leopards and Indian sandalwood (*Santalum album*). Invasive species, such as water hyacinth, are at risk on the site as well as the effects of urbanization and water abstraction. It is managed by the Forest Conservator Bureau.

Kabartal Wetland: It is known as the Kanwar Jheel, covering 2,620 hectares in the northern Indo-Gangetic plains of the Bihar State in the District of Begusarai. It is the number 2,436 Ramsar Designated Site in the 'Wetlands of International Significance' list. With 165 flora and 394 fauna of different types, this wetland provides an abundance of biodiversity. This wetland is a significant stop along the Central Asian flyway, and every year 58 migratory birds are a paradise destination. The wetland faces grave challenges from water management operations such as irrigation, water supply, drainage, and so forth.

Asan Conservation Reserve: The Ramsar Named Site No. 2.437 on the 'Wetlands of International Significance' list of 443 acres of the Asan River flowing across its confluence Yamuna River in Dehradun district of Uttarakhand. This wetland supports 330 bird species, among which critically endangered vulture species. It supports 49 species of fish.

Sur Sarovar: It is popularly known as Keetham Lake, a man-made reservoir mainly constructed to supply water to Agra city. It is now the Ramsar Site 2,440 on the designated list of 'Wetlands of International Importance' in the State of Uttar Pradesh. It is an important Central Asian Flyway which records 30,000 visiting birds every year.

Lonar Lake: This is one of the most recent Ramsar sites in Maharashtra state, No. 2441 on the Deccan plateau. It is a closed or endorheic basin, containing a circle of meteorites that impact the basalt base. The lake is exceptionally saline and alkaline, provided that there is no outflow leading to mineral accumulation as the water evaporates from the lake. In intense chemical environments, the Lake hosts specialist microorganisms such as cyanobacterium and phytoplankton. The site contains some 160 bird species, 46 reptile species, and 12 mammalian species. The site is severely vulnerable to household wastewater, municipal wastewater, and unsustainable tourism.

3. Son Beel wetland of Assam

Son Beel is considered to be one of the Asia's most extensive wetlands and is considered not only a significant wetland in Karimganj district of Assam but in the whole region. The house located in Southern Assam Karimganj district, India, between and 24°36′40″ 24°44′30″ N and 92°24′50″ 92°28′25″ E. The wetland of Son Beel is situated near the Ramkrishna Nagar district of Karimganj, just 10 km from the town of Ramkrishna Nagar. A particularly unique and fascinating Son Beel Reservoir feature is that it is turned into a rice farm during the winter season and that it is usually flooded with water after the winter season and is now a vast water body due to rainwater. As further rain rises, the lake overflows and excess water flows through the dam, eventually mixing with the Kushira Dam, which then flows into Bangladesh (Kar and Dey, 1990). This humid area's average depth is much smaller. Son Beel is famous for the fishing. Wetland is the center for all kinds of fish processing and is known for its small miniature fish, Son Beel's unique identity. Net fishing and marketing includes 500 Son Beel wetland communities. State government surveys and records of traditional fishing systems have shown that over 35,000 households rely directly or

Photo 17.1 View of Son Beel wetland during the monsoon season. (Photo Courtesy: Saanku Dash).

indirectly on the lake. Evergreen forest water bodies cover over 40% of Son Beel (Photo 17.1). In Son Beel, reptiles and other marine animals exist. The Son Beel measures in length and width at 13.2 and 4.2 km. The closest coast is 35 km inland (Photo 17.2).

4. Monetary valuation of Son Beel

The Economics of Ecosystems and Biodiversity (TEEB) is an international project to determine the monetary importance of the environmental resources and the relevance of biodiversity significance based on "Cost–Benefit Stewardship Model" for the earth and its living constitution. Although it was a challenging thought initially to have a TEEB initiative of Son Beel, but we studied the pattern of ecosystem services thoroughly around the wetland where 500 fisher-folk families directly involved and 35,000 families directly or indirectly accessing ecosystem services and benefits either. Son Beel offers various ecosystem services in the Millennium Ecosystem Assessment, which was first proposed in the Millennium Ecosystem Assessment Report (MA, 2005). Son Beel ecosystem offers a variety of products. This wetland offers fresh potable drinking water to 35,000 rural families. Apart from being the source of fresh water it provides genetic capital, fruit, fiber, electricity, natural medicinal products, etc.

Son Beel wetland is an important natural infrastructure to offer flood resistance as well as a disaster risk structure during drought days. Communities' access benefit in various ways viz water regulation, erosion regulation, cleaning up water, managing waste, environment and disaster risk reduced for flooding and drought in and around Son Beel, in particular. Local inhabitants and communities have deep sacred belief in Son Beel because it protects them from terrible risk of frequent flood and drought every year. People accessing spiritual stimulation, cognitive growth, reflection, exercise, and esthetical activities are nonmaterial benefits. People pay worship and trust it heavily that the Beel protects them in long run. It is an important natural structure integrating cultural diversity, knowledge resource, educational, research scope, social interface, cultural heritage, and ecotourism. There are other exceptional services that are not included under provisional, regulation, and cultural category of services and

Photo 17.2 View of Son Beel wetland premonsoon season.
(Photo Courtesy: Saanku Dash).

benefits. Such services and benefits also occur over very long time directly or indirectly for individuals.

Monetary value is an integration of a translational mechanism estimated and calculated by assessing qualitative and quantitative indicators, which get converted to develop a monetary figure. Waste water purification service provided by Son Beel has been valued in monetary transactions after having a cost analysis on the waste water management system by treatment plant. The revenues generated by tourism gives an indication of cultural ecosystem services this precious ecosystem is providing. Despite being a beautiful destination it is not being promoted as the site for ecotourism. Still this site is in precarious state as far as tourism is concerned because it lacks logistics and lodging amenities at large. Fish-folks and their fish catching services are monetized readily and easily alongside with their other livelihood activities together. Monetary valuation of Son Beel is a gross indication of ecosystem services local and indigenous people are extracting accordingly to their livelihood's preferences, which are an absolute baseline to access direct data for correct monetary valuation. It helps understanding express desires concealed from market values that are not expressed. Problem lies in the fact that we have limited access to policy decisions as far as Son Beel is concerned. Policy-making process favors provisioning services (food, forest, and fuel) because such services and benefits have visible effects. But there are so many other ecosystem services less noticeable, which have been underrated or largely ignored relatively, has yet been included in the decision-making process. The evaluation of the economic value of the ecosystem services is taken into account, which are traditionally less well covered, advances arguments of sustainability, rational use, and restoration.

Methodologies of monetary value of Son Beel: The structures are based on three major monetary appraisal categories. The first approach of this type shall be focused on markets. The calculation of values by the avoided expense of the pre-empted environmental harm shall also be used by consumer rates for commodities not on the consumer (nonmarketed fish, wood, forest goods, water, etc.).

The second monetary evaluation approach is displayed on the basis of preferences. For example, a hedonic type of value is used to estimate the value of the protected area by spending time and resources on the journey. Price fluctuations of properties are used as an indicator of landscape value due to changes in surrounding environment and habitat.

The third monetary valuation method is based on the preferences specified. For instance, using the contingent assessment process, individuals may voluntarily pay for improved environmental protection or accept compensation for environmental quality degradation.

Simplification of the economics of ecosystems and biodiversity valuation is our preliminary part of this entire exercise. We have limited baseline data and Son Beel is facing serious anthropogenic pressure. People are not well informed and unaware about its rich productivity and its vast range of ecosystem services and benefits. We try to an optimum level so that we could bring larger community's association and make them informed about the economic status of Son Beel. Adopting a six-step approach while taking into account of such international standard, we finally initiated to evolve a preliminary monetary valuation of Son Beel Wetland. We tried to evaluate value of nature by putting on indicators of ecosystem services.

Monetary valuation was done when US$ = $62.2

Final calculation:

Total number of used estimates = 34 + 30 + 9 + 13 = 86

Total minimum values (Rs/ha/year) = PS + RS + SS + CS = 2250 + 1986 + 595 + 861 = 5692

Total maximum values (Rs/ha/year) = PS + RS + SS + CS = 516,678 + 948,430 + 140,640 + 325,770 = 1,931,518

Final monetary valuation of Son Beel Wetland varies in the range of Rs. 5693/hectare/year to Rs. 1,931,518/hectare/year (Kumar et al., 2020).

This is a preliminary valuation within a limit of information. But it could have been a baseline for further extensive and advance valuation. Son Beel offers a fertile alluvial soil for highly productive field supporting abundant rice cultivation, vegetation cultivation, coconut cultivation, banana cultivation, livestock water, commercial water, domestic water, trapa-farming, and fisheries, etc (Tables 17.1 and 17.2).

Table 17.1 TEEB value.

Category of wetland	Ecosystem services	Estimate sum	Min. application value (int. $/hectare/year)	Max. application value (int. $/hectare/year)
Inland vegetative wetlands other than rivers and lakes	Provisioning services	34	34	9,709
	Regulating services	30	30	23,018
Floodplains, swamps/marshes, and peatlands	Habitat services	09	09	3,471
	Cultural services	13	13	8,399
	Total	86	86	44,597

Table 17.2 TEEB, 2010.

Inland vegetative wetland	Estimates used No.	Min. values (int. $/ha/year)	Max. values (int. $/ha/year)
Thorough	86	86	44,597
Related services	34	34	9,709
(1) Food	16	16	2,091
(2) Supply of fresh water	06	06	5,189
(3) Direct products	12	12	2,430
(4) Genetic resources	–	–	–
(5) Assets as medicine	–	–	–
(6) Decorative asset facilities	–	–	–
Services of oversight	30	30	23,018
(7) Air quality effect	–	–	–
(8) Legislation on climate change	05	05	351
(9) Extreme incident management	07	07	4,430
(10) Regulatory oversight of the flow of water	04	04	9,369
(11) Handling of waste and treatment of water	09	09	4,280
(12) Erosion resistance			
(13) Cycling of nutrients and soil fertility conservation	05	05	4,588
(14) Pollinating factor			
(15) Biophysiological audit			
Support and services	09	09	3,471
(16) Life cycle control	02	02	917
(17) Preservation of gene pools	07	07	2,554
Social and cultural events	13	13	8,399
(18) Functional and esthetic details	02	02	3,906
(19) Recreational and leisure incentives	09	09	3,700
(20) Community, art, and design inspiration	02	02	793
(21) Spire experiencing divine knowledge	–	–	–
(22) Learning and sciences (cognitive information)	–	–	–

5. Standard TEEB (2010) valuation of an inland vegetative wetland

Inland monetary value of services vegetative lake.

(Flood plains, marshes, peatlands, and swamps).

Int.$/ha/year—2007

TEEB, 2010, de Groot et al. (2010).

(A) Provisioning Ecosystem Services and monetary valuation of Son Beel (Table 17.3):

(B) Regulating ecosystem services and monetary valuation of Son Beel (Table 17.4):

(C) Supporting ecosystem services and monetary valuation of Son Beel (Table 17.5):

(D) Cultural ecosystem services and monetary valuation of Son Beel (Table 17.6):

Biodiversity resources of Son Beel: Owing to their importance in terms of plant community, there are some local plant species that recorded during the field survey near Son Beel area, which are used as medicinal preparatory by the locals.

Cassia sophera

Local name: Bonua gach

Part used: Leaves

Process of utilization: The leaves are used externally for ringworm, and a decoction of plant is beneficial in acute bronchitis.

Mimosa pudica

Local name: Chhoitemara/lajjaboti

The paste of leaves and roots is used in toothache.

Used in the treatment of boils

Table 17.3 Monetary evaluation of Son Beel.

Provisioning services	Number of figures used	Min. values (Rs./ha/year)	Max. values (Rs./ha/year)
1. Food	16	1,059	113,550
2. Fresh water supply	06	397	287,448
3. Raw materials	12	794	115,680
4. Genetic resources	Not available	Not available	Not available
5. Assets of medicine	Not available	Not available	Not available
6. Ornamental resources	Not available	Not available	Not available
Total	34	2,250	516,678

Table 17.4 Monetary evaluation of Son Beel.

Regulating services	Number of figures used	Min. values Rs./ha/year	Max. values Rs./ha/year
7. Air quality impact	Not available	Not available	Not available
8. Climate regulation	05	331	15,200
9. Moderation of extreme events	07	463	203,880
10. Regulation of water flows	04	264	448,500
11. Waste treatment/water purification	09	596	170,600
12. Erosion control	Not available	Not available	Not available
13. Nutrient cycling/ maintenance of soil fertility	05	331	110,250
14. Pollination	Not available	Not available	Not available
15. Biological regulation	Not available	Not available	Not available
Total	30	1,986	948,430

Table 17.5 Monetary valuation of Son Beel.

Supporting services (SS)	Number of figures used	Min values (Rs/ ha/year)	Max values (Rs/ ha/year)
16. Life cycle maintenance	02	132	25,300
17. Gene pool protection (conservation)	07	463	115,340
Total	09	5	140,640

Table 17.6 Monetary valuation of Son Beel.

Cultural services	Number of figures used	Min. value (Rs./ha/year)	Max. value (Rs./ha/year)
18. Esthetic information	02	132	115,225
19. Opportunities for recreation and tourism	09		180,345
20. Culture, production, and design inspiration	02	729	30,200
21. Spire experience divine knowledge	Not available	Not available	Not available
22. Cognitive information (education and research)	Not available	Not available	Not available
Total	13	861	325,770

Centella asiatica
Local name: Thankuni Pata
Used in the treatment of leprosy, thread worm, dysentery, etc.
Ageratum conyzoides
Local name: Uchandi
Part used: Leaf
Process of utilization: The paste made from the fresh leaf is used in cut injury.
The leaf juice is used in cut injury of human body or in animals is more effective than Dettol.
Mikania cordata
Local name: Refujilata
Part used: Leaf
Process of utilization: The juice extracted from the leaf is used in wounds and cuts.
Wadelia chinensis
Local name: Bhringaraj
Parts used: Leaves
Process of utilization: The juice of leaves is used as hair tonic in the preparation of hair oil.
Leucas aspera
Local name: Donkalosh/Shetdrone
Parts used: Leaves and seeds
Process of utilization: The pastes made from the leaves are used to cure wound burn etc. The oil extracted from the seeds is used to prepare different food items viz. curry pickles, etc.
Achyranthes aspera
Local name: Apang
Part used: Leaf
Process of utilization: The paste made from the leaf of the plants used to apply on hair to reduce dandruff.
Dose: The paste is used thrice in a week.
Polygonum hydropiper
Local name: Pani marich
Part used: The roots
Process of utilization: The roots are used to wrap on forehead which reduce headache.
Dose: It is used for whole day.
During the recent study medicinal plants and its value highlighted for Son Beel Wetland (Choudhury et al., 2020)

6. Challenges before Son Beel

Agenda 2030 provides the governments, civil society, the private sector, and other state/nonstate actors with a broader map of national and international policy action to reach sustainable development goals (SDGs) for our current and future generations. For the life and well-being of individuals and society, Son Beel offers a wide array of natural capital flows in ecosystem services. The planning and implementation of SDGs must be made essential to wetland protection, wise use, and restoration. Including wetlands resources and benefits in the Paris Climate Change Agreement nationally defined commitments is keys to the attainment of SDGs. To increase wetland intervention and minimize local conflicts at community level, adopting policies and practices in the conservation and wise use of wetlands to prevent further depletion and degradation of wetlands is important. When designated to improve understanding of values at a site and the subsequent suggested actions related to its conservation, safely use and restoration, identifying services and benefits such a wetland offers people and the environment. Son Beel is facing serious ecosystem marginalization. Main Channels of link up with rivers are silted heavily and illegal encroachments are a matter of grave concern.

The following steps can be a major breakthrough for the conservation of the wetlands. These steps are suggested for Son Beel wetland in particular and for any wetland in general.

Step 1: Join and agree with the stakeholders on wetland-related problems.
Step 2: Identify a prioritization list of relevant local ecosystem services.
Step 3: Baseline Information Identification.
Step 4: Assessment of planned improvements in the availability and delivery of ecosystem resources.
Step 5: Identification Analysis of expected ecosystem services and evaluation of policy measures with change.
Step 6: Evaluation of market and financial effects of choices for policies in terms of how changes to ecosystem services affect individuals.

In addition to the abovementioned steps general awareness among the masses is must to resolve the issue. Awareness among the habitants and farmer community about the significance of wetlands as pollution and chemical absorbers, natural defense against storms and surges, etc. should be widely disseminated. The problem of shrinkage of wetlands due to encroachment should be dealt by strict legal action. The administration of Ramsar Indian wetlands infringes several provisions of the Ramsar Convention, including those on "wise use" of wetlands as set out in Provisions 4.2 and 1.7.2. From the Ramsar Convention on Migratory Water Birds (Criterion 5 of the Convention), formulate and execute their planning to encourage the protection of the wetlands and the wise management of wetlands in their territories included in the List.

7. Conclusion

The wetlands are the kidneys of the earth. They act as natural water purifier. Coastal wetlands are the buffer zones between the high tides of the sea and agricultural fields. By acting as buffers, they serve the vital function of conserving and protecting the wetlands from the problem of salinization, which may otherwise occur in the absence of these buffer zones due to rising sea levels brought to us by the rising global temperatures. They are the saviors of the farmer community, which saves them from the loss, and their livelihood is protected from the deteriorating effects of saline waters of the sea. They also act as pollution absorbers, take away excess of pesticides, chemicals, and thus prevent the agricultural fields from the degrading effects of storms from time to time. The changing climate is imposing several harmful effects on these ecosystems, as discussed above. Undue human interference leads to the shrinkage of the space of these wetlands, thereby leading to the loss of unique flora and fauna. So their protection, improvement, preservation, and conservation require that all the related factors be well studied and researched. The data all the quarters should be collected, compiled, and coordinated to check the reasons for the deterioration of the wetlands from the changing climatic conditions. Of course, climate change is connected with many other factors like pollution from industries, transport, deforestation, and increasing population leading to a chain of events.

Deterioration of wetlands is one of those problems. Through the case study of Son Beel, we have highlighted the significance and challenges of any wetland. Although, climate change has brought with it many reasons for the degradation of these unique ecosystems. Enough has been said how the wetlands curb the menaces brought by the extremes of weather triggered by changing climatic conditions and how the wetlands curb the problem of climate change by acting as a sink for carbon dioxide. Overall, the lake Son Beel has excellent potential for designation as a Ramsar site. Thus, wetlands with their high potential of acting as the purifiers of the earth owe uniqueness. Their conservation and protection are of utmost importance from every perspective, be it pollution protection, stream bank, and coastal margin stabilization from erosion and cyclones arising as a consequence of climate change. So a particular management policy and awareness among the general public, mainly farming community, are essential for the overall development and protection of wetlands.

Acknowledgments

The authors are thankful to the local residents for their cooperation during the field study. The authors are also grateful to members of Voice of Environment (Youth Environmental Organization) for their spontaneous support in carrying out research study and field visit.

References

Albertson, F.W., Weaver, J.E., 1942. History of the native vegetation of western Kansas during seven years of continuous drought. Ecol. Monogr. 12, 23–51.

Barbier, E.B., Acreman, M.C., Knowler, D., 1997. Economic Valuation of Wetlands: A Guide for Policy Makers and Planners.

Burkett, V., Kusler, J., 2007. Climate change: potential impacts and interactions in wetlands of the United States. J. Am. Water Resour. Assoc. 36 (2), 313–320. https://doi.org/10.1111/j.1752-1688.2000.tb04270.x.

Choudhury, M., et al., 2020. Wetlands are in Peril - A Case Study of Son Beel Wetland of Assam, India Intercontinental Webinar Proceeding on “Ecosystem Services and UNSDG” Held on 5th June 2020, pp. 103–110.

Dillehay, T.D., Quivira, M.P., Bonzani, R., Silva, C., Wallner, J., Le Quesne, C., 2007. Cultivated wetlands and emerging complexity in south-central Chile and long distance effects of climate change. Antiquity 81 (314), 949–960.

Finlayson, B.A., 2013. The method of weighted residuals and variational principles. Society for Industrial and Applied Mathematics.

Johnson, W.C., Boettcher, S.E., Poiani, K.A., Guntenspergen, G.R., 2004. Influence of weather extremes on the water levels of glaciated prairie wetlands. Wetlands 24, 385–398.

Junk, W.J., An, S., Finlayson, C.M., 2013. Current state of knowledge regarding the world's wetlands and their future under global climate change: a synthesis. Aquat. Sci. 75, 151–167. https://doi.org/10.1007/s00027-012-0278-z.

Kar, D., Dey, S.C., 1990. Fish disease syndrome: a preliminary study from Assam Bangladesh. J. Zool. 18, 115–118.

Kumar, D., Choudhury, M., Rathore, A., 2020. Valuation of ecosystem services & benefits of Son Beel wetland in Assam, India: a case study of natural solutions to climate change & water. In: Climate Change Conference. Hamburg University, Germany.

Larson, D., 1995. Effects of climate on numbers of northern prairie wetlands. Clim. Change 30, 169–180.

MA (Millennium Ecosystem Assessment), 2005. Ecosystems and Human Well-Being: The Assessment Series (Four Volumes and Summary). Island Press, Washington, DC.

Mingkui, C., Keith, G., Steward, M., 1998. Global methane emission from wetlands and its sensitivity to climate change. Atmos. Environ. 32 (19), 3293–3299. https://doi/10.1016.S1352-2310(98)00105-8.

Mitsch, W.J., Bernal, B., Nahlik, A.M., 2013. Wetlands, carbon, and climate change. Landsc. Ecol. 28, 583–597. https://doi.org/10.1007/s10980-012-9758-8.

Poiani, K.A., Johnson, W.C., 1991. Global warming and prairie wetlands. Bioscience 41, 611–618.

Shapley, M.D., Johnson, W.C., Engstrom, D.R., Osterkamp, W.R., 2005. A 1000-year record of late Holocene flooding and drought from Day County, South Dakota, reconstructed from tree rings, lake sediments, and ancient shorelines. Holocene 15, 29–41.

TEEB, 2010. In: Kumar, P. (Ed.), The Economics of Ecosystems and Biodiversity: Ecological and Economic Foundations. Earthscan, London and Washington.

Van der Valk, A.G., Davis, C.B., 1978. The role of seed banks in the vegetation dynamics of prairie glacial marshes. Ecology 59, 322–335.

William, K.M., Elizabeth, R.B., Keith, L.B., Mark, M.B., Leonard, R.G., 1997. Climate change, hurricanes and tropical cyclones, and rising sea level in coastal wetlands. Ecol. Appl. 7 (3), 770–801. https://doi.org/10.1890/1051-0761(1997)007(0770:CCHATS)2.0.20;2.

Winter, T.C., 2007. The vulnerability of wetlands to climate change: a hydrologic landscape perspective. J. Am. Water Resour. Assoc. 36 (2), 305–311.

Winter, T.C., Rosenberry, D., 1998. Hydrology of prairie pothole wetlands during drought and deluge: a 17-year study of the Cottonwood Lake wetlands complex in North Dakota in the perspective of longer term measured and proxy hydrological records. Clim. Change 40, 189–209.

World Business Council for Sustainable Development, 2017. Incentives for Natural Infrastructure: Review of Existing Policies, Incentives and Barriers Related to Permitting, Finance and Insurance of Natural Infrastructure. Geneva.

Index

Note: 'Page numbers followed by "*f*" indicate figures and "*t*" indicate tables.'

D

FSC
www.fsc.org
MIX
Papier | Fördert
gute Waldnutzung
FSC® C083411

Kolibri 360 GmbH
Ferdinand-Jühlke-Straße 7
99095 Erfurt, Deutschland
produktsicherheit@kolibri360.de